Advanced Topics
in
Signal Processing

PRENTICE HALL SIGNAL PROCESSING SERIES

Alan V. Oppenheim, Editor

99/66
c.1

Advanced Topics
in
Signal Processing

Edited by

Jae S. Lim
Alan V. Oppenheim

PRENTICE HALL, Englewood Cliffs, New Jersey 07632

Library of Congress Cataloging-in-Publication Data

Advanced topics in signal processing.

(Prentice Hall signal processing series)
Bibliography: p.
Includes index.
1. Signal processing.   I. Lim, Jae S.   II. Oppenheim,
Alan V.   III. Series.
TK5102.5.A333 1987   621.38′043   87-11457
ISBN 0-13-013129-6   025

Editorial/production supervision and
   interior design: Barbara G. Flanagan
Cover design: 20/20 Services, Inc.
Manufacturing buyer: S. Gordon Osbourne

© 1988 by Prentice Hall
A Division of Simon & Schuster
Englewood Cliffs, New Jersey 07632

Printed in the United States of America

10 9 8 7 6 5 4 3 2

ISBN   0-13-013129-6   025

Prentice-Hall International (UK) Limited, *London*
Prentice-Hall of Australia Pty. Limited, *Sydney*
Prentice-Hall Canada Inc., *Toronto*
Prentice-Hall Hispanoamericana, S.A., *Mexico*
Prentice-Hall of India Private Limited, *New Delhi*
Prentice-Hall of Japan, Inc., *Tokyo*
Prentice-Hall of Southeast Asia Pte. Ltd., *Singapore*
Editora Prentice-Hall do Brasil, Ltda., *Rio de Janeiro*

Contents

3 Multirate Processing of Digital Signals   123

Ronald E. Crochiere
Lawrence R. Rabiner

AT&T Bell Laboratories

4 Efficient Fourier Transform and Convolution Algorithms   199

C. S. Burrus

Rice University

5 Fundamentals of Adaptive Signal Processing   246

Samuel D. Stearns
Sandia National Laboratories

6 Short-Time Fourier Transform   289

S. Hamid Nawab
Boston University

Thomas F. Quatieri
Lincoln Laboratory

7 Two-Dimensional Signal Processing   338

Jae S. Lim

Massachusetts Institute of Technology

8 Some Advanced Topics in Filter Design   416

Hans Wilhelm Schüssler
Peter Steffen

University of Erlangen-Nuremberg

Appendix Fundamentals of Digital Signal Processing   492

Index   511

Preface

The field of digital signal processing has maintained tremendous vitality for more than two decades, and there is every indication that this will continue and even accelerate. Advances in technology provide the capability in signal processing microprocessors that was associated with mainframe computers and relatively costly high-speed array processors only a short time ago. New applications from military, industrial, and consumer areas continue to be found and developed, and existing applications are becoming increasingly more cost-effective and important. In concert with the applications and technology, the theoretical foundations of the field have continued to expand, leading to deeper understanding of the basis for signal processing and to important new algorithms and applications to which they relate.

The fundamentals of digital signal processing are readily available in a variety of excellent text and reference books and form the basis for senior and graduate-level courses at virtually all engineering schools. Many schools also offer advanced topics courses in digital signal processing, as has been the case at M.I.T. for the past ten years. More advanced material appropriate for these courses has been accessible primarily through journal articles or through books dealing with a single specialized topic or area. The need for a principal reference book or textbook on advanced topics, both for advanced courses and for general reference, has been a prime motivation behind the development of this book.

Our goal is to provide in a single volume an introduction to a variety of advanced topics in signal processing. While an edited book has certain limitations and drawbacks, it is an ideal format for treating a broad spectrum of advanced topics because it provides the opportunity for each topic to be written by an expert in that field. Effort has been made to keep the notation as consistent as possible throughout and to cross-reference between chapters when appropriate.

Among the advanced topics currently being pursued in the field, we have chosen eight that we feel are among the most important. This book consists of eight chapters, with each chapter covering a specific advanced topic. The topics covered are parametric signal modeling, spectral estimation, multirate signal processing, Fourier transform and convolution algorithms, adaptive signal processing, short-time Fourier transform, two-dimensional signal processing, and filter design. Each chapter has been organized so that it can be covered in two to three weeks when this book is used as a principal reference or text in an advanced topics course at a university.

It is generally assumed that the reader has prior exposure to the fundamentals of digital signal processing. A brief summary of the basic concepts is included as an appendix, which also serves to establish much of the notation used throughout the book. The chapters have been written with an emphasis on a tutorial presentation so that the reader interested in pursuing a particular advanced topic further will be able to obtain a solid introduction to the topic through the appropriate chapter in this book. While the topics covered are related and chapters are cross-referenced, each chapter can be read and used independently of the others.

This book is primarily a result of the collective efforts of the chapter authors. We are very grateful for their enthusiastic support, timely responses, and willingness to incorporate suggestions from us, from other contributing authors, and from a number of our colleagues who served as reviewers.

Jae S. Lim
Alan V. Oppenheim

1

Parametric Signal Modeling

James H. McClellan
Schlumberger Well Services

1.0 INTRODUCTION

The concept of representing a complicated process in terms of a simple model under-lies work in many fields. In electronics, semiconductor devices are analyzed via lumped parameter equivalent circuits. Economic models for the national economy have been built to predict the impact of policy decisions made by the government and private industry. Speech production, an acoustic wave phenomenon, has been mod-eled successfully as a lumped parameter linear system excited by a pulse train or white noise. The list of applications for parametric modeling is extensive, but all cases share a common idea: *reduce a complicated process with many variables to a simpler one involving a small number of parameters*. This reduction often requires approximation, so some degree of error may be involved in modeling the original process. Even so, if the parameters of the model turn out to be physically meaningful, then one can gain insight into the behavior of the overall process by understanding the influence of each parameter.

In this chapter, we will concentrate on methods for modeling a discrete-time signal as the output of a linear time-invariant system driven by a fixed input signal (e.g., the unit-sample signal or white Gaussian noise). In the case of the unit sample input, the signal is modeled as the impulse response of a linear time-invariant (LTI) system. Since such systems have z-transforms that are rational functions, the term "rational" modeling is often employed. An equivalent name is "pole-zero" modeling, since both the numerator and denominator polynomials of a rational system function have a finite number of roots. The coefficients of the system function (or its poles and zeros) comprise the small set of parameters that represent the signal. With a white

James H. McClellan is with Schlumberger Well Services, P.O. Box 2175, Houston, TX 77252-2175.

noise input, the LTI model yields a rational power spectrum for the signal. In fact, this approach to spectrum analysis is now widely used in place of the traditional Fourier methods (see Chapter 2).

Although many signals are not stationary and cannot be modeled with just one LTI system, it is often possible to model small sections of a signal with a rational model and then account for nonstationarity by allowing the parameters of the model to change from section to section. This, in fact, is the approach that is usually taken in speech applications, where the fundamental block for analysis is one pitch period (≈ 10 ms). Another development along this line is that of adaptive filtering, in which the coefficients of the model evolve over time according to an update strategy that is always working to minimize an error measure (e.g., the LMS algorithm presented in Chapter 5).

A number of significant issues are related to rational modeling. First, the rational model imposes a special form on the types of time functions that can be represented exactly. It is well known that the impulse response $h(n)$ of a pth order LTI system is a linear combination of p complex exponentials:[†]

$$h(n) = \sum_{k=1}^{p} c_k \lambda_k^n u(n) \tag{1.1}$$

This form appears similar to the Fourier transform—both are sums of complex exponentials. There is, however, a profound difference. The Fourier representation is based on a *fixed* set of basis functions: complex sinusoids. In addition, the number of complex sinusoids is equal to the number of data points being transformed. By contrast, the exponentials of the rational model in Eq. (1.1) are adjustable. This adjustment, which amounts to computing the basis functions $\{\lambda_k^n u(n)\}$, is the essential step in the modeling process. A good rational model represents the signal with a small number of well-chosen exponentials.

Whenever a general signal is modeled with a fixed-order rational model, there is, of necessity, an approximation to be made. Exact modeling is possible in Eq. (1.1) only when the original signal happens to be representable as a linear combination of p exponentials. When the representation is not exact, a measure of "goodness of fit" must be assumed, so that the $2p$ parameters, $\{c_k\}$ and $\{\lambda_k\}$, can be selected in a systematic fashion by finding the "best" approximation. Many possible norms exist and may be used to evaluate the approximation error, but the least-squares-error norm, which measures the energy in the error, is the most popular, primarily because of its mathematical tractability. All of the methods described in this chapter are formulated as least-squares ($\mathcal{L}_2$ norm) minimization problems. Recent research has extended some of the modeling techniques to other norms (e.g., the Chebyshev $\mathcal{L}_\infty$ norm and the $\mathcal{L}_1$ norm).

Another important issue is efficient calculation of the best approximation. A complete pole-zero model leads to a set of nonlinear equations to be solved for the best approximation. While it is certainly possible to solve such nonlinear equations using iterative methods such as steepest descent or Newton's method, the computational requirements are severe. On the other hand, the *linear prediction* formulation for the

---

[†]We have implicitly assumed distinct roots for the system function of the model, but only to simplify the appearance of this formula; repeated roots would introduce terms of the form $n^r \lambda_k^n u(n)$.

special case of all-pole modeling involves an "indirect" least-squares approximation problem but yields a set of linear equations describing the best approximation. Furthermore, in one special case, called the *autocorrelation* method, the linear equations take a special form and a fast algorithm, known as Levinson's method, is available for their solution. Those cases where high-speed algorithms exist will be a main focus of this chapter. The interest in such techniques for real-time applications (e.g., speech communication) has led to the design of special hardware and, more recently, to VLSI implementations.

The aims of this presentation are many, then, but perhaps the most important is that the reader be able to develop sufficient skills in the technical details of several different modeling methods to appreciate the variety of options available. No single technique necessarily stands out as the best. Circumstances sometimes favor one technique over the others. Also, individual researchers have their own favorites. For this author, Prony's method has served as a workhorse in several applications and is, therefore, the centerpiece of the presentation. The view that one is approximating a signal with a linear combination of adjustable exponentials is quite appealing, even though it neglects the statistical nature of the problem. Interestingly, however, a very popular view from the statistical side is the approximate maximum likelihood method, which yields basically the same set of equations for optimality.

Another facet of this presentation is the significant amount of material devoted to fast algorithms. Even with several sections devoted to the subject, the treatment is far from complete. The intent of the presentation is to cover the basic manipulations of the recursive algorithms so that the reader will accumulate the background to understand research papers published on the subject.

Finally, the techniques described here as *parametric signal modeling* have been developed over the years in many diverse fields. One objective of this chapter is to give a unified treatment of several different viewpoints on the same signal modeling problem so the reader will be able to research the published literature in various fields and extend his or her own knowledge. The similarities and differences among methods are often subtle and not appreciated by the casual observer. For example, the equations obtained in the autocorrelation method of linear prediction can be derived in a stochastic or a deterministic setting. The deterministic case can, in turn, be formulated in the time or frequency domain. The resulting equations are identical to those of the *maximum entropy method* (MEM) of spectral estimation as well as the method of *predictive deconvolution*. Even further, the *Yule-Walker* equations discussed in Chapter 2 are the same. On the other hand, a slight change in the formulation could lead to an entirely different answer, as in the *covariance* method of linear prediction. Each of these formulations emphasizes a different characteristic about the signal being modeled, and each offers a different insight into the problem at hand.

1.1 DIRECT MODELING

In this section, we consider a problem that turns out to be rather difficult to solve and, therefore, not widely used in practice. We begin here, nevertheless, because this problem states accurately the goal of matching the impulse response of a rational system to an arbitrary signal. The resulting method is referred to here as "direct" modeling.[†] Even though it seems to be the obvious way to proceed, we find, after a short walk down this path, that we must retreat to find simpler methods. To do so, it will be necessary to change the problem formulation—the result being "indirect" methods whose solution can be computed from linear equations. For now we are content to show the difficulty of the "direct" problem by deriving its equations of optimality.

Suppose the causal signal $x(n)$ is to be modeled as the impulse response of a rational system as in Fig. 1.1. An error signal $e_D(n)$ is computed as the difference

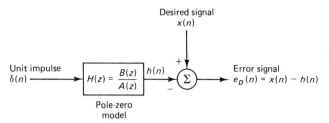

Figure 1.1 Block diagram for a direct modeling problem. The system function $H(z)$ is a rational function $B(z)/A(z)$ with p poles and q zeros. $A(z) = 1 + \Sigma_{k=1}^{p} a_k z^{-k}$; $B(z) = \Sigma_{k=0}^{q} b_k z^{-k}$.

between the impulse response $h(n)$ and the given signal $x(n)$. The parameters of the LTI system are to be chosen so that the error is small. Minimization of the least-squares norm (i.e., the total error energy) is the criterion used.

$$\min_{A(z),B(z)} E_D = \sum_{n=0}^{\infty} |e_D(n)|^2 \tag{1.2}$$

From Parseval's theorem, we can write a frequency-domain expression for the error in terms of $A(\omega)$ and $B(\omega)$:

$$E_D = \frac{1}{2\pi} \int_{-\pi}^{\pi} \left| X(\omega) - \frac{B(\omega)}{A(\omega)} \right|^2 d\omega \tag{1.3}$$

The global minimum is found at a stationary point of E_D:

$$\frac{\partial E_D}{\partial a_k} = 0 \quad \text{and} \quad \frac{\partial E_D}{\partial b_k} = 0 \tag{1.4}$$

The resulting equations turn out to be *nonlinear* in the parameters $\{a_k\}$ and, for that reason, the solution of Eq. (1.4) is a formidable task. Calculation of the partial derivatives will demonstrate the difficulty:

$$\frac{\partial E_D}{\partial a_k} = \frac{1}{2\pi} \int_{-\pi}^{\pi} 2\mathcal{R}e\left\{ \frac{-B(\omega)e^{-jk\omega}}{A^2(\omega)} \left[X^*(\omega) - \frac{B^*(\omega)}{A^*(\omega)} \right] \right\} d\omega$$

$$\frac{\partial E_D}{\partial b_k} = \frac{1}{2\pi} \int_{-\pi}^{\pi} 2\mathcal{R}e\left\{ \frac{e^{-jk\omega}}{A(\omega)} \left[X^*(\omega) - \frac{B^*(\omega)}{A^*(\omega)} \right] \right\} d\omega$$

[†] Some authors prefer the terminology "forward" and "inverse" to describe the "direct" and "indirect" methods of modeling, respectively (e.g., Chapter 5).

In addition, we must consider the possibility that these equations may not have a unique solution because the stationary points given in Eq. (1.4) are only necessary conditions for a local minimum (or maximum).

Since the equations for optimality are nonlinear, general iterative methods such as the gradient method or Newton's method are usually employed for their solution. The amount of computation needed is nontrivial. For example, each iteration of Newton's method would require evaluation of a $(p + q + 1)$-order gradient vector and the inversion of a $(p + q + 1) \times (p + q + 1)$ Hessian matrix of second derivatives of E_D; and the number of iterations might be large.

Because of the heavy computational requirements, direct modeling is not too useful. Fortunately, a number of indirect methods have been developed for modeling. Each of these indirect methods changes the error norm of the problem slightly, so that some or all of the parameters of the rational model can be obtained by solving *linear equations. It is important to keep in mind that none of these indirect methods will solve the problem stated above.* The direct modeling problem remains a difficult task. In the final section, on pole-zero modeling, we will reconsider the "direct" problem and attempt to solve it by a succession of these linear "indirect" problems.

1.2 INDIRECT MODELING

Now we reformulate the modeling problem so that the model coefficients can be computed by solving linear equations. We do this by minimizing a least-squares error generated by filtering the signal $x(n)$ with a finite impulse response (FIR) filter. Since the FIR filter will attempt to remove the poles of $X(z)$, it should be a reasonable estimate of the denominator polynomial in an all-pole model for the signal. Figure 1.2 shows a block diagram of this approach. Contrast this figure with that for direct modeling, Fig. 1.1. Loosely speaking, the error in the "indirect" method is $E_I(z) = X(z)A(z) - B(z)$, while that of the direct method was $E_D(z) = X(z) - B(z)/A(z)$. Since $E_I(z) = A(z)E_D(z)$, these are never the same. The denominator $A(z)$ weights the "indirect" error relative to the "direct" error. However, the "indirect" error has the advantage that it is linear in the coefficients of $A(z)$ and $B(z)$.

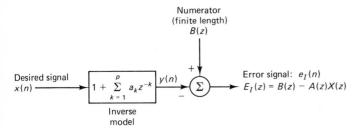

Figure 1.2   Block diagram for the indirect modeling problem. The system function $A(z)$ is an FIR filter with p zeros. $A(z) = 1 + \sum_{k=1}^{p} a_k z^{-k}$.

The indirect method is the basis of all the modeling techniques discussed in the remainder of this chapter. By far the most important one is Prony's method, which is presented in Section 1.2.2. As a prelude, we describe the Padè approximant to help motivate the concept of indirect modeling.

1.2.1 Padè Approximation

In the Padè approximation technique, the parameters of the pole-zero model are chosen so that its impulse response *exactly* matches the first $p + q + 1$ points of the signal. The name "approximation" is actually a misnomer because there is no approximation involved. In fact, the signal samples $x(n)$ for $n > p + q$ do not enter into the procedure at all. Padè *matching* would be a more accurate description. The reader is cautioned that we are presenting this method primarily as a motivational device—the Padè approximant is usually unsatisfactory in practice unless the signal just happens to be exactly the impulse response of some low-order rational system.

A matter of terminology also arises when referring to this method. Padè's work, published in 1892, addressed the problem of matching the terms in a power series with a rational model. Nearly 100 years earlier, in 1795, Prony published a method for exactly representing a set of data points with a model that was the linear combination of *damped* exponentials. As we have observed, the linear combination of complex exponentials is equivalent to a rational z-transform model. Thus Prony's method is equivalent to Padè approximation, although specialized to the case of purely real exponentials. It may seem that proper credit is not being given here to Prony's contribution. Instead, we reserve the name "Prony's method" for a wider set of techniques, in which a set of data samples is approximated by a linear combination of complex exponentials. Strictly speaking, this is still an abuse of terminology because the use of approximation together with Prony's exponential model is a recent development. Some would insist on calling such a technique an "extended Prony method." In any event, Prony's method is discussed in the next section.

Now we formulate the Padè approximation method. Suppose we have a causal signal $x(n)$ that we hope to model with a rational model

$$H(z) = \frac{B(z)}{A(z)} = \frac{\sum\limits_{k=0}^{q} b_k z^{-k}}{1 + \sum\limits_{k=1}^{p} a_k z^{-k}} = \sum_{n=0}^{\infty} h(n)z^{-n} \tag{1.4}$$

Since $H(z)$ depends on $p + q + 1$ free parameters, it seems plausible that we could force $h(n) = x(n)$ for $n = 0, 1, \ldots, p + q$ by proper choice of the a_k and b_k. A simple example with $p = 1$ and $q = 0$ will confirm this suspicion. For $H(z) = b_0/(1 + a_1 z^{-1})$, the impulse response is $h(n) = b_0(-a_1)^n u(n)$, and we match $x(0)$ and $x(1)$ by choosing

$$b_0 = x(0) \qquad \text{and} \qquad a_1 = -x(1)/x(0) \tag{1.5}$$

This solution assumes that $x(0) \neq 0$; otherwise, the time origin needs to be redefined so as to model the nonzero portion of the signal.

The foregoing example seems to suggest that the equations to be solved are, in general, nonlinear. However, it is possible to obtain the solution using a two-step process that requires the solution only of linear equations. This development involves some simple notions from linear algebra and will lead quite naturally to the least-squares approximation problem presented in the next section (the extended Prony method).

Cross-multiplying in Eq. (1.4), we obtain $B(z) = A(z)H(z)$. Equivalently, b_n is the convolution of $h(n)$ and a_n. This can be expressed in convolutional matrix form:

$$
\begin{bmatrix}
h(0) & 0 & 0 & \cdots & 0 \\
h(1) & h(0) & 0 & \ddots & 0 \\
h(2) & h(1) & h(0) & \ddots & 0 \\
h(3) & h(2) & h(1) & \ddots & \vdots \\
\vdots & \ddots & \ddots & \ddots & h(0) \\
\vdots & \ddots & \ddots & \ddots & h(1) \\
\vdots & \vdots & \vdots & & \vdots
\end{bmatrix}
\begin{bmatrix}
1 \\ a_1 \\ a_2 \\ \vdots \\ a_p
\end{bmatrix}
=
\begin{bmatrix}
b_0 \\ b_1 \\ \vdots \\ b_q \\ 0 \\ 0 \\ \vdots
\end{bmatrix}
\tag{1.6}
$$

After imposing the impulse response matching condition, $h(n) = x(n)$, for $n = 0, 1, 2, \ldots, p + q$, we retain the first $p + q + 1$ equations from Eq. (1.6):

$$
\begin{bmatrix}
x(0) & 0 & \cdots & 0 \\
x(1) & x(0) & \ddots & 0 \\
\vdots & \ddots & \ddots & \vdots \\
x(q) & x(q-1) & \cdots & x(q-p) \\
x(q+1) & x(q) & \cdots & \vdots \\
\vdots & \ddots & \ddots & \vdots \\
x(q+p) & x(q+p-1) & \cdots & x(q)
\end{bmatrix}
\begin{bmatrix}
1 \\ a_1 \\ a_2 \\ \vdots \\ a_p
\end{bmatrix}
=
\begin{bmatrix}
b_0 \\ b_1 \\ \vdots \\ b_q \\ 0 \\ 0 \\ \vdots
\end{bmatrix}
\tag{1.7}
$$

These equations are then partitioned and rewritten as[†]

$$
\begin{bmatrix}
\mathbf{X}^{p,0,q} \\
\mathbf{X}^{p,q+1,q+p}
\end{bmatrix}
\mathbf{a} =
\begin{bmatrix}
\mathbf{b} \\
\mathbf{0}_p
\end{bmatrix}
\tag{1.8}
$$

The matrix partition tells us that the solution for the vector $\mathbf{a}$ is independent of $\mathbf{b}$ and can be obtained from the lower partition of Eq. (1.8):

$$
\mathbf{X}^{p,q+1,q+p}\,\mathbf{a} = \mathbf{0}_p
\tag{1.9}
$$

Since $\mathbf{a}^T = [1 \ \mathbf{a}_1^T]$, the first column of $\mathbf{X}^{p,q+1,q+p}$ can be moved to the right-hand side of the equation, and we see that Eq. (1.9) is actually a set of $p \times p$ simultaneous linear equations:

$$
\mathbf{X}^{p-1,q,q+p-1}\,\mathbf{a}_1 = -\mathbf{x}_{q+1,q+p}
\tag{1.10}
$$

There are three cases to consider for the solution of Eq. (1.10).

Case 1. $\mathbf{X}^{p-1,q,q+p-1}$ nonsingular. The unique solution for $\mathbf{a}_1$ is given by

$$
\mathbf{a}_1 = -(\mathbf{X}^{p-1,q,q+p-1})^{-1}\mathbf{X}^{0,q+1,q+p}
$$

[†] Consult Appendix 1.1 at the end of this chapter for a description of the notation in Eq. (1.8).

Case 2. $\mathbf{X}^{p-1,q,q+p-1}$ singular, but Eq. (1.10) has a (nonunique) solution. If the rank of $\mathbf{X}^{p-1,q,q+p-1}$ is r, the solution of Eq. (1.10) with $p - r$ zero entries is preferred because the zero coefficients may reduce the order of the rational model found via the Padè technique.

Case 3. $\mathbf{X}^{p-1,q,q+p-1}$ singular, no solution for Eq. (1.10). In this case, the tacit assumption that $a_0 = 1$ is incorrect. If, instead, a_0 is set to 0, then we can find a (nonunique) solution to Eq. (1.9), although it may not be able to match $x(p + q)$.

In all cases, we use the denominator coefficients of our model, $\mathbf{a}$, to obtain the numerator coefficients $\{b_k\}$ from the upper partition of Eq. (1.9) by matrix multiplication.

$$\mathbf{X}^{p,0,q} \, \mathbf{a} = \mathbf{b} \qquad\qquad (1.11)$$

Example 1

Given $x(n) = n + 1$, $0 \leq n \leq 4$, find a second-order system (i.e., $p = 2$ and $q = 2$) to model $x(n)$. The matrix equations are

$$\begin{bmatrix} 1 & 0 & 0 \\ 2 & 1 & 0 \\ 3 & 2 & 1 \\ 4 & 3 & 2 \\ 5 & 4 & 3 \end{bmatrix} \begin{bmatrix} 1 \\ a_1 \\ a_2 \end{bmatrix} = \begin{bmatrix} b_0 \\ b_1 \\ b_2 \\ 0 \\ 0 \end{bmatrix}$$

The lower partition consists of two equations for which there is a unique solution (Case 1):

$$\begin{bmatrix} a_1 \\ a_2 \end{bmatrix} = \begin{bmatrix} -2 \\ 1 \end{bmatrix}$$

and

$$\begin{bmatrix} b_0 \\ b_1 \\ b_2 \end{bmatrix} = \begin{bmatrix} 1 & 0 & 0 \\ 2 & 1 & 0 \\ 3 & 2 & 1 \end{bmatrix} \begin{bmatrix} 1 \\ -2 \\ 1 \end{bmatrix} = \begin{bmatrix} 1 \\ 0 \\ 0 \end{bmatrix}$$

The model is $H(z) = 1/(1 - 2z^{-1} + z^{-2})$. Since we are seeking a causal signal, the inverse z-transform yields $h(n) = (n + 1)u(n)$ for all n. Thus, $h(n)$ does match $x(n)$ for $0 \leq n \leq 4$.

There are two notable defects in the Padè approximation philosophy. The first is illustrated in the preceding example: the model system is not necessarily stable. No mechanism is built into the Padè approximation technique to guarantee stability of the answer. Later, we will study techniques in which stability can be guaranteed *a priori*. The second is a much more serious problem: the order of the model is tied directly to the number of signal points being matched. Thus, we may need a high-order model if we are to match a significant portion of the signal $x(n)$. The one exception occurs when the signal can be represented *exactly* by a low-order rational model—an unlikely situation for real data. If we have a long signal record available for which we want to derive a low-order model, then we must use the approximation method of the next section.

1.2.2 Prony's Method: Approximate Matching

When constructing a model, it is important to be able to use all the available signal points and at the same time create a low-order model. To accomplish these goals, it is necessary to give up the exact matching imposed in the Padè method and introduce a notion of approximation. Thus we now reconsider the impulse response matching problem when the orders of the numerator and denominator of the model are fixed and finite, but an entire causal signal $x(n)$, $n = 0, 1, \ldots, \infty$ is to be modeled. There are now an infinite number of equations available for matching terms of $x(n)$:

$$\begin{bmatrix} x(0) & 0 & \ddots & 0 \\ x(1) & x(0) & \ddots & 0 \\ x(2) & x(1) & \ddots & 0 \\ \cdot & \cdot & \cdot & \cdot \\ & \ddots & \ddots & \ddots \end{bmatrix} \begin{bmatrix} 1 \\ a_1 \\ a_2 \\ \vdots \\ a_p \end{bmatrix} \overset{?}{=} \begin{bmatrix} b_0 \\ b_1 \\ \vdots \\ b_q \\ 0 \\ 0 \\ \vdots \end{bmatrix} \tag{1.12}$$

Once again these equations can be partitioned to decouple the effects of the numerator and denominator coefficients. The first $q + 1$ rows are separated and we obtain

$$\begin{bmatrix} \mathbf{X}^{p,0,q} \\ \mathbf{X}^{p,q+1,\infty} \end{bmatrix} \begin{bmatrix} 1 \\ \mathbf{a}_1 \end{bmatrix} \overset{?}{=} \begin{bmatrix} \mathbf{b} \\ \mathbf{0} \end{bmatrix} \tag{1.13}$$

The lower partition contains an infinite number of equations to be solved for $\mathbf{a}$:

$$\mathbf{X}^{p,q+1,\infty} \, \mathbf{a} = [\mathbf{x}_{q+1,\infty} \quad \mathbf{X}^{p-1,q,\infty}] \begin{bmatrix} 1 \\ \mathbf{a}_1 \end{bmatrix} \overset{?}{=} \mathbf{0} \tag{1.14}$$

In general, Eq. (1.14) describes a system of overdetermined linear equations that need not have an exact solution, even when the columns of $\mathbf{X}^{p-1,q,\infty}$ are linearly independent vectors. Restated in another way, the vector $\mathbf{x}_{q+1,\infty}$ can only be approximated by the columns of $\mathbf{X}^{p-1,q,\infty}$, so we are forced to choose $\mathbf{a}_1$ to minimize the "equation error"

$$\mathbf{e} = \mathbf{X}^{p,q+1,\infty} \, \mathbf{a} = \mathbf{x}_{q+1,\infty} + \mathbf{X}^{p-1,q,\infty} \, \mathbf{a}_1 \tag{1.15}$$

The least-squares-error norm (min $\mathbf{e}^T\mathbf{e}$) is chosen because of its mathematical simplicity. In summation form, this error can be expressed as

$$\mathbf{e}^T\mathbf{e} = \sum_{i=q+1}^{\infty} e^2(n) = \sum_{i=q+1}^{\infty} \left[x(i) + \sum_{k=1}^{p} a_k x(i - k) \right]^2 \tag{1.16}$$

The reader is reminded that *this is a different least-squares error from that in Eq. (1.2).*

In the special case where the error can be reduced to zero, $\mathbf{x}_{q+1,\infty}$ lies in the span of the columns of $\mathbf{X}^{p-1,q,\infty}$, and Eq. (1.14) holds with equality. This solution is the same as would be obtained via Padè approximation, so we have obtained a generalization of the Padè method.

Minimization of $\mathbf{e}^T\mathbf{e}$ with respect the a_k leads to a set of linear equations, called the *normal equations*. As described in Appendix 1.2, the solution to this linear least-squares minimization problem can be written compactly as[†]

$$(\mathbf{X}^{p,q+1,\infty})^T\mathbf{X}^{p,q+1,\infty}\,\mathbf{a} = \begin{bmatrix} E_a \\ \mathbf{0} \end{bmatrix} \tag{1.17}$$

Actually, Eq. (1.17) contains two equations (see Eqs. 1.2.4–1.2.7). The first is due to the orthogonality condition, $\mathbf{e}_{\min}^T\mathbf{X}^{p-1,q,\infty} = \mathbf{0}$, used in least-squares minimization:

$$(\mathbf{X}^{p-1,q,\infty})^T\mathbf{X}^{p-1,q,\infty}\,\mathbf{a}_1 = -(\mathbf{X}^{p-1,q,\infty})^T\mathbf{x}_{q+1,\infty} \tag{1.18}$$

The second is an expression for the minimal value of the least-squares error:

$$\begin{aligned} E_a = \min \mathbf{e}^T\mathbf{e} &= \mathbf{e}_{\min}^T(\mathbf{X}^{p-1,q,\infty}\,\mathbf{a}_1 + \mathbf{x}_{q+1,\infty}) \\ &= \mathbf{x}_{q+1,\infty}^T\mathbf{X}^{p,q+1,\infty}\,\mathbf{a} \end{aligned} \tag{1.19}$$

Of course, we knew beforehand that the optimal value of $\mathbf{a}_1$ would be found as the solution to a set of *linear* equations, but things are even better because the equations have a special form. The $p \times p$ coefficient matrix $(\mathbf{X}^{p-1,q,\infty})^T\mathbf{X}^{p-1,q,\infty}$ is always symmetric, or Hermitian if the data were complex-valued. It is also positive semidefinite and will be invertible (i.e., positive definite) if and only if the columns of $\mathbf{X}^{p-1,q,\infty}$ are linearly independent. Further examination of the form of these normal equations (1.17) in the case of all-pole modeling will reveal more special structure that can be exploited in the development of fast solution algorithms (e.g., Levinson's recursion).

The precise form of the matrix $\mathbf{R} = (\mathbf{X}^{p,q+1,\infty})^T\mathbf{X}^{p,q+1,\infty}$ is rather simple. First note that the ith column of $\mathbf{X}^{p,q+1,\infty}$ has elements $x(\ell - i)$ for $\ell = q + 1,\, q + 2,$ $\ldots,\, \infty$. Thus, the entries $r_x(i, j)$ of the matrix $\mathbf{R}$ are time correlations of the signal $x(n)$. For this reason, $\mathbf{R}$ is usually referred to as a correlation matrix.

$$r_x(i, j) = \sum_{\ell=q+1}^{\infty} x(\ell - i)x(\ell - j) = r_x(j, i), \qquad i, j = 0, 1, 2, \ldots, p \tag{1.20}$$

Comparison of Eq. (1.20) with a similar definition in Chapter 2 for correlation estimates will reveal a slight difference in that $r_x(i, j)$ is unscaled.

An important distinction to be made at this point lies in the interpretation of the infinite upper limit on the sum in Eq. (1.20). This limit must be chosen to be finite in the practical case where the signal $x(n)$ is necessarily of finite length. There are two ways to impose this condition, and they result in methods with quite different properties. The techniques are known as the *autocorrelation* method and the *covariance* method of all-pole modeling.[‡]

[†] Alternatively, Eq. (1.17) follows from Eq. (1.16) by differentiation with respect to the a_k.

[‡] These names are very misleading, but the terminology is so widespread now that we follow the general custom.

1.2.3 Autocorrelation Method: All-Pole Modeling

The first way to deal with the issue of finite signal length is to consider a finite-length signal to be an infinitely long signal in which most of the values happen to be zero. For example, a signal of length N might be zero-padded to form an infinitely long signal. The error in Eq. (1.16) can be rewritten to account for the fact that $x(n) = 0$ for $n < 0$ and $n \geq N$:

$$\sum_{n=q+1}^{N+p-1} \left[x(n) + \sum_{k=1}^{p} a_k x(n-k) \right]^2 \tag{1.21}$$

Notice the limits on the summation. They reflect the fact that the error is identically zero for $n \geq N + p$. A key observation is that this error sum involves terms where the signal $x(n)$ is zero. Over the range $[N, N + p - 1]$, $x(n)$ is identically zero but is being approximated by a linear combination of terms $x(n - k)$, some of which are nonzero. Thus there is no way to make $e(n) = 0$ over this entire range, except by choosing all the $a_k = 0$. The net result is that this "edge effect" causes a bias in the coefficients $\{a_k\}$. It is a bit like the effect of windowing with respect to the Fourier transform.

Other than the fact that $x(n)$ has many zero values, there is no essential difference between this case and the one where $x(n)$ would have infinitely many nonzero values. However, a major simplification is possible when we specialize further to the case of all-pole modeling (i.e., $q = 0$). Then $r_x(i, j)$ in Eq. (1.20) is only a function of the difference $i - j$. To see this, write out the $(i + 1, j + 1)$st entry:

$$
\begin{aligned}
r_x(i + 1, j + 1) &= \sum_{\ell=q+1}^{\infty} x(\ell - i - 1)x(\ell - j - 1) \\
&= x(q - i)x(q - j) + r_x(i, j)
\end{aligned}
\tag{1.22}
$$

The first term will be zero in the all-pole case ($q = 0$), when $1 \leq i, j \leq p$, because $x(n) = 0$ for $n < 0$. This is almost what we want, except that everything would be much tidier if the result were also to apply for $i = 0$ and $j = 0$. This can be done by modifying the error to include the constant term $e^2(0) = x^2(0)$. Then the sums in Eqs. (1.21) and (1.22) will start at 0 instead of 1, but minimization of the error still gives the same solution, and we have the desired result:

$$r_x(i + 1, j + 1) = r_x(i, j) \stackrel{\text{def}}{=} r_x(i - j), \qquad i, j = 0, 1, \ldots, p \tag{1.23}$$

Thus the matrix $\mathbf{R} = (\mathbf{X}^{p,0,\infty})^T \mathbf{X}^{p,0,\infty}$ has equal-valued entries along each diagonal. This type of matrix is called Toeplitz; the following is a 5×5 symmetric example.

$$\mathbf{R} = \begin{bmatrix} r_x(0) & r_x(1) & r_x(2) & r_x(3) & r_x(4) \\ r_x(1) & r_x(0) & r_x(1) & r_x(2) & r_x(3) \\ r_x(2) & r_x(1) & r_x(0) & r_x(1) & r_x(2) \\ r_x(3) & r_x(2) & r_x(1) & r_x(0) & r_x(1) \\ r_x(4) & r_x(3) & r_x(2) & r_x(1) & r_x(0) \end{bmatrix} \tag{1.24}$$

The name *autocorrelation* method comes from the observation that the entries in the matrix $\mathbf{R}$ are the first $p + 1$ autocorrelation coefficients of the signal $x(n)$. The special Toeplitz structure greatly simplifies the solution of the autocorrelation normal equations (ACNE) by virtue of the Levinson algorithm (Section 1.3). It also happens that we can guarantee that $\mathbf{R}$ is invertible because the p columns of $\mathbf{X}^{p,q+1,\infty}$ are always linearly independent for the all-pole case.

Now we ask an important question. Suppose that $x(n)$ was produced by windowing an infinitely long signal $x_\infty(n)$:

$$x(n) = \begin{cases} w(n)x_\infty(n), & n = 0, 1, \ldots, N - 1 \\ 0, & \text{elsewhere} \end{cases} \tag{1.25}$$

How does the model for $x(n)$ compare with the z-transform $X_\infty(z)$ when the latter is exactly all-pole? The answer is that the model from the autocorrelation method can *never* be correct. A simple example illustrates this shortcoming of the autocorrelation method.

Example 2

Suppose that $x_\infty(n) = \gamma^n u(n)$, with z-transform $X_\infty(z) = 1/(1 - \gamma z^{-1})$ and $|\gamma| < 1$. To build a first-order all-pole model from N points of $x(n)$, we must compute $r_x(0)$ and $r_x(1)$:

$$r_x(0) = \sum_{n=0}^{N-1} \gamma^n \gamma^n = \frac{1 - \gamma^{2N}}{1 - \gamma^2}$$

$$r_x(1) = \sum_{n=0}^{N-2} \gamma^n \gamma^{n+1} = \gamma \frac{1 - \gamma^{2N-2}}{1 - \gamma^2} \tag{1.26}$$

The solution for a_1 should be $-\gamma$, but it is always biased for finite N. Of course, as $N \to \infty$, a_1 converges to $-\gamma$:

$$a_1 = -\frac{r_x(1)}{r_x(0)} = -\gamma \frac{1 - \gamma^{2N-2}}{1 - \gamma^{2N}} \tag{1.27}$$

Incidentally, you might want to try computing the minimum squared error for this problem. The result should be $(1 - \gamma^{4N-2})/(1 - \gamma^{2N})$, which tends to its minimum value of $x^2(0) = 1$ as $N \to \infty$.

The ACNE solution is incorrect because the autocorrelation coefficients produced from $x(n)$ differ from those of $x_\infty(n)$. Although Example 2 presents a deterministic case, the same statements hold true when $x(n)$ is a random process. The implicit rectangular window distorts the autocorrelation estimates. The use of other windows such as the Hamming window can improve the quality of the model. This is the case in the spectral estimation of narrowband signals closely spaced in frequency. The next technique considered, the covariance method, does not suffer this same difficulty—it can compute the correct model from a finite segment of data.

As a final note, there has been no mention yet of the numerator coefficient b_0 for the all-pole model. Since certain properties of the autocorrelation method depend on the choice of this coefficient the discussion is deferred to Section 1.3. In any event, b_0 is merely a gain and is clearly of secondary importance in all-pole modeling.

1.2.4 Covariance Method

A second way to deal with a finite-length signal, $x(n)$, is to approximate only over the range available, say $[0, N - 1]$. Equivalently, we restrict the number of equations in Eq. (1.10) to be finite, with the result that, in the normal equations (1.17), the matrix $\mathbf{X}^{p,q+1,N-1}$ is used in place of $\mathbf{X}^{p,q+1,\infty}$:

$$\mathbf{X}^{p,q+1,N-1} = \begin{bmatrix} x(q+1) & x(q) & \cdots & x(q-p+1) \\ x(q+2) & x(q+1) & \cdots & x(q-p+2) \\ \vdots & \vdots & \vdots & \vdots \\ x(N-1) & \cdots & \cdots & x(N-p-1) \end{bmatrix} \tag{1.28}$$

Stated another way, the minimization of the error is restricted to the range $[q + 1, N - 1]$. The error signal is never evaluated outside the finite range of the data, in particular, for indices greater than $N - 1$. This property is often summarized by stating that "the covariance error operator does not run off the end of the data." Figure 1.3 summarizes the situation schematically.

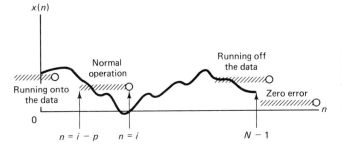

Figure 1.3 Schematic representation of the error calculation operator $e(n) = x(n) - \sum_{j=1}^{p} (-a_j)x(n - j)$. The error is the difference between $x(n)$ indicated by the open circle and a linear combination of the previous p signal points, shown by the shaded region.

Consider the all-pole case ($q = 0$), where the signal $x(n)$ extends over $[0, N - 1]$. The autocorrelation method takes the error over $[1, \infty]$, whereas the covariance method uses the interval $[1, N - 1]$. In fact, $e(n)$ is identically zero for $n \geq N + p$, so that the autocorrelation method actually extends over $[1, N + p - 1]$. However, the inclusion of $e(n)$ for $N \leq n < N + p$ makes a significant difference because $x(n) = 0$ for $n \geq N$. As already noted for the autocorrelation method, this "edge effect" will bias the model coefficients. The covariance method avoids this circumstance because its error operator never "runs off the data."

It is interesting to note that including $e(n)$ in the interval $[1, p - 1]$ does not cause a bias for either method, even though one is "running onto the data." This may seem puzzling, but the explanation boils down to the fact that a model's impulse response is assumed to be produced with zero initial conditions. Incidentally, this case is called "prewindowed" by several authors and offers some simplification in the development of fast algorithms.

Unlike the autocorrelation method, the covariance method can determine the exact model from a finite data segment when the given signal is truly all-pole. This also holds true in the pole-zero case. Recall the example from the last section, $X_\infty(z) = 1/(1 - \gamma z^{-1})$. The first-order all-pole covariance model is obtained by solving the normal equations.

$$\begin{bmatrix} \gamma & \gamma^2 & \cdots & \gamma^{N-1} \\ 1 & \gamma & \cdots & \gamma^{N-2} \end{bmatrix} \begin{bmatrix} \gamma & 1 \\ \gamma^2 & \gamma \\ \vdots & \vdots \\ \gamma^{N-1} & \gamma^{N-2} \end{bmatrix} \begin{bmatrix} 1 \\ a_1 \end{bmatrix} = \begin{bmatrix} E_{CV} \\ 0 \end{bmatrix} \tag{1.29}$$

The solution, $a_1 = -r_x(1, 0)/r_x(1, 1) = -\gamma$, is the correct model coefficient, provided $N \geq 2$. The squared error can be computed to be

$$E_{CV} = r_x(0,0) + a_1 r_x(0, 1) = 0$$

A further advantage of the covariance method is that it works just as well for the pole-zero case ($q \neq 0$). The poles of the model are determined by minimizing the error starting at index $q + 1$. More generally, any range of signal values can be used. If L and U are any arbitrary lower and upper limits on the range of a signal, then poles for a covariance-type model can be calculated by minimizing the error over the range $[L + p, U]$. (This is sometimes called the "unwindowed" case.)

$$E_{CV} = \sum_{n=L+p}^{U} \left[x(n) + \sum_{k=1}^{p} a_k x(n - k) \right]^2 \tag{1.30}$$

A version of pole-zero modeling based on this idea is described in the final section of this chapter.

In the covariance method, the normal equations do *not* have the Toeplitz matrix structure, even in the all-pole case. The entries $r_x(i, j)$ of the covariance matrix $\mathbf{R} = (\mathbf{X}^{p,q+1,N-1})^T \mathbf{X}^{p,q+1,N-1}$ are

$$r_x(i, j) = r_x(j, i) = \sum_{\ell=q+1}^{N-1} x(\ell - i) x(\ell - j) \tag{1.31}$$

In an attempt to imitate the previous derivation, we can write

$$r_x(i + 1, j + 1) = r_x(i, j) + x(q - i)x(q - j) - x(N - 1 - i)x(N - 1 - j) \tag{1.32}$$

In this case, $r_x(i, j)$ depends on both indices i and j. It still appears similar to a covariance estimate, although for a nonstationary signal. Thus we have a possible rationalization for the name "covariance" method.

1.3 SOLUTION OF THE AUTOCORRELATION NORMAL EQUATIONS

The special form of the equations arising in the autocorrelation method, namely a symmetric Toeplitz matrix, can be exploited to obtain a fast method of solution. The term "fast" has a precise meaning in this context: the linear equations can be solved using a number of operations proportional to p^2 rather than p^3, as would be the case for a general method such as Gaussian elimination. This procedure, first published by Norman Levinson in 1947, is well known as the Levinson recursion.

The autocorrelation solution also possesses a number of elegant properties. The most amazing is the fact that the polynomial $A(z)$ formed from the coefficients $\{a_k\}$ is guaranteed to have all its roots inside the unit circle. This and other properties are discussed in section 1.3.2.

1.3.1 Levinson's Recursion

The original Levinson algorithm was a method for solving a Wiener filtering problem. The equations were a bit more general than the ACNE in that right-hand side of Eq. (1.33) was an arbitrary vector. A refinement, due to Durbin, simplifies the recursion when the right-hand side is composed of correlations. This is the case to be derived here; the more general case is left as an exercise.

Since we are now interested in the order of the $\mathbf{R}$ matrix, a superscript is added to all vectors and matrices. In addition, we suppress the subscript on r_x that indicates that the autocorrelation is for $x(n)$. Repeating the ACNE, we have

$$\mathbf{R}^j \mathbf{a}^j = \begin{bmatrix} E_a^j \\ \mathbf{0}_j \end{bmatrix} \tag{1.33}$$

where $\mathbf{R}^j$ is a symmetric $(j+1) \times (j+1)$ Toeplitz matrix.

$$\mathbf{R}^j = \begin{bmatrix} r(0) & r(1) & \cdots & r(j-1) & r(j) \\ r(1) & r(0) & \ddots & \ddots & r(j-1) \\ \vdots & \ddots & \ddots & \ddots & \ddots \\ r(j-1) & \ddots & \ddots & \ddots & r(1) \\ r(j) & r(j-1) & \ddots & r(1) & r(0) \end{bmatrix} \tag{1.34}$$

and the solution of the jth-order problem is E_a^j and $\mathbf{a}^j$:

$$\mathbf{a}^j = \begin{bmatrix} 1 & a_1^j & a_2^j & \cdots & a_j^j \end{bmatrix}^T \tag{1.35}$$

For a pth-order model, we need the solution at $j = p$. The basic idea is to build an "order recursion" in which the solution for $\mathbf{a}^{j+1}$ and E_a^{j+1} can be found from the jth-order solution $\mathbf{a}^j$, E_a^j. When $j = 1$, Eq. (1.33) is 2×2:

$$\begin{bmatrix} r(0) & r(1) \\ r(1) & r(0) \end{bmatrix} \begin{bmatrix} 1 \\ a_1^1 \end{bmatrix} = \begin{bmatrix} E_a^1 \\ 0 \end{bmatrix} \tag{1.36}$$

so the solution is $k_1 = a_1^1 = -r(1)/r(0)$ and $E_a^1 = r(0) + k_1 r(1) = (1 - k_1^2)r(0)$, and we are under way.

First, we assume that the jth-order solution, $\mathbf{a}^j$, is known.

$$\mathbf{R}^j \mathbf{a}^j = \begin{bmatrix} E_a^j \\ \mathbf{0}_j \end{bmatrix} \tag{1.37}$$

Then the crucial observation is that the solution to the equation

$$\mathbf{R}^j \boldsymbol{\alpha}^j = \begin{bmatrix} \mathbf{0}_j \\ E_a^j \end{bmatrix} \tag{1.38}$$

is merely the reversed version of the solution to Eq. (1.37). This provides the key to the update derived in Eqs. (1.44) and (1.50).

$$\boldsymbol{\alpha}^j = [a_j^j \quad a_{j-1}^j \quad \cdots \quad a_1^j \quad 1]^T \tag{1.39}$$

Now consider the $(j + 1)$st-order problem.

$$\mathbf{R}^{j+1}\mathbf{a}^{j+1} = \begin{bmatrix} E_a^{j+1} \\ \mathbf{0}_{j+1} \end{bmatrix} \tag{1.40}$$

Two partitions of $\mathbf{R}^{j+1}$ can be made in terms of the lower-order problem:

$$\mathbf{R}^{j+1} = \begin{bmatrix} \mathbf{R}^j & \mathbf{r}^j \\ (\mathbf{r}^j)^T & r(0) \end{bmatrix} = \begin{bmatrix} r(0) & (\boldsymbol{\rho}^j)^T \\ \boldsymbol{\rho}^j & \mathbf{R}^j \end{bmatrix} \tag{1.41}$$

where

$$\mathbf{r}^j = [r(1) \quad r(2) \quad \cdots \quad r(j + 1)]^T \tag{1.42}$$

and

$$\boldsymbol{\rho}^j = [r(j + 1) \quad r(j) \quad \cdots \quad r(1)]^T \tag{1.43}$$

Another key is that $\boldsymbol{\rho}^j$ is merely $\mathbf{r}^j$ turned around.

It is always true that the new solution $\mathbf{a}^{j+1}$ can be written in the form

$$\mathbf{a}^{j+1} = \begin{bmatrix} \mathbf{a}^j \\ 0 \end{bmatrix} + k_{j+1} \begin{bmatrix} 0 \\ \boldsymbol{\epsilon}^j \end{bmatrix} \tag{1.44}$$

where the last entry of $\boldsymbol{\epsilon}^j$ is taken to be 1. The quantity k_{j+1} is called a "reflection coefficient" and will be shown to have many special properties. To determine k_{j+1} and the $(j + 1)$-vector $\boldsymbol{\epsilon}^j$, we expand Eq. (1.40) using Eqs. (1.41) and (1.44):

$$\begin{bmatrix} \mathbf{R}^j & \mathbf{r}^j \\ (\mathbf{r}^j)^T & r(0) \end{bmatrix} \begin{bmatrix} \mathbf{a}^j \\ 0 \end{bmatrix} + k_{j+1} \begin{bmatrix} r(0) & (\boldsymbol{\rho}^j)^T \\ \boldsymbol{\rho}^j & \mathbf{R}^j \end{bmatrix} \begin{bmatrix} 0 \\ \boldsymbol{\epsilon}^j \end{bmatrix} = \begin{bmatrix} E_a^{j+1} \\ \mathbf{0}_{j+1} \end{bmatrix} \tag{1.45}$$

Then we perform the matrix multiplications:

$$\begin{bmatrix} \mathbf{R}^j \mathbf{a}^j \\ (\mathbf{r}^j)^T \mathbf{a}^j \end{bmatrix} + k_{j+1} \begin{bmatrix} (\boldsymbol{\rho}^j)^T \boldsymbol{\epsilon}^j \\ \mathbf{R}^j \boldsymbol{\epsilon}^j \end{bmatrix} = \begin{bmatrix} E_a^{j+1} \\ \mathbf{0}_{j+1} \end{bmatrix} \tag{1.46}$$

Since $\mathbf{R}^j \mathbf{a}^j$ is known from Eq. (1.37), we have

$$\begin{bmatrix} E_a^j \\ \mathbf{0}_j^j \\ (\mathbf{r}^j)^T \mathbf{a}^j \end{bmatrix} + k_{j+1} \begin{bmatrix} (\boldsymbol{\rho}^j)^T \boldsymbol{\epsilon}^j \\ \mathbf{R}^j \boldsymbol{\epsilon}^j \end{bmatrix} = \begin{bmatrix} E_a^{j+1} \\ \mathbf{0}_{j+1} \end{bmatrix} \tag{1.47}$$

and we can now write two equations:

$$E_a^j + k_{j+1}(\boldsymbol{\rho}^j)^T \boldsymbol{\epsilon}^j = E_a^{j+1} \tag{1.48}$$

$$k_{j+1} \mathbf{R}^j \boldsymbol{\epsilon}^j = -\begin{bmatrix} \mathbf{0}_j \\ (\mathbf{r}^j)^T \mathbf{a}^j \end{bmatrix} \tag{1.49}$$

The solution to Eq. (1.49) is found from Eq. (1.38) by noting that $\boldsymbol{\epsilon}^j$ must be a scalar multiple of $\boldsymbol{\alpha}^j$:

$$k_{j+1}E_a^j\boldsymbol{\epsilon}^j = -[(\mathbf{r}^j)^T\mathbf{a}^j]\boldsymbol{\alpha}^j \tag{1.50}$$

In fact, the condition that the last entry of $\boldsymbol{\epsilon}^j$ be 1 means that $\boldsymbol{\epsilon}^j = \boldsymbol{\alpha}^j$, from which we calculate k_{j+1}:

$$k_{j+1} = \frac{-[(\mathbf{r}^j)^T\mathbf{a}^j]}{E_a^j} \tag{1.51}$$

Since $(\boldsymbol{\rho}^j)^T\mathbf{a}^j = (\mathbf{r}^j)^T\mathbf{a}^j$, the reflection coefficient k_{j+1} can also be expressed as

$$k_{j+1} = \frac{-[(\boldsymbol{\rho}^j)^T\boldsymbol{\alpha}^j]}{E_a^j} \tag{1.52}$$

The denominator is the minimum squared error for the jth-order problem, so we need not worry about dividing by zero. In fact, zero error means that the exact solution has been obtained and that the recursion can be terminated. In general, we are interested in monitoring the error. This is possible if we write a recursive update for the error using Eq. (1.48).

$$
\begin{aligned}
E_a^{j+1} &= E_a^j + k_{j+1}(\boldsymbol{\rho}^j)^T\boldsymbol{\alpha}^j \\
&= E_a^j - k_{j+1}^2 E_a^j \\
&= (1 - k_{j+1}^2)E_a^j
\end{aligned}
\tag{1.53}
$$

Equations (1.53), (1.51), and (1.44) constitute the core of the algorithm. Collecting the vital equations together, we write a complete specification of the Levinson recursion:

$$a_1^1 = k_1 = \frac{-r(1)}{r(0)} \tag{1.54a}$$

$$E_a^1 = (1 - k_1^2)r(0) \tag{1.54b}$$

$$k_{j+1} = -\frac{r(j+1) + \sum_{i=1}^{j} a_i^j r(j+1-i)}{E_a^j} \tag{1.54c}$$

$$a_{j+1}^{j+1} = k_{j+1} \tag{1.54d}$$

$$a_i^{j+1} = a_i^j + k_{j+1}a_{j+1-i}^j, \qquad i = 1, 2, \ldots, j \tag{1.54e}$$

$$E_a^{j+1} = (1 - k_{j+1}^2)E_a^j \tag{1.54f}$$

The Levinson recursion can be generalized to solve any system of equations of the form $\mathbf{Ra} = \mathbf{c}$. When the column vector $\mathbf{c}$ on the right-hand side is arbitrary, an additional recursion is required, but the derivation is similar to the one above. When the $\mathbf{R}$ matrix is not Toeplitz but is close to Toeplitz as in the covariance method, the normal equations can still be solved by a "fast" algorithm, as shown in Section 1.4.

The approach is similar, although complicated by the fact that several additional updates are necessary.

One significant feature of the Levinson recursion is that it reduces the amount of computation necessary to solve the $(p + 1) \times (p + 1)$ normal equations from something proportional to p^3 (as with Gaussian elimination) to an amount proportional to p^2. The solution of a general set of p equations in p unknowns with Gaussian elimination requires $\frac{1}{3}p^3 + O(p^2)$ operations (i.e., multiplications and divisions) plus p^2 storage locations. Since the matrix $\mathbf{R}^p$ is a symmetric positive semidefinite matrix, and is usually positive definite, the normal equations can also be solved via Cholesky's method. The amount of computation is then $\frac{1}{6}p^3 + O(p^2)$, the total storage is $\frac{1}{2}p^2$, and the method is known to be numerically well behaved.

By comparison, the Levinson recursion requires $2j + 2$ multiplications, $2j + 1$ additions, and 1 division to update from the jth-order solution to the $(j + 1)$st. Summing over p iterations we get a total of $\sum_{j=0}^{p-1} 2j + 3 = p(p + 2) = p^2 + 2p$ multiplications. For p in the range of 10 to 14 (as is the case for speech modeling), the savings can be significant. The storage is also much less, namely $2p + 2$ memory locations for $\{a_1^p, \ldots, a_p^p, r(0), r(1), r(2), \ldots, r(p), E_a^p\}$.

It is important to point out that in the overall modeling process the solution of the normal equations may be a small fraction of the total calculation. The computation of the correlation coefficients $r_x(i, j)$ requires pN multiplications, where N is the length of the signal $x(n)$. If $N \gg p$, then pN will dominate the p^2 operations of Levinson's method or the p^3 operations of Cholesky's method.

A final consideration is the numerical stability of the calculation. When the normal equations of the least-squares problem are ill-conditioned, recursive methods for solving these linear equations are quite sensitive to rounding errors. The problem can be attributed to errors in forming the correlation matrix. An alternate least-squares computational algorithm, known as Golub's method, avoids explicit calculation of the correlation coefficients. It works directly with the signal matrix and should be more robust when the normal equations are ill conditioned. The number of operations for Golub's method is about the same as for Cholesky's method.

The form of Levinson's recursion not only gives an efficient means of calculation for the ACNE, but it also leads to a number of elegant properties for the ACNE solution. The next section discusses these properties, especially those depending on the reflection coefficients.

1.3.2 Properties of the Autocorrelation Solution

In addition to the fast solution of the autocorrelation equations using Levinson's method, many elegant properties of the autocorrelation solution apply to the all-pole model $H(z)$:

$$H(z) = \frac{b_0}{1 + \sum_{k=1}^{p} a_k z^{-k}} = \frac{b_0}{A(z)} \tag{1.55}$$

Most of these properties relate to the reflection coefficients generated as a by-product of Levinson's recursion. It should not be assumed that these properties are of theoretical interest only. Many have important practical implications. First, stability of the all-pole model can be guaranteed *a priori*.

Property.   The polynomial $A(z)$ from the autocorrelation method is minimum phase (i.e., its p roots lie strictly inside the unit circle).

We demonstrate the minimum-phase condition directly from the error minimization property of Prony's method. This method of proof can be extended to certain instances of the covariance method, so we formulate it with that generalization in mind. We consider that the energy in the error signal $e(n)$ is minimized over the range $[L, U]$ and specialize to the autocorrelation method by taking $L = 0$ and $U = \infty$.

The proof is accomplished by contradiction. Suppose the optimal $A(z)$ has at least one root q such that $|q| \geq 1$. Then $A(z)$ can written as

$$A(z) = (1 - qz^{-1})A_0(z), \tag{1.56}$$

where $A_0(z)$ is monic (i.e., the leading coefficient is equal to 1). In other words, $x(n)$ is filtered by the cascade of the two systems $A_0(z)$ and $(1 - qz^{-1})$, as shown in Fig. 1.4.

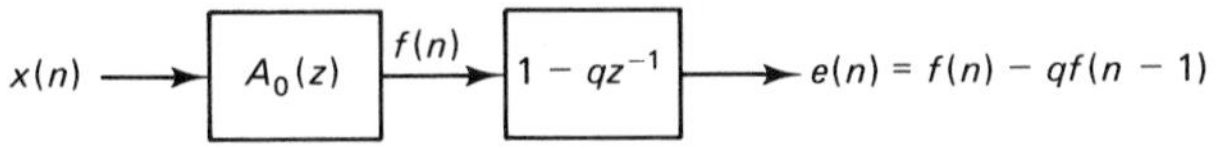

Figure 1.4   Removing one zero from $A(z)$.

If $f(n)$ denotes the output signal after $x(n)$ is filtered by $A_0(z)$, then the output error energy can be written as

$$E = \sum_{n=L}^{U} |e(n)|^2 = \sum_{n=L}^{U} [f(n) - qf(n-1)][f(n) - qf(n-1)]^* \tag{1.57}$$

Letting $q = re^{j\theta}$, we obtain the error energy

$$E = \sum_{n=L}^{U} |f(n)|^2 + r^2|f(n-1)|^2 - 2\Re e\{re^{j\theta} f(n-1)f^*(n)\} \tag{1.58}$$

We now consider $\partial E/\partial r$, which should be zero for the optimal predictor.

$$\frac{\partial E}{\partial r} = 2r \sum_{n=L}^{U} |f(n-1)|^2 - 2\Re e \sum_{n=L}^{U} e^{j\theta} f(n-1)f^*(n) \tag{1.59}$$

The Cauchy-Schwartz inequality tells us that

$$\Re e \sum_{n=L}^{U} e^{j\theta} f(n-1)f^*(n) \leq \left[\sum_{n=L}^{U} |f(n)|^2\right]^{1/2} \left[\sum_{n=L}^{U} |e^{j\theta} f(n-1)|^2\right]^{1/2} \tag{1.60}$$

with equality if and only if $f(n) = e^{j\theta} f(n-1)$ for $n = L, L+1, \ldots, U$.

If we denote the quantity $(\sum_{n=L}^{U} |f(n-1)|^2)^{1/2}$ by F, then Eq. (1.59) is

$$\frac{\partial E}{\partial r} \geq 2rF^2 - 2F[F^2 + |f(U)|^2 - |f(L-1)|^2]^{1/2} \tag{1.61}$$

with the same condition for equality as before. We now apply Eq. (1.61) to the autocorrelation method, where $L = 0$ and $U = \infty$. Since the input signal $x(n)$ is causal and has finite energy, it follows that $f(-1) = 0$ and $f(\infty) = 0$, and Eq. (1.61) becomes

$$\frac{\partial E}{\partial r} \geq 2(r - 1)F^2 \tag{1.62}$$

If the root lies outside the unit circle ($r \geq 1$), the right-hand side of Eq. (1.62) is always greater than or equal to zero. In fact, the derivative $\partial E/\partial r$ can be zero if and only if $r = 1$ and $f(n) = e^{j\theta} f(n - 1)$ for $n \geq 0$. But then $f(n)$ and $x(n)$ must be identically zero. Thus $\partial E/\partial r > 0$ for $n \geq 1$, contradicting the assumption that r was a stationary point of E, and we conclude that all the roots of $A(z)$ must lie strictly within the unit circle.

The foregoing proof relies solely on the error minimization property of Prony's method. Most other methods of proof have been based on the following property of the reflection coefficients, $\{k_j\}$, generated as a by-product in Levinson's recursion.

Property. The reflection coefficients are all less than 1 in magnitude; $|k_j| < 1$.

Proof. From the Levinson's recursion (Eqs. 1.54a–f), $E_a^j = (1 - k_j^2)E_a^{j-1}$ and $E_a^j > 0$ for all j. Thus, $1 - k_j^2 > 0$, which implies $|k_j| < 1$.

This property of the reflection coefficients has led to their use in special digital filter implementations known as *lattice filters*. The lattice filters are nothing more than an interpretation of Levinson's recursion, rewritten in polynomial form. Let $A_j(z)$ denote the jth-order model. The following equation is equivalent to the Levinson update (Eqs. 1.54d,e).

$$A_{j+1}(z) = A_j(z) + k_{j+1}z^{-(j+1)}A_j(z^{-1}) \tag{1.63}$$

The initial condition is $A_0(z) = 1$. Defining $\alpha_j(z) = z^{-j}A_j(z^{-1})$, we obtain a companion recursion from Eq. (1.63) by replacing z with z^{-1}:

$$\alpha_{j+1}(z) = z^{-1}\alpha_j(z) + k_{j+1}A_j(z) \tag{1.64}$$

Also, Eq. (1.63) can be written as

$$A_{j+1}(z) = A_j(z) + k_{j+1}z^{-1}\alpha_j(z) \tag{1.65}$$

The polynomial $\alpha_j(z)$ is just $A_j(z)$ flipped around and, therefore, corresponds to the solution α^j used in the Levinson recursion.

Equations (1.64) and (1.65) serve as the basis of two types of lattice filters: the all-zero lattice for $A_p(z)$ in Fig. 1.5 and the all-pole lattice for $1/A_p(z)$ in Fig. 1.6. In the all-zero lattice, each section is an implementation of the Levinson recursion in Eqs. (1.64) and (1.65). The all-pole lattice follows from rewriting Eqs. (1.64) and (1.65) as

$$A_j(z) = A_{j+1}(z) - k_{j+1}z^{-1}\alpha_j(z) \quad \text{and} \quad \alpha_{j+1}(z) = k_{j+1}A_j(z) + z^{-1}\alpha_j(z) \tag{1.66}$$

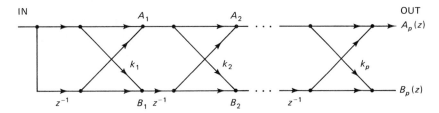

Figure 1.5 All-zero lattice structure for an FIR filter (no feedback).

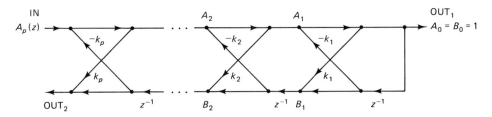

Figure 1.6 All-pole lattice filter implementation.

Each section has inputs $A_{j+1}(z)$ and $\alpha_j(z)$ and outputs $A_j(z)$ and $\alpha_{j+1}(z)$. The transfer function from IN to OUT_1 is $1/A_p(z)$. Interestingly, the transfer function to the second output OUT_2 is an all-pass filter $z^{-p}A_p(z^{-1})/A_p(z)$. Lattice filters are finding increased use in filter implementations and adaptive filtering applications. The reasons are twofold: first, the stability or minimum-phase nature of the filter can be controlled by keeping $|k_j| < 1$; and second, the filter is modular—increasing the order is accomplished by merely cascading additional sections without recomputing the existing k_j. Although it may not be obvious, any minimum-phase all-pole filter can be implemented in lattice form. Indeed, we will show that the reflection coefficients can be calculated from the filter coefficients $\{a_k\}$ by a reverse Levinson recursion (see Eq. 1.77).

Up to this point we have not considered the calculation of the numerator constant b_0 used in the all-pole model (Eq. 1.55). From the previous discussion of Prony's method, we might pick $b_0 = x(0)$ in order to match the first term of $x(n)$ with that of the model's impulse response, $h(0)$. However, a much better choice is to select b_0 such that the energy in the two signals $x(n)$ and $h(n)$ is the same. This second criterion leads to the so-called correlation matching property.

Property.   The autocorrelation sequences for $x(n)$ and $h(n)$ match term by term over the range $-p, -p + 1, \ldots, -1, 0, 1, 2, \ldots, p$.

Furthermore, we can show that the correlation matching requires that b_0 be calculated from the Levinson recursion as $b_0^2 = E_a^p$, where E_a^p is the minimum error energy found at the pth step. Proof of the correlation matching property is rather difficult, but the argument is roughly as follows:

Multiplying both sides of Eq. (1.55) by $H(z^{-1})A(z)$, we obtain

$$H(z)H(z^{-1})A(z) = b_0 H(z^{-1}) \tag{1.67}$$

Let $H(z)H(z^{-1}) = \sum_{\ell=-\infty}^{\infty} r_h(\ell)z^{-\ell}$ be the z-transform of the autocorrelation sequence for $h(n)$. Expanding Eq. (1.67) and recalling that $h(n)$ is right-sided gives

$$\left(\sum_{\ell=-\infty}^{\infty} r_h(\ell)z^{-\ell} \right)\left(1 + \sum_{k=1}^{p} a_k z^{-k} \right) = b_0 \sum_{n=0}^{\infty} h(n)z^{+n} \tag{1.68}$$

Comparing powers of z in Eq. (1.68) yields

$$r_h(\ell) + a_1 r_h(\ell - 1) + \cdots + a_p r_h(\ell - p) = \begin{cases} b_0 h(0), & \ell = 0 \\ 0, & \ell = 1, 2, \ldots, p \end{cases} \tag{1.69}$$

Using symmetry, $r_h(\ell) = r_h(-\ell)$, we obtain

$$\sum_{k=1}^{p} a_k r_h(\ell - k) = -r_h(\ell), \qquad \ell = 1, \ldots, p \tag{1.70}$$

and using $h(0) = b_0$, we have an expression for the gain:

$$b_0^2 = r_h(0) + \sum_{k=1}^{p} a_k r_h(k) \tag{1.71}$$

In matrix form,

$$\begin{bmatrix} r_h(0) & r_h(1) & \cdot & r_h(p) \\ r_h(1) & r_h(0) & \cdot & \cdot \\ \cdot & \cdot & \cdot & \cdot \\ \cdot & \cdot & \cdot & \cdot \\ r_h(p) & r_h(p-1) & \cdot & r_h(0) \end{bmatrix} \begin{bmatrix} 1 \\ a_1 \\ a_2 \\ \vdots \\ a_p \end{bmatrix} = -\begin{bmatrix} b_0^2 \\ 0 \\ 0 \\ \vdots \\ 0 \end{bmatrix} \tag{1.72}$$

Therefore, the following two matrix equations have the same solution

$$\mathbf{Ra} = \begin{bmatrix} E_a^p \\ \mathbf{0}_p \end{bmatrix} \tag{1.73}$$

and

$$\mathbf{R}_h \mathbf{a} = \begin{bmatrix} b_0^2 \\ \mathbf{0}_p \end{bmatrix} \tag{1.74}$$

Using $r(0) = r_h(0)$ (by virtue of the definition of b_0), we can argue that $r(\ell) = r_h(\ell)$ for $|\ell| \leq p$. This follows when both Eqs. (1.73) and (1.74) are solved with Levinson's method. The first-order solutions must match, then the second-order ones, etc. The details of the proof are left as an exercise. As a result, the expression for b_0 in Eq. (1.71) becomes

$$b_0^2 = r(0) + \sum_{k=1}^{p} a_k r(k) = E_a^p \tag{1.75}$$

Analogously, the gain for the all-pole covariance method is usually chosen to be

$$b_0^2 = r(0, 0) + \sum_{k=1}^{p} a_k r(0, k) \tag{1.76}$$

The foregoing argument also supports the following property.

Property.   There is a one-to-one relationship between the two sets of parameters $\{a_1, a_2, \ldots, a_p, b_0\}$ and $\{r(0), r(1), \ldots, r(p)\}$.

It is also the case that knowledge of the reflection coefficients k_j, $j = 1, 2,$ $\ldots, p$ and the signal energy $r(0)$ is equivalent to either of these two sets. This is a consequence of the reverse Levinson algorithm in which Eqs. (1.54a–f) are turned around to compute the reflection coefficients from the model parameters:

$$k_j = a_j^j \tag{1.77a}$$

$$a_i^{j-1} = \left(\frac{1}{1 - k_j^2}\right)(a_i^j - k_j a_{j-1}^j), \qquad i = 1, 2, \ldots, j - 1 \tag{1.77b}$$

The reverse Levinson algorithm starts with $\mathbf{a}^p$ known and generates each of the lower-order models from Eqs. (1.77).

Property.   There is a one-to-one relationship between the two sets of parameters $\{a_1, a_2, \ldots, a_p, b_0\}$ and $\{k_1, k_2, \ldots, k_p, E_a^p\}$.

The equivalence between these two sets means that *any* minimum-phase all-pole filter can be realized in lattice form. Taking the last two properties together, we have demonstrated an equivalence among three different sets of parameters: coefficients for a stable model, positive-definite correlations, and reflection coefficients that are less than 1 in magnitude.

1.4 FAST SOLUTION OF COVARIANCE EQUATIONS

It is also possible to derive a fast $O(p^2)$ algorithm for the solution of the normal equations from the covariance method. The derivation is considerably more detailed than that leading to Levinson's recursion, but the approach is similar in spirit. In the covariance case, the new twist is that two types of updating must be done. The first, *order updating,* is like that found in the Levinson recursion. The second, *time updating,* arises because of the finite extent of the signal used in the covariance method. It turns out that there is a simple relationship between the covariance solution over the interval $[L, U]$ and the solution over the interval $[L + 1, U]$ or $[L, U - 1]$. To incorporate time updating into a fast covariance algorithm, two auxiliary least-squares problems must be solved and updated recursively. Time updating is also the basis of a powerful adaptive filtering technique known as *recursive least squares,* presented in Sections 1.4.8 and 1.4.9.

A word of warning: the derivations presented in this section are rather intricate. The reader may wish to skip most of this material during a first reading. Before continuing, all readers are urged to study Appendices 1.1 and 1.2 at the end of the chapter to become comfortable with the matrix notations used here and to master the essentials of the least-squares problems treated here. In the course of deriving the fast covariance algorithm, we must solve four different least-squares problems, each requiring a time and order update. Thus we must derive eight recursions. For each, we have a recursion for both the model coefficients and the error energy.

1.4.1 Notation and Preliminaries

Before we begin the derivation of the fast covariance algorithm, it is necessary to introduce a number of quantities that occur frequently in the recursions. If we assume that the signal $x(n)$ is available only over the range $L \le n \le U$, then the (forward) error can be written as

$$e_a^{j,L,U}(n) = \tilde{\mathbf{x}}_j^T(n)\mathbf{a}^{j,L,U} = x(n) + \sum_{k=1}^{j} a_k^{j,L,U} x(n-k), \qquad n \in [L+j, U]$$

(1.78)

where

$$\mathbf{a}^{j,L,U} = [1 \quad a_a^{j,L,U} \quad \cdots \quad a_j^{j,L,U}]^T$$

The error vector can also be expressed in a matrix-vector form (see Appendix 1.1):

$$\mathbf{e}_a^{j,L,U} = \tilde{\mathbf{X}}^{j,L,U} \mathbf{a}^{j,L,U}$$

(1.79)

The superscripts j, L, and U are carried throughout this development because the covariance solution depends not only on the order j but also the interval of the signal $[L, U]$. At first glance, this notation seems incredibly cumbersome. Every vector is loaded with numerous subscripts and superscripts. But, after some familiarity, you will notice that most updates involve changes in only one or two indices and are not nearly as complicated as they appear at first.

Minimization of the error energy over the range $[L+j, U]$ leads to the normal equations for the covariance method solution (see Appendix 1.2):

$$(\tilde{\mathbf{X}}^{j,L,U})^T \tilde{\mathbf{X}}^{j,L,U} \mathbf{a}^{j,L,U} = \mathbf{R}^{j,L,U} \mathbf{a}^{j,L,U} = \begin{bmatrix} E_a^{j,L,U} \\ \mathbf{0}_j \end{bmatrix}$$

(1.80)

where $\mathbf{R}^{j,L,U}$ is a $(j+1) \times (j+1)$ covariance matrix, i.e., it is symmetric and positive semidefinite, and $E_a^{j,L,U}$ is the minimum value of the error energy.

Three other least-squares problems must be solved simultaneously as a part of the fast covariance algorithm. First is the backward predictor, $\mathbf{b}^{j,L,U}$, which minimizes the energy in the error vector

$$\mathbf{e}_b^{j,L,U} = \tilde{\mathbf{X}}^{j,L,U} \mathbf{b}^{j,L,U}$$

(1.81)

This vector has 1 as its last element: $(b^{j,L,U})^T = [b_0^{j,L,U} \quad \cdots \quad b_{j-1}^{j,L,U} \quad 1]$. The normal equations for $\mathbf{b}^{j,L,U}$ are

$$\mathbf{R}^{j,L,U} \mathbf{b}^{j,L,U} = \begin{bmatrix} \mathbf{0}_j \\ E_b^{j,L,U} \end{bmatrix}$$

(1.82)

Two comments are in order. First, $\mathbf{b}^{j,L,U}$ is not the coefficient for a numerator polynomial in a pole-zero model. Second, we might recall a similar vector, $\boldsymbol{\alpha}^j$ in the Levinson recursion, which solved an equation like Eq. (1.82). In that case, $\boldsymbol{\alpha}^j$ was $\mathbf{a}^j$ reversed. In the covariance case, no such simplification happens.

The other two auxiliary least-squares problems involve minimization of the errors:

$$\mathbf{e}_c^{j,L,U} = \begin{bmatrix} \mathbf{0}_{U-L-j+1} \\ 1 \end{bmatrix} + \tilde{\mathbf{X}}^{j,L,U}\,\mathbf{c}^{j,L,U} \tag{1.83}$$

$$\mathbf{e}_d^{j,L,U} = \begin{bmatrix} 1 \\ \mathbf{0}_{U-L-j+1} \end{bmatrix} + \tilde{\mathbf{X}}^{j,L,U}\,\mathbf{d}^{j,L,U} \tag{1.84}$$

where

$$(\mathbf{c}^{j,L,U})^T = [c_0^{j,L,U} \quad c_1^{j,L,U} \quad \cdots \quad c_j^{j,L,U}]$$

and

$$(\mathbf{d}^{j,L,U})^T = [d_0^{j,L,U} \quad d_1^{j,L,U} \quad \cdots \quad d_j^{j,L,U}]$$

The resulting normal equations are

$$\mathbf{R}^{j,L,U}\,\mathbf{c}^{j,L,U} + \tilde{\mathbf{x}}_j(U) = \mathbf{0}_j \tag{1.85}$$

$$\tilde{\mathbf{x}}_j^T(U)\mathbf{c}^{j,L,U} + 1 = E_c^{j,L,U} \tag{1.86}$$

and

$$\mathbf{R}^{j,L,U}\,\mathbf{d}^{j,L,U} + \tilde{\mathbf{x}}_j(L+j) = \mathbf{0}_j \tag{1.87}$$

$$\tilde{\mathbf{x}}_j^T(L+j)\mathbf{d}^{j,L,U} + 1 = E_d^{j,L,U} \tag{1.88}$$

1.4.2 Recursions for the Covariance Matrix

The recursive updating required in the fast covariance algorithm is derived by examining partitions of the covariance matrix $\mathbf{R}^{j,L,U}$. First, it is easy to verify that

$$\mathbf{R}^{j,L,U} = (\tilde{\mathbf{X}}^{j,L,U})^T\tilde{\mathbf{X}}^{j,L,U} = \sum_{i=L+j}^{U} \tilde{\mathbf{x}}_j(i)\tilde{\mathbf{x}}_j^T(i) \tag{1.89}$$

From this outer product identity, it is straightforward to derive each of the following four properties:

Time partitions:

$$\mathbf{R}^{j,L,U} = \mathbf{R}^{j,L+1,U} + \tilde{\mathbf{x}}_j(L+j)\tilde{\mathbf{x}}_j^T(L+j) \tag{1.90}$$

$$\mathbf{R}^{j,L,U} = \mathbf{R}^{j,L,U-1} + \tilde{\mathbf{x}}_j(U)\tilde{\mathbf{x}}_j^T(U) \tag{1.91}$$

Order partitions:

$$\mathbf{R}^{j+1,L,U} = \begin{bmatrix} \mathbf{R}^{j,L+1,U} & \mathbf{q}^{j,L,U} \\ (\mathbf{q}^{j,L,U})^T & \displaystyle\sum_{i=L}^{U-j-1} x^2(i) \end{bmatrix} \tag{1.92}$$

$$\mathbf{R}^{j+1,L,U} = \begin{bmatrix} \displaystyle\sum_{i=L+j+1}^{U} x^2(i) & (\mathbf{r}^{j,L,U})^T \\ \mathbf{r}^{j,L,U} & \mathbf{R}^{j,L,U-1} \end{bmatrix} \tag{1.93}$$

where the vectors $\mathbf{q}$ and $\mathbf{r}$ are defined by

$$\mathbf{q}^{j,L,U} = \sum_{i=L+j+1}^{U} x(i-j-1)\tilde{\mathbf{x}}_j(i) \tag{1.94}$$

$$\mathbf{r}^{j,L,U} = \sum_{i=L+j+1}^{U} x(i)\tilde{\mathbf{x}}_j(i-1) \tag{1.95}$$

These vectors contain covariance values and can, themselves, be generated recursively:

$$\mathbf{q}^{j+1,L,U} = \begin{bmatrix} 0 \\ \mathbf{q}^{j,L,U} \end{bmatrix} + \begin{bmatrix} \sum_{i=L+j+2}^{U} x(i)x(i-j-2) \\ -x(U-j-1)\tilde{\mathbf{x}}_j(U) \end{bmatrix} \tag{1.96}$$

$$\mathbf{r}^{j+1,L,U} = \begin{bmatrix} \mathbf{r}^{j,L,U} \\ 0 \end{bmatrix} + \begin{bmatrix} -x(L+j+1)\tilde{\mathbf{x}}_j(L+j) \\ \sum_{i=L+j+2}^{U} x(i)x(i-j-2) \end{bmatrix} \tag{1.97}$$

With all these definitions made, we are finally ready to begin the task of generating the recursions for $\mathbf{a}$, $\mathbf{b}$, $\mathbf{c}$, and $\mathbf{d}$. For each vector there are two recursions: a time update and an order update. First we consider the order updates for $\mathbf{a}$ and $\mathbf{b}$.

1.4.3 Order Update for a and b

With hindsight, we can express $\mathbf{a}^{j+1,L,U}$ as

$$\mathbf{a}^{j+1,L,U} = \begin{bmatrix} \mathbf{a}^{j,L+1,U} \\ 0 \end{bmatrix} + k_a^{j+1} \begin{bmatrix} 0 \\ \mathbf{b}^{j,L,U-1} \end{bmatrix} \tag{1.98}$$

Note that $k_a^{j+1} = a_{j+1}^{j+1,L,U}$. It is analogous to the reflection coefficients generated during the Levinson recursion, although the property that $|k_a^{j+1}| < 1$ need not be true. The update of Eq. (1.98) is easily verified by substituting into the normal equations and expanding via the order partitions of $\mathbf{R}^{j+1,L,U}$.

$$\mathbf{R}^{j,L+1,U} \mathbf{a}^{j+1,L,U} = \begin{bmatrix} E_a^{j+1,L,U} \\ \mathbf{0}_{j+1} \end{bmatrix} \tag{1.99}$$

$$\begin{bmatrix} \mathbf{R}^{j,L+1,U} \mathbf{a}^{j,L+1,U} \\ (\mathbf{q}^{j,L,U})^T \mathbf{a}^{j,L+1,U} \end{bmatrix} + k_a^{j+1} \begin{bmatrix} (\mathbf{r}^{j,L,U})^T \mathbf{b}^{j,L,U-1} \\ \mathbf{R}^{j,L,U-1} \mathbf{b}^{j,L,U-1} \end{bmatrix} = \begin{bmatrix} E_a^{j+1,L,U} \\ \mathbf{0}_{j+1} \end{bmatrix} \tag{1.100}$$

Invoking the known solutions for $\mathbf{a}^{j,L,U}$ and $\mathbf{b}^{j,L,U}$, we have

$$\begin{bmatrix} E_a^{j,L+1,U} \\ \mathbf{0}_j \\ (\mathbf{q}^{j,L,U})^T \mathbf{a}^{j,L+1,U} \end{bmatrix} + k_a^{j+1} \begin{bmatrix} (\mathbf{r}^{j,L,U})^T \mathbf{b}^{j,L,U-1} \\ \mathbf{0}_j \\ E_b^{j,L,U-1} \end{bmatrix} = \begin{bmatrix} E_a^{j+1,L,U} \\ \mathbf{0}_j \\ 0 \end{bmatrix} \tag{1.101}$$

we obtain

$$k_a^{j+1} = -\frac{(\mathbf{q}^{j,L,U})^T \mathbf{a}^{j,L+1,U}}{E_b^{j,L,U-1}} \tag{1.102}$$

and

$$E_a^{j+1,L,U} = E_a^{j,L+1,U} + k_a^{j+1}(\mathbf{r}^{j,L,U})^T \mathbf{b}^{j,L,U-1}$$
$$= E_a^{j,L+1,U}(1 - k_a^{j+1}k_b^{j+1}) \tag{1.103}$$

The recursion for $\mathbf{b}^{j,L,U}$ is derived in exactly the same manner and yields

$$\mathbf{b}^{j+1,L,U} = \begin{bmatrix} 0 \\ \mathbf{b}^{j,L,U-1} \end{bmatrix} + k_b^{j+1}\begin{bmatrix} \mathbf{a}^{j,L+1,U} \\ 0 \end{bmatrix} \tag{1.104}$$

$$k_b^{j+1} = -\frac{(\mathbf{r}^{j,L,U})^T \mathbf{b}^{j,L,U-1}}{E_a^{j,L+1,U}} \tag{1.105}$$

$$E_b^{j+1,L,U} = E_b^{j,L,U-1} + k_b^{j+1}(\mathbf{q}^{j,L,U})^T \mathbf{a}^{j,L+1,U}$$
$$= E_b^{j,L,U-1}(1 - k_b^{j+1}k_a^{j+1}) \tag{1.106}$$

Further simplification can be performed because the following identity holds:

$$k_a^{j+1}E_b^{j+1,L,U} = (\mathbf{b}^{j+1,L,U})^T \mathbf{R}^{j+1,L,U}\mathbf{a}^{j+1,L,U} = k_b^{j+1}E_a^{j+1,L,U} \tag{1.107}$$

After some manipulation, it follows that $(\mathbf{q}^{j,L,U})^T\mathbf{a}^{j,L+1,U} = (\mathbf{r}^{j,L,U})^T\mathbf{b}^{j,L,U-1}$, so only one of these inner products needs to be evaluated.

1.4.4 Time Update for a and b

It is important to note that the order updates for $\mathbf{a}^{j+1,L,U}$ and $\mathbf{b}^{j+1,L,U}$ involve lower-order solutions over *different time intervals*, namely $[L + 1, U]$ and $[L, U - 1]$. Thus we now turn our attention to the need for a *time update*. We propose a time update for $\mathbf{a}^{j,L+1,U}$ in the form

$$\mathbf{a}^{j,L+1,U} = \mathbf{a}^{j,L,U} + \begin{bmatrix} 0 \\ \boldsymbol{\epsilon} \end{bmatrix} \tag{1.108}$$

where $\boldsymbol{\epsilon}$ is to be determined. Note that the zero in the first entry of the unknown correction vector is necessary because $\mathbf{a}^{j,L,U}$ always contains 1 as its first entry. Now we can expand the equation for $\mathbf{a}^{j,L+1,U}$:

$$\mathbf{R}^{j,L+1,U}\mathbf{a}^{j,L+1,U} = [\mathbf{R}^{j,L,U} - \tilde{\mathbf{x}}_j(L+j)\tilde{\mathbf{x}}_j^T(L+j)]\left[\mathbf{a}^{j,L,U} + \begin{bmatrix} 0 \\ \boldsymbol{\epsilon} \end{bmatrix}\right] \tag{1.109}$$

Thus

$$\mathbf{R}^{j,L,U}\mathbf{a}^{j,L,U} + \begin{bmatrix} (\mathbf{r}^{j-1,L,U})^T\boldsymbol{\epsilon} \\ \mathbf{R}^{j-1,L,U-1}\boldsymbol{\epsilon} \end{bmatrix} - \lambda_a^j\begin{bmatrix} x(L+p) \\ \tilde{\mathbf{x}}_{j-1}(L+j-1) \end{bmatrix} = \begin{bmatrix} E_a^{j,L+1,U} \\ 0_j \end{bmatrix} \tag{1.110}$$

where

$$\lambda_a^j = \tilde{\mathbf{x}}_j^T(L+j)\mathbf{a}^{j,L,U} + \tilde{\mathbf{x}}_{j-1}^T(L+j-1)\boldsymbol{\epsilon} \tag{1.111}$$

Using the fact that the first term of Eq. (1.110) contains an already solved least-squares problem, we obtain the two equations

$$E_a^{j,L,U} + (\mathbf{r}^{j-1,L,U})^T\boldsymbol{\epsilon} - \lambda_a^j x(L+j) = E_a^{j,L+1,U} \tag{1.112}$$

$$0_j + \mathbf{R}^{j-1,L,U-1}\boldsymbol{\epsilon} - \lambda_a^j\tilde{\mathbf{x}}_{j-1}^T(L+j-1) = 0_j \tag{1.113}$$

Equation (1.113) can be compared to the equation for $\mathbf{d}^{j,L,U}$ and we can conclude that

$$\boldsymbol{\epsilon} = -\lambda_a^j \mathbf{d}^{j-1,L,U-1} \tag{1.114}$$

The constant λ_a^j can be evaluated from Eq. (1.111):

$$\lambda_a^j = \tilde{\mathbf{x}}_j^T(L+j)\mathbf{a}^{j,L,U} - \lambda_a^j \tilde{\mathbf{x}}_{j-1}^T(L+j-1)\mathbf{d}^{j-1,L,U-1} \tag{1.115}$$

which leads to

$$\lambda_a^j = \frac{\tilde{\mathbf{x}}_j^T(L+j)\mathbf{a}^{j,L,U}}{1 + \tilde{\mathbf{x}}_{j-1}^T(L+j-1)\mathbf{d}^{j-1,L,U-1}} = \frac{e_a^{j,L,U}(L+j)}{E_d^{j-1,L,U-1}} \tag{1.116}$$

The update for the error energy becomes, from Eq. (1.112),

$$\begin{aligned} E_a^{j,L+1,U} &= E_a^{j,L,U} - \lambda_a^j[x(L+j) + (\mathbf{r}^{j-1,L,U})^T\mathbf{d}^{j-1,L,U-1}] \\ &= E_a^{j,L,U} - \lambda_a^j e_a^{j,L,U}(L+j) \end{aligned} \tag{1.117}$$

The use of $e_a^{j,L,U}(L+j)$ in Eq. (1.117) is justified as follows:

$$\begin{aligned} e_a^{j,L,U}(L+j) &= \tilde{\mathbf{x}}_j^T(L+j)\mathbf{a}^{j,L,U} \\ &= x(L+j) + \tilde{\mathbf{x}}_{j-1}^T(L+j-1)(\mathbf{R}^{j-1,L,U-1})^{-1}\mathbf{r}^{j-1,L,U} \\ &= x(L+j) + (\mathbf{d}^{j-1,L,U-1})^T\mathbf{r}^{j-1,L,U} \end{aligned} \tag{1.118}$$

This fact is also needed later in the order update for $\mathbf{c}$ and $\mathbf{d}$. To summarize, we rewrite the time update for $\mathbf{a}^{j,L+1,U}$:

$$\mathbf{a}^{j,L+1,U} = \mathbf{a}^{j,L,U} - \lambda_a^j\begin{bmatrix} 0 \\ \mathbf{d}^{j-1,L,U-1} \end{bmatrix} \tag{1.119}$$

Likewise, the time update for $\mathbf{b}^{j,L,U-1}$ can be derived to obtain

$$\mathbf{b}^{j,L,U-1} = \mathbf{b}^{j,L,U} - \lambda_b^j\begin{bmatrix} \mathbf{c}^{j-1,L+1,U} \\ 0 \end{bmatrix} \tag{1.120}$$

$$\lambda_b^j = \frac{\tilde{\mathbf{x}}_j^T(U)\mathbf{b}^{j,L,U}}{E_c^{j-1,L+1,U}} = \frac{e_b^{j,L,U}(U)}{E_c^{j-1,L+1,U}} \tag{1.121}$$

$$\begin{aligned} E_b^{j,L,U-1} &= E_b^{j,L,U} - \lambda_b^j[x(U-j) + (\mathbf{q}^{j-1,L,U})^T\mathbf{c}^{j-1,L+1,U}] \\ &= E_b^{j,L,U} - \lambda_b^j e_b^{j,L,U}(U) \end{aligned} \tag{1.122}$$

Since the auxiliary equations for $\mathbf{c}$ and $\mathbf{d}$ arise naturally in the time update for $\mathbf{a}$ and $\mathbf{b}$, it is necessary to develop recursive updating (in both order and time) for the solutions to these auxiliary equations.

1.4.5 Order Update for c and d

Again, with hindsight we propose an update of the form

$$\mathbf{c}^{j+1,L,U} = \begin{bmatrix} \mathbf{c}^{j,L+1,U} \\ 0 \end{bmatrix} + \boldsymbol{\epsilon} \tag{1.123}$$

Substituting into the equation for $\mathbf{c}^{j+1,L,U}$

$$\mathbf{R}^{j+1,L,U}\mathbf{c}^{j+1,L,U} + \tilde{\mathbf{x}}_{j+1}(U) = \mathbf{0}_{j+2} \tag{1.124}$$

we obtain

$$\begin{bmatrix} \mathbf{R}^{j,L+1,U}\mathbf{c}^{j,L+1,U} \\ (\mathbf{q}^{j,L,U})^T\mathbf{c}^{j,L+1,U} \end{bmatrix} + \mathbf{R}^{j+1,L,U}\boldsymbol{\epsilon} + \begin{bmatrix} \tilde{\mathbf{x}}_j(U) \\ x(U-j-1) \end{bmatrix} = \mathbf{0}_{j+2} \tag{1.125}$$

Using the solution for $\mathbf{c}^{j,L+1,U}$, we obtain

$$\begin{bmatrix} -\tilde{\mathbf{x}}_j(U) \\ (\mathbf{q}^{j,L,U})^T\mathbf{c}^{j,L+1,U} \end{bmatrix} + \mathbf{R}^{j+1,L,U}\boldsymbol{\epsilon} + \begin{bmatrix} \tilde{\mathbf{x}}_j(U) \\ x(U-j-1) \end{bmatrix} = \mathbf{0}_{j+2} \tag{1.126}$$

or

$$\mathbf{R}^{j+1,L,U}\boldsymbol{\epsilon} = -\begin{bmatrix} \mathbf{0}_{j+1} \\ x(U-j-1) + (\mathbf{q}^{j,L,U})^T\mathbf{c}^{j,L+1,U} \end{bmatrix} \tag{1.127}$$

Thus we see that $\boldsymbol{\epsilon}$ must be a scalar multiple of $\mathbf{b}^{j+1,L,U}$.

$$\mathbf{c}^{j+1,L,U} = \begin{bmatrix} \mathbf{c}^{j,L+1,U} \\ 0 \end{bmatrix} + k_c^{j+1}\mathbf{b}^{j+1,L,U} \tag{1.128}$$

Note that $k_c^{j+1} = c_{j+1}^{j+1,L,U}$ can be evaluated from Eq. (1.127):

$$k_c^{j+1} = -\frac{x(U-j-1) + (\mathbf{q}^{j,L,U})^T\mathbf{c}^{j,L+1,U}}{E_b^{j+1,L,U}} \tag{1.129}$$

$$= -\frac{e_b^{j+1,L,U}(U)}{E_b^{j+1,L,U}}$$

This follows from the identity

$$k_c^{j+1} E_b^{j+1,L,U} = (\mathbf{c}^{j+1,L,U})^T\mathbf{R}^{j+1,L,U}\mathbf{b}^{j+1,L,U} = -e_b^{j+1,L,U}(U) \tag{1.130}$$

Also we can obtain a recursion for $E_c^{j+1,L,U}$:

$$E_c^{j+1,L,U} = 1 + \tilde{\mathbf{x}}_{j+1}^T(U)\begin{bmatrix} \mathbf{c}^{j,L+1,U} \\ 0 \end{bmatrix} + k_c^{j+1}\tilde{\mathbf{x}}_{j+1}^T(U)\mathbf{b}^{j+1,L,U}$$

$$= E_c^{j,L+1,U} + k_c^{j+1}e_b^{j+1,L,U}(U) \tag{1.131}$$

$$= E_c^{j,L+1,U} - (k_c^{j+1})^2 E_b^{j+1,L,U}$$

Similar development leads to the following equations for $\mathbf{d}$:

$$\mathbf{d}^{j+1,L,U} = \begin{bmatrix} 0 \\ \mathbf{d}^{j,L,U-1} \end{bmatrix} + k_d^{j+1}\mathbf{a}^{j+1,L,U} \tag{1.132}$$

$$k_d^{j+1} = -\frac{x(L+j+1) + (\mathbf{r}^{j,L,U})^T\mathbf{d}^{j,L,U-1}}{E_d^{j+1,L,U}} = -\frac{e_a^{j+1,L,U}(L+j+1)}{E_a^{j+1,L,U}} \tag{1.133}$$

$$E_d^{j+1,L,U} = E_d^{j,L,U-1} + k_d^{j+1}\tilde{\mathbf{x}}_{j+1}^T(L+j)\mathbf{a}^{j+1,L,U}$$

$$= E_d^{j,L,U-1} + k_d^{j+1}e_a^{j+1,L,U}(L+j+1) \tag{1.134}$$

$$= E_d^{j,L,U-1} - (k_d^{j+1})^2 E_a^{j+1,L,U}$$

Once again, a simplification was made based on the identity

$$k_d^{j+1} E_a^{j+1,L,U} = (\mathbf{d}^{j+1,L,U})^T \mathbf{R}^{j+1,L,U} \mathbf{a}^{j+1,L,U} = -e_a^{j+1,L,U}(L + j + 1) \qquad (1.135)$$

1.4.6 Time Update for c and d

These equations indicate that one further recursion (a time update) is needed to complete the solution. For $\mathbf{c}^{j,L+1,U}$ the update is of the form

$$\mathbf{c}^{j,L+1,U} = \mathbf{c}^{j,L,U} + \boldsymbol{\epsilon} \qquad (1.136)$$

Expanding the normal equations for $\mathbf{c}^{j,L+1,U}$ gives

$$[\mathbf{R}^{j,L,U} - \tilde{\mathbf{x}}_j(L + j)\tilde{\mathbf{x}}_j^T(L + j)][\mathbf{c}^{j,L,U} + \boldsymbol{\epsilon}] + \tilde{\mathbf{x}}_j(U) = \mathbf{0}_{j+1} \qquad (1.137)$$

$$\mathbf{R}^{j,L,U}\mathbf{c}^{j,L,U} + \mathbf{R}^{j,L,U}\boldsymbol{\epsilon} - \lambda_c^j\tilde{\mathbf{x}}_j(L + j) + \tilde{\mathbf{x}}_j(U) = \mathbf{0}_{j+1} \qquad (1.138)$$

where

$$\lambda_c^j = \tilde{\mathbf{x}}_j^T(L + j)\mathbf{c}^{j,L,U} + \tilde{\mathbf{x}}_j^T(L + j)\boldsymbol{\epsilon} \qquad (1.139)$$

Thus

$$(\mathbf{R}^{j,L,U}\boldsymbol{\epsilon} - \lambda_c^j\tilde{\mathbf{x}}_j(L + j) = \mathbf{0}_{j+1} \qquad (1.140)$$

and we can conclude that $\boldsymbol{\epsilon}$ must be a scalar multiple of $\mathbf{d}^{j,L,U}$. Hence the update for $\mathbf{c}^{j,L+1,U}$ is

$$\mathbf{c}^{j,L+1,U} = \mathbf{c}^{j,L,U} - \lambda_c^j\mathbf{d}^{j,L,U} \qquad (1.141)$$

Evaluation of λ_c^j follows from Eq. (1.139) and $\boldsymbol{\epsilon} = \lambda_c^j\mathbf{d}^{j,L,U}$.

$$\lambda_c^j = \frac{\tilde{\mathbf{x}}_j^T(L + j)\mathbf{c}^{j,L,U}}{1 + \tilde{\mathbf{x}}_j^T(L + j)\mathbf{d}^{j,L,U}} = \frac{e_c^{j,L,U}(L + j)}{E_d^{j,L,U}} \qquad (1.142)$$

The update for the least-squares error is

$$E_c^{j,L+1,U} = 1 + \tilde{\mathbf{x}}_j^T(U)\mathbf{c}^{j,L,U} - \lambda_c^j\tilde{\mathbf{x}}_j^T(U)\mathbf{d}^{j,L,U} \qquad (1.143)$$

$$E_c^{j,L+1,U} = E_c^{j,L,U} - \lambda_c^j e_d^{j,L,U}(U)$$

The corresponding equations for $\mathbf{d}^{j,L,U-1}$ are

$$\mathbf{d}^{j,L,U-1} = \mathbf{d}^{j,L,U} - \lambda_d^j\mathbf{c}^{j,L,U} \qquad (1.144)$$

$$\lambda_d^j = \frac{e_d^{j,L,U}(U)}{E_c^{j,L,U}} \qquad (1.145)$$

and

$$E_d^{j,L,U-1} = E_d^{j,L,U} - \lambda_d^j e_c^{j,L,U}(L + j) \qquad (1.146)$$

1.4.7 Summary of Fast Covariance Recursion

Now we can summarize the main loop of the recursion. At the beginning of each iteration we have the four solution vectors $\mathbf{a}^{j,L,U}$, $\mathbf{b}^{j,L,U}$, $\mathbf{c}^{j-1,L+1,U}$, and $\mathbf{d}^{j-1,L,U-1}$ plus

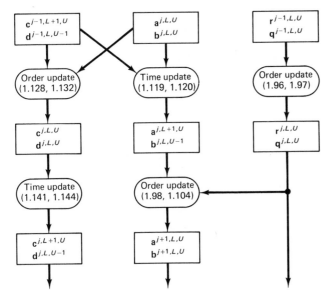

Figure 1.7   Fast covariance recursion dependencies among the time and order updates.

the two correlation vectors $\mathbf{q}^{j-1,L,U}$ and $\mathbf{r}^{j-1,L,U}$. The sequence for generating the next higher order solutions is given in Fig. 1.7.

The only remaining piece of information is the initial conditions for starting the recursion.

$$\mathbf{q}^{0,L,U} = \mathbf{r}^{0,L,U} = \rho_{01} = \rho_{10} = \sum_{i=L+1}^{U} x(i)x(i-1) \qquad (1.147a)$$

$$\mathbf{a}^{1,L,U} = \begin{bmatrix} 1 \\ -\rho_{01}/\rho_{11} \end{bmatrix} \qquad (1.147b)$$

$$E_a^{1,L,U} = \rho_{00} - \rho_{01}^2/\rho_{11} \qquad (1.147c)$$

$$\mathbf{b}^{1,L,U} = \begin{bmatrix} -\rho_{01}/\rho_{00} \\ 1 \end{bmatrix} \qquad (1.147d)$$

$$E_b^{1,L,U} = \rho_{11} - \rho_{01}^2/\rho_{00} \qquad (1.147e)$$

$$\mathbf{c}^{0,L+1,U} = \frac{-x(U)}{\rho_{00}} \qquad (1.147f)$$

$$E_c^{0,L+1,U} = 1 - \frac{x^2(U)}{\rho_{00}} \qquad (1.147g)$$

$$\mathbf{d}^{0,L,U+1} = \frac{-x(L)}{\rho_{11}} \qquad (1.147h)$$

$$E_d^{0,L,U+1} = 1 - \frac{x^2(L)}{\rho_{11}} \qquad (1.147i)$$

where

$$E_s = \sum_{i=L}^{U} x^2(i)$$

$$\rho_{00} = E_s - x^2(L)$$

$$\rho_{11} = E_s - x^2(U)$$

$$\mathbf{R}^{1,L,U} = \begin{bmatrix} \rho_{00} & \rho_{01} \\ \rho_{10} & \rho_{11} \end{bmatrix}$$

From the complete specification of the fast covariance algorithm, it is a simple matter to show that the amount of computation (number of multiplications and divisions) is proportional to p^2. There are five steps in the update. We find that going from the jth-order solution to the $(j + 1)$st requires the following:

0. Initial conditions: $2N + 7$ multiplications.
1. Order update for **c** and **d**: $2j + 4$ multiplications.
2. Time update for **a** and **b**: $4j + 4$ multiplications.
3. Order update for **q** and **r**: $2N - 2$ multiplications, where N is the total number of signal points, i.e., $N = U - L + 1$. Note that the computation of the covariance values is generally not included in the number of operations for Levinson's recursion. For large data sets it is the dominant factor, although we consider it separately here.
4. Time update for **c** and **d**: $4j + 8$ multiplications.
5. Order update for **a** and **b**: $3j + 7$ multiplications.

The total for steps 1, 2, 4, and 5 is $13j + 23$ multiplications. Thus the total for generating the pth-order solution is

$$\text{Number of mults} = 13 \sum_{j=1}^{p-1} j + 23(p - 1) = \frac{13p(p - 1)}{2} + 23(p - 1) \approx \frac{13}{2}p^2$$

$$(1.148)$$

If we compare this with the $\frac{1}{3}p^3$ multiplications of a general Gaussian elimination algorithm, we can see that the crossover point for more efficient calculation via the *fast* covariance algorithm is around $p \approx 20$. Of course, the recursive algorithm does have one major advantage: all lower-order problems are also solved. This can be useful when one is searching for a good model order and would like to examine the error energy at each step.

1.4.8 Recursive Least Squares (RLS)

In the covariance method, the error is summed over a finite (but fixed) range of the signal. In situations where data are being acquired sequentially, it would be convenient to be able to update the covariance solution each time an additional signal value becomes available. Such a technique would fall into the general category of adaptive signal processing. For example, the problem of tracking a nonstationary signal requires an adaptive system. Several approaches are popular for adaptive filtering. One is the least-mean-square (LMS) method described in Chapter 5. This technique, however, only approximates the least-squares solution to the modeling problem, converging to the ACNE model in the limit where the signal length becomes infinite. On the other hand, the RLS method described here will maintain the model coefficients at their optimal values. It operates by holding the model order fixed and optimizing the coefficients over successively larger blocks of data. We will discuss two equivalent versions of the RLS update. The first is an $O(p^2)$ algorithm that amounts to an update scheme for the inverse of the covariance matrix. The second is an $O(p)$ algorithm reminiscent of the fast covariance algorithm.

To begin, we assume that the existing covariance solution has been computed when the signal is available over the time interval $[L, U]$. The error, taken over the range $[L + p, U]$, was minimized by solving the normal equations for the vector $\mathbf{a}^{p,L,U}$ and $E_a^{p,L,U}$.

$$(\tilde{\mathbf{X}}^{p,L,U})^T \tilde{\mathbf{X}}^{p,L,U} \mathbf{a}^{p,L,U} = \mathbf{R}^{p,L,U} \mathbf{a}^{p,L,U} = \begin{bmatrix} E_a^{p,L,U} \\ \mathbf{0}_p \end{bmatrix} \tag{1.149}$$

When an additional datum, $x(U + 1)$, becomes available, we would like to update the optimal coefficients of Eq. (1.149) without recomputing and inverting the entire covariance matrix $\mathbf{R}^{p,L,U+1}$. With the acquisition of a new datum, $x(U + 1)$, the normal equations can be decomposed using the time partition of the covariance matrix.

$$\mathbf{R}^{p,L,U+1} \mathbf{a}^{p,L,U+1} = (\mathbf{R}^{p,L,U} + \tilde{\mathbf{x}}_p(U + 1)\tilde{\mathbf{x}}_p^T(U + 1))\mathbf{a}^{p,L,U+1}$$

$$= \begin{bmatrix} E_a^{p,L,U+1} \\ \mathbf{0}_p \end{bmatrix} \tag{1.150}$$

One easy way to solve this equation is to apply the following matrix identity:

$$(\lambda \mathbf{R} + \mathbf{x}\mathbf{y}^T)^{-1} = \frac{1}{\lambda}\left(\mathbf{R}^{-1} - \frac{\mathbf{R}^{-1}\mathbf{x}\mathbf{y}^T\mathbf{R}^{-1}}{\lambda + \mathbf{y}^T\mathbf{R}^{-1}\mathbf{x}}\right) \tag{1.151}$$

to the covariance matrix and its inverse $\mathbf{Q}^{p,L,U} = (\mathbf{R}^{p,L,U})^{-1}$.

$$\mathbf{a}^{p,L,U+1} = \left[\mathbf{Q}^{p,L,U} - \frac{\mathbf{Q}^{p,L,U}\tilde{\mathbf{x}}_p(U + 1)\tilde{\mathbf{x}}_p^T(U + 1)\mathbf{Q}^{p,L,U}}{1 + \tilde{\mathbf{x}}_p^T(U + 1)\mathbf{Q}^{p,L,U}\tilde{\mathbf{x}}_p(U + 1)}\right]\begin{bmatrix} E_a^{p,L,U+1} \\ \mathbf{0}_p \end{bmatrix} \tag{1.152}$$

This expression, in turn, can be broken into simpler parts by defining a quantity, $\mathbf{k}^{p,L,U+1}$, called the vector gain,

$$\mathbf{k}^{p,L,U+1} = -\mathbf{Q}^{p,L,U}\tilde{\mathbf{x}}_p(U + 1), \tag{1.153}$$

and by giving a notation for the denominator in Eq. (1.152):

$$\delta^{p,L,U+1} = 1 - \tilde{\mathbf{x}}_p^T(U + 1)\mathbf{k}^{p,L,U+1} \tag{1.154}$$

Recognizing the solution for $\mathbf{a}^{p,L,U}$, we obtain for Eq. (1.152)

$$\mathbf{a}^{p,L,U+1} = \frac{E_a^{p,L,U+1}}{E_a^{p,L,U}}\left[\mathbf{a}^{p,L,U} + \frac{\mathbf{k}^{p,L,U+1}\tilde{\mathbf{x}}_p^T(U + 1)\mathbf{a}^{p,L,U}}{\delta^{p,L,U+1}}\right] \tag{1.155}$$

It is worth noting that $\tilde{\mathbf{x}}_p^T(U + 1)\mathbf{a}^{p,L,U}$ is just the error the old operator makes at $n = U + 1$. To finish the derivation, we must evaluate $E_a^{p,L,U+1}$. This is accomplished by observing that the leading entry in any coefficient vector $\mathbf{a}^{p,L,U}$ must be 1.

$$
1 = \frac{E_a^{p,L,U+1}}{E_a^{p,L,U}}\left[1 + \frac{k_0^{p,L,U+1} e_a^{p,L,U}(U + 1)}{\delta^{p,L,U+1}} \right]
$$

$$
\Rightarrow E_a^{p,L,U+1} = \frac{\delta^{p,L,U+1} E_a^{p,L,U}}{\delta^{p,L,U+1} + k_0^{p,L,U+1} e_a^{p,L,U}(U + 1)}
\tag{1.156}
$$

where $k_0^{p,L,U+1}$ is the first entry of the vector $\mathbf{k}^{p,L,U+1}$.

Collecting the important equations together in Eq. (1.157), we see that three quantities are carried along in the recursion: $\mathbf{a}^{p,L,U}$, the vector of optimal model coefficients; $E_a^{p,L,U}$, the optimal value of the error; and $\mathbf{Q}^{p,L,U}$, the inverse covariance matrix. The number of multiplications is given to the left of each formula. The computational expense of the RLS update for a pth order model is $\frac{3}{2}p^2 + \frac{17}{2}p + 7$ multiplications, 4 divisions, and $\frac{3}{2}p^2 + \frac{11}{2}p + 4$ additions. The symmetry of the inverse covariance matrix helps to reduce the total by approximately $\frac{1}{2}p^2$. The storage required for the symmetric matrix $\mathbf{Q}^{p,L,U}$ will be $\frac{1}{2}(p + 1)(p + 2)$ values.

$$
e_a^{p,L,U}(U + 1) = \tilde{\mathbf{x}}_p^T(U + 1)\mathbf{a}^{p,L,U} \qquad\qquad p \qquad (1.157a)
$$

$$
\mathbf{k}^{p,L,U+1} = -\mathbf{Q}^{p,L,U}\tilde{\mathbf{x}}_p(U + 1) \qquad\qquad (p + 1)^2 \quad (1.157b)
$$

$$
\delta^{p,L,U+1} = 1 - \tilde{\mathbf{x}}_p^T(U + 1)\mathbf{k}^{p,L,U+1} \qquad\qquad p + 1 \quad (1.157c)
$$

$$
E_a^{p,L,U+1} = \frac{\delta^{p,L,U+1} E_a^{p,L,U}}{\delta^{p,L,U+1} + k_0^{p,L,U+1} e_a^{p,L,U}(U + 1)} \qquad\qquad 2 \quad (1.157d)
$$

$$
\mathbf{a}^{p,L,U+1} = \frac{E_a^{p,L,U+1}}{E_a^{p,L,U}}\left[\mathbf{a}^{p,L,U} + \frac{e_a^{p,L,U}(U + 1)\mathbf{k}^{p,L,U+1}}{\delta^{p,L,U+1}} \right] \qquad 2p \quad (1.157e)
$$

$$
\mathbf{Q}^{p,L,U+1} = \mathbf{Q}^{p,L,U} - \frac{\mathbf{k}^{p,L,U+1}(\mathbf{k}^{p,L,U+1})^T}{\delta^{p,L,U+1}} \qquad\qquad \tfrac{1}{2}(p + 1)(p + 4) \quad (1.157f)
$$

A closer examination of Eqs. (1.157) would reveal that $E_a^{p,L,U+1}$ and $E_a^{p,L,U}$ are not necessary because the scaling for $\mathbf{a}^{p,L,U+1}$ can always be determined relative to its first element. In addition, Eq. (1.150) implies that $\mathbf{a}^{p,L,U+1}$ is just a scaled version of the first column of $\mathbf{Q}^{p,L,U+1}$, so the calculation of $e_a^{p,L,U}(U + 1)$ is also unnecessary. These simplifications would reduce slightly the $O(p)$ term in the number of operations, but the RLS update, Eqs. (1.157), remains $O(p^2)$. There is a good reason for computing the errors—we can monitor the approximation error of the model. Even so, the alert student will notice that the value of $E_a^{p,L,U+1}$ is just the inverse of the first element in the first column of $\mathbf{Q}^{p,L,U+1}$.

An interesting modification of Eqs. (1.157) will accommodate an exponentially weighted least-squares method. If the weighted error is taken to be

$$
E_w = \sum_{i=L}^{U} \gamma^{U-i} e_a^2(i)
\tag{1.158}
$$

then we can show that the order update for the covariance matrix is

$$\mathbf{R}^{p,L,U+1} = \sum_{i=L}^{U+1} \gamma^{U+1-i}\tilde{\mathbf{x}}_p(i)\tilde{\mathbf{x}}_p^T(i) = \gamma\mathbf{R}^{p,L,U} + \tilde{\mathbf{x}}_p(U)\tilde{\mathbf{x}}_p^T(U) \qquad (1.159)$$

In this case the general version of the matrix inversion identity, Eq. (1.151), will still apply. Such a weighted solution is attractive in an adaptive filtering problem where the input signal is time-varying. The exponential weight applied to the error will place more importance on recent errors and will eventually discount errors in the distant past.

The initial conditions to begin the recursion are not necessarily easy to compute, especially $\mathbf{Q}^{p,L,U}$. A quick look back at the fast covariance method would reveal that the inverse covariance matrix is never explicitly computed. It is possible to compute this matrix inverse with an $O(p^2)$ algorithm, although the algorithm is not presented in this chapter. Another approach might be to start with the solution for $L = U$, but this fails because $\mathbf{R}^{p,L,U}$ is singular when $U - L + 1 < p$. A variation on this theme is to use an approximate inverse formed by adding a small positive number to the diagonal of $\mathbf{R}^{p,L,U}$, until one is sure the covariance matrix is nonsingular. This works especially well in conjunction with weighted RLS because the long-term effect of the incorrect covariance matrix diminishes.

It should be obvious that a similar updating strategy applies to the case where a datum is deleted from the beginning of the interval $[L, U]$. There is a remote possibility that this would fail should the covariance matrix become singular; adding data never destroys invertibility, but deleting data might. In any event, it would be possible to recursively update a model by minimizing the error over a sliding window. If the signal statistics were nonstationary, such a finite-memory RLS would provide a means of tracking the time evolution of the signal characteristics. The computational cost of $O(p^2)$ operations per update appears rather high compared with that of other adaptive filtering methods, such as the LMS algorithm, with $O(p)$ operations per point (see Chapter 5). However, as we now show, there is an $O(p)$ update for RLS.

1.4.9 Fast Algorithm for RLS

This section summarizes an $O(p)$ algorithm for updating the coefficients of the RLS problem. The derivation of the algorithm follows similar lines to the fast covariance method. Time and order partitions of the covariance matrix (Section 1.4.2) are exploited to build the recursions. Almost all details are omitted; readers interested in derivations can either consult the published literature for references on the subject or apply their knowledge from the derivation of the fast covariance method to this problem.

Five quantities must be updated during the recursion: $\mathbf{a}^{p,L,U}$, $\mathbf{b}^{p,L,U}$, $\mathbf{c}^{p-1,L,U}$, $\mathbf{d}^{p-1,L,U}$, and $E_a^{p,L,U}$. These are the same vectors that were defined previously for the fast covariance algorithm (see Section 1.4.1). The reason we have a different recursion here is that we concentrate on the time update in going from $[L, U]$ to $[L, U + 1]$; there is no order update. The complete fast RLS algorithm is summarized in Eqs. (1.160). The number of multiplications is given to the left of each equation number.

$$e_a^{p,L,U}(U+1) = \tilde{\mathbf{x}}_p^T(U+1)\mathbf{a}^{p,L,U} \qquad\qquad p \quad (1.160\text{a})$$

$$\mathbf{a}^{p,L,U+1} = \mathbf{a}^{p,L,U} + e_a^{p,L,U}(U+1)\begin{bmatrix} 0 \\ \mathbf{c}^{p-1,L,U} \end{bmatrix} \qquad p \quad (1.160\text{b})$$

$$e_a^{p,L,U+1}(U+1) = \tilde{\mathbf{x}}_p^T(U+1)\mathbf{a}^{p,L,U+1} \qquad\qquad p \quad (1.160\text{c})$$

$$E_a^{p,L,U+1} = E_a^{p,L,U} + e_a^{p,L,U}(U+1)e_a^{p,L,U+1}(U+1) \qquad 1 \quad (1.160\text{d})$$

$$\mathbf{c}^{p,L,U+1} = \begin{bmatrix} 0 \\ \mathbf{c}^{p-1,L,U} \end{bmatrix} - \frac{e_a^{p,L,U+1}(U+1)}{E_a^{p,L,U+1}}\mathbf{a}^{p,L,U+1} \qquad p \quad (1.160\text{e})$$

$$e_b^{p,L,U}(U+1) = \tilde{\mathbf{x}}_p^T(U+1)\mathbf{b}^{p,L,U} \qquad\qquad p \quad (1.160\text{f})$$

$$\begin{bmatrix} \mathbf{c}^{p-1,L+1,U+1} \\ 0 \end{bmatrix} = \frac{\mathbf{c}^{p,L,U+1} - c_p^{p,L,U+1}\mathbf{b}^{p,L,U}}{1 + c_p^{p,L,U+1}e_b^{p,L,U}(U+1)} \qquad 2p+1 \quad (1.160\text{g})$$

$$\mathbf{b}^{p,L,U+1} = \mathbf{b}^{p,L,U} + e_b^{p,L,U}(U+1)\begin{bmatrix} \mathbf{c}^{p-1,L+1,U+1} \\ 0 \end{bmatrix} \qquad p \quad (1.160\text{h})$$

$$e_c^{p-1,L+1,U+1}(L+p-1) = \tilde{\mathbf{x}}_{p-1}^T(L+p-1)\mathbf{c}^{p-1,L+1,U+1} \qquad p \quad (1.160\text{i})$$

$$e_d^{p-1,L,U}(U+1) = \tilde{\mathbf{x}}_{p-1}^T(U+1)\mathbf{d}^{p-1,L,U} \qquad p \quad (1.160\text{j})$$

$$\mathbf{c}^{p-1,L,U+1} = \frac{\mathbf{c}^{p-1,L+1,U+1} + e_c^{p-1,L+1,U+1}(L+p-1)\mathbf{d}^{p-1,L,U}}{1 - e_c^{p-1,L+1,U+1}(L+p-1)e_d^{p-1,L,U}(U+1)} \qquad 2p+1 \quad (1.160\text{k})$$

$$\mathbf{d}^{p-1,L,U+1} = \mathbf{d}^{p-1,L,U} + e_d^{p-1,L,U}(U+1)\mathbf{c}^{p-1,L,U+1} \qquad p \quad (1.160\text{l})$$

The relationship (1.160d) is not obvious. It can be obtained from the following identity after multiplication of Eq. (1.160b) by $\mathbf{R}^{p,L,U+1}$.

$$\begin{aligned}
e_a^{p,L,U+1}(U+1) &= \tilde{\mathbf{x}}_p^T(U+1)\mathbf{a}^{p,L,U+1} \\
&= x(U+1) + \tilde{\mathbf{x}}_{p-1}^T(U)(\mathbf{R}^{p-1,L,U})^{-1}\mathbf{r}^{p-1,L,U+1} \\
&= x(U+1) + (\mathbf{c}^{p-1,L,U})^T\mathbf{r}^{p-1,L,U+1}
\end{aligned}$$

The update (1.160f) for $\mathbf{c}^{p-1,L+1,U+1}$ requires the solution of two simultaneous equations involving $\mathbf{b}^{p,L,U+1}$. Both of these follow from order partitions. The first is

$$\mathbf{c}^{p,L,U+1} = \begin{bmatrix} \mathbf{c}^{p-1,L+1,U+1} \\ 0 \end{bmatrix} + c_p^{p,L,U+1}\mathbf{b}^{p,L,U+1}$$

and the second is the update (1.160h) for $\mathbf{b}^{p,L,U}$.

$$\mathbf{b}^{p,L,U+1} = \mathbf{b}^{p,L,U} + \tilde{\mathbf{x}}_p^T(U)\mathbf{b}^{p,L,U}\begin{bmatrix} \mathbf{c}^{p-1,L+1,U+1} \\ 0 \end{bmatrix}$$

The updates (1.160k, l) for $\mathbf{d}^{p-1,L,U}$ and $\mathbf{c}^{p-1,L,U}$ are derived by solving the following two simultaneous equations:

$$\mathbf{c}^{p-1,L,U+1} = \mathbf{c}^{p-1,L+1,U+1} + [\tilde{\mathbf{x}}_{p-1}^T(L+p-1)\mathbf{c}^{p-1,L+1,U+1}]\mathbf{d}^{p-1,L,U+1}$$

$$\mathbf{d}^{p-1,L,U+1} = \mathbf{d}^{p-1,L,U} + [\tilde{\mathbf{x}}_{p-1}^T(U+1)\mathbf{d}^{p-1,L,U}]\mathbf{c}^{p-1,L,U+1}$$

The total number of operations is $13p + 3$ multiplications and $11p + 1$ additions and 3 divisions per update. One special case of the fast RLS algorithm has received extra attention because there is a simplification that reduces the computation even more. This is the *prewindowed* case, where the signal is assumed to start at L, but the error summation is also begun at L. The signal values are assumed to be zero for $n < L$, and it follows that $\tilde{\mathbf{x}}_{p-1}(L + p - 1) = \mathbf{0}$. Thus $\mathbf{d}^{p-1,L,U} = \mathbf{0}$ and $\mathbf{c}^{p-1,L,U+1} = \mathbf{c}^{p-1,L+1,U+1}$. The net result is that the amount of computation is reduced to approximately $8p$ multiplications per update.

The exponentially weighted least-squares error can also be incorporated into the fast RLS algorithm. As shown in the preceding section, this type of weighting introduces a scalar multiplier into the time update of the covariance matrix $\mathbf{R}^{p,L,U}$. Derivation of the details of this case are left to the interested reader.

As a final comment, you should realize that there are many other versions of the fast RLS algorithm. For example, it is possible to derive a different updating scheme if the vector $\mathbf{k}^{p,L,U}$ from the last section were used instead of $\mathbf{c}^{p,L,U}$. Even a casual search of the published literature will reveal that this topic has been an extremely active area of research during the past several years. Numerous papers have been published for the fast RLS and fast covariance methods, all aimed at reducing the number of operations in the updates.

1.5 DIFFERENT MODELS

The autocorrelation and covariance methods are not the only two modeling methods in use. Numerous other techniques have been proposed. In this section, we discuss two that have seen widespread use, primarily for model-based spectral estimation.

The motivation for considering these additional techniques is as follows. The autocorrelation method is a successful modeling technique due, in part, to the fact that the resulting all-pole model is guaranteed to be stable. However, this stability property has a cost; one must window the signal $x(n)$ and pad with zeros before calculating the autocorrelation coefficients. In many situations, the effect of such windowing is minimal. However, one application where it is quite harmful occurs in high-resolution spectral estimation, where the resolution of closely spaced spectral lines is rendered impossible by the application of a short data window. The covariance method works much better in these situations because it treats the edge effects as if there were no window (i.e., it doesn't run off the data). As we have seen, it can exactly match an all-pole spectrum from a short section of data; but it has no stability guarantee.

So we consider two new ways to compute the coefficients of an all-pole model, one of which can even guarantee stability. Both are based on the minimization of yet another error norm: the sum of the forward and backward errors. Assuming that signal values $x(n)$ are available over the range $[L, U]$, this forward-backward error is

$$E_{FB} = E_{a+} + E_{a-} \qquad (1.161a)$$

where

$$E_{a+} = \sum_{n=L+p}^{U} \left[x(n) + \sum_{k=1}^{p} a_k x(n - k) \right]^2 \tag{1.161b}$$

$$E_{a-} = \sum_{n=L}^{U-p} \left[x(n) + \sum_{k=1}^{p} a_k x(n + k) \right]^2 \tag{1.161c}$$

Note that the range of the summation in each case has been chosen carefully so that only signal points within the interval $[L, U]$ are required to define E_{FB}. Also note that the forward error E_{a+} is exactly the same as that defined for the covariance method. The backward error E_{a-} should not be confused with the backward error operator $\mathbf{b}^{p,L,U}$ encountered in the fast covariance algorithm. E_{a-} is the error generated by taking the forward operator $\mathbf{a}^{p,L,U}$ and running it backward over the data.

Unconstrained minimization of this error norm gives the first method—the modified covariance method. The second method, Burg's recursion, results from a constrained minimization of E_{FB} such that the solution can be constructed iteratively via an order update of exactly the same form as the Levinson recursion.

1.5.1 Modified Covariance Method

The unconstrained minimization of E_{FB} with respect to the parameters $\{a_k\}$ would lead to a set of normal equations similar in form to those of the covariance method.

$$[(\tilde{\mathbf{X}}^{p,L,U})^T \quad (\tilde{\mathbf{X}}^{p,L,U})^T \mathbf{J}] \begin{bmatrix} \tilde{\mathbf{X}}^{p,L,U} \\ \tilde{\mathbf{X}}^{p,L,U} \mathbf{J} \end{bmatrix} \mathbf{a}^{p,L,U} = \begin{bmatrix} E_{FB}^{p,L,U} \\ \mathbf{0}_p \end{bmatrix} \tag{1.162}$$

where the matrix $\mathbf{J}$ has 1's along its antidiagonal; thus $\mathbf{Jx}$ will reverse the vector $\mathbf{x}$.

$$\mathbf{J} = \begin{bmatrix} 0 & 0 & \cdots & 1 \\ \vdots & \vdots & \cdots & \vdots \\ 0 & 1 & \cdots & 0 \\ 1 & 0 & \cdots & 0 \end{bmatrix}$$

It is possible to derive a fast algorithm for these normal equations by following the derivation already given for the covariance method. The covariance matrix has similar structure, so the same types of updates are necessary.

An advantage of the modified covariance method is that the error is summed over twice as many points as in the ordinary covariance method. This can be important when the duration of the available signal is extremely short. However, there must be some rationale for expecting the signal to appear the same in the forward and backward directions. One situation where this seems justified is in analyzing multiple sinusoids in noise. A sinusoid does look the same in both directions. Likewise, for a random process the covariance function is independent of direction, being the inverse transform of a magnitude-squared Fourier spectrum. On the other hand, for a decaying exponential, one would expect poor results because, in the reverse direction, the exponential appears unstable. Indeed, in a first-order example we observe a bias in the pole location computed from a finite data segment of γ^n.

Example 3

Given the signal $x(n) = \gamma^n u(n)$ over the range $[0, N - 1]$, the first-order model computed via the modified covariance method is obtained by substituting in Eq. (1.165):

$$a_1 = \frac{-2 \sum_{n=1}^{N-1} \gamma^n \gamma^{n-1}}{\sum_{n=1}^{N-1} \gamma^{2(n-1)} + \gamma^{2n}} = -\gamma \frac{2}{1 + \gamma^2}$$

This parameter estimate is biased and, surprisingly, independent of N. The true value is $-\gamma$. Assuming that $|\gamma| < 1$, the value for a_1 will always be greater than γ. In other words, the pole of the first-order model will be closer to the unit circle than the true pole is. When γ is close to 1, as for a sinusoid, the bias is small and even serves to "sharpen" the spectral line by placing the pole closer to the unit circle than it has to be. This might explain why these techniques perform so well for high-resolution spectral estimation. For small values of γ the pole estimate could differ from the true value by up to a factor of 2.

   As happens for the covariance method, the modified covariance method may also produce unstable models. However, in the first-order case, $p = 1$, the model is always stable. We prove this fact because the Burg recursion uses this first-order model as a starting point. The forward-backward error is

$$E_{FB} = \sum_{n=L+1}^{U} [x(n) + a_1 x(n - 1)]^2 + \sum_{n=L}^{U-1} [x(n) + a_1 x(n + 1)]^2$$

$$= \sum_{n=L+1}^{U} [x(n) + a_1 x(n - 1)]^2 + [x(n - 1) + a_1 x(n)]^2 \tag{1.163}$$

Differentiating with respect to a_1, we obtain

$$\frac{dE_{FB}}{da_1} = \sum_{n=L+1}^{U} 2[x(n) + a_1 x(n - 1)]x(n - 1) + 2[x(n - 1) + a_1 x(n)]x(n)$$

$$\tag{1.164}$$

Setting the derivative to 0 and solving for a_1 yields

$$a_1 = \frac{-2 \sum_{n=L+1}^{U} x(n)x(n - 1)}{\sum_{n=L+1}^{U} [x^2(n - 1) + x^2(n)]} \tag{1.165}$$

The proof that $|a_1| \le 1$ follows from the fact that $-2ab \le a^2 + b^2$. Unfortunately, for higher-order models, stability is no longer guaranteed, although Burg's clever modification of the problem does lead to a stable answer.

1.5.2 Burg's Recursion

Burg proposed a constrained minimization of E_{FB} in which the model coefficients had to be constructed using a "Levinson-like" order update. The rationale for this approach is that stability can be characterized by reflection coefficients satisfying $k_j < 1$. Thus, if the constrained minimization yields reflection coefficients that are less than 1 in magnitude, then stability is assured because we will start from a stable first-order solution (which is identical to that of the modified covariance method).

We now plunge into the details. From the solution of the jth-order problem (denoted by a_i^j), the solution of the $(j + 1)$st problem must be constructed in the form

$$a_{j+1}^{j+1} = k_{j+1}$$
$$a_i^{j+1} = a_i^j + k_{j+1}a_{j+1-i}^j, \qquad i = 1, 2, \ldots, j \tag{1.166}$$

The forward-backward error is a function of the coefficients $\{a_k^{j+1}\}$ or, equivalently, of k_{j+1} and the coefficients $\{a_k^j\}$. The reflection coefficient k_{j+1} is the only free parameter to be optimized, so the error is actually a function of one variable and will be denoted as the Burg error $E_B(k_{j+1})$.

Since the minimization is always done with respect to only one free parameter, we prefer a summation notation rather than matrices for this derivation. The forward and backward error signals at the jth step are

$$e_{a+}^j(n) = x(n) + \sum_{k=1}^{j} a_k^j x(n - k)$$

$$e_{a-}^j(n) = x(n) + \sum_{k=1}^{j} a_k^j x(n + k) \tag{1.167}$$

Then the total error can be written as

$$E_B^j = \sum_{n=L+j}^{U} [e_{a+}^j(n)]^2 + \sum_{n=L}^{U-j} [e_{a-}^j(n)]^2 \tag{1.168}$$

The range of each sum depends on j, such that the error operator does not run off the data. A recursive expression for $e_{a+}^{j+1}(n)$ and $e_{a-}^{j+1}(n)$ can be obtained from the update (1.166) assumed for a_i^j:

$$e_{a+}^{j+1}(n) = x(n) + \sum_{k=1}^{j} (a_k^j + k_{j+1}a_{j+1-k}^j)x(n - k) + k_{j+1}x(n - j - 1)$$

$$= x(n) + \sum_{k=1}^{j} a_k^j x(n - k)$$

$$+ k_{j+1}[x(n - j - 1) + \sum_{k=1}^{j} a_k^j x(n - j - 1 + k)] \tag{1.169}$$

$$= e_{a+}^j(n) + k_{j+1}e_{a-}^j(n - j - 1)$$

Likewise, $e_{a-}^{j+1}(n)$ can be expressed in terms of the reflection coefficients k_{j+1}:

$$e_{a-}^{j+1}(n) = x(n) + \sum_{k=1}^{j} (a_k^j + k_{j+1}a_{j+1-k}^j)x(n + k) + k_{j+1}x(n + j + 1)$$

$$= e_{a-}^j(n) + k_{j+1}e_{a+}^j(n + j + 1) \tag{1.170}$$

The minimization of E_B^{j+1} with respect to k_{j+1} leads to

$$\frac{dE_B^{j+1}}{dk_{j+1}} = \sum_{n=L+j+1}^{U} 2e_{a+}^{j+1}(n)e_{a-}^{j}(n-j-1) + \sum_{n=L}^{U-j-1} 2e_{a-}^{j+1}(n)e_{a+}^{j}(n+j+1) = 0$$

$$(1.171)$$

Changing variables so that the sums run over the same range of indices and substituting for e^{j+1}, we obtain

$$\sum_{n=L+j+1}^{U} [e_{a+}^{j}(n) + k_{j+1}e_{a-}^{j}(n-j-1)]e_{a-}^{j}(n-j-1)$$

$$+ [e_{a-}^{j}(n-j-1) + k_{j+1}e_{a+}^{j}(n)]e_{a+}^{j}(n) = 0 \qquad (1.172)$$

Finally, the reflection coefficient k_{j+1} is

$$k_{j+1} = -2 \frac{\displaystyle\sum_{n=L+j+1}^{U} e_{a+}^{j}(n)e_{a-}^{j}(n-j-1)}{\displaystyle\sum_{n=L+j+1}^{U} [e_{a+}^{j}(n)]^2 + [e_{a-}^{j}(n-j-1)]^2} \qquad (1.173)$$

Again, proof that $|k_{j+1}| < 1$ follows easily from the fact that $-2ab \le a^2 + b^2$. Since the first-order model is stable, all higher-order models generated from the Burg recursion will be stable.

As a summary, we collect together the important equations that describe Burg's algorithm. Given the signal points $x(L), x(L+1), \ldots, x(U)$, the Burg update is

$$e_{a+}^{0}(n) = x(n) \qquad (1.174a)$$

$$e_{a-}^{0}(n) = x(n) \qquad (1.174b)$$

$$k_{j+1} = -2 \frac{\displaystyle\sum_{n=L+j+1}^{U} e_{a+}^{j}(n)e_{a-}^{j}(n-j-1)}{\displaystyle\sum_{n=L+j+1}^{U} [e_{a+}^{j}(n)]^2 + [e_{a-}^{j}(n-j-1)]^2} \qquad (1.174c)$$

$$e_{a+}^{j+1}(n) = e_{a+}^{j}(n) + k_{j+1}e_{a-}^{j}(n-j-1),$$

$$n = L+j+1, \ldots, U \qquad (1.174d)$$

$$e_{a-}^{j+1}(n) = e_{a-}^{j}(n) + k_{j+1}e_{a+}^{j}(n+j+1),$$

$$n = L, L+1, \ldots, U-j-1 \qquad (1.174e)$$

$$a_{j+1}^{j+1} = k_{j+1} \qquad (1.174f)$$

$$a_{i}^{j+1} = a_{i}^{j} + k_{j+1}a_{j+1-i}^{j}, \qquad i = 1, 2, \ldots, j \qquad (1.174g)$$

One significant feature of this recursion is that it works directly from the data. There is no intermediate computation of autocorrelation coefficients as in the Levinson recursion.

The updating in Eqs. (1.174f, g) is identical to that in the Levinson recursion; but the reflection coefficients have different values. As you might guess, there are many other ways to compute reflection coefficients for models. Equation (1.174c) is the harmonic mean of the forward and backward errors; another possibility is to use the geometric mean of these errors. In all cases, a lattice implementation is possible due to the structure of the updating in Eqs. (1.174f, g).

1.6 DIFFERENT INTERPRETATIONS

In this section, we discuss three other approaches to the modeling problem: Wiener filtering, linear prediction, and predictive deconvolution. Each one leads to exactly the same set of equations as found already for the autocorrelation method. As such, there are no new techniques presented here. However, the different viewpoints offered by each of these approaches can provide insight into different applications for all-pole models. Wiener filtering and predictive deconvolution are important techniques in seismic signal processing. The linear prediction formulation is attractive when considering bandwidth compression for speech.

1.6.1 Digital Wiener Filters

Wiener shaping filters are now commonly used to deconvolve seismic traces (See section 1.6.3). The problem addressed by this type of Wiener filtering is that of converting a given input signal $x(n)$ into a desired output signal $d(n)$. Naively, we might declare that a filter with transfer function $D(z)/X(z)$ is the exact solution and we are finished. There are many difficulties with this instant solution. First, $D(z)/X(z)$ may not be a stable, causal system. Second, $D(z)/X(z)$ may be a very-high-order system and thus not realizable in any practical sense. Finally, this solution is predicated on exact knowledge of the signals $x(n)$ and $d(n)$, but these might be random, described only by some underlying probability law.

Hence, we formulate a system design problem that guarantees a realizable answer. We do so under a stochastic formulation to illustrate that the results from Prony's method (Section 1.2) can be derived for random processes. In restating the least-squares problem for random processes, it is crucial to define an error norm and a notion of orthogonality (i.e., inner product). Then all the previous results apply to the stochastic case. The norm is taken to be the variance $E\{|e(n)|^2\}$ of the error signal, and the inner product to be the cross-correlation of $x(n)$ and $y(n)$, $E\{x(n)y(n)\}$.[†]

The net result is that time correlations will be replaced by statistical correlations, or vice versa. The expected value $E\{\cdot\}$ replaces the summation used previously for time correlation. In fact, one way to view the deterministic formulation is that it gives an explicit way to estimate, from the given data, the autocorrelation and cross-correlation functions needed in the normal equations. Even if one sticks exclusively to the stochastic formulation, this problem of estimating the correlations must be taken into account.

In the block diagram of Fig. 1.8, we want to find the FIR filter $A(z)$ whose output

---

[†] All random signals are assumed to be zero-mean and wide-sense stationary.

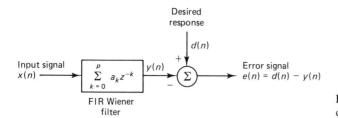

Figure 1.8   Wiener filter block diagram.

$y(n)$ best approximates $d(n)$ in the sense that the mean-square norm of the error is minimized. The constraint that the filter be finite impulse response (FIR) guarantees realizability but requires an approximation in matching the statistics of the output $y(n)$ to those of the desired signal $d(n)$.

In contrast to the modeling problem, we are not really interested in $1/A(z)$ for the Wiener shaping filter. In fact, the interesting quantity is $y(n)$—the filtered (or shaped) response. If modeling is the aim, a special case of the FIR Wiener filter can be solved: namely, let $d(n)$ be unit-variance white Gaussian noise. Then $|X(\omega)A(\omega)|^2 \approx 1$ and $A(z)$ is suitable for use as the poles of a model.

The optimal Wiener filter is obtained by solving the normal equations for the least-squares problem. The error signal can be written

$$e(n) = y(n) - d(n) = \sum_{k=0}^{p} a_k x(n-k) - d(n) \tag{1.175}$$

As always, the normal equations are obtained by applying the projection theorem, which states that the optimal error signal must be orthogonal to the basis signals $x_j(n) = x(n-j)$, $j = 0, 1, \ldots, p$. Expressing this orthogonality condition in equation form gives the following set of $p + 1$ equations for the optimal filter:

$$E\{e(n)x_j(n)\} = 0, \qquad j = 0, 1, \ldots, p \tag{1.176}$$

$$E\left\{\left[\sum_{k=0}^{p} a_k x(n-k) - d(n)\right]x(n-j)\right\} = 0, \qquad j = 0, 1, \ldots, p \tag{1.177}$$

$$\sum_{k=0}^{p} a_k E\{x(n-k)x(n-j)\} = E\{d(n)x(n-j)\}, \qquad j = 0, 1, \ldots, p \tag{1.178}$$

Defining $r_x(j) = E\{x(n)x(n-j)\}$ and $r_{dx}(j) = E\{d(n)x(n-j)\}$, the statistical auto-correlation of $x(n)$ and the cross-correlation of $d(n)$ and $x(n)$, respectively, we obtain

$$\sum_{k=0}^{P} a_k r_x(j-k) = r_{dx}(j), \qquad j = 0, 1, \ldots, p \tag{1.179}$$

In matrix form,

$$\mathbf{R_x}\begin{bmatrix} a_0 \\ a_1 \\ \vdots \\ \vdots \\ a_p \end{bmatrix} = \begin{bmatrix} r_{dx}(0) \\ r_{dx}(1) \\ \vdots \\ r_{dx}(p) \end{bmatrix} \tag{1.180}$$

where $\mathbf{R}_x$ is a $(p + 1) \times (p + 1)$ Toeplitz matrix with entries $r_x(j - k)$. The solution to Eq. (1.180) can be obtained efficiently with Levinson's method, generalized for an arbitrary right-hand side. Using the optimal filter solution, we obtain the minimum value of the error:

$$
\begin{aligned}
E\{e^2(n)\} &= E\left\{\left[\sum_{k=0}^{p} a_k x(n - k) - d(n)\right] e(n)\right\} \\
&= -E\left\{d(n)\left[\sum_{k=0}^{p} a_k x(n - k) - d(n)\right]\right\} \\
&= r_d(0) - \sum_{k=0}^{p} a_k r_{dx}(k)
\end{aligned}
\tag{1.181}
$$

The normal equations for the optimal Wiener filter are similar to Eq. (1.33) for the autocorrelation method. However, the Wiener filtering problem is more general. The autocorrelation normal equations are obtained by taking $d(n)$ to be unit-variance white Gaussian noise, such that $r_{dx}(0) = 1$ and $r_{dx}(j) = 0$ for $j = 0$:

$$
\mathbf{R}_x \begin{bmatrix} a_0 \\ a_1 \\ \vdots \\ a_p \end{bmatrix} = \begin{bmatrix} 1 \\ 0 \\ \vdots \\ 0 \end{bmatrix}
\tag{1.182}
$$

Then Eq. (1.182) is nearly identical (to within the scale factor a_0) to the autocorrelation normal equations, and the solution can be obtained in two steps. First, let $\{1, \alpha_1, \alpha_2, \ldots, \alpha_p\}$ be the solution for

$$
\mathbf{R}_x \begin{bmatrix} 1 \\ \alpha_1 \\ \vdots \\ \alpha_p \end{bmatrix} = \begin{bmatrix} E_\alpha^p \\ 0 \\ \vdots \\ 0 \end{bmatrix}
\tag{1.183}
$$

Then $a_j = \alpha_j / E_\alpha^p$ for $j = 1, 2, \ldots, p$. This particular Wiener filter is usually called a *whitening* filter because its output $X(z)A(z)$ is the best mean-square approximation to white noise. The deterministic counterpart is named the *inverse* filter because the desired signal is chosen to be the unit impulse and $X(z)A(z) \approx 1$.

1.6.2 Linear Prediction

Yet another formulation of the all-pole modeling problem is available with linear prediction. While the problem is often formulated in a stochastic setting, we will consider a deterministic form. Suppose we are given a signal $x(n)$ that is known to be the impulse response of an all-pole system $G/A(z)$. Then $x(n)$ must satisfy the time-domain difference equation

$$
x(n) = -\sum_{k=1}^{p} a_k x(n - k) + G\delta(n)
\tag{1.184}
$$

Hence, for $n > 0$ it is possible to express $x(n)$ as a linear combination of the previous p values of the signal, $x(n-1), x(n-2), \ldots, x(n-p)$. To put it another way, it must be possible to predict, using a linear combination of p samples of the signal, the next value of the signal.

For an arbitrary signal (not necessarily all-pole), such an attempt at exact linear prediction must fail, so we are forced to adopt a notion of approximation to find the optimal p-point linear predictor for $x(n)$. As shown in Fig. 1.9, the prediction error $e(n)$ is the difference between the actual signal value $x(n)$ and the predicted signal value $\hat{x}(n) = \sum_{k=1}^{P}(-a_k)x(n-k)$. The coefficients of the optimal linear predictor are chosen so as to minimize the energy in the error signal over the range $[L, U]$.

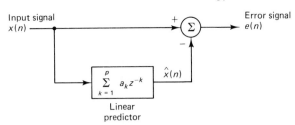

Figure 1.9   Block diagram for linear prediction.

Invoking the projection theorem for least-squares approximation, we must have the error $e(n)$ orthogonal to the basis functions $x_j(n) = x(n-j), j = 1, 2, \ldots, p$.

$$\sum_{n=L}^{U} e(n)x_j(n) = 0 \tag{1.185}$$

$$\sum_{n=L}^{U} \left[x(n) + \sum_{k=1}^{p} a_k x(n-k) \right] x(n-j) = 0, \quad j = 1, 2, \ldots, p \tag{1.186}$$

$$\sum_{k=1}^{P} a_k \left[\sum_{n=L}^{U} x(n-k)x(n-j) \right] = -\sum_{n=L}^{U} x(n)x(n-j), \\ j = 1, 2, \ldots, p \tag{1.187}$$

These are, of course, the familiar normal equations. Depending on the range of the signal, $[L, U]$, and how we treat the end conditions, we obtain either the autocorrelation method or the covariance method.

The basic philosophy of linear prediction is that if we can predict future values of the signal, then, in effect, we know the structure of the signal and, hence, can model the signal. In fact, the prediction error filter of Fig. 1.9 can be regarded as a filter that attempts to remove the predictable part of the signal and produce an output, $e(n)$, that is completely unpredictable. When $x(n)$ is a random signal, this last viewpoint means that the prediction error filter is a "whitening" filter because it attempts to produce an uncorrelated (i.e., white) error signal.

The viewpoint of the model as a linear predictor is a common one in the field of speech communication. Linear predictive coding (LPC) is a popular bandwidth compression scheme for speech signals. The system works as follows: At the coder the speech signal is processed by a prediction error filter. Since speech is a quasi-periodic signal, one set of predictor coefficients is needed for each pitch period of the signal. Over the communication channel, one sends the predictor coefficients and a very low bit rate version of the error signal, which should be random. Side information about the pitch period is also needed, but that part of the problem is separate from LPC. At the receiver, an all-pole model formed from the predictor coefficients is driven by the error signal to resynthesize the speech. Bandwidth compression relative to pulse code modulation (PCM) is obtained if the error signal can be quantized much more coarsely than the original speech signal.

The linear prediction viewpoint is also used in seismic signal processing, commonly for deconvolution. The next section presents details of the predictive deconvolution technique.

1.6.3 Predictive Deconvolution

An application of parametric signal modeling that has gained widespread use, especially in the seismic exploration industry, is *predictive deconvolution*, which is used to remove reverberations from a seismic trace.[†] It requires a simple modification of the linear prediction concept to allow prediction distances greater than one time sample ahead. In this section, a general description of the deconvolution problem is given, along with specific comments directed at predictive deconvolution. Some aspects of seismic deconvolution may be oversimplified or omitted so that the essence of the problem (from the point of view of modeling) can be conveyed.

First we discuss the deconvolution problem in general. Deconvolution is the process of removing the distorting effect of a linear filter (i.e., convolution) from a signal. Consider a recorded signal $t(n)$ that was formed by passing a source signal $w(n)$ through an LTI system with impulse response $r(n)$; i.e., $t(n) = w(n) * r(n)$. The objective of deconvolution is to process the recorded trace $t(n)$ to recover either the source wavelet $w(n)$ or the filter response $r(n)$. From the point of view of the output signal, it is impossible to say whether the distorting filter is $w(n)$ or $r(n)$. In any event, one signal must be considered distortion and the other assigned the role of desired signal. For example, in seismic processing, the desired signal is the impulse response of a layered earth model. The distortion comes from the nonimpulsive nature of the wavelet generated by a source such as an airgun or from reverberations in the water layer for marine seismic recordings.

Without further information, the deconvolution problem is ill-posed. Something extra must be known or assumed about either $w(n)$ or $r(n)$. One possibility is that $w(n)$ can be measured directly. Then the problem statement becomes to determine $r(n)$, given $t(n)$ and $w(n)$. In effect, one has to build an inverse filter for $w(n)$. As already stated, the naive approach of dividing Fourier transforms, $R(\omega) = T(\omega)/W(\omega)$, rarely works because $W(\omega)$ may be zero at some frequencies. One way to deal effectively with the problem of spectral zeros is use the Wiener inverse filter, $H(\omega)$, which gives the best least-squares approximation for $W(\omega)H(\omega) \approx 1$. Deconvolution then amounts to application of the Wiener FIR filter $h(n)$ to the recorded signal $t(n)$.

The foregoing approach is feasible when the distorting signal can be observed directly. In seismic processing, one must resort to a more subtle assumption, namely, that the earth impulse response $r(n)$ is an uncorrelated (white) time series. In effect,

---

[†]In seismic applications, the terms "trace" and "wavelet" refer to particular types of signals. Specifically, a seismic trace is the convolution of a source wavelet with the impulse response of the earth.

we are saying that the layered earth model is uncorrelated in the sense that knowledge of the first ℓ layers does not tell us anything about the $(\ell + 1)$st layer. While this assumption might seem a bit rash from a physical point of view, it has allowed the development of a class of very successful deconvolution techniques that depend on signal modeling. The net result of assuming that $r(n)$ is white is that its correlation function is an impulse, and thus the correlation function for $w(n)$ is the same as that of $t(n)$, to within a scale factor. Thus a modeling technique that depends solely on the autocorrelation function can be used to design a deconvolution filter that will remove the effect of $w(n)$. We will have more to say about this type of deconvolution, after a brief description of the application scenario for seismic processing.

A simplified model of a seismic reflection survey is illustrated in Fig. 1.10. Energy from a seismic source propagates within the earth, where it is reflected and refracted at the interfaces between layers. It is then recorded by receivers at the surface in the form of seismic traces. Typically, the source-receiver offset distance is much smaller than the depth to the various reflectors. Thus, it is usually a good first approximation to think of the local earth as a set of plane layers, each presenting a different acoustic impedance to the seismic wave. Waveform stacking procedures used prior to the deconvolution filters effectively map the source and receiver to the same surface position, making the problem appear one-dimensional. Thus we need only think about normally incident and reflected waves. The received waveform contains primary reflections (single-bounce echoes) and multiple reflections, as illustrated in Fig. 1.11. This figure is not a ray path diagram, even though it appears as such; rather, it shows how multiple reflections and returns from deeper reflectors arrive later in time.

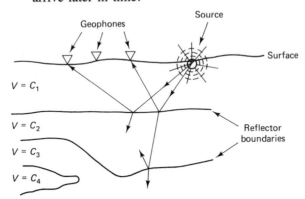

Figure 1.10   Seismic prospecting scenario.

The response of the earth that produces the various received events can be modeled by an impulse response that is an impulse train, called the reflector series. The location of the impulses is dictated by the two-way travel time to the reflectors; the strength is determined by the magnitudes of the reflection coefficient at each interface. The arrival times of the primary reflections are used to determine the depths to the reflectors once the subsurface velocities are also known. Multiple reflections usually constitute a form of interference. One objective of the deconvolution filter is to remove multiples from the portion of the trace containing primaries.

Due to the nonimpulsive nature of most seismic sources, the received seismic trace is smeared, and the locations of the impulses are obscured. What is needed is

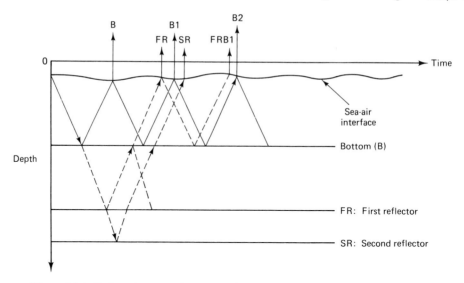

Figure 1.11 Reflected returns shown as a function of depth and time. The time axis is provided to show how the reflections at depth produce the earth impulse response as a function of time.

a wavelet-shaping filter to sharpen the trace and restore the impulsive nature of the reflector series. Then the resolution of thin layers whose primary reflections have travel-time intervals shorter than the seismic source duration becomes possible. A major hurdle for this type of processing is the determination of the shape of the seismic source pulse. Direct measurement of the source waveform is not feasible. Even if it were attempted, it would neglect changes experienced by the source wavelet as it propagates through the earth.

This problem seems tailor-made for Wiener shaping filters, but it requires knowledge of $r_w(\ell)$. If the autocorrelation function of the reflector series $r_r(\ell)$ is assumed to be an impulse, then

$$r_r(\ell) = r_r(0)\delta(\ell) \tag{1.188}$$

Thus the autocorrelation function of the total seismic trace $t(n)$

$$r_t(\ell) = r_w(\ell) * r_r(\ell) = r_r(0)r_w(\ell) \tag{1.189}$$

is equal to the autocorrelation function of the reverberated seismic source $w(n)$ multiplied by the scale factor $r_r(0)$. The significance of Eq. (1.189) is that the received trace can be used to estimate the autocorrelation of the reverberated seismic wavelet $w(n)$.

Thus the design of a Wiener inverse filter for $w(n)$ can be carried out from the autocorrelation function alone. In fact, it is independent of any scaling of the autocorrelation function. Levinson's method is applied to solve the autocorrelation normal equations for the FIR wave-shaping deconvolution filter. Note that we are limited to the autocorrelation method because the measurement is restricted to the autocorrelation function of $w(n)$. The effect of the Wiener shaping filter is to perform pulse compression of the seismic source pulse and, thereby, increase the resolution of the reflector boundaries.

This technique is quite powerful, but it does have some limits. First, the attempt to design a Wiener filter that approximates an impulse fails to recognize the band-limited nature of the signals involved. Both $W(\omega)$ *and* $T(\omega)$ are bandlimited, so it is better to take a bandlimited pulse as the desired signal $d(n)$ for the Wiener filter design. One common form for $d(n)$ is the impulse response of a bandpass filter. A second issue arises because the FIR inverse filter is always minimum phase. If the wavelet $w(n)$ is minimum phase, the inverse filter will cancel the effects of $w(n)$. However, seismic sources are not inherently minimum phase. One example of a decidedly non-minimum-phase source is the airgun. Since the all-pole model must be minimum phase, there will always be a residual error when modeling a nonminimum-phase source, and the wave-shaping deconvolution filter will perform rather poorly. A fix for this problem is to let $d(n)$ be a nonminimum-phase signal. One popular approach is to convert $d(n)$ into a zero-phase signal, $d(n) * d(-n)$, with the same bandpass spectrum; Gaussian pulse shapes have also been used.

Now we turn to another problem faced in processing seismic records: rever-beration. This problem is most annoying on marine seismic records, where the water layer acts effectively as a waveguide for seismic energy. Consider the situation diagrammed in Fig. 1.11. The sequence of returns B1, B2, B3, FRB1, etc., represents the water reverberation. For a shallow water layer with a hard bottom (i.e., a good reflector), the reverberation overlaps the primary returns from both the shallow and deep earth layers, which, in addition, are illuminated by only a tiny fraction of the total energy of the seismic source. Such a situation was typical of early exploration in the Gulf of Mexico. In deep ocean water the reverberation does not overlay the first reflection, but it may obscure the deeper reflections, whose low energy makes them very hard to find, even in the absence of reverberation. A similar interference results from interlayer multiple reverberations.

The impulse response of the water layer can be determined for our model (see Fig. 1.11). Suppose that the two-way travel time in the water layer is n_w samples. Then it is possible to represent the water reverberation effect as a convolution operator of the form

$$C(z) = 1 - cz^{-n_w} + c^2 z^{-2n_w} - c^3 z^{-3n_w} + \cdots$$

$$= \frac{1}{1 + cz^{-n_w}} \tag{1.190}$$

where c is the reflectivity of the bottom. Physically, $|c| < 1$, so $C(z)$ is a recursive minimum-phase operator. If the primary earth reflector series is $p(n)$ (single bounce echoes, including the water bottom), then the total earth-water system function is given by $C(z)P(z)C(z)$. $C(z)$ is used twice because the water layer acts on the signal when the energy propagates down and again when it returns, as illustrated in Fig. 1.11, for the reflection FRB1. $P(z)$ contains the information about the reflectors within the earth that we are seeking.

If we had exact knowledge of the velocity c and the two-way travel time n_w, then the three-point filter $(1 + cz^{-n_w} + c^2 z^{-2n_w})$ would completely dereverberate the seis-mic return. However, such a deterministic approach is not very practical because the reverberation pattern encountered in practice is not so simple. Fortunately, even very complicated reverberation patterns seem to be minimum-phase operators.

The total received seismic trace $t(n)$ can be written as

$$t(n) = w(n) * c(n) * c(n) * p(n) = w_r(n) * p(n) \qquad (1.191)$$

where $w_r(n)$ represents the reverberated seismic source. If the wave-shaping filter design is performed on Eq. (1.191), then the inverse filter will remove the part of the total seismic trace that is predictable (the reverberated seismic source signal $w_r(n)$) and leave the unpredictable part (the primary reflector series $p(n)$). One problem with this approach is that it requires extremely long filters because the water reverberation component contains the two-way water delay n_w. Thus, this approach is not suited for deep water seismograms.

A reexamination of the three-point dereverberation operator $(1 + cz^{-n_w} + c^2 z^{-2n_w})$ suggests an obvious modification to the Wiener filter design procedure. Since the three-point operator has zero values for its impulse response in the range from $n = 1$ to $n = n_w - 1$, a similar constraint can be imposed on the deconvolution filter. Thus, we seek an all-pole model for $w_r(n)$ of the form

$$W_r(z) = \frac{G}{1 + z^{-\alpha} \sum\limits_{k=1}^{p} a_k z^{-k}} \qquad (1.192)$$

where α is the prediction distance. In effect, we are modifying the linear prediction filter (Fig. 1.12) to incorporate a delay $z^{-\alpha}$, which is a function of the two-way travel time in the water layer.

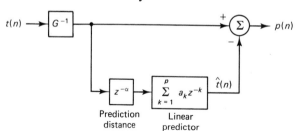

Figure 1.12   Predictive deconvolution with prediction distance α.

The extension of the least-squares technique to the model of Eq. (1.192) is straightforward. The normal equations are derived in exactly the same manner and can be solved with the Levinson recursion generalized for an arbitrary right-hand side.

$$\mathbf{R} \begin{bmatrix} a_1 \\ a_2 \\ \vdots \\ a_p \end{bmatrix} = \begin{bmatrix} r(\alpha + 1) \\ r(\alpha + 2) \\ \vdots \\ r(\alpha + p) \end{bmatrix} \qquad (1.193)$$

The only new consideration is the determination of a good value for the prediction distance α. If we assume that α is known *a priori*, we merely solve Eq. (1.193) for $A(z)$. A better strategy is to search for a good value of α. This can be done by computing the optimal filter for several trial values of α and then choosing the one that gives the smallest least-squares error. As one might expect, there is a variant of Levinson's recursion that provides a fast algorithm to step from a prediction distance of α to $\alpha + 1$.

1.7 POLE-ZERO MODELING

In this final section, we reconsider the "direct" problem first raised in Section 1.2. We have already concluded that the equations to be solved for the optimal pole-zero model in the direct problem are nonlinear and, therefore, quite difficult to solve. This led us to develop the various indirect methods presented in this chapter. However, there are situations where the solution from an "indirect" method is inadequate, and the difference between the model output and the desired signal must be as small as possible. What is needed is a way to improve the solution from the "indirect" method. We will show that an improved solution can be constructed from a linear problem and, further, that a succession of these linear problems can be solved to iteratively refine the pole-zero estimate. This is the method of *iterative prefiltering*.

First we must discuss ways to obtain zeros for a modeling problem. The strategy dictated by Prony's method is to calculate the poles first and then obtain the zeros by a simple matrix multiplication. This requires that the number of zeros q be known in advance. Since the error function for $n > q$ does not depend on the zeros, the poles are determined by minimizing the energy in the error signal for $n > q$, giving the denominator polynomial $A'(z)$. The superscript denotes the fact that this polynomial was computed via an "indirect" model. The $q + 1$ free parameters of $B(z)$ are then chosen to make the error exactly zero over the range $[0, q]$. If the z-transform of the original signal were $X(z)$ and if $A'(z)$ were the perfect estimate of the poles, then the correct numerator polynomial $B'(z)$ would be obtained by polynomial multiplication $X(z)A'(z) = B'(z)$. In general, both assumptions break down: $X(z)$ is not representable (with zero error) by a pole-zero model, and $A'(z)$ from an "indirect" problem is not the correct denominator. In practice, such a simple cross-multiplication idea will not yield good results. As always, some notion of (least-squares) approximation must be introduced into the problem.

One possible method of approximation is based on the observation that the error in the "direct" problem is linear in the coefficients of $B(z)$. Thus we can formulate a linear least-squares problem to solve for $B(z)$ given a fixed denominator, denoted as $A'(z)$. If $h'(n)$ is the impulse response of the known system $1/A'(z)$, then the error is

$$e_1(n) = x(n) - b_n * h'(n) \tag{1.194}$$

In this form, the problem looks just like that of Wiener filtering, with $x(n)$ playing the role of the desired signal and $h'(n)$ the input to the Wiener filter. Minimization of the error energy $\sum_{n=0}^{\infty} |e_1(n)|^2$ leads to a set of normal equations for the b_n, involving the autocorrelation of $h'(n)$.

An equivalent approach results from expanding $B(z)/A(z)$ in its partial fraction form

$$\frac{B(z)}{A'(z)} = \sum_{k=1}^{p} \frac{c_k}{(1 - \lambda_k z^{-1})} \tag{1.195}$$

Since the poles λ_k are fixed, the error is linear in the coefficients c_k and (for the special case of nonrepeated roots) can be expressed as

$$e_2(n) = x(n) - \sum_{k=1}^{p} c_k (\lambda_k)^n \tag{1.196}$$

The advantage of this form is that one can operate on the basis signals $(\lambda_k)^n$ analytically to compute correlations. In any event, the answers should be nearly the same; Eq. (1.194) is just direct form and Eq. (1.196) is parallel form for filter $B(z)/A^I(z)$.

Now we introduce a notion of iteration that will allow us to compute a new set of zeros *and* poles based on a given set of poles. It also has the very desirable property of doing the update by solving a *linear* problem. Consider the following error function, which is a linear function of the unknowns $A^{i+1}(z)$ and $B^{i+1}(z)$

$$E^{i+1}(z) = \frac{A^{i+1}(z)X(z) - B^{i+1}(z)}{A^i(z)} \tag{1.197}$$

$A^i(z)$ is the "old" denominator, which is assumed known. If $h^i(n)$ is the inverse z-transform of $1/A^i(z)$, then the error can be written as a linear function of the finite-length signals a_n^{i+1} and b_n^{i+1}, $a_0^{i+1} = 1$.

$$e^{i+1}(n) = a_n^{i+1} * x(n) * h^i(n) - b_n^{i+1} * h^i(n) \tag{1.198}$$

The coefficients $\{a_n^{i+1}\}$ and $\{b_n^{i+1}\}$ are chosen by minimizing the error energy $e^{i+1}(n)$ summed over the range of indices available for $x(n)$. In effect, the filter $1/A^i(z)$ acts as a weighting function on the least-squares error.

There is not much that one can prove about the convergence of the iteration implied by Eq. (1.198). No general conditions have been given to guarantee convergence. However, if the iteration does converge, it must give the same answer as the "direct" method. To prove this, assume that the sequence of $A^{i+1}(z)$ tends to a limiting value, $A^*(z)$. In this limiting condition, $A^{i+1}(z)/A^i(z) \to 1$ and

$$E^{i+1}(z) = \frac{A^{i+1}(z)X(z) - B^{i+1}(z)}{A^i(z)} \longrightarrow X(z) - \frac{B^*(z)}{A^*(z)}$$

Since the energy in the error $E^{i+1}(z)$ is minimized at each step, $B^*(z)/A^*(z)$ must be the solution to the "direct" problem. In practice, we usually don't need complete convergence, and often the answer is good enough within 5–10 iterations.

The precise form of the normal equations can be exhibited by writing Eq. (1.198) in matrix form and applying the projection theorem:

$$[\tilde{\mathbf{X}}_h^i \quad \mathbf{H}^i] \begin{bmatrix} \mathbf{a}^{i+1} \\ \mathbf{b}^{i+1} \end{bmatrix} \overset{?}{=} \mathbf{0} \tag{1.199}$$

The submatrix $\tilde{\mathbf{X}}_h^i$ contains entries that are filtered versions of the original signal $x(n)$, $x_h^i(n) = x(n) * h^i(n)$:

$$\tilde{\mathbf{X}}_h^i = \begin{bmatrix} x_h^i(L+p) & x_h^i(L+p-1) & \ddots & x_h^i(L) \\ x_h^i(L+p+1) & x_h^i(L+p) & \ddots & x_h^i(L+1) \\ \vdots & & \ddots & \vdots \\ x_h^i(U) & x_h^i(U-1) & \cdots & x_h^i(U-p) \end{bmatrix}$$

and $\mathbf{H}^i$ contains entries from the impulse response of $1/A(z)$.

$$\mathbf{H}^i = \begin{bmatrix} h^i(L+p) & h^i(L+p-1) & \ddots & h^i(L+p-q) \\ h^i(L+p+1) & h^i(L+p) & \ddots & h^i(L+p+1-q) \\ \vdots & \ddots & \ddots & \vdots \\ h^i(U) & h^i(U-1) & \cdots & h^i(U-q) \end{bmatrix}$$

Both $x_h^i(n)$ and $h^i(n)$ can be generated from a recursive filter implementation of $1/A^i(z)$.

The normal equations for this problem can be expressed as

$$\begin{bmatrix} (\tilde{\mathbf{X}}_h^i)^T\tilde{\mathbf{X}}_h^i & (\tilde{\mathbf{X}}_h^i)^T\mathbf{H}^i \\ (\mathbf{H}^i)^T\tilde{\mathbf{X}}_h^i & (\mathbf{H}^i)^T\mathbf{H}^i \end{bmatrix} \begin{bmatrix} \mathbf{a}^{i+1} \\ \mathbf{b}^{i+1} \end{bmatrix} = \begin{bmatrix} E^{i+1} \\ \mathbf{0} \end{bmatrix} \tag{1.200}$$

The error sum is taken over the range $[L + p, U]$ because the signal $x(n)$ is available only over the range $[L, U]$. Likewise, it does no good to extend beyond the interval $[L, U]$ because edge effects then become a problem. There is no problem filling the entries of $\mathbf{H}^i$ because the necessary values of the impulse response, $h^i(n)$, can be generated analytically.

With this type of iteration, we can approach the solution of the "direct" pole-zero modeling problem. Although there is no guarantee of convergence, most practical applications of this technique have been successful. Iterative prefiltering allows us to introduce zeros and to improve the all-pole model obtained by an "indirect" method.

1.8 SUMMARY

In this chapter we have discussed various methods of signal modeling. We concentrated on "indirect" methods, in general, and Prony's method, in particular. Two cases of Prony's method were treated in detail: the covariance method and the auto-correlation method for all-pole modeling. The covariance method has the desirable property of matching a noise-free all-pole signal exactly. The autocorrelation solution does not have this same property but does possess a number of elegant properties. In both cases, the model coefficients can be calculated with a fast $O(p^2)$ algorithm.

The rest of the chapter discussed other related methods. Some give exactly the same solution as the autocorrelation method and thus offer different interpretations of the same modeling technique. Others are extensions of the basic methods and result in different model coefficients. In a closing section, we presented an iterative method for pole-zero modeling based on the "indirect" methods developed earlier.

Despite the length of the chapter, there were still several topics omitted. No treatment of the parametric modeling of sinusoids in noise was given, even though this topic is closely related to the rational modeling techniques presented in this chapter. Also, very little was said about the problem of selecting the correct model order, a subject that has received a fair amount of attention from several researchers. With the iterative algorithms for solving the normal equations via order updates, it would be possible to monitor the error of a least-squares fit and then pick the model order where this error is acceptable. The research question is how to define "acceptable."

Modeling remains a very active area for research and development. New methods are constantly being discovered and old ones adapted for new applications. This chapter attempts to present enough background to demonstrate the power of these techniques and the differences among the various methods.

REFERENCES

It would be virtually impossible to provide a complete bibliography for the topics presented in this chapter. Such a compilation would be longer than the material of the chapter itself. Therefore, we provide an indirect referencing scheme. The following five references should be regarded as pointers into the literature, where one could begin to search for particular papers of interest. Each of the references cited has an extensive bibliography containing hundreds of references.

S. L. Marple, *Modern Spectrum Analysis*, Prentice-Hall, Englewood Cliffs, NJ, 1987.

S. M. Kay and S. L. Marple, "Spectrum Analysis: A Modern Perspective." *Proc. IEEE*, Vol. 69, pp. 1380–1419, Nov. 1981.

J. Makhoul, "Linear Prediction: A Tutorial Review," *Proc. IEEE*, Vol. 63, pp. 561–580, April 1975.

D. G. Childers, Ed., *Modern Spectrum Analysis*, IEEE Press (Selected Reprint Series), New York, 1978.

S. B. Kesler, Ed. *Modern Spectrum Analysis: II*, IEEE Press (Selected Reprint Series), New York, 1986.

In addition, we provide a sampling of important names and publications related to this subject, so the reader can gain some appreciation for the long history of research devoted to the topic of parametric signal modeling.

G. R. B. Prony, "Essai expérimental et analytique sur les lois de la dilatabilité de fluides elastiques et sur celles de la force expansion de la vapeur de l'alcool, à différentes températures," *Journal de l'Ecole Polytechnique* (Paris), Vol. 1, no. 2, pp. 24–76, 1795.

H. E. Padè, "Sur la représentation approchée d'une fonction par des fractions rationelles," *Annales Scientifique de l'Ecole Normale Supérieure*, Vol. 9, no. 3 (supplement), pp. 1–93, 1892.

G. U. Yule, "On a Method of Investigating Periodicities in Disturbed Series, with Special Reference to Wolfer's Sun-spot Numbers," *Philos. Trans. Royal Society: London*, Vol. 226, Series A, pp. 267–298, 1927.

G. Walker, "On Periodicity in Series of Related Terms," *Proc. Royal Society: London*, Vol. 131, Series A, pp. 518–532, 1931.

N. Levinson, "The Wiener RMS Error Criterion in Filter Design and Prediction," *J. Math. Phys.*, Vol. 25, pp. 261–278, Jan. 1947.

N. Wiener, *Interpolation and Smoothing of Stationary Time Series*, MIT Press, Cambridge, MA, 1949.

E. Robinson, "Predictive Decomposition of Time Series with Application to Seismic Exploration," *Geophysics*, Vol. 32, no. 3, pp. 418–484, June 1967.

R. N. McDonough and W. H. Huggins, "Best Least-Squares Representation of Signals by Exponentials," *IEEE Trans. Automatic Control*, Vol. 13, pp. 408–412, Aug. 1968.

J. Burg, *Maximum Entropy Spectral Analysis*, Ph.D. Thesis, Dept. Geophysics, Stanford University, Stanford, CA, 1975.

M. Morf, B. Dickinson, T. Kailath, A. Vieira, "Efficient Solution of Covariance Equations for Linear Prediction," *IEEE Trans. Acoustics, Speech, and Signal Processing*, Vol. 25, pp. 429–433, Oct. 1977.

K. Steiglitz, "On the Simultaneous Estimation of Poles and Zeros in Speech Analysis," *IEEE Trans. Acoustics, Speech, and Signal Processing*, Vol. 25, pp. 229–234, June 1977.

APPENDIX 1.1

NOTATION FOR SIGNAL MATRICES

The dependence of various matrices on the range of signal values available and the model order requires a general notation to express such dependencies (see Section 1.4). In fact, two notations are needed because of differences in the way end conditions are handled for finite signal lengths. These notations are, of necessity, cumbersome because two or three superscripts (or subscripts) must be used to denote all the information. After some familiarity, it becomes easy to work with formulas expressed in this notation.

First we define a notation for the signal matrix that arises naturally in the discussion of Padè approximation and Prony's method:

$$\mathbf{X}^{p,L,U} \stackrel{\text{def}}{=} \begin{bmatrix} x(L) & x(L-1) & \cdots & x(L-p) \\ x(L+1) & x(L) & \ddots & x(L+1-p) \\ \vdots & \ddots & \ddots & \vdots \\ x(U) & x(U-1) & \cdots & x(U-p) \end{bmatrix} \tag{1.1.1}$$

This matrix has $(U - L + 1)$ rows and $p + 1$ columns. The superscripts on $\mathbf{X}$ indicate that this signal matrix is associated with a pth-order modeling problem and that the error vector is evaluated over the range $[L, U]$. Note that this requires signal values from the range $[L - p, U]$. By convention, the signal values for negative indices are taken to be zero.

The $\mathbf{X}^{p,L,U}$ notation for the signal matrix would be sufficient, except that the signal range depends on the model order. A second notation is needed when the signal range is fixed and the model order varied, as happens in the fast covariance algorithm (Section 1.4).

$$\tilde{\mathbf{X}}^{p,L,U} \stackrel{\text{def}}{=} \begin{bmatrix} x(L+p) & x(L+p-1) & \cdots & x(L) \\ x(L+p+1) & x(L+p) & \ddots & x(L+1) \\ \vdots & \ddots & \ddots & \vdots \\ x(U) & x(U-1) & \cdots & x(U-p) \end{bmatrix} \tag{1.1.2}$$

This $\tilde{\mathbf{X}}^{p,L,U}$ matrix still has $p + 1$ columns but only $(U - L - p + 1)$ rows. In this case, the range of the error evaluation depends on p because the available signal range is fixed. Obviously, the following equalities hold:

$$\tilde{\mathbf{X}}^{p,L,U} = \mathbf{X}^{p,L+p,U} \tag{1.1.3}$$

$$\mathbf{X}^{p,L,U} = \tilde{\mathbf{X}}^{p,L-p,U} \tag{1.1.4}$$

The difference between these two notations is perhaps minor, but you are warned to be alert to the different situations that require one or the other.

It is also worthwhile to have a separate notation for the rows and columns of these matrices. For example, in the derivation of Prony's method, the first column of $\mathbf{X}^{p,L,U}$ is treated separately. Thus we define a column to be

$$\mathbf{x}_{L,U}^T = [x(L) \quad x(L + 1) \quad \cdots \quad x(U)] \tag{1.1.5}$$

Likewise, we can describe the rows of these matrices by defining

$$\tilde{\mathbf{x}}_p^T(i) = [x(i) \quad x(i - 1) \quad \cdots \quad x(i - p)] \tag{1.1.6}$$

Note that within a column the index for $x(i)$ is increasing, whereas along a row it is decreasing. Furthermore, the notation for the column emphasizes the beginning and end of the index range for the column, whereas for the row, the argument gives the starting index, and the subscript denotes the model order (equal to vector length minus 1).

Finally, we can express the $\mathbf{X}$ and $\tilde{\mathbf{X}}$ matrices in terms of the columns

$$\mathbf{X}^{p,L,U} = [\mathbf{x}_{L,U} \quad \mathbf{x}_{L-1,U-1} \quad \cdots \quad \mathbf{x}_{L-p,U-p}] \tag{1.1.7}$$

$$\tilde{\mathbf{X}}^{p,L,U} = [\mathbf{x}_{L+p,U} \quad \mathbf{x}_{L+p-1,U-1} \quad \cdots \quad \mathbf{x}_{L,U-p}] \tag{1.1.8}$$

or in terms of the rows

$$\mathbf{X}^{p,L,U} = \begin{bmatrix} \tilde{\mathbf{x}}_p^T(L) \\ \tilde{\mathbf{x}}_p^T(L + 1) \\ \vdots \\ \tilde{\mathbf{x}}_p^T(U) \end{bmatrix} \tag{1.1.9}$$

$$\tilde{\mathbf{X}}^{p,L,U} = \begin{bmatrix} \tilde{\mathbf{x}}_p^T(L + p) \\ \tilde{\mathbf{x}}_p^T(L + p + 1) \\ \vdots \\ \tilde{\mathbf{x}}_p^T(U) \end{bmatrix} \tag{1.1.10}$$

APPENDIX 1.2

LEAST-SQUARES NORMAL EQUATIONS

One particular least-squares problem arises often in the discussion of signal modeling—that of finding the least-squares solution to a set of overdetermined linear equations $\mathbf{Xa_1} = -\mathbf{y}$ (note the minus sign). This set of equations can also be expressed as

$$\mathbf{Ya} = [\mathbf{y}\ \mathbf{X}]\begin{bmatrix} 1 \\ \mathbf{a_1} \end{bmatrix} \overset{?}{=} \mathbf{0} \tag{1.2.1}$$

Since Eq. (1.2.1) may not have an exact solution, we are interested in finding the solution vector $\mathbf{a}$ such that the error vector $(\mathbf{e} = \mathbf{Ya})$ has minimum length $(\mathbf{e}^T\mathbf{e})$. The value of this minimum length (E_a) is also of interest:

$$E_a = \min_{\mathbf{a}} (\mathbf{Ya})^T \mathbf{Ya} \tag{1.2.2}$$

The projection theorem of linear least-squares optimization theory tells us that, at the optimum value of $\mathbf{a}$, the error $(\mathbf{e} = \mathbf{Ya})$ is orthogonal to the linear subspace spanned by the basis vectors, in this case the columns of $\mathbf{X}$.

$$\mathbf{X}^T\mathbf{e} = \mathbf{X}^T \mathbf{Ya} = \mathbf{0} \tag{1.2.3}$$

This orthogonality condition leads to the normal equations of the least-squares problem

$$(\mathbf{X}^T\mathbf{X})\mathbf{a_1} = -\mathbf{X}^T\mathbf{y} \tag{1.2.4}$$

The minimum value of the error is found by using the orthogonality condition from Eq. (1.2.3):

$$\begin{aligned} E_a &= \min (\mathbf{e}^T\mathbf{e}) \\ &= (\mathbf{y} + \mathbf{Xa_1})^T\mathbf{e} \\ &= \mathbf{y}^T\mathbf{e} \\ &= \mathbf{y}^T\mathbf{Ya} \end{aligned} \tag{1.2.5}$$

Further reduction is possible:

$$\begin{aligned} E_a &= \mathbf{y}^T(\mathbf{y} + \mathbf{Xa_1}) \\ &= \mathbf{y}^T\mathbf{y} + \mathbf{y}^T\mathbf{Xa_1} \\ &= \mathbf{y}^T\mathbf{y} + \mathbf{a_1}^T\mathbf{X}^T\mathbf{Xa_1} \end{aligned} \tag{1.2.6}$$

The desired result is obtained by combining Eqs. (1.2.4) and (1.2.6) into one expression:

$$\begin{bmatrix} \mathbf{y}^T \\ \mathbf{X}^T \end{bmatrix}[\mathbf{y}\ \mathbf{X}]\begin{bmatrix} 1 \\ \mathbf{a_1} \end{bmatrix} = \begin{bmatrix} E_a \\ \mathbf{0} \end{bmatrix} \tag{1.2.7}$$

In other words, the least-squares solution for $\mathbf{Ya} \overset{?}{=} \mathbf{0}$ is calculated from

$$\mathbf{Y}^T\mathbf{Ya} = \begin{bmatrix} E_a \\ \mathbf{0} \end{bmatrix} \tag{1.2.8}$$

where the solution vector $\mathbf{a}$ has a 1 in its first entry and E_a is the minimum value of the least-squares error.

2

Spectral Estimation

Steven Kay
University of Rhode Island

2.0 INTRODUCTION

The problem of spectral estimation is that of determining the distribution in frequency of the power of a random process. Questions such as "Does most of the power of the signal reside at low or high frequencies?" or "Are there resonances in the spectrum?" are often answered as a result of a spectral analysis. As one might expect, spectral analysis finds wide use in such diverse fields as radar, sonar, speech, biomedicine, economics, geophysics, and others in which signals of unknown or questionable origin are of interest.

Estimation of the power spectral density is usually complicated by the lack of a sufficiently long duration data record on which to base the spectral estimate. The shortness of the record may be due to a genuine lack of data, as in the seismic patterns of an erupting volcano, or due to an artificially imposed restriction necessary to ensure that the spectral characteristics of a signal do not change over the duration of the data record, as in speech processing. For reliable spectral estimates we would wish for large amounts of data, but for many practical cases of interest, such as the two just mentioned, the data set is limited. The principles of statistical inference allow us to make the most of the available data. Tradeoffs can be expected in spectral estimation; the paramount one is bias versus variance. As we will see, if the spectral estimator yields good estimates on the average (low bias), then we can expect much variability from one data realization to the next (high variance). On the other hand, if we choose a spectral estimator with low variability, then on the average the spectral estimate may be poor. The only way out of this dilemma is to increase the data record length.

Spectral estimators may be classified as either nonparametric or parametric. The nonparametric ones such as the periodogram, Blackman-Tukey, and minimum variance spectral estimators require no assumptions about the data other than wide-sense

Steven Kay is with the Electrical Engineering Department, University of Rhode Island, Kingston, RI 02881.

stationarity. The parametric spectral estimators, on the other hand, are based on rational transfer function or time series models of the data. Hence, their application is more restrictive. The time series models are the autoregressive, moving average, and autoregressive moving average types. The advantage of the parametric spectral estimator is that when applicable it yields a more accurate spectral estimate. Without having to increase the data record length, we can simultaneously reduce the bias and the variance over the nonparametric spectral estimator. Of course, the improvement is due to the use of *a priori* knowledge afforded by the modeling assumption. Such adjectives as *high resolution* [1], *maximum entropy* [2], and *linear prediction* [3] are all synonymous with *parametric*.

2.1 DEFINITIONS

It will be assumed that $x(n)$ is a real discrete-time random process that is wide-sense stationary. To be wide-sense stationary, the mean of $x(n)$ for any n must be the same and the autocorrelation function must depend only on the lag between samples. Mathematically, it is assumed that

$$E[x(n)] = m_x \tag{2.1}$$

$$E[x(n)x(n + k)] = r_x(k) \tag{2.2}$$

where m_x is the mean and $r_x(k)$ is the autocorrelation function evaluated at lag k, both of which are independent of n. For convenience we will further assume that $m_x = 0$. This assumption is not restrictive in that we may equivalently consider $y(n) = x(n) - m_x$ with the result that $m_y = 0$.

The power spectral density (PSD) of $x(n)$ is defined as

$$P_x(\omega) = \sum_{k=-\infty}^{\infty} r_x(k) \exp(-j\omega k) \tag{2.3}$$

over the range $-\pi \leq \omega \leq \pi$. Equation (2.3) is sometimes called the Wiener-Khinchin theorem. The interpretation of $P_x(\omega)d\omega$ is as the average power of $x(n)$, which resides in the frequency band from ω to $\omega + d\omega$. Although this interpretation may not be evident from Eq. (2.3), an alternative but equivalent definition of the PSD will more clearly illustrate this property. The PSD may also be defined as

$$P_x(\omega) = \lim_{M \to \infty} E\left[\frac{1}{2M + 1} \left| \sum_{n=-M}^{M} x(n) \exp(-j\omega n) \right|^2 \right] \tag{2.4}$$

Eq. (2.4) says that the PSD at frequency ω is found by first taking the magnitude squared of the Fourier transform of $x(n)$ and then dividing by the data record length to yield power. Since the power will be a random variable (a different value will be obtained for each realization of $x(n)$), the expected value is taken. Finally, since the random process is in general of infinite duration, a limiting operation is required.

Eq. (2.4) is identical to Eq. (2.3) as we will now show. Starting with Eq. (2.4), we have

$$P_x(\omega) = \lim_{M \to \infty} E\left[\frac{1}{2M + 1} \sum_{n=-M}^{M} \sum_{m=-M}^{M} x(n)x(m) \exp[-j\omega(n - m)]\right]$$

$$= \lim_{M \to \infty} \frac{1}{2M + 1} \sum_{n=-M}^{M} \sum_{m=-M}^{M} r_x(n - m) \exp[-j\omega(n - m)] \tag{2.5}$$

But

$$\sum_{n=-M}^{M} \sum_{m=-M}^{M} f(n - m) = \sum_{k=-2M}^{2M} (2M + 1 - |k|)f(k) \tag{2.6}$$

which may be verified by considering $f(n - m)$ as the (n, m) element of a matrix of dimension $(2M + 1) \times (2M + 1)$. Applying Eq. (2.6) to Eq. (2.5), we obtain

$$P_x(\omega) = \lim_{M \to \infty} \frac{1}{2M + 1} \sum_{k=-2M}^{2M} (2M + 1 - |k|)r_x(k) \exp(-j\omega k)$$

$$= \lim_{M \to \infty} \sum_{k=-2M}^{2M} \left(1 - \frac{|k|}{2M + 1}\right)r_x(k) \exp(-j\omega k)$$

$$= \sum_{k=-\infty}^{\infty} r_x(k) \exp(-j\omega k)$$

The last step assumes that $r_x(k) \to 0$ as $k \to \infty$ at a sufficiently rapid rate, which is violated for random processes with nonzero means or sinusoidal components. Excluding these processes, Eq. (2.4) is equivalent to Eq. (2.3). Note that Eq. (2.3) is able to accommodate nonzero means and sinusoidal components but only via the use of Dirac delta functions. In practice, the data records will of necessity be finite and so this slight discrepancy is a moot point. Equations (2.3) and (2.4) will provide starting points for the class of nonparametric spectral estimators.

2.2 THE PROBLEM OF SPECTRAL ESTIMATION

The problem of PSD estimation or, more succinctly, spectral estimation is the following. Given a *finite* segment of a realization of a random process, i.e., $\{x(0), x(1), \ldots, x(N - 1)\}$, estimate the PSD $P_x(\omega)$ for $|\omega| \le \pi$. Because $r_x(-k) = r_x(k)$ the PSD will be an even function, so we really need to estimate it only over the frequency interval $0 \le \omega \le \pi$. It is clear from the definition of the PSD given by Eq. (2.4) that the stated problem is difficult if not impossible. To find the PSD requires knowledge of $x(n)$ for all n as well as all possible realizations so that the expectation operation can be applied. Therefore, we should not expect perfect estimates. Another viewpoint is that, according to Eq. (2.3), spectral estimation is equivalent to autocorrelation estimation. Given N samples of $x(n)$ we must estimate an infinite number of autocorrelation lags, again an impossible task. One way out of this dilemma is to assume *a priori* that the autocorrelation function has certain properties. These properties would allow us to determine exactly the autocorrelation function for $k \geq p + 1$ if we knew the values for $k = 0, 1, \ldots, p$. (Recall that $r_x(-k) = r_x(k)$.) As an example, we might assume that

$$r_x(k) = -a_1 r_x(k - 1) - \cdots - a_p r_x(k - p) \qquad \text{for } k \geq p + 1 \qquad (2.7)$$

so that given the initial conditions of the difference equation $\{r_x(1), r_x(2), \ldots, r_x(p)\}$ and the coefficients $\{a_1, a_2, \ldots, a_p\}$ we could generate the remaining lags using Eq. (2.7). Add $r_x(0)$ to this information and we now have enough information to determine $P_x(\omega)$. The salient feature of this approach is that a model has been assumed for the autocorrelation function or, equivalently, for the PSD. The general spectral estimation problem in which we must estimate a continuous function has been reduced to a parameter estimation problem in which we estimate a finite set of parameters. For the example of Eq. (2.7) we will see in Section 2.5.2 that the a_k coefficients are easily determined from $r_x(k)$ for $0 \leq k \leq p$. Hence, knowledge of $r_x(k)$ for $0 \leq k \leq p$ is equivalent to knowledge of the PSD. Basic concepts in statistics tell us that if $N \gg p + 1$, our estimates of $r_x(k)$ for $0 \leq k \leq p$ and hence of the PSD will be good. This modeling approach forms the basis of the parametric techniques of Section 2.5.

A word of caution concerning the parametric techniques is in order. Since they rely on a model, it is critical that the model be correct. Any departure from the model will result in a systematic or bias error in the spectral estimate. As an example, if we model the PSD by a Gaussian function

$$P_x(\omega) = A \exp[-\tfrac{1}{2}(\omega - \omega_0)^2/\sigma^2], \qquad |\omega| \leq \pi$$

then we need only estimate $\{A, \omega_0, \sigma\}$ to estimate the PSD. If N is large, our estimates of these parameters presumably will not vary greatly from realization to realization. Hence, the spectral estimate will have low variability. If, however, the true PSD does not fit the model, the spectral estimator will be highly biased, an example of which is shown in Fig. 2.1. Clearly, the accuracy of the model is of utmost importance.

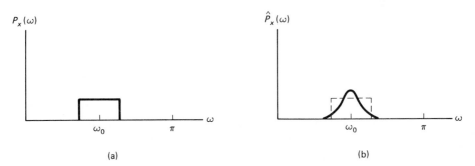

Figure 2.1   Example of difficulty of modeling approach using Gaussian model. (a) True power spectral density; (b) spectral estimate.

Since spectral estimators are random functions, it is necessary to base them on sound statistical techniques of estimation. Furthermore, we will need to describe the accuracy of a spectral estimate in statistical terms. To accomplish these goals, the next section briefly reviews the basic concepts of statistics employed in spectral estimation.

2.3 REVIEW OF STATISTICS

2.3.1 Properties of Estimators

The field of statistics deals with inferring information from random data. The theory of *statistical estimation* is particularly important for spectral estimation and hence we will restrict our review to this area. To illustrate the concepts, we will consider the problem of the estimation of the mean m_x of a wide-sense stationary Gaussian random process. We assume that the PSD $P_x(\omega)$ is white

$$P_x(\omega) = \sigma_x^2 + m_x^2 \delta_c(\omega), \qquad |\omega| \le \pi \tag{2.8}$$

so that the variance is σ_x^2. Here $\delta_c(\omega)$ is the Dirac delta function. The observed data are $\{x(0), x(1), \ldots, x(N-1)\}$. We wish to find an estimator of m_x, which we will denote by $\hat{m}_x$. To be useful the estimator should be a function of only the observed data. A reasonable estimator might be the sample mean

$$\hat{m}_x = \frac{1}{N} \sum_{n=0}^{N-1} x(n) \tag{2.9}$$

Note that $\hat{m}_x$ is a random variable since it is a function of several random variables. As such its value will change for each realization of the observed data. The *rule* of assigning a value to $\hat{m}_x$ is called the *estimator* of m_x, while the *actual value* of $\hat{m}_x$ computed for a realization of data is called the *estimate*. A desirable property of an estimator is that on the average it should yield the correct value or

$$E[\hat{m}_x] = m_x \tag{2.10}$$

The sample mean has this property since

$$E[\hat{m}_x] = E\left[\frac{1}{N} \sum_{n=0}^{N-1} x(n)\right] = \frac{1}{N} \sum_{n=0}^{N-1} E[x(n)] = m_x$$

Such an estimator is said to be *unbiased*. If this is not the case, then the bias of the estimator is defined to be

$$B[\hat{m}_x] = m_x - E[\hat{m}_x] \tag{2.11}$$

and is referred to as a *bias error*. An estimator may be unbiased but fluctuate wildly from realization to realization. For a reliable estimate we would also like the variance of the estimator to be small. We will denote the variance of an estimator by $\text{VAR}[\hat{m}_x]$. The variance of the sample mean estimator is easily found. Noting that $x(n)$ is a white noise process, we see that samples of $x(n)$ are uncorrelated. This is easily verified by taking the inverse Fourier transform of $P_x(\omega)$, which is given by Eq. (2.8), to yield the autocorrelation function

$$r_x(k) = \sigma_x^2 \delta(k) + m_x^2$$

Since $r_x(k) - m_x^2 = 0$ for $m \ne 0$, the data samples are uncorrelated. It follows that

$$\text{VAR}[\hat{m}_x] = \frac{1}{N^2} \sum_{n=0}^{N-1} \text{VAR}[x(n)] = \frac{1}{N^2}(N\sigma_x^2) = \frac{1}{N}\sigma_x^2 \tag{2.12}$$

It is comforting that the variance approaches zero as the number of data samples tends to infinity. In a sense, then, the estimator is consistent in that

$$\lim_{N \to \infty} \hat{m}_x = m_x \tag{2.13}$$

and we should probably always require our estimators to have this property. Looking at it in another way we may rewrite Eq. (2.13) as

$$\lim_{N \to \infty} \frac{1}{N} \sum_{n=0}^{N-1} x(n) = E[x(m)] \tag{2.14}$$

The estimator $\hat{m}_x$ may be thought of as a temporal mean since we are averaging in time the samples of one realization, while $E[x(m)]$ represents an ensemble mean. The ensemble mean is computed by effectively averaging over many realizations at the identical instant of time. Since the temporal average as $N \to \infty$ equals the ensemble average, we refer to $x(n)$ as being *ergodic* in the mean. For arbitrary processes, ergodicity in the mean is attained if

$$\lim_{k \to \infty} r_x(k) = m_x^2 \tag{2.15}$$

Consistency of the sample mean estimator requires that the samples of the process be uncorrelated if separated by large enough lags. Otherwise, not enough temporal averaging will be accomplished to cancel out the random fluctuations of the samples.

It frequently occurs in practice that estimators are biased. Furthermore, the bias may depend on the parameter to be estimated. (If the bias did not depend on the unknown parameter, then in most cases we could remove the bias. As an example, consider the mean estimator $\sum_{n=0}^{N-1} x(n)$.) In such cases it is better to describe the performance of the estimator using the mean-square error (MSE)

$$\text{MSE}[\hat{m}_x] = E[(\hat{m}_x - m_x)^2] \tag{2.16}$$

In the case of unbiased estimators, we strive to use an estimator having a low variance, but here an estimator with a low MSE is desirable. The MSE is easily related to the variance since

$$\text{MSE}[\hat{m}_x] = E[[(\hat{m}_x - E[\hat{m}_x]) + (E[\hat{m}_x] - m_x)]^2]$$

$$= E[(\hat{m}_x - E[\hat{m}_x])^2 + 2(\hat{m}_x - E[\hat{m}_x])(E[\hat{m}_x] - m_x) + (E[\hat{m}_x] - m_x)^2]$$

$$= \text{VAR}[\hat{m}_x] + 2[E[\hat{m}_x] - E[\hat{m}_x]][E[\hat{m}_x] - m_x] + B^2[\hat{m}_x]$$

$$\text{MSE}[\hat{m}_x] = \text{VAR}[\hat{m}_x] + B^2[\hat{m}_x] \tag{2.17}$$

It is apparent from Eq. (2.17) that if an estimator is unbiased, then the MSE is identical to the variance. The question arises as to whether we can trade bias for variance in an attempt to yield a smaller MSE. For example, if $\hat{m}_x$ is an unbiased estimator and we define a new estimator as

$$\bar{m}_x = a\hat{m}_x$$

then from Eq. (2.11),

$$B[\bar{m}_x] = m_x - am_x = (1 - a)m_x \tag{2.18}$$

and also

$$\text{VAR}[\tilde{m}_x] = a^2 \, \text{VAR}[\hat{m}_x] \tag{2.19}$$

Clearly, for $|a| < 1$ the variance is decreased but the bias is increased over that for $\hat{m}_x$. Also, from Eqs. (2.17)–(2.19),

$$\text{MSE}[\tilde{m}_x] = a^2 \, \text{VAR}[\hat{m}_x] + (1 - a)^2 \, m_x^2 \tag{2.20}$$

which is minimized over a by the optimal choice

$$a_{\text{OPT}} = \frac{m_x^2}{m_x^2 - \text{VAR}[\hat{m}_x]}. \tag{2.21}$$

If $\hat{m}_x$ is the sample mean estimator, then

$$a_{\text{OPT}} = \frac{m_x^2}{m_x^2 - \sigma_x^2/N}$$

Unfortunately, the optimal value of a depends on m_x, which is exactly what we are attempting to estimate. It is interesting to note that if we knew *a priori* that m_x was close to zero, we could choose a close to zero to reduce the variance but still not increase the bias significantly. This type of operation is termed a *bias-variance tradeoff,* and we frequently encounter it in spectral estimation.

The final performance measure of an estimator, which will prove useful, is the confidence interval. Instead of describing the performance of an estimator by its mean and variance, it is more complete to state the probability density function (PDF) of the estimator. For our example, $\hat{m}_x$ is Gaussian since it is a sum of jointly Gaussian random variables. Hence,

$$\hat{m}_x \sim N(m_x, \sigma_x^2/N) \tag{2.22}$$

where $\sim$ means "is distributed according to" and $N(\mu, \sigma^2)$ denotes a Gaussian (normal) distribution with mean μ and variance σ^2. The PDF of the estimator itself may be conveniently summarized by giving an interval on the real line where m_x will lie with high probability. For instance, we might say that m_x lies within the interval $\hat{m}_x \pm \Delta$ with probability 0.9. The interval $(\hat{m}_x - \Delta, \hat{m}_x + \Delta)$ is called a 90% confidence interval for m_x. We now derive this confidence interval.

First consider the probability statement

$$\text{Prob}\left[-\alpha \le \frac{\hat{m}_x - m_x}{\sqrt{\sigma_x^2/N}} \le \alpha \right] = 0.9 \tag{2.23}$$

where "Prob" denotes probability. We must choose α so that Eq. (2.23) is true. According to Eq. (2.22),

$$\frac{\hat{m}_x - m_x}{\sqrt{\sigma_x^2/N}} \sim N(0, 1)$$

so $\alpha = 1.645$. Manipulating Eq. (2.23), we have

$$\text{Prob}[-\alpha\sqrt{\sigma_x^2/N} - \hat{m}_x \le -m_x \le \alpha\sqrt{\sigma_x^2/N} - \hat{m}_x] = 0.9$$
$$\text{Prob}[\hat{m}_x - \alpha\sqrt{\sigma_x^2/N} \le m_x \le \hat{m}_x + \alpha\sqrt{\sigma_x^2/N}] = 0.9 \tag{2.24}$$

The 90% confidence interval is $\hat{m}_x \pm 1.645\sigma_x/\sqrt{N}$. Although it appears from Eq. (2.24) that m_x will fall within the confidence interval with 90% probability, this is an incorrect interpretation because m_x is not a random variable. We should say that the *random* interval $\hat{m}_x \pm 1.645\sigma_x/\sqrt{N}$ will *cover* the true value of m_x with 90% probability. An illustration is given in Fig. 2.2 in which the random interval covers the true value of the mean 90% of the time.

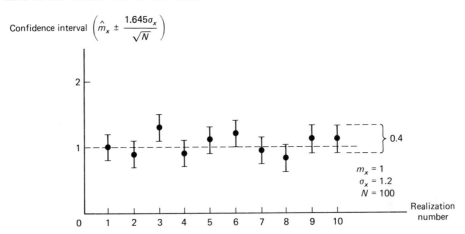

Figure 2.2   Illustration of concept of 90% confidence interval for mean.

2.3.2 Finding Good Estimators

Although we have discussed the properties of estimators and in particular examined the sample mean estimator, it is not yet clear how we actually find the estimator. The technique that has proven to be most useful in practice is the maximum likelihood estimator (MLE). The principle says that given the PDF $p(\mathbf{x}; \theta)$ of $\mathbf{x} = [x(0)x(1) \ldots x(N - 1)]^T$, which depends on an unknown parameter θ, a good estimate of θ is found by choosing the value that maximizes $p(\mathbf{x}; \theta)$. The data sample values $\mathbf{x}_0$ are substituted for $\mathbf{x}$ so that the PDF is only a function of θ. The rationale for the approach is that by maximizing the PDF over θ we are finding the value of θ that results in the highest probability of $\mathbf{x}_0$ being observed. Since $\mathbf{x}_0$ was indeed observed, that value of θ is a reasonable one. Furthermore, it can be shown that as $N \to \infty$ the MLE is unbiased and has the smallest variance (and hence the smallest MSE) of all unbiased estimators [4]. As an example, consider the previous problem of mean estimation. To find the MLE we need to maximize $p(\mathbf{x}; m_x)$ over m_x. Since the data samples are jointly Gaussian and uncorrelated, they are also independent. Hence,

$$p(\mathbf{x}; m_x) = \prod_{n=0}^{N-1} \frac{1}{\sqrt{2\pi}\sigma_x} \exp\left[-\frac{1}{2\sigma_x^2}[x(n) - m_x]^2\right]$$

$$= \frac{1}{(2\pi\sigma_x^2)^{N/2}} \exp\left[-\frac{1}{2\sigma_x^2} S(m_x)\right]$$

where $S(m_x) = \sum_{n=0}^{N-1} [x(n) - m_x]^2$. Maximizing $p(\mathbf{x}; m_x)$ is equivalent to minimizing S. Hence,

$$\frac{\partial S}{\partial m_x} = -2 \sum_{n=0}^{N-1} [x(n) - m_x]$$

Setting the derivative equal to zero and replacing m_x by $\hat{m}_x$, we have

$$\sum_{n=0}^{N-1} [x(n) - \hat{m}_x] = 0 \qquad \text{or} \qquad \hat{m}_x = \frac{1}{N} \sum_{n=0}^{N-1} x(n)$$

which we have already seen to be an intuitively pleasing estimator. As a second example, consider the problem of estimating the linear trend in a time series, as illustrated in Fig. 2.3. A reasonable model for the data of Fig. 2.3 is

$$x(n) = \theta_1 n + \theta_2 + e(n), \qquad n = 0, 1, \ldots, N - 1 \qquad (2.25)$$

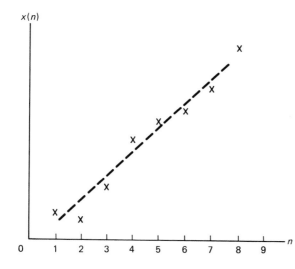

Figure 2.3 Data with apparent linear trend.

where θ_1, θ_2 are the slope and intercept, respectively, of a line. Both parameters of the line are unknown and $e(n)$ is modeled as a zero mean wide-sense stationary white Gaussian noise process with variance σ_e^2. The MLE of θ_1, θ_2 can be determined by maximizing $p(\mathbf{x}; \theta_1, \theta_2)$ over θ_1 and θ_2. In a similar fashion to the previous example we need to minimize

$$S'(\theta_1, \theta_2) = \sum_{n=0}^{N-1} [x(n) - \theta_1 n - \theta_2]^2 \qquad (2.26)$$

The problem and solution are recast more conveniently in matrix notation. Let

$$\mathbf{x} = [x(0) \, x(1) \, \ldots \, x(N - 1)]^T$$

$$\boldsymbol{\theta} = [\theta_1 \, \theta_2]^T$$

$$\mathbf{e} = [e(0) \, e(1) \, \ldots \, e(N - 1)]^T$$

$$\mathbf{H} = \begin{bmatrix} 0 & 1 \\ 1 & 1 \\ 2 & 1 \\ \cdot & \cdot \\ \cdot & \cdot \\ \cdot & \cdot \\ N-1 & 1 \end{bmatrix}$$

Then Eq. (2.25) becomes

$$\mathbf{x} = \mathbf{H}\boldsymbol{\theta} + \mathbf{e} \tag{2.27}$$

where $\mathbf{H}$ is a known matrix, $\boldsymbol{\theta}$ is a vector of unknown parameters, and $\mathbf{e} \sim N(\mathbf{0}, \sigma_e^2 \mathbf{I})$. $N(\boldsymbol{\mu}, \mathbf{K})$ denotes a multivariate Gaussian PDF with mean $\boldsymbol{\mu}$ and covariance matrix $\mathbf{K}$, and $\mathbf{I}$ is the identity matrix. Equation (2.26) now becomes

$$S'(\theta) = (\mathbf{x} - \mathbf{H}\boldsymbol{\theta})^T (\mathbf{x} - \mathbf{H}\boldsymbol{\theta}) \tag{2.28}$$

To find the MLE, we take the gradient of S' by making use of the following identities:

$$\frac{\partial \boldsymbol{\theta}^T \mathbf{A}\boldsymbol{\theta}}{\partial \boldsymbol{\theta}} = 2\mathbf{A}\boldsymbol{\theta} \qquad \text{if } \mathbf{A}^T = \mathbf{A}$$

$$\frac{\partial \boldsymbol{\theta}^T \mathbf{b}}{\partial \boldsymbol{\theta}} = \mathbf{b}$$

Hence,

$$S'(\boldsymbol{\theta}) = \mathbf{x}^T\mathbf{x} - 2\boldsymbol{\theta}^T\mathbf{H}^T\mathbf{x} + \boldsymbol{\theta}^T\mathbf{H}^T\mathbf{H}\boldsymbol{\theta}$$

$$\frac{\partial S'(\boldsymbol{\theta})}{\partial \boldsymbol{\theta}} = -2\mathbf{H}^T\mathbf{x} + 2\mathbf{H}^T\mathbf{H}\boldsymbol{\theta} = 0$$

which results in

$$\hat{\boldsymbol{\theta}} = (\mathbf{H}^T\mathbf{H})^{-1}\mathbf{H}^T\mathbf{x} \tag{2.29}$$

$\hat{\boldsymbol{\theta}}$ is the MLE of $\boldsymbol{\theta}$ under the conditions stated above.

It frequently occurs in practice that an estimation problem can be expressed in the form of Eq. (2.27) but that $\mathbf{e}$ is not composed of uncorrelated zero-mean Gaussian random variables. The least-squares modified Yule-Walker equation estimator of Section 2.5.8 is one such example. In such a case $\hat{\boldsymbol{\theta}}$ can still be used as an estimator of $\boldsymbol{\theta}$ although there are no optimality properties associated with the estimator; $\hat{\boldsymbol{\theta}}$ is called the *least-squares estimator* since it minimizes a sum of squares as given by Eq. (2.28). That is, it picks the $\hat{\boldsymbol{\theta}}$ such that the difference of the sum of the squares of the output produced by $\hat{\boldsymbol{\theta}}$ (i.e., $\mathbf{H}\hat{\boldsymbol{\theta}}$) and the output actually observed (i.e., $\mathbf{x}$) is minimized over all possible $\hat{\boldsymbol{\theta}}$.

2.3.3 A Useful PDF

The PDF for a sum of squares of independent $N(0, 1)$ random variables is of interest in spectral estimation. If $y = \sum_{n=0}^{N-1} x^2(n)$ where $x(n) \sim N(0, 1)$ for $n = 0, 1, \ldots, N - 1$, and all the $x(n)$'s are independent, then the PDF of y is

$$p(y) = \begin{cases} \dfrac{1}{2^{N/2}\Gamma(N/2)} y^{N/2-1} \exp(-y/2) & \text{if } y \geq 0 \\ 0 & \text{if } y < 0 \end{cases} \qquad (2.30)$$

where $\Gamma(u)$ is the gamma integral. The random variable y is said to be distributed according to a chi-squared distribution with N degrees of freedom or

$$y \sim \chi_N^2.$$

It is easily shown that the mean and variance are

$$E[y] = N \qquad (2.31)$$

$$\text{VAR}[y] = 2N \qquad (2.32)$$

2.4 NONPARAMETRIC SPECTRAL ESTIMATION

In this section we will discuss several popular methods of nonparametric spectral estimation. The Fourier or classical methods are the periodogram, which makes use of Eq. (2.4), and the Blackman-Tukey approach, which relies on Eq. (2.3). A recently proposed method termed the minimum variance spectral estimator will also be described. In all cases it will become apparent that for a fixed data record length we can reduce either the bias or the variance of the estimator but not both simultaneously.

2.4.1 Periodogram Spectral Estimator

The periodogram spectral estimator relies on Eq. (2.4):

$$P_x(\omega) = \lim_{M \to \infty} E\left[\frac{1}{2M+1} \left| \sum_{n=-M}^{M} x(n) \exp(-j\omega n) \right|^2 \right] \qquad (2.4)$$

Recall that the observed data set is $\{x(0), x(1), \ldots, x(N-1)\}$. By neglecting the expectation operator and by using the available data, we define the periodogram spectral estimator as

$$\hat{P}_{\text{PER}}(\omega) = \frac{1}{N} \left| \sum_{n=0}^{N-1} x(n) \exp(-j\omega n) \right|^2 \qquad (2.33)$$

An interesting interpretation of the periodogram estimator becomes apparent if we replace ω by ω_0 to emphasize that we are estimating the PSD at a particular frequency and rewrite Eq. (2.33) as

$$\hat{P}_{\text{PER}}(\omega_0) = \left[N \left| \sum_{k=0}^{N-1} h(n-k)x(k) \right|^2 \right]\Bigg|_{n=0}$$

where

$$h(n) = \begin{cases} \dfrac{1}{N} \exp(j\omega_0 n) & \text{for } n = -(N-1), \ldots, -1, 0 \\ 0 & \text{otherwise} \end{cases}$$

and $h(n)$ is the impulse response of a linear shift-invariant filter with frequency response

$$H(\omega) = \sum_{n=-(N-1)}^{0} h(n) \exp(-j\omega n)$$

$$= \frac{\sin[N(\omega - \omega_0)/2]}{N \sin[(\omega - \omega_0)/2]} \exp\left[j\left(\frac{N-1}{2}\right)(\omega - \omega_0) \right]$$

(2.34)

This is a bandpass filter with center frequency at $\omega = \omega_0$. Hence, the periodogram estimates the power at frequency ω_0 by filtering the data with a bandpass filter, sampling the output at $n = 0$, and computing the magnitude squared. The N factor is necessary to account for the bandwidth of the filter, which can be shown to be approximately $1/N$ [5], assuming a 3-dB bandwidth. The power when divided by $1/N$ yields the spectral estimate.

It might be supposed that if enough data were available, say $N \rightarrow \infty$, then

$$\hat{P}_{PER}(\omega) \rightarrow P_x(\omega)$$

or that the periodogram is a consistent estimator of the PSD. This was the case for estimation of the mean. To test this hypothesis we will consider the periodogram of zero-mean white Gaussian noise. In Fig. 2.4 we have computed the periodogram for several data record lengths. Each N-point data record was obtained by taking the first N points of a 1024-point data record. It appears that the random fluctuation or variance of the periodogram does not decrease with increasing N and hence that the periodogram is *not* a consistent estimator.

To verify this observation we will derive the PDF of the periodogram for white Gaussian noise. We will assume for simplicity that $\omega = \omega_k = k2\pi/N$ for $k = 0, 1, \ldots, N/2$ (N even). Note that $\hat{P}_{PER}(-\omega_k) = \hat{P}_{PER}(\omega_k)$, so we need not consider $k = -N/2 + 1, \ldots, -1$. The general results for an arbitrary frequency may be found in [6]. We show in the Appendix at the end of this chapter that

$$\frac{2\hat{P}_{PER}(\omega_k)}{P_x(\omega)} \sim \chi_2^2, \qquad k = 1, 2, \ldots, N/2 - 1$$

$$\frac{\hat{P}_{PER}(\omega_k)}{P_x(\omega)} \sim \chi_1^2, \qquad k = 0, N/2$$

(2.35)

It immediately follows from Eqs. (2.31) and (2.32) that

$$E[\hat{P}_{PER}(\omega_k)] = P_x(\omega_k), \qquad k = 0, 1, \ldots, N/2 \qquad (2.36)$$

$$\text{VAR}[\hat{P}_{PER}(\omega_k)] = \begin{cases} P_x^2(\omega_k), & k = 1, 2, \ldots, N/2 - 1 \\ 2P_x^2(\omega_k), & k = 0, N/2 \end{cases} \qquad (2.37)$$

We see that the periodogram is an unbiased estimator of the PSD but that it is *not* consistent in that the variance does not decrease with increasing data record length. This accounts for the appearance of Fig. 2.4. The periodogram estimator is unreliable because the standard deviation, which is the square root of the variance, is as large

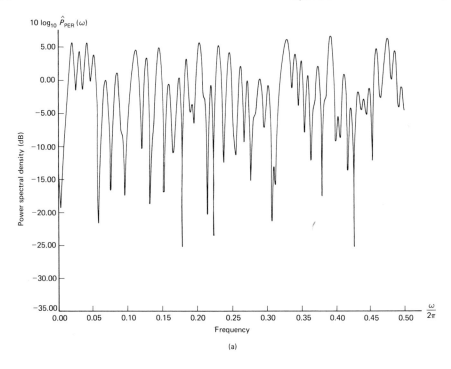

(a)

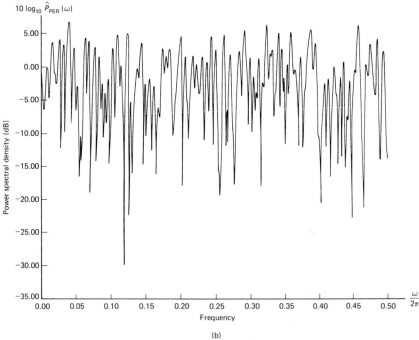

(b)

Figure 2.4 Illustration of the inconsistency of the periodogram for white Gaussian noise ($\sigma_x^2 = 1$). (a) $N = 128$, (b) $N = 256$, (c) $N = 512$, (d) $N = 1024$.

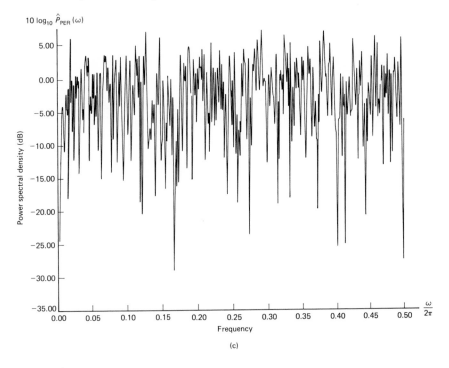

(c)

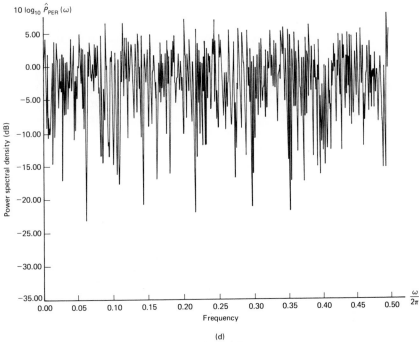

(d)

Figure 2.4 (*cont.*)

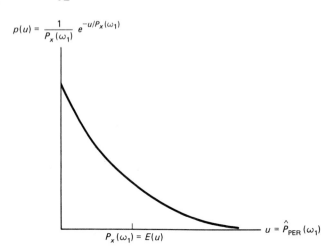

$$p(u) = \frac{1}{P_x(\omega_1)} e^{-u/P_x(\omega_1)}$$

$$u = \hat{P}_{\mathrm{PER}}(\omega_1)$$

$$P_x(\omega_1) = E(u)$$

Figure 2.5 Probability density function of periodogram estimator.

as the mean, the quantity to be estimated. The PDF of $\hat{P}_{\mathrm{PER}}(\omega_1)$ is given in Fig. 2.5 (see Section 2.3.3), which shows that the probability of obtaining an estimate near $P_x(\omega_1)$ is small.

The reason the periodogram is a poor estimator is that, given N data points, we are attempting to estimate about $N/2$ unknown parameters, i.e., $\{P_x(\omega_0), P_x(\omega_1), \dots, P_x(\omega_{N/2})\}$. As N increases, our estimator does not improve because we are estimating proportionally more parameters. In the case of mean estimation the number of unknown parameters was fixed at one. Similar conclusions about the performance of the periodogram can be drawn for arbitrary frequencies and processes with arbitrary PSDs.

The way out of this dilemma is to use an averaged periodogram estimator in an attempt to approximate the expectation operator of Eq. (2.4). Assume we are given K data records uncorrelated with each other and all for the interval $0 \le n \le L - 1$. Also assume that they are drawn from the same random process. The data are $\{x_0(n), 0 \le n \le L - 1; x_1(n), 0 \le n \le L - 1; \dots; x_{K-1}(n), 0 \le n \le L - 1\}$. Then the averaged periodogram estimator is

$$\hat{P}_{\mathrm{AVPER}}(\omega) = \frac{1}{K} \sum_{m=0}^{K-1} \hat{P}_{\mathrm{PER}}^{(m)}(\omega) \qquad (2.38)$$

where

$$\hat{P}_{\mathrm{PER}}^{(m)}(\omega) = \frac{1}{L} \left| \sum_{n=0}^{L-1} x_m(n) \exp(-j\omega n) \right|^2$$

The expected value of the averaged periodogram for white Gaussian noise will be the true PSD as before, but the variance will be decreased by a factor of K, the number of periodograms averaged. Since the data records are uncorrelated and hence independent, the individual periodograms are independent and hence uncorrelated. It follows from Eq. (2.37) that for $k \ne 0, N/2$,

$$\mathrm{VAR}[\hat{P}_{\mathrm{AVPER}}(\omega_k)] = \frac{1}{K^2} \sum_{m=0}^{K-1} \mathrm{VAR}[\hat{P}_{\mathrm{PER}}^{(m)}(\omega_k)]$$

$$= \frac{1}{K^2} K P_x^2(\omega_k) \tag{2.39}$$

$$= \frac{1}{K} P_x^2(\omega_k)$$

As an example, consider the averaged periodogram estimate for white noise with $K = 8$ and $L = 128$ as shown in Fig. 2.6. A comparison with Fig. 2.4(a) illustrates the reduction in variance. In practice we seldom have uncorrelated data sets, but only one data record of length N on which to base the spectral estimator. A frequent approach is to segment the data into K nonoverlapping blocks of length L, where $N = KL$. In this manner we can use Eq. (2.38) with

$$x_m(n) = x(n + mL), \qquad n = 0, 1, \ldots, L - 1; \quad m = 0, 1, \ldots, K - 1$$

Since the blocks are contiguous, they cannot be uncorrelated for any other process but white noise. The variance reduction factor will in general be less than K. For processes not exhibiting sharp resonances, the autocorrelation function will damp out rapidly so that the use of Eq. (2.39) will be a good approximation.

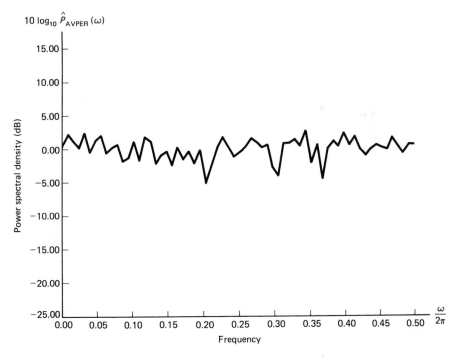

Figure 2.6 Averaged periodogram for white noise with $\sigma_x^2 = 1$, $L = 128$, and $K = 8$.

The astute reader may now ask why we could not just segment the data record into more and more subrecords in an effort to reduce the variance. The problem with this approach is that as the number of subrecords increases, the bias of the averaged periodogram estimator will increase for *any other process but white noise*. To see this we now derive the expected value of the averaged periodogram:

$$E[\hat{P}_{\text{AVPER}}(\omega)] = E\left[\frac{1}{K}\sum_{m=0}^{K-1}\hat{P}_{\text{PER}}^{(m)}(\omega)\right] = E[\hat{P}_{\text{PER}}^{(0)}(\omega)] \tag{2.40}$$

where we have used the fact that all the individual periodograms have the same PDF or are identically distributed. It can be shown, however, that the periodogram defined in Eq. (2.33) can also be written as

$$\hat{P}_{\text{PER}}(\omega) = \sum_{k=-(N-1)}^{N-1}\hat{r}_x(k)\exp(-j\omega k) \tag{2.41}$$

where

$$\hat{r}_x(k) = \frac{1}{N}\sum_{n=0}^{N-1-|k|}x(n)x(n+|k|) \tag{2.42}$$

and $\hat{r}_x(k)$ is observed to be an estimator of the autocorrelation function. Using Eq. (2.41) with N replaced by L in Eq. (2.40), we have

$$E[\hat{P}_{\text{AVPER}}(\omega)] = E\left[\sum_{k=-(L-1)}^{L-1}\hat{r}_x^{(0)}(k)\exp(-j\omega k)\right]$$

where

$$\hat{r}_x^{(0)}(k) = \frac{1}{L}\sum_{n=0}^{L-1-|k|}x(n)x(n+|k|) \tag{2.43}$$

so that

$$E[\hat{P}_{\text{AVPER}}(\omega)] = \sum_{k=-(L-1)}^{L-1}E[\hat{r}_x^{(0)}(k)]\exp(-j\omega k)$$

But from Eq. (2.43),

$$E[\hat{r}_x^{(0)}(k)] = \left(1-\frac{|k|}{L}\right)r_x(k) \qquad \text{for } |k| \le L-1$$

so

$$E[\hat{P}_{\text{AVPER}}(\omega)] = \sum_{k=-(L-1)}^{L-1}\left(1-\frac{|k|}{L}\right)r_x(k)\exp(-j\omega k)$$

If we define

$$w_B(k) = \begin{cases} 1-|k|/L, & |k| \le L-1 \\ 0, & |k| > L \end{cases}$$

then we see that

$$E[\hat{P}_{\text{AVPER}}(\omega)] = \mathcal{F}[w_B(k)r_x(k)] = \int_{-\pi}^{\pi}W_B(\omega-\xi)P_x(\xi)\frac{d\xi}{2\pi} \tag{2.44}$$

where $\mathscr{F}$ denotes the Fourier transform operator and $W_B(\omega)$ is the Fourier transform of $w_B(k)$. The sequence $w_B(k)$ is sometimes referred to as a *triangular* or *Bartlett window*. Its Fourier transform is easily shown to be

$$W_B(\omega) = \frac{1}{L} \left(\frac{\sin \omega L/2}{\sin \omega/2} \right)^2 \tag{2.45}$$

and is plotted in Fig. 2.7. The result of the convolution operation is to produce an average spectral estimate that is smeared. An example is shown in Fig. 2.8. To avoid the smearing, the Bartlett window length L must be chosen so that the width of the main lobe of $W_B(\omega)$ is much less than the width of the narrowest peak in $P_x(\omega)$. Since the 3-dB bandwidth of the main lobe of $W_B(\omega)$ is about $\Delta\omega = 2\pi/L$, we cannot resolve details in the PSD finer than $2\pi/L$. The spectral estimator is then said to have a resolution of $1/L$ cycles/sample. Clearly, for maximum resolution we should choose L as large as possible or $L = N - 1$, which results in the standard periodogram. However, we know that for good variance reduction we should choose $K = N/L$ large according to Eq. (2.39) or L small. Since both goals cannot be met simultaneously, we are forced to trade off bias (or, equivalently, resolution) for variance by adjusting L. In practice, a good strategy is to compute several averaged periodogram spectral estimates, each successive one having a larger L. If the spectral

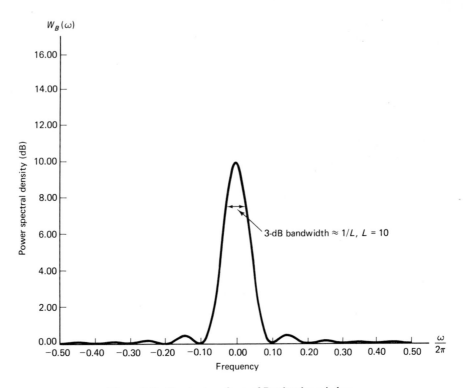

Figure 2.7   Fourier transform of Bartlett lag window.

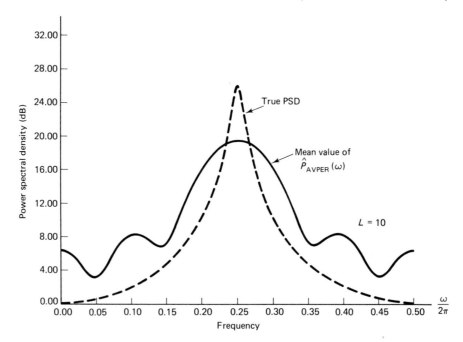

Figure 2.8 Example of mean value of averaged periodogram.

estimate does not change significantly as L is increased, then all the spectral detail has been found. This technique, known as *window closing* [6], may suffer from statistical instability problems. Another technique that is useful in practice is that of pre-whitening the data prior to spectral estimation. Recall from Eq. (2.44) that the expected value of the averaged periodogram is a convolution of the Fourier transform of the Bartlett window with the PSD. If the process is white noise so that $P_x(\omega) = \sigma_x^2$, then

$$E[\hat{P}_{\text{AVPER}}(\omega)] = \sigma_x^2 \int_{-\pi}^{\pi} W_B(\omega - \xi) \frac{d\xi}{2\pi}$$

$$= \sigma_x^2 \int_{-\pi}^{\pi} W_B(\xi) \frac{d\xi}{2\pi} \tag{2.46}$$

$$= \sigma_x^2 w_B(0) = \sigma_x^2 = P_x(\omega)$$

Hence, regardless of the value of L, the estimator is unbiased, due of course to the lack of peaks and valleys in the PSD. If this is the case, then L can be made small to reduce the variance. In practice, if we have some idea of the general shape of the PSD but possibly not the details, a good technique is to filter the data with a linear shift-invariant filter to yield a PSD that is flatter at the output of the filter than at the input. To do so, we find the spectral estimate of the filter output and then divide it by $|H(\omega)|^2$, where $H(\omega)$ is the frequency response of the prewhitener. Such an approach is termed *prewhitening the data*.

It is extremely important in interpreting spectral estimates to be able to ascertain whether spectral detail is due to statistical fluctuation or is actually present. In other words, we need some measure of confidence in the spectral estimate. Assuming that the spectral estimator is approximately unbiased, we can derive a confidence interval for the estimator. (See Section 2.3.1 for a discussion of confidence intervals.) The unbiased assumption requires the bandwidth of the narrowest peak or valley of the PSD to be much larger than the bandwidth of the Bartlett spectral window, $W_B(\omega)$. If we recall the bandpass filtering interpretation of the periodogram, then at least within the vicinity of the frequency under consideration, we can replace the data by a white noise process. It is then possible to use the previous results to yield the approximation

$$\frac{2\hat{P}_{\text{PER}}(\omega)}{P_x(\omega)} \sim \chi_2^2 \tag{2.47}$$

(We omit $\omega = 0$ and $\omega = \pi$ from further consideration.) From Eq. (2.38),

$$\hat{P}_{\text{AVPER}}(\omega) = \frac{1}{K} \sum_{m=0}^{K-1} \hat{P}_{\text{PER}}^{(m)}(\omega)$$

so

$$\frac{2K\hat{P}_{\text{AVPER}}(\omega)}{P_x(\omega)} = \sum_{m=0}^{K-1} \frac{2\hat{P}_{\text{PER}}^{(m)}(\omega)}{P_x(\omega)}$$

But according to Eq. (2.47), each random variable in the summation is a χ_2^2 random variable or the sum of the squares of two independent $N(0, 1)$ random variables. Furthermore, the blocks of data used to form each periodogram are approximately uncorrelated and hence independent due to the Gaussian nature of the data. Each random variable in the summation is then independent. The PDF for the sum of squares $2K$ $N(0, 1)$ random variables is χ_{2K}^2 or

$$\frac{2K\hat{P}_{\text{AVPER}}(\omega)}{P_x(\omega)} \sim \chi_{2K}^2 \tag{2.48}$$

We now define the α percentage point of a χ_{2K}^2 cumulative distribution function as

$$\text{Prob}[\chi_{2K}^2 \leq \chi_{2K}^2(\alpha)] = \alpha$$

Then, from Eq. (2.48),

$$\text{Prob}\left[\chi_{2K}^2(\alpha/2) \leq \frac{2K\hat{P}_{\text{AVPER}}(\omega)}{P_x(\omega)} \leq \chi_{2K}^2(1 - \alpha/2)\right] = 1 - \alpha$$

$$\text{Prob}\left[\frac{2K\hat{P}_{\text{AVPER}}(\omega)}{\chi_{2K}^2(\alpha/2)} \geq P_x(\omega) \geq \frac{2K\hat{P}_{\text{AVPER}}(\omega)}{\chi_{2K}^2(1 - \alpha/2)}\right] = 1 - \alpha$$

so that a $(1 - \alpha) \times 100\%$ confidence interval is

$$\left(\frac{2K\hat{P}_{\text{AVPER}}(\omega)}{\chi_{2K}^2(1 - \alpha/2)}, \frac{2K\hat{P}_{\text{AVPER}}(\omega)}{\chi_{2K}^2(\alpha/2)}\right) \tag{2.49}$$

If we plot the PSD in decibels, the confidence interval becomes a constant-length interval for all ω, or

$$10 \log_{10} \hat{P}_{\text{AVPER}}(\omega) \begin{cases} +10 \log_{10} \dfrac{2K}{\chi^2_{2K}(\alpha/2)} \\[4mm] -10 \log_{10} \dfrac{\chi^2_{2K}(1 - \alpha/2)}{2K} \end{cases} \text{dB} \qquad (2.50)$$

As an example, for a 95% confidence interval with $K = 10$, $\alpha = 0.05$, $\chi^2_{20}(0.025) = 10.85$, $\chi_{20}(0.975) = 31.41$, and the confidence interval is

$$10 \log_{10} \hat{P}_{\text{AVPER}}(\omega) \begin{cases} +2.65 \\ -1.96 \end{cases} \text{dB}$$

This means that the interval

$$(10 \log_{10} \hat{P}_{\text{AVPER}}(\omega) - 1.96, \ 10 \log_{10} \hat{P}_{\text{AVPER}}(\omega) + 2.65)$$

will cover the true value of $10 \log_{10} P_x(\omega)$ with a probability of 0.95. Hence, spectral peaks and valleys of more than a few decibels should be considered as actually being present and not due to statistical fluctuation.

Before concluding our discussion of periodogram spectral estimators it is worthwhile to note the use of the fast Fourier transform (FFT) in computing them. Since we cannot expect to compute $\hat{P}_{\text{PER}}(\omega)$ for a continuum of frequencies, we are forced to sample it. Typically, one uses $\omega_k = 2\pi k/N$ for $k = 0, 1, \ldots, N - 1$, so

$$\hat{P}_{\text{PER}}(\omega_k) = \frac{1}{N} \left| \sum_{n=0}^{N-1} x(n) \exp(-j\omega_k n) \right|^2$$

$$= \frac{1}{N} \left| \sum_{n=0}^{N-1} x(n) \exp(-j2\pi kn/N) \right|^2 \qquad (2.51)$$

The periodogram for $0 \le \omega \le \pi$ is found from the samples $k = 0, 1, \ldots, N/2$, assuming N is even. Equation (2.51) is in the form of a discrete Fourier tranform (DFT) and hence the FFT may be used to efficiently perform the computation. To approximate $\hat{P}_{\text{PER}}(\omega)$ more closely, we may need to have a finer frequency spacing. This is accomplished by zero padding the data, i.e., by defining

$$x'(n) = \begin{cases} x(n), & n = 0, 1, \ldots, N - 1 \\ 0, & n = N, N + 1, \ldots, N' - 1 \end{cases} \qquad (2.52)$$

Then, letting $\omega'_k = 2\pi k/N'$ for $k = 0, 1, \ldots, N' - 1$, we have

$$\hat{P}_{\text{PER}}(\omega'_k) = \frac{1}{N} \left| \sum_{n=0}^{N'-1} x'(n) \exp(-j2\pi kn/N') \right|^2$$

$$= \frac{1}{N} \left| \sum_{n=0}^{N-1} x(n) \exp(-j2\pi kn/N') \right|^2 \qquad (2.53)$$

so the effective frequency spacing of the periodogram samples is $2\pi/N' < 2\pi/N$. No extra resolution is afforded by zero padding, but we achieve a better evaluation of the periodogram. See Fig. 2.9 for an example.

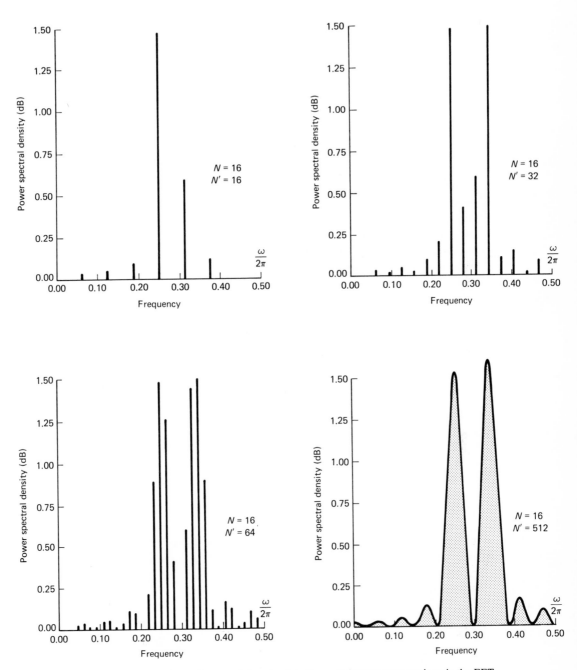

Figure 2.9   Effect of zero padding the data for periodogram computation via the FFT.

2.4.2 Blackman-Tukey Spectral Estimator

We saw in the previous section that the periodogram estimator could be expressed as

$$\hat{P}_{PER}(\omega) = \sum_{k=-(N-1)}^{N-1} \hat{r}_x(k) \exp(-j\omega k) \tag{2.41}$$

where

$$\hat{r}_x(k) = \frac{1}{N} \sum_{n=0}^{N-1-|k|} x(n)x(n + |k|) \tag{2.42}$$

Here $\hat{r}_x(k)$ is a biased estimator of the autocorrelation function, and in this form the periodogram is seen to be an estimator based on the Wiener-Khinchin theorem. The poor performance of the periodogram may be attributed to the poor performance of the autocorrelation function estimator. In fact, from Eq. (2.41), $r_x(N - 1)$ is estimated by $(1/N)x(0)x(N - 1)$ no matter how large N is. This estimator will be highly variable because of the lack of averaging of lag products, and it will be biased as well. The higher lags of the autocorrelation function will be poorer estimates since they involve fewer lag products. One way to avoid this problem is to weight the higher lags less, or

$$\hat{P}_{BT}(\omega) = \sum_{k=-(N-1)}^{N-1} w(k)\hat{r}_x(k) \exp(-j\omega k) \tag{2.54}$$

where $w(k)$ is a *lag window* with the following properties:

1. $0 \leq w(k) \leq w(0) = 1$

2. $w(-k) = w(k)$ (2.55)

3. $w(k) = 0$ for $|k| > M$

where $M \leq N - 1$. Equation (2.54) is called the Blackman-Tukey spectral estimator. It is equivalent to the periodogram if $w(k) = 1$ for $|k| \leq M = N - 1$. The Blackman-Tukey spectral estimator is sometimes called the *weighted covariance* estimator. The approach of reducing the variance of a random variable by weighting it by a factor less than 1 was discussed in Section 2.3.1. Here, the weighting is applied to the autocorrelation function estimator and, as we saw previously, we can expect an increase in the bias. Many windows are available. Table 2.1 lists a few of them. We must be careful to ensure that the chosen window will always lead to a nonnegative spectral estimate. To see how a negative spectral estimate might occur as a result of windowing, note from Eq. (2.54) that

$$\hat{P}_{BT}(\omega) = \mathcal{F}\{w(k)\hat{r}_x(k)\}$$

$$= \int_{-\pi}^{\pi} W(\omega - \xi)\hat{P}_{PER}(\xi)\frac{d\xi}{2\pi} \tag{2.56}$$

since $\mathcal{F}\{\hat{r}_x(k)\} = \hat{P}_{PER}(\omega)$. Although $\hat{P}_{PER}(\omega) \geq 0$, $W(\omega)$ may be negative enough to cause $\hat{P}_{BT}(\omega)$ to be negative. To ensure that this will not be the case, $w(k)$ should have

TABLE 2.1   COMMON LAG WINDOWS

Name	Definition	Fourier Transform
Rectangular	$w(k) = \begin{cases} 1, & |k| \le M \\ 0, & |k| > M \end{cases}$	$W(\omega) = W_R(\omega)$ $= \dfrac{\sin\frac{\omega}{2}(2M+1)}{\sin \omega/2}$
Bartlett	$w(k) = \begin{cases} 1 - \dfrac{|k|}{M}, & |k| \le M \\ 0, & |k| > M \end{cases}$	$W(\omega) = W_B(\omega)$ $= \dfrac{1}{M}\left(\dfrac{\sin M\omega/2}{\sin \omega/2}\right)^2$
Hanning	$w(k) = \begin{cases} \dfrac{1}{2} + \dfrac{1}{2}\cos\dfrac{\pi k}{M}, & |k| \le M \\ 0, & |k| > M \end{cases}$	$W(\omega) = \dfrac{1}{4}W_R(\omega - \pi/M)$ $\quad + \dfrac{1}{2}W_R(\omega)$ $\quad + \dfrac{1}{4}W_R(\omega + \pi/M)$
Hamming	$w(k) = \begin{cases} 0.54 + 0.46\cos\dfrac{\pi k}{M}, & |k| \le M \\ 0, & |k| > M \end{cases}$	$W(\omega) = 0.23\,W_R(\omega - \pi/M)$ $\quad + 0.54\,W_R(\omega)$ $\quad + 0.23\,W_R(\omega + \pi/M)$
Parzen	$w(k) = \begin{cases} 2\left(1 - \dfrac{|k|}{M}\right)^3 - \left(1 - 2\dfrac{|k|}{M}\right)^3, & |k| \le M/2 \\ 2\left(1 - \dfrac{|k|}{M}\right)^3, & \dfrac{M}{2} < k \le M \\ 0, & |k| > M \end{cases}$	$W(\omega) = \dfrac{8}{M^3}\left(\dfrac{3}{2}\dfrac{\sin^4 M\omega/4}{\sin^4 \omega/2}\right.$ $\left. - \dfrac{\sin^4 M\omega/4}{\sin^2 \omega/2}\right)$

a Fourier transform that is nonnegative. A window that satisfies these requirements is the Bartlett window. Of the other windows given in Table 2.1 only the Parzen window has this property.

We now examine the bias and variance of the Blackman-Tukey spectral estimator. From Eq. (2.56) the mean is

$$E[\hat{P}_{BT}(\omega)] = \int_{-\pi}^{\pi} W(\omega - \xi)E[\hat{P}_{PER}(\xi)]\frac{d\xi}{2\pi}$$

If we assume that the periodogram is approximately an unbiased estimator, then

$$E[\hat{P}_{BT}(\omega)] \approx \int_{-\pi}^{\pi} W(\omega - \xi)P_x(\xi)\frac{d\xi}{2\pi} \tag{2.57}$$

The unbiased assumption will be valid if the data record is long enough so that $P_x(\omega)$ is smooth over any $2\pi/N$ interval. This follows from Eq. (2.44) if we consider $L = N$.

Note from Eq. (2.57) that the mean of the Blackman-Tukey spectral estimator is a smeared version of the true PSD. It is said that $W(\omega)$ acts as a *spectral window*. The PSDs that will be heavily biased are those with nonflat spectra.

The variance of the Blackman-Tukey spectral estimator may be shown [6] to be

$$\text{VAR}[\hat{P}_{\text{BT}}(\omega)] \approx \frac{P_x^2(\omega)}{N} \int_{-\pi}^{\pi} W^2(\xi) \frac{d\xi}{2\pi} \tag{2.58}$$

Equation (2.58) is derived by making the assumptions that $P_x(\omega)$ is smooth over the main lobe of the spectral window ($\approx 4\pi/M$) and that $N \gg M$. Using Parseval's theorem, we can rewrite Eq. (2.58) as

$$\text{VAR}[\hat{P}_{\text{BT}}(\omega)] \approx \frac{P_x^2(\omega)}{N} \sum_{k=-M}^{M} w^2(k) \tag{2.59}$$

As an example, for the Bartlett window,

$$w_B(k) = \begin{cases} 1 - |k|/M, & |k| \le M - 1 \\ 0, & |k| > M \end{cases}$$

$$\text{VAR}[\hat{P}_{\text{BT}}(\omega)] \approx \frac{2M}{3N} P_x^2(\omega) \tag{2.60}$$

Again a bias-variance tradeoff is evident if we examine Eqs. (2.57) and (2.60). For a small bias we would like M large to make the spectral window in Eq. (2.57) behave as an impulse function. On the other hand, for a small variance, M should be small according to Eq. (2.60). Much of the art in nonparametric spectral estimation is in choosing an appropriate window, both in type and in length.

A confidence interval for the Blackman-Tukey spectral estimator may be derived in a similar fashion to that for the averaged periodogram. The results are

$$10 \log_{10} \hat{P}_{\text{BT}}(\omega) \begin{cases} +10 \log_{10} \dfrac{\nu}{\chi_\nu^2(\alpha/2)} \\[2ex] -10 \log_{10} \dfrac{\chi_\nu^2(1 - \alpha/2)}{\nu} \end{cases} \text{dB} \tag{2.61}$$

where $\nu = 2N/\Sigma_{k=-M}^{M} w^2(k)$ = degrees of freedom [6]. As an example, for the Bartlett window, $\nu \approx 3N/M$.

2.4.3 Minimum Variance Spectral Estimation

The minimum variance spectral estimator (MVSE) estimates the PSD by effectively measuring the power out of a set of narrowband filters [7, 8]. The popularly used name maximum likelihood method (MLM) is actually a misnomer in that the spectral estimator is not a maximum likelihood spectral estimator *nor does it possess any of the properties of a maximum likelihood estimator*. Even the name MVSE that has been chosen to describe this estimator is not meant to imply that the spectral estimator is one that possesses minimum variance but is used only to describe the origins of the estimator. The MVSE is also referred to as the Capon spectral estimator [9]. In the MVSE the shapes of the narrowband filters are, in general, dependent on the frequency under consideration, in contrast to the periodogram, for which the shapes of the narrowband filters are the same for all frequencies. In the MVSE the filters adapt to the process for which the PSD is sought with the advantage that the filter sidelobes can be adjusted to reduce the response to spectral components outside the band of interest.

Recall from the discussion of Section 2.4.1 that the periodogram estimates the PSD by forming a bank of narrowband filters with frequency responses given by Eq. (2.34). The frequency response of each filter is unity at $\omega = \omega_0$, the frequency of interest. For a good spectral estimate the filter output power should be due only to the PSD near $\omega = \omega_0$. However, because of the high sidelobes of Eq. (2.34), it is possible to observe a large output power owing to the PSD outside the band of interest. In an attempt to combat this *leakage* it is desirable to design a bank of filters that will adaptively adjust its sidelobes to minimize the power at the filter outputs due to "out of band" spectral components. To do so we can design the filters to minimize the power at their output or we choose the filter to minimize

$$\rho = \int_{-\pi}^{\pi} |H(\omega)|^2 P_x(\omega) \frac{d\omega}{2\pi} \tag{2.62}$$

subject to the constraint $H(\omega_0) = 1$. The filter frequency response is of the form

$$H(\omega) = \sum_{n=-(N-1)}^{0} h(n) \exp(-j\omega n)$$

in accordance with Eq. (2.34). Note that the filter coefficients will in general be complex as was the case for the periodogram. To minimize ρ, note that it may be expressed as

$$
\begin{aligned}
\rho &= \int_{-\pi}^{\pi} \sum_{k=-(N-1)}^{0} h(k) \exp(-j\omega k) \sum_{\ell=-(N-1)}^{0} h^*(\ell) \exp(j\omega \ell) P_x(\omega) \frac{d\omega}{2\pi} \\
&= \sum_{k=-(N-1)}^{0} \sum_{\ell=-(N-1)}^{0} h(k) h^*(\ell) \int_{-\pi}^{\pi} P_x(\omega) \exp[j\omega(\ell - k)] \frac{d\omega}{2\pi} \\
&= \sum_{k=-(N-1)}^{0} \sum_{\ell=-(N-1)}^{0} h(k) h^*(\ell) r_x(\ell - k) \\
&= \mathbf{h}^H \mathbf{R}_x \mathbf{h}
\end{aligned}
\tag{2.63}
$$

where $\mathbf{h}^* = [h(0)\, h(-1)\, \ldots\, h(-(N-1))]^T$, $\mathbf{R}_x$ is the $N \times N$ autocorrelation matrix with (i, j) element $r_x(i - j)$, and H denotes the complex conjugate transpose. The constraint of unity frequency response at $\omega = \omega_0$ can also be rewritten as

$$\mathbf{h}^H \mathbf{e} = 1$$

where $\mathbf{e} = [1\, \exp(j\omega_0)\, \ldots\, \exp[j(N - 1)\omega_0]^T$. This constrained minimization may be accomplished by making use of the identity

$$\mathbf{h}^H \mathbf{R}_x \mathbf{h} = (\mathbf{h} - \tilde{\mathbf{h}})^H \mathbf{R}_x (\mathbf{h} - \tilde{\mathbf{h}}) + \tilde{\mathbf{h}}^H \mathbf{R}_x \tilde{\mathbf{h}} \tag{2.64}$$

where

$$\tilde{\mathbf{h}} = \frac{\mathbf{R}_x^{-1}\mathbf{e}}{\mathbf{e}^H\mathbf{R}_x^{-1}\mathbf{e}}$$

and which holds if $\mathbf{h}^H\mathbf{e} = 1$. To verify this identity, note that

$$(\mathbf{h} - \tilde{\mathbf{h}})^H\mathbf{R}_x(\mathbf{h} - \tilde{\mathbf{h}}) + \tilde{\mathbf{h}}^H\mathbf{R}_x\tilde{\mathbf{h}} = \mathbf{h}^H\mathbf{R}_x\mathbf{h} + (2\tilde{\mathbf{h}}^H\mathbf{R}_x\tilde{\mathbf{h}} - \tilde{\mathbf{h}}^H\mathbf{R}_x\mathbf{h} - \mathbf{h}^H\mathbf{R}_x\tilde{\mathbf{h}})$$

$$= \mathbf{h}^H\mathbf{R}_x\mathbf{h} + \frac{1}{\mathbf{e}^H\mathbf{R}_x^{-1}\mathbf{e}}(2 - \mathbf{e}^H\mathbf{h} - \mathbf{h}^H\mathbf{e})$$

Since $\mathbf{h}^H\mathbf{e} = 1$, it follows that $\mathbf{e}^H\mathbf{h} = 1$, and hence the identity is proved. To minimize the variance, observe from Eq. (2.64) that $\tilde{\mathbf{h}}^H\mathbf{R}_x\tilde{\mathbf{h}}$ does not depend on $\mathbf{h}$ and $(\mathbf{h} - \tilde{\mathbf{h}})^H\mathbf{R}_x(\mathbf{h} - \tilde{\mathbf{h}}) \geq 0$ because $\mathbf{R}_x$ is a positive-definite matrix. Therefore, the minimum value is obtained by letting $\mathbf{h} = \tilde{\mathbf{h}}$, or

$$\mathbf{h} = \frac{\mathbf{R}_x^{-1}\mathbf{e}}{\mathbf{e}^H\mathbf{R}_x^{-1}\mathbf{e}} \tag{2.65}$$

Finally, substituting Eq. (2.65) into Eq. (2.63), we obtain the minimum power

$$\rho = \frac{1}{\mathbf{e}^H\mathbf{R}_x^{-1}\mathbf{e}} \tag{2.66}$$

As an example, assume that $x(n)$ is a first-order autoregressive process with parameter $a_1 = -r$. The PSD is that of a lowpass process with its power concentrated at $\omega = 0$ (see Fig. 2.11b in the next section for an example). The autocorrelation function is shown in Section 2.5.2 to be

$$r_x(k) = \frac{\sigma^2}{1 - a_1^2}(-a_1)^{|k|}$$

Using this, we can easily verify that

$$\mathbf{R}_x^{-1} = \frac{1}{\sigma^2}\begin{bmatrix} 1 & a_1 & 0 & 0 & \cdots & 0 \\ a_1 & 1 + a_1^2 & a_1 & 0 & \cdots & 0 \\ \vdots & & \ddots & \ddots & & \vdots \\ & & & & & \\ 0 & \cdots & 0 & a_1 & 1 + a_1^2 & a_1 \\ 0 & \cdots & 0 & 0 & a_1 & 1 \end{bmatrix}$$

so that

$$\mathbf{R}_x^{-1}\mathbf{e} = \frac{1}{\sigma^2}\begin{bmatrix} A^*(\omega_0) \\ \exp(j\omega_0)|A(\omega_0)|^2 \\ \vdots \\ \exp[j\omega_0(N-2)]|A(\omega_0)|^2 \\ \exp[j\omega_0(N-1)]A(\omega_0) \end{bmatrix}$$

where $A(\omega_0) = 1 - r \exp(-j\omega_0)$. It follows that

$$\mathbf{e}^H \mathbf{R}_x^{-1} \mathbf{e} = \frac{1}{\sigma^2}[N - 2(N - 1)r \cos \omega_0 + (N - 2)r^2]$$

The filter coefficients and hence the frequency response can be found by using these two expressions in Eq. (2.65). The magnitude of the frequency response is plotted in Fig. 2.10 for $N = 10$, $\omega_0/2\pi = 0.25$ for various values of r. For $r = 0$, i.e., $a_1 = 0$, the noise is white and the filter has the usual sinc-type response as given in Eq. (2.34) (shown as a dashed curve in the figure). For other values of r, the optimal filter attempts to reject the noise by adjusting the response so as to attenuate that region of the frequency band where the noise PSD is the largest. In this case the PSD of the noise is (see Section 2.5.2)

$$P_{AR}(\omega) = \frac{\sigma^2}{|1 + a_1 \exp(-j\omega)|^2} = \frac{\sigma^2}{|1 - r \exp(-j\omega)|^2} \tag{2.67}$$

The PSD has a peak at $\omega/2\pi = 0$ and the sharpness of the peak increases with r. This is reflected in the frequency response, in which the attenuation in the frequency band centered about $\omega/2\pi = 0$ increases with r. In fact, as $r \to 1$, so that the PSD approaches a Dirac delta function located at $\omega/2\pi = 0$, it can be shown that $H(0) \to 0$. In effect, a null is placed at the frequency location where the noise power is greatest. As expected, the gain of the filter at $\omega/2\pi = \omega_0/2\pi = 0.25$ is unity.

It is now apparent that the frequency response of the optimal filter will depend on the noise background near the center frequency $\omega = \omega_0$. For a given PSD $\mathbf{h}$ will form a different filter for each assumed value of ω_0. The filter will adjust itself to reject noise components with frequencies not near $\omega = \omega_0$ and to pass noise components at and near $\omega = \omega_0$. The power out of the filter is therefore a good indication of the power in the process in the vicinity of ω_0. A spectral estimator can thus be defined as

$$\hat{P}_{MV}(\omega) = \frac{p}{\mathbf{e}^H \hat{\mathbf{R}}_x^{-1} \mathbf{e}} \tag{2.68}$$

where $\mathbf{e} = [1 \ \exp(j\omega) \ \exp(j2\omega) \ \ldots \ \exp(j\omega(p - 1))]^T$. The factor p is included to account for the bandwidth of the filter, i.e., to yield a power spectral *density* estimate by dividing the estimate of the power by $1/p$, which is approximately the 3-dB bandwidth of the optimal filter. To compute the spectral estimate we need only an estimate of the $p \times p$ autocorrelation matrix $\mathbf{R}_x$. Note that the dimensions of the autocorrelation matrix should be much less than $N \times N$ to allow the higher order autocorrelation samples to be reliably estimated. The spectral estimator $\hat{P}_{MV}(\omega)$ has been referred to as the MLM or Capon method.

In practice, the MVSE exhibits more resolution than the periodogram and Blackman-Tukey spectral estimators but less than an autoregressive (AR) spectral estimator [8]. Also, the variance of the MVSE appears to be less than that of the AR spectral estimator when both are based on the same number of autocorrelation lags. The critical choice in the MVSE is in the value of p, the dimension of the estimated autocorrelation matrix. For large p, the filter bandwidth will be small, which will yield high-resolution spectral estimates. However, the variance may also be large due to the

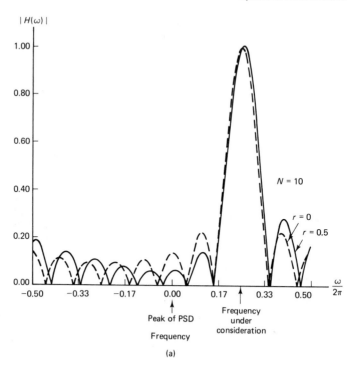

(a)

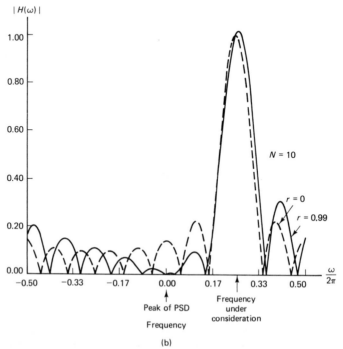

(b)

Figure 2.10 Frequency response magnitude of optimal filter: (a) $r = 0.5$, (b) $r = 0.99$.

large number of autocorrelation lags estimated. As usual, then, a tradeoff must be effected between bias and variance.

In summary, nonparametric spectral estimation based on a limited data set involves trading off bias for variance. A basic problem is that the PSD depends on an infinite number of autocorrelation function lags, all of which need to be estimated to obtain a good spectral estimate. Noting the impossibility of the task, one may reasonably ask whether it might be better to assume a model for the PSD or autocorrelation function that depends on only a finite set of parameters. If the number of data points is large relative to the number of PSD parameters, then good estimates of the parameters and hence the PSD would be expected. This approach is termed *parametric spectral estimation* and is based on classical models of time series analysis. We will study this alternative spectral estimation approach in the next section.

2.5 PARAMETRIC SPECTRAL ESTIMATION

2.5.1 Time Series Models

The most general time series model is one in which the random process is assumed to have been generated by exciting a linear shift-invariant causal pole-zero filter with white noise, or

$$x(n) = -\sum_{k=1}^{p} a_k x(n-k) + \sum_{k=0}^{q} b_k \epsilon(n-k) \tag{2.69}$$

where $b_0 = 1$ and $\epsilon(n)$ is a white noise process with zero mean and variance σ^2. Such a model is termed an autoregressive moving average (ARMA) model and is given the abbreviation ARMA(p, q). We see that $x(n)$ is the output of the linear shift-invariant filter with transfer function

$$
H(z) = \frac{B(z)}{A(z)}
$$
$$
= \frac{1 + \sum\limits_{k=1}^{q} b_k z^{-k}}{1 + \sum\limits_{k=1}^{p} a_k z^{-k}} \tag{2.70}
$$

There are no restrictions on the number of poles p or zeros q. It is assumed that both $A(z)$ and $B(z)$ have their zeros inside the unit circle of the z-plane so that $H(z)$ as well as $1/H(z)$ are stable and causal filters. The latter requirements are reasonable since the time series models are usually justified by physical considerations.

The PSD of $x(n)$ follows from Eq. (2.70) as

$$
P_{\text{ARMA}}(\omega) = |H(\exp(j\omega))|^2 P_\epsilon(\omega)
$$
$$
= \frac{\sigma^2 |1 + \sum\limits_{k=1}^{q} b_k \exp(-j\omega k)|^2}{|1 + \sum\limits_{k=1}^{p} a_k \exp(-j\omega k)|^2} \tag{2.71}
$$

Equation (2.71) represents a model for the PSD based on the time series model of Eq. (2.69) for the data. Note that the PSD depends *only on the filter coefficients and white noise variance*. To estimate the PSD we need only estimate $\{b_1, b_2, \ldots, b_q, a_1, a_2, \ldots, a_p, \sigma^2\}$ and substitute the estimated values into Eq. (2.71). We hope that $N \gg p + q + 1$ so that we will obtain good estimates of the unknown parameters. Equivalently, the autocorrelation function must also be a function of the same model parameters and, hence, we have achieved our objective of replacing estimation of an infinite set of autocorrelation function lags by estimation of a finite set of parameters.

It should be observed that the time series model of Eq. (2.69) is fundamentally different from the deterministic signal model discussed in Chapter 1. For the deterministic signal it was appropriate to model it as an impulse response, with the error being the difference between the actual signal and the modeled signal. For the spectral estimation problem the "signal" itself is inherently random, and the error is the difference between the true PSD and the estimated one. Alternately, for our problem there does not exist a single "signal" that is to be modeled in terms of its *sequence values* but rather an ensemble of sequences for which the *average distribution of power with frequency* is sought. Curiously, though, some of the estimation algorithms are nearly identical, but this equivalence is coincidental, attributable to the need for easily implementable algorithms; in general the *optimal algorithms* for the two cases will be quite different. Finally, assessing the performance of an estimation algorithm must be done in the context of the fundamental data assumptions made. Failure to do so results in a comparison of "apples and oranges."

Less general, although more useful, models in practice are found by setting either $b_k = 0$ for $k = 1, 2, \ldots, q$ or $a_k = 0$ for $k = 1, 2, \ldots, p$ in Eq. (2.69). Adopting the latter leads to the moving average (MA) times series model

$$x(n) = \sum_{k=0}^{q} b_k \epsilon(n - k) \tag{2.72}$$

and a corresponding PSD model

$$P_{MA}(\omega) = \sigma^2 \left| 1 + \sum_{k=1}^{q} b_k \exp(-j\omega k) \right|^2 \tag{2.73}$$

The abbreviation MA(q) is used and q is termed the MA model order. The b_k's and σ^2 are called the MA parameters. If we set the b_k's equal to zero in the ARMA model, we then have the autoregressive (AR) time series model

$$x(n) = -\sum_{k=1}^{p} a_k x(n - k) + \epsilon(n) \tag{2.74}$$

and a corresponding PSD model

$$P_{AR}(\omega) = \frac{\sigma^2}{\left| 1 + \sum_{k=1}^{p} a_k \exp(-j\omega k) \right|^2} \tag{2.75}$$

The abbreviation AR(p) is used, and p is termed the AR model order. The a_k's and σ^2 are called the AR parameters.

Usually, the appropriate model is chosen based on physical modeling, as in vocal tract modeling for speech [10]. Frequently, in practice one does not know which of the given models yields an accurate representation of the PSD. An important result from the Wold decomposition and Kolmogorov theorems is that any ARMA or MA process can be represented by an AR process of possibly infinite order; likewise, any ARMA or AR process can be represented by an MA process of possibly infinite order [11,12]. Hence, if we choose the wrong model among the three, we may still obtain a reasonable approximation by using a high enough model order.

An important relationship for parametric spectral estimation is that between the parameters of the model and the autocorrelation function. Considering the general ARMA model, we now derive the celebrated Yule-Walker equations. First, multiply Eq. (2.69) by $x(n - k)$ and take the expectation:

$$E[x(n)x(n - k)] = -\sum_{\ell=1}^{p} a_\ell E[x(n - \ell)x(n - k)] + \sum_{\ell=0}^{q} b_\ell E[\epsilon(n - \ell)x(n - k)]$$

$$r_x(k) = -\sum_{\ell=1}^{p} a_\ell r_x(k - \ell) + \sum_{\ell=0}^{q} b_\ell r_{x\epsilon}(k - \ell) \tag{2.76}$$

where $r_{x\epsilon}(k) = E[x(n)\epsilon(n + k)]$. To evaluate $r_{x\epsilon}(k)$ note that $x(n)$ is the output of a filter with impulse response $h(n)$ excited at the input by $\epsilon(n)$ so that

$$r_{x\epsilon}(k) = E\left[\sum_{\ell=-\infty}^{\infty} h(n - \ell)\epsilon(\ell)\epsilon(n + k)\right]$$

$$= \sum_{\ell=-\infty}^{\infty} h(n - \ell)\sigma^2\delta(n + k - \ell) \tag{2.77}$$

$$= h(-k)\sigma^2$$

Since the filter is causal, $r_{x\epsilon}(k) = 0$ for $k > 0$. Equation (2.76) becomes

$$r_x(k) = \begin{cases} -\sum_{\ell=1}^{p} a_\ell r_x(k - \ell) + \sigma^2 \sum_{\ell=0}^{q-k} h(\ell)b_{\ell+k} & k = 0, 1, \ldots, q \\ -\sum_{\ell=1}^{p} a_\ell r_x(k - \ell) & k \geq q + 1 \end{cases} \tag{2.78}$$

Equations (2.78) are termed the Yule-Walker equations. They relate the auto-correlation function to the ARMA parameters and are useful for estimating the ARMA parameters given estimates of the autocorrelation function lags.

We now give examples of the autocorrelation function and PSD for low-order AR, MA, and ARMA models. By doing so we will observe the types of spectra that are representable by these models.

2.5.2 Examples of AR Processes

Consider first an AR(p) model. From Eq. (2.78), on letting $q = 0$, we have

$$r_x(k) = \begin{cases} -\sum_{\ell=1}^{p} a_\ell r_x(k - \ell) + \sigma^2 h(k), & k = 0 \\ -\sum_{\ell=1}^{p} a_\ell r_x(k - \ell), & k \geq 1 \end{cases}$$

But using Eqs. (2.70), we have

$$h(0) = \lim_{z \to \infty} H(z) = 1$$

so

$$r_x(k) = \begin{cases} -\sum_{\ell=1}^{p} a_\ell r_x(\ell) + \sigma^2, & k = 0 \\ -\sum_{\ell=1}^{p} a_\ell r_x(k - \ell), & k \geq 1 \end{cases} \tag{2.79}$$

Equations (2.79) allow us to compute the autocorrelation function recursively given the AR parameters, an example of which will be given shortly. The process is reversible since given $\{r_x(0), r_x(1), \ldots, r_x(p)\}$ we can determine the AR parameters using Eq. (2.79) for $k = 1, 2, \ldots, p$ as follows

$$\begin{bmatrix} r_x(0) & r_x(1) & \cdots & r_x(p-1) \\ r_x(1) & r_x(0) & \cdots & r_x(p-2) \\ \vdots & \vdots & \ddots & \vdots \\ r_x(p-1) & r_x(p-2) & \cdots & r_x(0) \end{bmatrix} \begin{bmatrix} a_1 \\ a_2 \\ \vdots \\ a_p \end{bmatrix} = - \begin{bmatrix} r_x(1) \\ r_x(2) \\ \vdots \\ r_x(p) \end{bmatrix} \tag{2.80}$$

$$\sigma^2 = r_x(0) + \sum_{\ell=1}^{p} a_\ell r_x(\ell)$$

The AR parameters can be found by solving this set of linear equations. As explained in Chapter 1, due to the special nature of the matrix and right-hand vector, the Levinson algorithm may be used to solve Eq. (2.80). The algorithm is summarized below.

$$a_{11} = -\frac{r_x(1)}{r_x(0)}$$

$$\sigma_1^2 = (1 - a_{11}^2) r_x(0)$$

For $k = 2, 3, \ldots, p$,

$$a_{kk} = -\frac{r_x(k) + \sum_{\ell=1}^{k-1} a_{k-1,\ell} r_x(k - \ell)}{\sigma_{k-1}^2}$$

$$a_{ki} = a_{k-1,i} + a_{kk} a_{k-1,k-i}, \qquad i = 1, 2, \ldots, k - 1 \tag{2.81}$$

$$\sigma_k^2 = (1 - a_{kk}^2) \sigma_{k-1}^2$$

The solution of Eq. (2.80) is given by $a_i = a_{pi}$, $i = 1, 2, \ldots, p$, and $\sigma^2 = \sigma_p^2$. For further details see Chapter 1.

As an example of the utility of Eq. (2.79) consider an AR(1) process. Then, for $p = 1$, Eq. (2.79) yields

$$r_x(k) = -a_1 r_x(k - 1), \qquad k \geq 1$$

which leads to the solution

$$r_x(k) = r_x(0)(-a_1)^{|k|}$$

To find $r_x(0)$ we use Eq. (2.79) for $k = 0$ to obtain

$$\sigma^2 = r_x(0) + a_1 r_x(1)$$
$$= r_x(0) - a_1^2 r_x(0)$$

which results in

$$r_x(0) = \frac{\sigma^2}{1 - a_1^2}$$

or finally

$$r_x(k) = \frac{\sigma^2}{1 - a_1^2}(-a_1)^{|k|} \tag{2.82}$$

Note that $|a_1| < 1$ to guarantee stability of $1/A(z)$. The autocorrelation function $r_x(k)$ is plotted in Fig. 2.11 for $a_1 < 0$ and $a_1 > 0$. The corresponding PSDs are given by

$$P_{AR}(\omega) = \frac{\sigma^2}{|1 + a_1 \exp(-j\omega)|^2} \tag{2.83}$$

and are also plotted in decibels in Fig. 2.11. Note that $a_1 < 0$ yields a lowpass process while $a_1 > 0$ yields a highpass process. Thus, an AR(1) process cannot model a bandpass process. To do so we must use an AR(2) process with complex conjugate poles, i.e., poles at $z = r \exp(\pm j\omega_0)$. It can be shown [13] that for this case

$$r_x(k) = \frac{\sigma^2 \dfrac{1 + r^2}{1 - r^2} \sqrt{1 + \left(\dfrac{1 - r^2}{1 + r^2}\right)^2 \cot^2(\omega_0)}}{1 - 2r^2\cos 2\omega_0 + r^4} r^{|k|} \cos(|k|\omega_0 - \phi) \tag{2.84}$$

where

$$\phi = \arctan\left[\frac{1 - r^2}{1 + r^2} \cot \omega_0\right]$$

$$a_1 = -2r \cos \omega_0, \qquad a_2 = r^2$$

The PSD is

$$P_{AR}(\omega) = \frac{\sigma^2}{|1 + a_1 \exp(-j\omega) + a_2 \exp(-j2\omega)|^2}$$

$$= \frac{\sigma^2}{|1 - r \exp[-j(\omega - \omega_0)]|^2 |1 - r \exp[-j(\omega + \omega_0)]|^2} \tag{2.85}$$

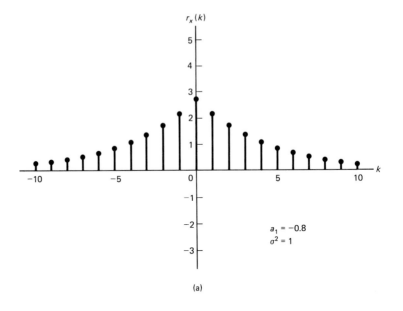

(a)

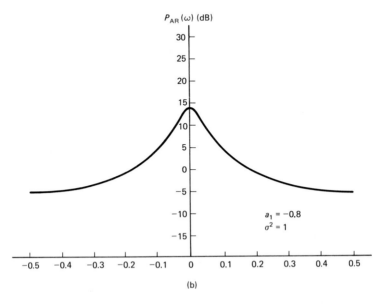

(b)

Figure 2.11 (a) Autocorrelation of AR(1) process; (b) power spectral density of AR(1) process; (c) autocorrelation of AR(1) process; (d) power spectral density of AR(1) process.

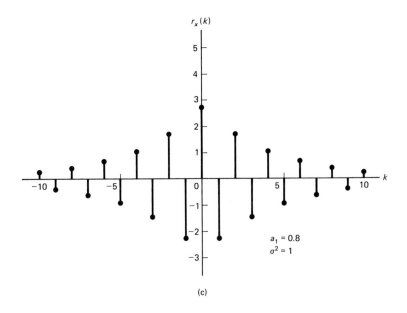

(c)

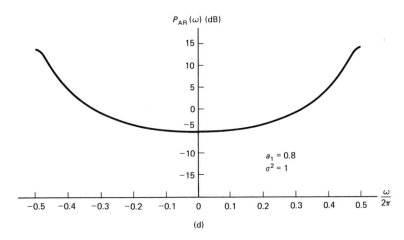

(d)

Figure 2.11 (*cont.*)

Examples of the autocorrelation function and PSD for complex poles at $r \exp(\pm j\pi/2)$ are given in Fig. 2.12. As $r \rightarrow 1$ the PSD becomes more peaked about $\omega = \omega_0$ and the autocorrelation function becomes more sinusoidal. In general, to represent L spectral peaks none of which are at $\omega = 0$ or π requires a model order of $p = 2L$.

2.5.3 Examples of MA Processes

To find the autocorrelation function for an MA(q) process we use Eq. (2.78) with $p = 0$. Then

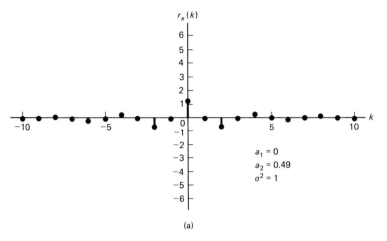

(a)

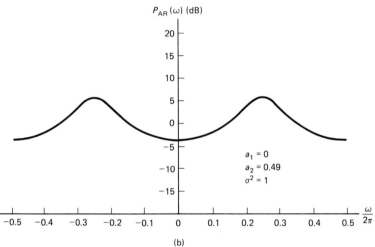

(b)

Figure 2.12 (a) Autocorrelation of AR(2) process with complex conjugate poles at $0.7 \exp[\pm j2\pi(0.25)]$; (b) power spectral density of AR(2) process with complex conjugate poles at $0.7 \exp[\pm j2\pi(0.25)]$; (c) autocorrelation of AR(2) process with complex conjugate poles at $0.95 \exp[\pm j2\pi(0.25)]$; (d) power spectral density of AR(2) process with complex conjugate poles at $0.95 \exp[\pm j2\pi(0.25)]$.

$$r_x(k) = \begin{cases} \sigma^2 \sum_{\ell=0}^{q-k} h(\ell)b_{\ell+k}, & k = 0, 1, \ldots, q \\ 0, & k \geq q + 1 \end{cases}$$

But

$$h(k) = \begin{cases} b_k, & k = 0, 1, \ldots, q \\ 0, & \text{otherwise} \end{cases}$$

which results in

$$r_x(k) = \begin{cases} \sigma^2 \sum_{\ell=0}^{q-|k|} b_{\ell}b_{\ell+k}, & |k| \leq q \\ 0, & |k| > q \end{cases} \tag{2.86}$$

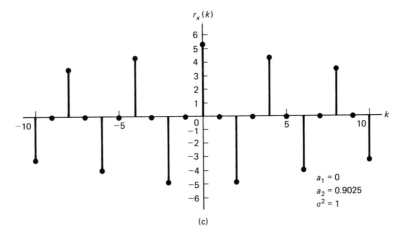

(c)

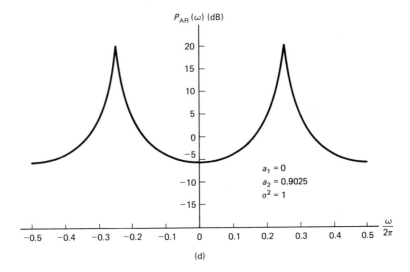

(d)

Figure 2.12 (*cont.*)

For an MA(1) process we have, on using Eq. (2.86),

$$r_x(k) = \begin{cases} \sigma^2(1 + b_1^2), & k = 0 \\ \sigma^2 b_1, & k = 1 \\ 0, & k \geq 2 \end{cases} \qquad (2.87)$$

with a corresponding PSD

$$P_{\text{MA}}(\omega) = \sigma^2 |1 + b_1 \exp(-j\omega)|^2 \qquad (2.88)$$

An example is given in Fig. 2.13 in which we see that the PSD exhibits a dip at $\omega = 0$, i.e., at the zero location of $B(z)$. For spectral valleys at other frequency locations we

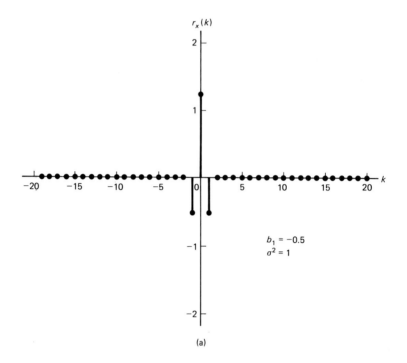

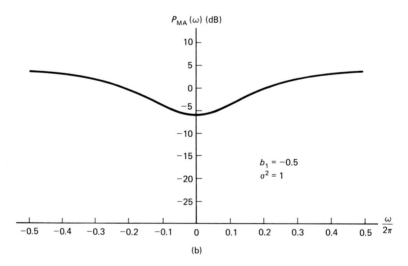

Figure 2.13 (a) Autocorrelation of MA(1) process; (b) power spectral density of MA(1) process; (c) autocorrelation of MA(1) process; (d) power spectral density of MA(1) process.

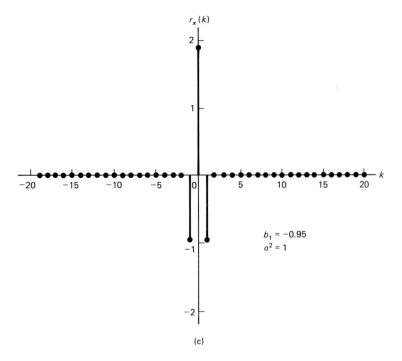

(c)

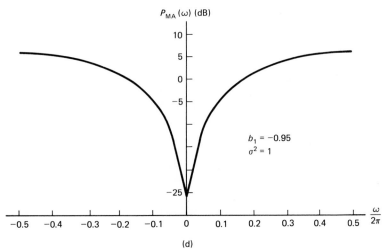

(d)

Figure 2.13 (*cont.*)

must use a higher-order MA model. For an MA(2) process,

$$r_x(k) = \begin{cases} \sigma^2(1 + b_1^2 + b_2^2), & k = 0 \\ \sigma^2(b_1 + b_1 b_2), & k = 1 \\ \sigma^2 b_2, & k = 2 \\ 0, & k \geq 3 \end{cases} \qquad (2.89)$$

and

$$P_{MA}(\omega) = \sigma^2 |1 + b_1 \exp(-j\omega) + b_2 \exp(-j2\omega)|^2 \qquad (2.90)$$

Assuming complex zeros at $r \exp(\pm j\omega_0)$ so that $b_1 = -2r \cos \omega_0$, $b_2 = r^2$, the PSD becomes

$$P_{MA}(\omega) = \sigma^2 |1 - r \exp[-j(\omega - \omega_0)]|^2 |1 - r \exp[-j(\omega + \omega_0)]|^2 \qquad (2.91)$$

Examples are shown in Fig. 2.14. Note that the PSD for MA processes tends to be broadband but may exhibit nulls if the zeros are close to the unit circle.

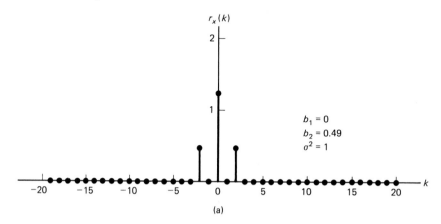

(a)

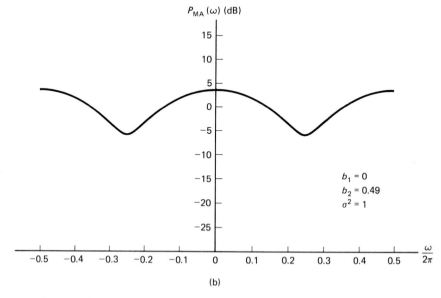

(b)

Figure 2.14 (a) Autocorrelation of MA(2) process with complex conjugate zeros at $0.7 \exp[\pm j2\pi(0.25)]$; (b) power spectral density of MA(2) process with complex conjugate zeros at $0.7 \exp[\pm j2\pi(0.25)]$; (c) autocorrelation of MA(2) process with complex conjugate zeros at $0.95 \exp[\pm j2\pi(0.25)]$; (d) power spectral density of MA(2) process with complex conjugate zeros at $0.95 \exp[\pm j2\pi(0.25)]$.

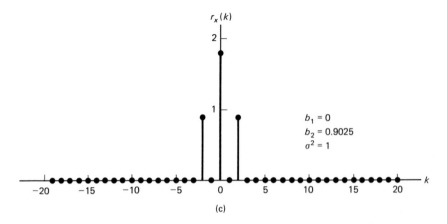

(c)

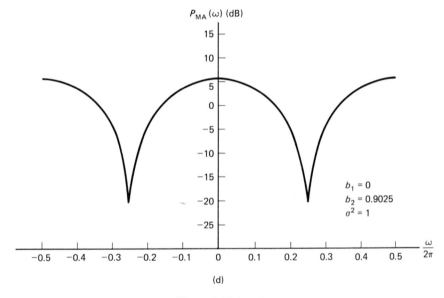

(d)

Figure 2.14 (*cont.*)

2.5.4 ARMA Processes

ARMA processes have PSDs that in general exhibit both spectral peaks and spectral valleys. If the zeros of the process are not near the unit circle, then an AR model may be appropriate. Likewise, if the poles of the process are not near the unit circle, then an MA process may suffice. If neither of these conditions is satisfied, then the general ARMA model will be necessary. The autocorrelation function and PSD for an ARMA process are similar in nature to those for the MA and AR processes and so will be omitted here. Examples may be found in [14].

2.5.5 Mixed Processes

In some cases the actual process may be modeled well by the sum of two uncorrelated time series. As an example, consider an AR(2) process plus white noise, or

$$y(n) = x(n) + w(n) \tag{2.92}$$

where $w(n)$ is zero-mean white noise with variance σ_w^2, and $x(n)$ is an AR(2) process. A process given by Eq. (2.92) is termed a mixed process. If we were to use an AR(2) model for $y(n)$ and solve for the AR parameters via the Yule-Walker equations using $r_y(k) = r_x(k) + \sigma_w^2 \delta(k)$, we would obtain unsatisfactory results. For a signal-to-noise ratio (SNR) of SNR $= 10 \log_{10} r_x(0)/\sigma_w^2 = 5$ dB the PSD that is obtained is shown in Fig. 2.15(a). If $p = 5$ is used, a better spectral fit is obtained, as shown in Fig. 2.15(b). This is to be expected since according to the Wold decomposition an AR(∞) model can represent any time series. The true model for an AR(p) process in white noise can be found in the following manner. Let $P_x(z)$ denote the z-transform of the autocorrelation function of $x(n)$. Then it follows from Eq. (2.92) that

$$P_y(z) = P_x(z) + P_w(z)$$

$$= \frac{\sigma^2}{A(z)A(z^{-1})} + \sigma_w^2 \tag{2.93}$$

$$= \frac{\sigma^2 + \sigma_w^2 A(z)A(z^{-1})}{A(z)A(z^{-1})}$$

It can be shown [15] that the numerator of Eq. (2.93) may be factored as

$$\sigma^2 + \sigma_w^2 A(z)A(z^{-1}) = \sigma_\eta^2 B(z)B(z^{-1}) \tag{2.94}$$

where $B(z) = 1 + \sum_{k=1}^{p} b_k z^{-k}$ has its zeros inside the unit circle and $\sigma_\eta^2 > 0$. Hence,

$$P_y(z) = \frac{\sigma_\eta^2 B(z)B(z^{-1})}{A(z)A(z^{-1})}$$

We see that $y(n)$ is actually an ARMA(p, p) process with its AR and MA parameters linked according to Eq. (2.94). Estimation of the PSD of $y(n)$ should be based on the appropriate ARMA model. This is an important consideration in that many time series of interest are actually composed of signals in noise.

2.5.6 Estimation of AR Power Spectral Densities

For good estimates of the AR parameters an MLE is usually employed (see Section 2.3.2). Assuming $x(n)$ is a Gaussian random process, we now show that an approximate MLE of the AR parameters is found by solving a set of simultaneous linear equations. This estimator is identical in form to the covariance method for signal modeling, which was discussed in Chapter 1. Assuming that $\mathbf{x} = [x(0)x(1) \cdots x(N-1)]^T$ is observed, we wish to estimate $\mathbf{a} = [a_1 a_2 \cdots a_p]^T$ and σ^2. The MLE

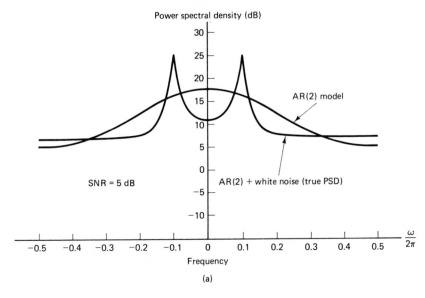

(a)

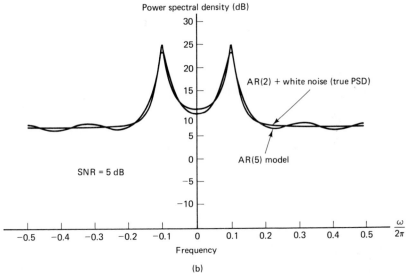

(b)

Figure 2.15 (a) AR(2) modeling of noise-corrupted AR(2) power spectral density; (b) AR(5) modeling of noise-corrupted AR(2) power spectral density.

is found by maximizing the joint PDF $p(\mathbf{x}; \mathbf{a}, \sigma^2)$ over $\mathbf{a}$ and σ^2 when $\mathbf{x}$ is replaced by the observed data samples. The PDF can also be written conditioned on the first p data samples as

$$p(x(p), x(p+1), \ldots, x(N-1) \mid x(0), x(1), \ldots, x(p-1); \mathbf{a}, \sigma^2)$$
$$p(x(0), x(1), \ldots, x(p-1); \mathbf{a}, \sigma^2) \tag{2.95}$$

For large data records the maximization of the PDF can be effected by maximizing only the conditional PDF in Eq. (2.95) as long as the poles are not too close to the unit circle [16]. We thus seek to maximize the conditional PDF. From Eq. (2.74),

$$\epsilon(n) = x(n) + \sum_{k=1}^{p} a_k x(n - k)$$

or, for $n = p, p + 1, \ldots, N - 1$,

$$\epsilon(p) = x(p) + a_1 x(p - 1) + \cdots + a_p x(0)$$
$$\epsilon(p + 1) = x(p + 1) + a_1 x(p) + \cdots + a_p x(1)$$
$$\vdots \tag{2.96}$$
$$\epsilon(N - 1) = x(N - 1) + a_1 x(N - 2) + \cdots + a_p x(N - p - 1)$$

Noting that the $\epsilon(n)$'s are uncorrelated Gaussian random variables and hence independent, the PDF of $\boldsymbol{\epsilon} = [\epsilon(p)\epsilon(p + 1) \cdots \epsilon(N - 1)]^T$ is

$$p(\boldsymbol{\epsilon}) = \prod_{n=p}^{N-1} \frac{1}{\sqrt{2\pi}\sigma} \exp\left[-\frac{1}{2\sigma^2}\epsilon^2(n)\right]$$
$$= \frac{1}{(2\pi\sigma^2)^{(N-p)/2}} \exp\left[-\frac{1}{2\sigma^2}\sum_{n=p}^{N-1}\epsilon^2(n)\right] \tag{2.97}$$

We now transform the PDF of $\{\epsilon(p), \epsilon(p + 1), \ldots, \epsilon(N - 1)\}$ to the PDF of $\{x(p), x(p + 1), \ldots, x(N - 1)\}$ by using the transformation of Eq. (2.96). The transformation may be rewritten as

$$\boldsymbol{\epsilon} = \underbrace{\begin{bmatrix} 1 & 0 & 0 & 0 & \cdots & 0 \\ a_1 & 1 & 0 & 0 & \cdots & 0 \\ a_2 & a_1 & 1 & 0 & \cdots & 0 \\ \vdots & & & & & \vdots \\ 0 & \cdots & a_p & a_{p-1} & \cdots & 1 \end{bmatrix}}_{\mathbf{J}} \mathbf{x}' + \begin{bmatrix} \sum_{i=1}^{p} a_i x(p - i) \\ \sum_{i=2}^{p} a_i x(p + 1 - i) \\ \vdots \\ a_p x(p - 1) \\ 0 \\ \vdots \\ 0 \end{bmatrix}$$

where $\mathbf{x}' = [x(p)x(p + 1) \cdots x(N - 1)]^T$. The Jacobian of the transformation $\partial\boldsymbol{\epsilon}/\partial\mathbf{x}'$ is just $\mathbf{J}$ and hence $|\det(\mathbf{J})| = 1$. Then, using Eq. (2.97), we have

$$p(\mathbf{x}'|x(0), x(1), \ldots, x(p - 1); \mathbf{a}, \sigma^2) = p(\boldsymbol{\epsilon}(\mathbf{x}'))\left|\det\left(\frac{\partial\boldsymbol{\epsilon}}{\partial\mathbf{x}'}\right)\right| = p(\boldsymbol{\epsilon}(\mathbf{x}'))$$

$$= \frac{1}{(2\pi\sigma^2)^{(N-p)/2}} \exp\left[-\frac{1}{2\sigma^2}\sum_{n=p}^{N-1}\left[x(n) + \sum_{j=1}^{p} a_j x(n - j)\right]^2\right] \tag{2.98}$$

To maximize Eq. (2.98) over **a** we need only minimize

$$S_1(\mathbf{a}) = \sum_{n=p}^{N-1} \left[x(n) + \sum_{j=1}^{p} a_j x(n-j) \right]^2 \tag{2.99}$$

Since S_1 is a quadratic function of **a**, differentiating Eq. (2.99) will produce a global minimum. Performing the differentiation, we have

$$\sum_{j=1}^{p} \hat{a}_j \sum_{n=p}^{N-1} x(n-j)x(n-k) = -\sum_{n=p}^{N-1} x(n)x(n-k), \quad k = 1, 2, \ldots, p \tag{2.100}$$

or in matrix form the MLE of **a** is found by solving

$$\underbrace{\begin{bmatrix} c_{11} & c_{12} & \cdots & c_{1p} \\ c_{21} & c_{22} & \cdots & c_{2p} \\ \vdots & \vdots & \ddots & \vdots \\ c_{p1} & c_{p2} & \cdots & c_{pp} \end{bmatrix}}_{\mathbf{C}} \underbrace{\begin{bmatrix} \hat{a}_1 \\ \hat{a}_2 \\ \vdots \\ \hat{a}_p \end{bmatrix}}_{\hat{\mathbf{a}}} = -\underbrace{\begin{bmatrix} c_{10} \\ c_{20} \\ \vdots \\ c_{p0} \end{bmatrix}}_{\mathbf{c}} \tag{2.101}$$

where

$$c_{jk} = \frac{1}{N-p} \sum_{n=p}^{N-1} x(n-j)x(n-k)$$

The factor $1/(N-p)$ has been introduced by dividing both sides of Eq. (2.100) by $N - p$. In this form c_{jk} is seen to be an estimate of $r_x(j-k)$ and hence *the MLE is obtained by replacing the true ACF in the Yule-Walker equations of Eq. (2.80) by suitable estimates*. It should be noted that **C** is symmetric and positive semidefinite. To solve Eq. (2.101) we may use a Cholesky decomposition or the fast solution for the covariance equations as discussed in Chapter 1.

To find $\hat{\sigma}^2$ we first substitute $\hat{\mathbf{a}}$ into Eq. (2.98) and then differentiate with respect to σ^2. Equivalently, we may take the logarithm of Eq. (2.98) since it is a monotonic function and differentiate to yield

$$\frac{\partial p}{\partial \sigma^2} = \frac{\partial}{\partial \sigma^2} \left[-\frac{N-p}{2} \ln 2\pi - \frac{N-p}{2} \ln \sigma^2 - \frac{1}{2\sigma^2} S_1(\hat{\mathbf{a}}) \right] = 0$$

which yields

$$\hat{\sigma}^2 = \frac{1}{N-p} S_1(\hat{\mathbf{a}})$$

$$= \frac{1}{N-p} \sum_{n=p}^{N-1} \left[x(n) + \sum_{j=1}^{p} \hat{a}_j x(n-j) \right]^2 \tag{2.102}$$

An alternate expression for $\hat{\sigma}^2$ is found by observing from Eq. (2.100) that

$$\sum_{n=p}^{N-1} \left[x(n) + \sum_{j=1}^{p} \hat{a}_j x(n-j) \right] \sum_{k=1}^{p} \hat{a}_k x(n-k) = 0$$

Hence, Eq. (2.102) becomes

$$\hat{\sigma}^2 = \frac{1}{N-p} \sum_{n=p}^{N-1} x^2(n) + \sum_{j=1}^{p} \hat{a}_j \frac{1}{N-p} \sum_{n=p}^{N-1} x(n)x(n-j)$$

or, finally,

$$\hat{\sigma}^2 = c_{00} + \sum_{j=1}^{p} \hat{a}_j c_{0j} \tag{2.103}$$

The estimation of the AR parameters via Eqs. (2.101) and (2.103) is called the *covariance method*. Once the estimates have been computed, the PSD estimate is given by

$$\hat{P}_{AR}(\omega) = \frac{\hat{\sigma}^2}{|1 + \hat{a}_1 \exp(-j\omega) + \cdots + \hat{a}_p \exp(-j\omega p)|^2} \tag{2.104}$$

Note that the approximate MLE was obtained by minimizing a sum of squared errors given by Eq. (2.99). If we consider $-\Sigma_{j=1}^{p} a_j x(n-j)$ as a linear prediction of $x(n)$ based on $\{x(n-1), x(n-2), \ldots, x(n-p)\}$ and $x(n) - [-\Sigma_{j=1}^{p} a_j x(n-j)]$ as a prediction error, then it may be said that the MLE of the AR parameters is found by finding the best pth-order linear predictor of $x(n)$ based on the previous p samples. In this manner, the name *linear prediction* is often associated with AR spectral estimation [3].

In addition to the covariance method, there are numerous other techniques that also are approximate MLEs. As an example it is easily shown that if $N \gg p$, then

$$c_{jk} \approx \hat{r}_x(j-k)$$

where $\hat{r}_x(k)$ is given by Eq. (2.42). When this autocorrelation function estimator is substituted into Eq. (2.101), a set of Yule-Walker equations results, with the theoretical autocorrelation function replaced by the biased autocorrelation function estimator. This approach is termed the *autocorrelation method* (see Chapter 1). Some other methods that have been found to work well are the forward-backward or modified covariance method [17,18] and the Burg method [2]. We will summarize these estimators. For further details see Chapter 1.

In the forward-backward method we minimize

$$S_2(\mathbf{a}) = \sum_{n=p}^{N-1} \left[x(n) + \sum_{j=1}^{p} a_j x(n-j) \right]^2 + \sum_{n=0}^{N-1-p} \left[x(n) + \sum_{j=1}^{p} a_j x(n+j) \right]^2 \tag{2.105}$$

which is the sum of forward and backward prediction error energies. The minimization proceeds analogously to that of the covariance method with the results given by Eq. (2.101) and Eq. (2.103) but with c_{jk} defined as

$$c_{jk} = \frac{1}{2(N-p)} \left[\sum_{n=p}^{N-1} x(n-j)x(n-k) + \sum_{n=0}^{N-1-p} x(n+j)x(n+k) \right] \tag{2.106}$$

One can show that $\mathbf{C}$ for the forward-backward method is symmetric and positive semidefinite. The inversion of $\mathbf{C}$ can be avoided by using a Cholesky decomposition or the recursive algorithm in [19].

The second method, the Burg algorithm, makes use of the Levinson algorithm and an estimate for the *reflection coefficients*, which are defined as $k_i = a_{ii}$ in Eq. (2.81). The Burg algorithm is initialized by

$$e_0(n) = x(n)$$

$$b_0(n) = x(n)$$

$$\sigma_0^2 = \frac{1}{N} \sum_{n=0}^{N-1} x^2(n)$$

For $i = 1, 2, \ldots, p,$

$$k_i = \frac{-2 \sum\limits_{n=i}^{N-1} e_{i-1}(n)b_{i-1}(n-1)}{\sum\limits_{n=i}^{N-1} [e_{i-1}^2(n) + b_{i-1}^2(n-1)]}$$

$$a_{ij} = \begin{cases} k_1 & \text{for } i = j = 1 \\ a_{i-1,j} + k_i a_{i-1,i-j} & \text{for } j = 1, 2, \ldots, i-1; \\ & \qquad\qquad i = 2, 3, \ldots, p \end{cases} \qquad (2.107)$$

$$\sigma_i^2 = (1 - k_i^2)\sigma_{i-1}^2$$

$$e_i(n) = e_{i-1}(n) + k_i b_{i-1}(n-1)$$

$$b_i(n) = b_{i-1}(n-1) + k_i e_{i-1}(n)$$

The estimates of the AR parameters are $\hat{a}_i = a_{pi}$, $i = 1, 2, \ldots, p$, and $\hat{\sigma}^2 = \sigma_p^2$.

2.5.7 Estimation of MA Power Spectral Densities

The MA spectral estimator was defined to be

$$P_{MA}(\omega) = \sigma^2 \left| 1 + \sum_{k=1}^{q} b_k \exp(-j\omega k) \right|^2 \qquad (2.108)$$

This can also be written as

$$P_{MA}(\omega) = \sum_{k=-q}^{q} r_x(k) \exp(-j\omega k) \qquad (2.109)$$

by expanding the factors in Eq. (2.108) with the autocorrelation function $r_x(k)$ given by Eq. (2.86). A natural estimator of $P_{MA}(\omega)$ would then seem to be

$$\hat{P}_{MA}(\omega) = \sum_{k=-q}^{q} \hat{r}_x(k) \exp(-j\omega k) \qquad (2.110)$$

where $\hat{r}_x(k)$ is some suitable estimator of the autocorrelation function. The MA spectral estimator bears a strong resemblance to the Blackman-Tukey spectral estimator. A subtle difference between the two estimators is that the MA spectral estimator is based on the MA(q) model and hence by assumption $\hat{r}_x(k) = 0$ for $|k| > q$. The Blackman-Tukey spectral estimator, on the other hand, can be applied to any process. Furthermore, in the Blackman-Tukey spectral estimator the autocorrelation function estimator is truncated at $k = M$ (as well as weighted) due to a finite-length data record.

To estimate the MA PSD we need to find the MLE of the MA parameters. Durbin has derived them for large data records. The exact derivation [20,14] relies heavily on advanced statistical theory and so will not be presented here. Instead we offer an intuitive justification for Durbin's method. The PSD of an MA process generalized to the z-plane is

$$P_{\mathrm{MA}}(z) = \sigma^2 B(z)B(z^{-1})$$

We can approximate $B(z)$ by $1/A(z)$, where $A(z) = 1 + \sum_{k=1}^{L} a_k z^{-k}$ to yield

$$P_{\mathrm{MA}}(z) = \frac{\sigma^2}{A(z)A(z^{-1})} \tag{2.111}$$

As an example, consider an MA(1) process. Then

$$B(z) \approx \frac{1}{A(z)}$$

$$\tag{2.112}$$

$$1 + b_1 z^{-1} \approx \frac{1}{A(z)}$$

or

$$A(z) \approx \frac{1}{1 + b_1 z^{-1}} = \sum_{k=0}^{\infty} (-b_1)^k z^{-k}$$

If we choose $a_k = (-b_1)^k$ for $k = 1, 2, \ldots, L$, and if $(-b_1)^k \approx 0$ for $k > L$, then the approximation will be a good one. Clearly, the approximation will be better when the zero of $B(z)$ is not near the unit circle or when $|b_1|$ is small. If L can be chosen such that the approximation is a good one, then the MA(q) process is equivalent to an AR(L) process. Consequently, the MLE of the AR parameters can be found using any of the methods of Section 2.5.6. Call these estimates $\{\hat{a}_1, \hat{a}_2, \ldots, \hat{a}_L, \hat{\sigma}^2\}$. To find the estimates of the b_k's, note from Eq. (2.112) that

$$B(z)B(z^{-1}) \approx \frac{1}{A(z)A(z^{-1})}$$

or

$$A(z)A(z^{-1}) \approx \frac{1}{B(z)B(z^{-1})} \tag{2.113}$$

The right-hand side of Eq. (2.113) represents the PSD of an AR(q) process. To estimate the b_k's we need only estimate the autocorrelation lags for $0 \le k \le q$, assuming we use the autocorrelation method of AR parameter estimation. But these lag estimates can be found from Eq. (2.113) by taking the inverse z-transform of $\hat{A}(z)\hat{A}(z^{-1})$, or

$$\hat{r}_a(k) = \frac{1}{L+1} \sum_{n=0}^{L-|k|} \hat{a}_n \hat{a}_{n+|k|} \tag{2.114}$$

The $1/(L+1)$ scale factor has been added to allow $\hat{r}_a(k)$ to be interpreted as an autocorrelation function estimator.

Durbin's method can be summarized as follows.

1. Fit a large-order AR model to the data. Specifically, using the Levinson algorithm solve

$$\begin{bmatrix} \hat{r}_x(0) & \hat{r}_x(1) & \cdots & \hat{r}_x(L-1) \\ \hat{r}_x(1) & \hat{r}_x(0) & \cdots & \hat{r}_x(L-2) \\ \vdots & \vdots & \ddots & \vdots \\ \hat{r}_x(L-1) & \hat{r}_x(L-2) & \cdots & \hat{r}_x(0) \end{bmatrix} \begin{bmatrix} \hat{a}_1 \\ \hat{a}_2 \\ \vdots \\ \hat{a}_L \end{bmatrix} = - \begin{bmatrix} \hat{r}_x(1) \\ \hat{r}_x(2) \\ \vdots \\ \hat{r}_x(L) \end{bmatrix} \tag{2.115}$$

where

$$\hat{r}_x(k) = \frac{1}{N} \sum_{n=0}^{N-1-k} x(n)x(n+k)$$

The estimate of σ^2 is given by

$$\hat{\sigma}^2 = \hat{r}_x(0) + \sum_{k=1}^{L} \hat{a}_k \hat{r}_x(k) \tag{2.116}$$

2. Using the data sequence $\{1, \hat{a}_1, \hat{a}_2, \ldots, \hat{a}_L\}$, fit an AR($q$) model. As before, solve

$$\begin{bmatrix} \hat{r}_a(0) & \hat{r}_a(1) & \cdots & \hat{r}_a(q-1) \\ \hat{r}_a(1) & \hat{r}_a(0) & \cdots & \hat{r}_a(q-2) \\ \vdots & \vdots & \ddots & \vdots \\ \hat{r}_a(q-1) & \hat{r}_a(q-2) & \cdots & \hat{r}_a(0) \end{bmatrix} \begin{bmatrix} \hat{b}_1 \\ \hat{b}_2 \\ \vdots \\ \hat{b}_q \end{bmatrix} = - \begin{bmatrix} \hat{r}_a(1) \\ \hat{r}_a(2) \\ \vdots \\ \hat{r}_a(q) \end{bmatrix} \tag{2.117}$$

where

$$\hat{r}_a(k) = \frac{1}{L+1} \sum_{n=0}^{L-k} \hat{a}_n \hat{a}_{n+k}$$

The estimation of the AR parameters in steps 1 and 2 may be effected by any of the AR estimation methods. The choice of the autocorrelation method allows a simple solution to the linear equations via the Levinson algorithm.

The only difficulty in using Durbin's method is in choosing L. From the intuitive justification it is clear that L should be equal to the effective impulse response length of $1/B(z)$. Unfortunately, this is unknown *a priori* since it is the MA parameters we wish to estimate. Also, we require $L \ll N$ for the AR parameter estimates to be statistically accurate.

2.5.8 Estimation of ARMA Power Spectral Densities

The ARMA(p, q) spectral estimator was defined in Section 2.5.1 to be

$$\hat{P}_{\text{ARMA}}(\omega) = \frac{\hat{\sigma}^2 \left| 1 + \sum\limits_{k=1}^{q} \hat{b}_k \exp(-j\omega k) \right|^2}{\left| 1 + \sum\limits_{k=1}^{p} \hat{a}_k \exp(-j\omega k) \right|^2} \tag{2.118}$$

For reliable estimates of the ARMA parameters we would once again like to use an MLE. Unfortunately, in this case the MLE is exceedingly difficult to obtain. Even with several simplifying approximations to the PDF *the equations obtained by differentiating are extremely nonlinear*. To illustrate the problem we will examine the function that needs to be minimized. First we return to the problem of AR estimation. There we found that the function we needed to minimize was

$$S_1(\mathbf{a}) = \sum_{n=p}^{N-1} \left[x(n) + \sum_{j=1}^{p} a_j x(n - j) \right]^2 \tag{2.119}$$

If we consider $x(n) = 0$ outside of the $0 \le n \le N - 1$ interval and replace the limits of the summation with $n = 0$ to $n = N + p - 1$, which for N large will not significantly alter the sum, then

$$S_1(\mathbf{a}) = \sum_{n=-\infty}^{\infty} \left[x(n) + \sum_{j=1}^{p} a_j x(n - j) \right]^2 \tag{2.120}$$

Note that $S_1(\mathbf{a})$ represents the energy out of the inverse or whitening filter $A(z) = 1 + \sum_{k=1}^{p} a_k z^{-k}$ for the input $x(n)$. For the true AR parameters, $x(n) + \sum_{j=1}^{p} a_j x(n - j)$ becomes $\epsilon(n)$, or we have whitened the $x(n)$ time series. It is not surprising then that for an ARMA process the approximate MLE is found by finding the inverse filter $A(z)/B(z)$, which whitens the ARMA time series. Whitening can also be shown to be equivalent to minimizing the output energy out of the filter. An explicit expression for the energy for the ARMA case can be obtained by transforming to the frequency domain. First for an AR process Eq. (2.120) can be rewritten using Parseval's theorem as

$$S_1(\mathbf{a}) = N \int_{-\pi}^{\pi} I(\omega) |A(\omega)|^2 \frac{d\omega}{2\pi}$$

where

$$I(\omega) = \frac{1}{N} \left| \sum_{n=0}^{N-1} x(n) \exp(-j\omega n) \right|^2$$

which is just the periodogram. The different notation is meant to emphasize that $I(\omega)$ is to be regarded as a function of the data and not necessarily as a spectral estimator. Likewise, the approximate MLE of the ARMA parameters is found by minimizing

$$S_2(\mathbf{a}, \mathbf{b}) = N \int_{-\pi}^{\pi} I(\omega) \frac{|A(\omega)|^2}{|B(\omega)|^2} \frac{d\omega}{2\pi} \tag{2.121}$$

As an example, consider the ARMA(1, 1) case. Then, differentiating Eq. (2.121) to obtain a set of necessary conditions for the approximate MLE, we have

$$\frac{\partial S_2}{\partial a_1} = N \int_{-\pi}^{\pi} \frac{I(\omega)}{|B(\omega)|^2} \frac{\partial}{\partial a_1} \{[1 + a_1 \exp(-j\omega)][1 + a_1 \exp(-j\omega)]\} \frac{d\omega}{2\pi}$$

$$= N \int_{-\pi}^{\pi} \frac{I(\omega)}{|B(\omega)|^2} \{[1 + a_1 \exp(-j\omega)] \exp(j\omega)$$

$$+ \exp(-j\omega)[1 + a_1 \exp(j\omega)]\} \frac{d\omega}{2\pi} \tag{2.122}$$

$$= N \int_{-\pi}^{\pi} \frac{I(\omega)}{|B(\omega)|^2} (2 \cos \omega + 2a_1) \frac{d\omega}{2\pi}$$

$$= 0$$

so

$$a_1 = - \frac{\displaystyle\int_{-\pi}^{\pi} \frac{I(\omega) \cos \omega}{|B(\omega)|^2} \frac{d\omega}{2\pi}}{\displaystyle\int_{-\pi}^{\pi} \frac{I(\omega)}{|B(\omega)|^2} \frac{d\omega}{2\pi}} \tag{2.123}$$

In a similar fashion we obtain

$$\frac{\partial S_2}{\partial b_1} = -N \int_{-\pi}^{\pi} \frac{I(\omega)|A(\omega)|^2}{|B(\omega)|^4} (2 \cos \omega + 2b_1) \frac{d\omega}{2\pi} = 0 \tag{2.124}$$

Substituting Eq. (2.123) for a_1 into Eq. (2.124) results in a very nonlinear equation in b_1. In general, we would obtain a set of nonlinear equations that need to be solved using iterative techniques. The standard Newton-Raphson technique has been proposed for this problem but suffers from convergence problems; furthermore, even if convergence is obtained, the solution may correspond to only a local and not a global minimum [21,14].

Due to the difficulty of computing the approximate MLE, several suboptimal approaches have been proposed. They rely on the theoretical relationship between the autocorrelation function and the ARMA parameters as embodied in the Yule-Walker equations. Recall from Eq. (2.78) that

$$r_x(k) = -\sum_{\ell=1}^{p} a_\ell r_x(k - \ell), \qquad k \geq q + 1 \tag{2.125}$$

If we rewrite this in matrix notation for $k = q + 1, q + 2, \ldots, q + p$, then

$$
\begin{bmatrix}
r_x(q) & r_x(q-1) & \cdots & r_x(q-p+1) \\
r_x(q+1) & r_x(q) & \cdots & r_x(q-p+2) \\
\vdots & \vdots & \ddots & \vdots \\
r_x(q+p-1) & r_x(q+p-2) & \cdots & r_x(q)
\end{bmatrix}
\underbrace{}_{\mathbf{R}_x'}
\begin{bmatrix}
a_1 \\ a_2 \\ \vdots \\ a_p
\end{bmatrix}
= -
\begin{bmatrix}
r_x(q+1) \\ r_x(q+2) \\ \vdots \\ r_x(q+p)
\end{bmatrix}
\tag{2.126}
$$

These equations are called the *modified Yule-Walker equations* and can be used to estimate the AR parameters of an ARMA(p, q) model if the theoretical autocorrelation lags are replaced by estimates. It is termed the modified Yule-Walker equation estimator. Once the AR parameter estimates have been obtained, the remaining parameters $\{b_1, b_2, \ldots, b_q, \sigma^2\}$ can be found as follows. Filter $x(n)$ with $\hat{A}(z)$, where $\hat{A}(z) = 1 + \sum_{k=1}^{p} \hat{a}_k z^{-k}$ and the $\hat{a}_k$'s have been found from Eq. (2.126). If $\hat{A}(z) \approx A(z)$, then the filter output will be very nearly an MA(q) process with parameters $\{b_1, b_2, \ldots, b_q, \sigma^2\}$. Now use Durbin's method on the output sequence to estimate the remaining parameters. Due to the memory of the $\hat{A}(z)$ filter the output sequence should be used only for $n = p, p + 1, \ldots, N - 1$.

The use of the modified Yule-Walker equations followed by Durbin's method produces good results at times. The technique can, however, yield highly variable spectral estimates, especially when the matrix $\mathbf{R}_x'$ is nearly singular [22]. The problem may be likened to attempting to fit a straight line through two data points that are subject to error. To overcome this deficiency it is better to use more than p equations. Since there will now be more equations than unknowns, we solve for the AR parameters in a least-squares sense, as explained in Section 2.3.2. Choosing the equations in Eq. (2.125) corresponding to $k = q + 1, q + 2, \ldots, M$ and rewriting in matrix notation, we have

$$
\begin{bmatrix}
r_x(q+1) \\ r_x(q+2) \\ \vdots \\ r_x(M)
\end{bmatrix}
= -
\begin{bmatrix}
r_x(q) & r_x(q-1) & \cdots & r_x(q-p+1) \\
r_x(q+1) & r_x(q) & \cdots & r_x(q-p+2) \\
\vdots & \vdots & \ddots & \vdots \\
r_x(M-1) & r_x(M-2) & \cdots & r_x(M-p)
\end{bmatrix}
\begin{bmatrix}
a_1 \\ a_2 \\ \vdots \\ a_p
\end{bmatrix}
\tag{2.127}
$$

By replacing $r_x(k)$ by an estimate, Eq. (2.127) will no longer hold exactly. An error term will need to be introduced so that Eq. (2.127) becomes

$$
\begin{bmatrix}
\hat{r}_x(q+1) \\ \hat{r}_x(q+2) \\ \vdots \\ \hat{r}_x(q+M)
\end{bmatrix}
$$

$$
=
\begin{bmatrix}
-\hat{r}_x(q) & -\hat{r}_x(q-1) & \cdots & -\hat{r}_x(q-p+1) \\
-\hat{r}_x(q+1) & -\hat{r}_x(q) & \cdots & -\hat{r}_x(q-p+2) \\
\vdots & \vdots & \ddots & \vdots \\
-\hat{r}_x(M-1) & -\hat{r}_x(M-2) & \cdots & -\hat{r}_x(M-p)
\end{bmatrix}
\begin{bmatrix}
a_1 \\ a_2 \\ \vdots \\ a_p
\end{bmatrix}
+
\begin{bmatrix}
e(q+1) \\ e(q+2) \\ \vdots \\ e(M)
\end{bmatrix}
\tag{2.128}
$$

But this is exactly in the form of Eq. (2.27), or

$$\mathbf{r} = \mathbf{Ra} + \mathbf{e} \tag{2.129}$$

so that a least-squares estimator of $\mathbf{a}$ would minimize

$$S_3(\mathbf{a}) = (\mathbf{r} - \mathbf{Ra})^T(\mathbf{r} - \mathbf{Ra})$$

The solution that is found from Eq. (2.29) is

$$\hat{\mathbf{a}} = (\mathbf{R}^T\mathbf{R})^{-1}\mathbf{R}^T\mathbf{r} \tag{2.130}$$

It has been found empirically that the estimator given by Eq. (2.130), which is termed the *least-squares modified Yule-Walker equation* (LSMYWE) estimator, yields better estimates than the modified Yule-Walker equation estimator. Neither technique, however, works well when the poles are sufficiently displaced from the unit circle. This is because the autocorrelation function damps out rapidly, leading to autocorrelation lag estimates dominated by the estimation error. It should be noted that the LSMYWE is not optimal in any reasonable sense. The relationship of the least-squares estimator with the MLE is not valid in this case since $\mathbf{R}$ is not a constant matrix but is random, and $\mathbf{e}$ is not distributed according to $N(\mathbf{0}, \sigma^2\mathbf{I})$. The LSMYWE minimizes the error in a theoretical equation, i.e., Eq. (2.127), when estimates are used. For this reason the technique is also referred to as an *equation error modeling* approach. Finally, once the AR parameter estimates have been obtained, Durbin's method can be used to estimate the remaining parameters based on the data at the output of $\hat{A}(z)$.

A difficulty of the LSMYWE is in choosing M. M should be chosen large to take advantage of the information in the higher-order lags but not so large that the auto-correlation function estimates are unreliable. Unless one has some *a priori* knowledge about the pole positions, this choice is difficult and is a problem that has not been resolved.

2.5.9 Model Order Determination

So far we have not addressed the question of how one actually chooses the appropriate model order for the time series model. The best way to make this decision is to base it on the physics of the process generating the data. For instance, in speech processing it is known that the vocal tract can be modeled as an all-pole filter having about 4 resonances in a 4-Khz band [3]. Hence, at least 8 complex conjugate poles are necessary and typically $p = 12$ is chosen in an AR model. When no such information is available, statistical tests can be used to estimate order. One such test is the Akaike information criterion (AIC) [23]. The AIC computes a measure over all possible model orders and chooses the model order that minimizes the measure. For an ARMA(p, q) process the AIC is defined as

$$\text{AIC}(i, j) = N \ln \hat{\sigma}_{ij}^2 + 2(i + j) \tag{2.131}$$

where $\hat{\sigma}_{ij}^2$ is the MLE of the white noise variance for an assumed ARMA(i, j) process. The AIC attempts to balance the modeling error (or bias) that is manifested by $\hat{\sigma}_{ij}^2$ and that generally decreases with increasing model order and the need to maintain a small number of model parameters to be estimated (or variance), which is embodied in the

$2(i + j)$ term and which increases with increasing model order. In practice $\hat{\sigma}_{ij}^2$ is usually any good estimate of the white noise variance. For an AR or MA process the AIC is defined as

$$\text{AIC}(i) = N \ln \hat{\sigma}_i^2 + 2i \tag{2.132}$$

where i is the assumed AR or MA model order. As an example consider the AR(2) process given by

$$x(n) = 1.34x(n - 1) - 0.9025x(n - 2) + \epsilon(n)$$

To estimate the AR parameters we use the Burg algorithm, so

$$\hat{\sigma}_i^2 = (1 - k_i^2)\hat{\sigma}_{i-1}^2$$

according to Eq. (2.107). Thus, the AIC becomes

$$\text{AIC}(i) = N \ln[(1 - k_i^2)\hat{\sigma}_{i-1}^2] + 2i$$
$$= \text{AIC}(i - 1) + N \ln(1 - k_i^2) + 2$$

Note that the term $N \ln(1 - k_i^2)$ causes the AIC to decrease with increasing i since $k_i^2 < 1$ while the constant term 2 causes it to increase with increasing i. For $N = 100$ the AIC yields the curve in Fig. 2.16. We can see that the minimum is attained at $i = 4$ even though the true model order is $p = 4$. This tendency to overestimate the true model order is characteristic of the AIC.

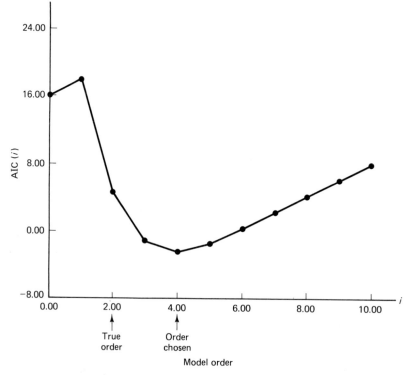

Figure 2.16 Example of Akaike information criterion in model order selection.

2.5.10 Maximum Entropy Spectral Estimation

Maximum entropy spectral estimation (MESE) is based on an explicit extrapolation of a segment of a known autocorrelation function for the samples that are not known [2]. In this way the characteristic smearing of the estimated PSD due to truncation of the autocorrelation function is alleviated. If $\{r_x(0), r_x(1), \ldots, r_x(p)\}$ is known, then the question arises as to how $\{r_x(p + 1), r_x(p + 2), \ldots\}$ should be specified in order to guarantee that the entire autocorrelation function is valid or that its Fourier transform is nonnegative. In general, there are an infinite number of possible extrapolations, all of which yield valid autocorrelation functions. In the MESE it is argued that the extrapolation should be chosen so that the time series characterized by the extrapolated autocorrelation function has maximum entropy. The time series will then be the most random one that has the known autocorrelation samples for its first $p + 1$ lag values. Alternatively, the PSD will be the one with the flattest (whitest) spectrum of all spectra for which the first $p + 1$ autocorrelation samples are equal to the known ones. The resultant spectral estimator is termed the MESE. The rationale for choosing the maximum entropy criterion is that it imposes the fewest constraints on the unknown time series by maximizing its randomness, thereby producing a minimum bias solution.

In particular, if one assumes a Gaussian random process, then the entropy per sample is proportional to

$$\int_{-\pi}^{\pi} \ln P_x(\omega)d\omega \tag{2.133}$$

The MESE is found by maximizing Eq. (2.133) subject to the constraints that the autocorrelation function corresponding to $P_x(\omega)$ has as its first $p + 1$ lags the known samples of the autocorrelation function, or

$$\int_{-\pi}^{\pi} P_x(\omega) \exp(j\omega k)\frac{d\omega}{2\pi} = r_x(k) \qquad \text{for } k = 0, 1, \ldots, p \tag{2.134}$$

The solution that results from applying the technique of Lagrangian multipliers is

$$P_x(\omega) = \frac{\sigma^2}{\left|1 + \sum_{k=1}^{p} a_k \exp(-j\omega k)\right|^2} \tag{2.135}$$

where $\{a_1, a_2, \ldots, a_p, \sigma^2\}$ are found by solving the Yule-Walker equations using the known samples of the autocorrelation function. Hence, with knowledge of $\{r_x(0), r_x(1), \ldots, r_x(p)\}$, the MESE is equivalent to the AR spectral estimator. This equivalence, however, is maintained only for Gaussian random processes and *known* autocorrelation function samples.

2.6 COMPUTER SIMULATION EXAMPLES

In an effort to illustrate the typical characteristics of the various spectral estimators that have been discussed and *not as a means of a quantitative comparison,* a test case was developed. The PSD is shown in Fig. 2.17(a). The process consists of narrowband

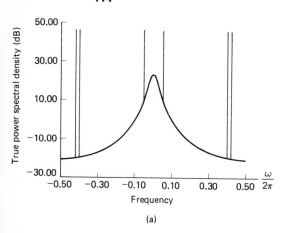

(a)

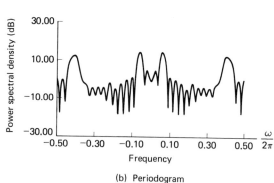

(b) Periodogram

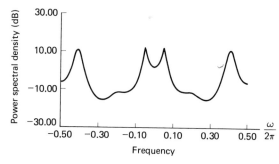

(c) Blackman–Tukey

(d) Minimum variance spectral estimator

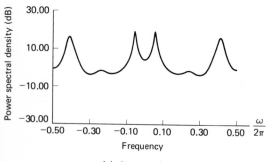

(e) Autocorrelation

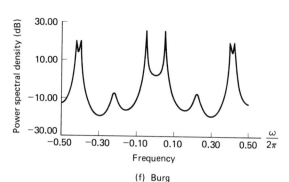

(f) Burg

Figure 2.17 Computer simulation results for test case data: (a) True power spectral density; (b) periodogram; (c) Blackman-Tukey; (d) MVSE; (e) autocorrelation; (f) Burg; (g) covariance; (h) modified covariance; (i) Durbin; (j) MYWE; (k) LSMYWE.

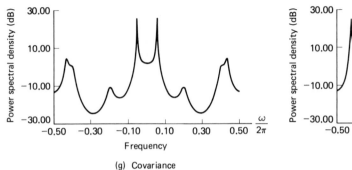

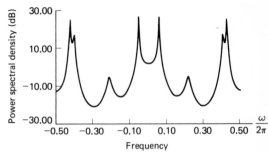

(g) Covariance

(h) Modified covariance

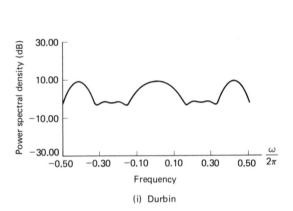

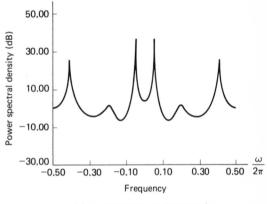

(i) Durbin

(j) Modified Yule–Walker equations

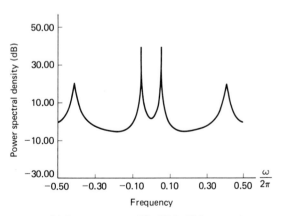

(k) Least-squares modified Yule–Walker equations

Figure 2.17 (*cont.*)

components or sinusoids as well as a broadband component. Specifically, 32 data points have been generated from the process:

$$x(n) = 2 \cos(\omega_1 n) + 2 \cos(\omega_2 n) + 2 \cos(\omega_3 n) + z(n) \qquad (2.136)$$

for $n = 0, 1, \ldots, 31$. In Eq. (2.136) $\omega_1/2\pi = 0.05$, $\omega_2/2\pi = 0.40$, $\omega_3/2\pi = 0.42$, and $z(n)$ is an AR process of order 1, or

$$z(n) = -a_1 z(n-1) + \epsilon(n) \qquad (2.137)$$

The values of a_1 and σ^2 are -0.85 and 0.1, respectively. The PSD of $z(n)$ is given by

$$P_{zz}(\omega) = \frac{\sigma^2}{|1 + a_1 \exp(-j\omega)|^2} \qquad (2.138)$$

The sinusoidal components at $\omega_2/2\pi = 0.40$ and $\omega_3/2\pi = 0.42$ will not be resolved by a Fourier spectral estimator since their separation is less than the resolution limit of $1/N = 0.03$. The component at $\omega_1/2\pi = 0.05$ is well resolved since it is 0.1 cycle/sample apart from its nearest neighbor. The SNR is defined as the power of any sinusoidal component to the broadband noise power. Specifically, it is given by

$$\text{SNR} = 10 \log_{10} \frac{2}{\sigma^2/(1 - a_1^2)} \text{ dB}$$

and equals 7.4 dB.

Each spectral estimation method, other than the periodogram, is constrained to estimate the same number of parameters so that a qualitative comparison is fair. The exact choices for the model orders, lag windows, etc., were not made to yield the best results but only to illustrate the typical characteristics of each estimator. Other choices will yield similar, although not identical, results.

The Fourier spectral estimates are displayed in Fig. 2.17(b) and (c). As expected, the periodogram is unable to resolve the closely spaced sinusoids, which are less than $1/N = 0.03$ cycle/sample apart, and exhibits the usual sidelobe structure. The Blackman-Tukey estimate in Fig. 2.17(c) was based on a biased autocorrelation estimator for lags $k = 0, 1, \ldots, M = 10$ and employed a Bartlett window. Being a smoothed version of the periodogram (see Eq. 2.56), the spectral estimate displays less detail than the periodogram.

The minimum variance spectral estimate based on an autocorrelation matrix of dimension $p \times p = 11 \times 11$ is shown in Fig. 2.17(d). The closely spaced sinusoids are not discernible. A larger-dimension autocorrelation matrix results in a spectral estimate that resolves the closely spaced sinusoids but that also gives rise to many spurious peaks.

The AR spectral estimates based on a model order of $p = 10$ are shown in Fig. 2.17(e)–(h). At most 10 peaks may be present in the spectral estimates. It is observed that the sinusoidal components are resolved by all the estimators except the autocorrelation method, Fig 2.17(e), which has the lowest resolution. All the methods produce a spurious peak at about $\omega/2\pi = 0.2$ although it is less pronounced for the autocorrelation method (Fig. 2.17e). Note that a peak is visible near $\omega/2\pi = 0.2$ in the periodogram, which may be responsible for the spurious peak observed in the AR spectral estimates. In general, the AR methods are able to resolve closely spaced narrowband components but tend to exhibit peaky spectra even for broadband processes.

The MA spectral estimate based on a model order of $q = 10$ and using Durbin's algorithm with a large AR model order of $L = 15$ is shown in Fig. 2.17(i). It is unable to resolve any of the narrowband spectral components. The choice of the large AR model order L did not significantly affect the spectral estimate. It is observed to be nearly identical to the Blackman-Tukey estimate (Fig. 2.17c). This is also expected since the forms of the PSD are identical (see Section 2.5.7) with only the estimates of the autocorrelation function being different.

The ARMA spectral estimates based on model orders of $p = 7$ and $q = 3$ are shown in Fig. 2.17(j) and (k). Neither the modified Yule-Walker equation spectral estimator (part j) nor the LSMYWE spectral estimator (part k) is able to resolve the spectral components centered about $\omega/2\pi = 0.4$. If the AR model order p is increased, it is possible that the peaks may have been resolved. Both estimates are similar in appearance. The LSMYWE spectral estimator (part k) used $M = 15$ but did not appear to be oversensitive to the number of equations $M - q$ used.

2.7 FURTHER TOPICS

The spectral estimation methods discussed in this chapter are only some of the many approaches that have been proposed. Many other approximate maximum likelihood estimators for AR, MA, and ARMA processes are available. Also, the important problem of estimation of the frequencies of sinusoidal signals in white noise has not been addressed. The spectral estimation methods discussed in this chapter have been adapted to the frequency estimation problem although it can be more appropriately termed a parameter estimation problem. Such algorithms as the Pisarenko method, MUSIC method, the iterative filtering method, and the singular value decomposition or principal component AR method have been widely investigated. All these topics as well as numerous references may be found in [14].

2.8 SUMMARY

The principal methods of spectral estimation are the nonparametric approaches of the periodogram, the Blackman-Tukey, the minimum variance spectral estimators, and the parametric approaches based on the time series models. A general comparison of the various approaches is given in Table 2.2. For the nonparametric approaches a tradeoff must always be effected between the bias and the variance of the estimator because of the finite number of autocorrelation lags that can be estimated. The parametric models do not suffer from a truncated autocorrelation function since the model implicitly extrapolates it. Hence, when the model is accurate, spectral estimates with higher resolution and lower variability are obtained. However, when the model is incorrect, then no amount of data will yield a good spectral estimate.

TABLE 2.2 COMPARISON OF SPECTRAL ESTIMATION APPROACHES

Spectral Estimator Type	Assumed Model	PSD Estimator Form	Representable PSDs	Resolution Capability
Fourier	None	$$\frac{1}{N}\left\lvert \sum_{n=0}^{N-1} x(n)e^{-j\omega n} \right\rvert^2$$ $$\sum_{k=-M}^{M} w(k)\hat{r}_x(k)e^{-j\omega k}$$	Broadband only	Low
Minimum variance	None	$$\frac{p}{e^H \hat{\mathbf{R}}_x^{-1} \mathbf{e}}$$	Broadband or narrowband but not both	Medium
Autoregressive	AR	$$\frac{\hat{\sigma}^2}{\left\lvert 1 + \sum_{k=1}^{p} \hat{a}_k e^{-j\omega k} \right\rvert^2}$$	Broadband or narrowband but not both	High
Autoregressive moving average	ARMA	$$\frac{\hat{\sigma}^2 \left\lvert 1 + \sum_{k=1}^{q} \hat{b}_k e^{-j\omega k} \right\rvert^2}{\left\lvert 1 + \sum_{k=1}^{p} \hat{a}_k e^{-j\omega k} \right\rvert^2}$$	Broadband or narrowband or both	High
Moving average	MA	$$\hat{\sigma}^2 \left\lvert 1 + \sum_{k=1}^{q} \hat{b}_k e^{-j\omega k} \right\rvert^2$$	Broadband only	Low

Confidence intervals are available for the Fourier spectral estimators. Unfortunately, the statistics of the minimum variance and parametric spectral estimators are unknown owing to the highly nonlinear relationship between the data and the spectral estimator. Consequently, no such confidence intervals for the parametric spectral estimators are available, making interpretation of the spectral estimate difficult.

In summary, nonparametric approaches are more robust due to the lack of restrictive assumptions, but where appropriate, parametric spectral estimators offer the promise of good spectral estimates even for short data records. The practitioner of spectral estimation will need to intelligently assess the characteristics of the data before applying an appropriate spectral estimation method (or methods).

REFERENCES

1. S. Kay and S. L. Marple, Jr., "Spectrum Analysis: A Modern Perspective," *Proc. IEEE,* Vol. 69, pp. 1380–1419, Nov. 1981.

2. J. P. Burg, "Maximum Entropy Spectral Analysis," Ph.D. Thesis, Dept. Geophysics, Stanford University, Stanford, CA, May 1975.

3. J. Makhoul, "Linear Prediction: A Tutorial Review," *Proc. IEEE*, Vol. 63, pp. 561–580, April 1975.

4. M. Kendall and A. Stuart, *The Advanced Theory of Statistics*, Vol. II, Macmillan, New York, 1979.

5. F. J. Harris, "On the Use of Windows for Harmonic Analysis with the Discrete Fourier Transform," *Proc. IEEE*, Vol. 66, pp. 51–83, Jan. 1978.

6. G. M. Jenkins and D. G. Watts, *Spectral Analysis and Its Applications*, Holden-Day, San Francisco, 1968.

7. J. Capon, "High-Resolution Frequency-Wavenumber Spectrum Analysis," *Proc. IEEE*, Vol. 57, pp. 1408–1418, Aug. 1969.

8. R. T. Lacoss, "Data Adaptive Spectral Analysis Methods," *Geophysics*, Vol. 36, pp. 661–675, Aug. 1971.

9. R. N. McDonough, "Application of the Maximum-Likelihood and the Maximum-Entropy Method to Array Processing," in *Nonlinear Methods of Spectral Analysis*, S. Haykin, Ed., Springer-Verlag, New York, 1983.

10. J. D. Markel and A. H. Gray, Jr., *Linear Prediction of Speech*, Springer-Verlag, New York, 1976.

11. H. Wold, *A Study in the Analysis of Stationary Time Series*, Almqvist and Wiksell, Uppsala, 1954.

12. A. N. Kolmogorov, "Interpolation and Extrapolation von Stationaren Zufalligen Folgen," *Bull. Acad. Sci. U.S.S.R., Ser. Math.*, Vol. 5, pp. 3–14, 1941.

13. S. Kay, "Autoregressive Spectral Analysis of Narrowband Processes in White Noise with Application to Sonar Signals," Ph.D. Thesis, Georgia Institute of Technology, 1980.

14. S. Kay, *Modern Spectral Estimation: Theory and Application*, Prentice-Hall, Englewood Cliffs, NJ, 1987.

15. M. Pagano, "Estimation of Models of Autoregressive Signal Plus White Noise," *Ann. Statistics*, Vol. 2, pp. 99–108, 1974.

16. S. Kay, "More Accurate Autoregressive Parameter and Spectral Estimates for Short Data Records," *Rec. 1st IEEE Workshop on Spectral Analysis*, Hamilton, Ont., Aug. 1981.

17. A. H. Nuttall, "Spectral Analysis of a Univariate Process with Bad Data Points, via Maximum Entropy and Linear Predictive Techniques," Naval Underwater Systems Center, Tech. Rep. 5303, New London, CT, March 26, 1976.

18. T. J. Ulrych and R. W. Clayton, "Time Series Modelling and Maximum Entropy," *Phys. Earth and Planetary Interiors*, Vol. 12, pp. 188–200, Aug. 1976.

19. S. L. Marple, Jr., "A New Autoregressive Spectrum Analysis Algorithm," *IEEE Trans. Acoustics, Speech, and Signal Processing*, Vol. ASSP-28, pp. 441–454, Aug. 1980.

20 J. Durbin, "Efficient Estimation of Parameters in Moving-Average Models," *Biometrika*, Vol. 46, pp. 306–316, 1959.

21. H. Akaike, "Maximum Likelihood Identification of Gaussian Autoregressive Moving Average Models," *Biometrika*, Vol. 60, pp. 255–265, 1973.

22. S. Kay, "Noise Compensation for Autoregressive Spectral Estimates," *IEEE Trans. Acoustics, Speech, and Signal Processing*, Vol. ASSP-28, pp. 292–303, June 1980.

23. H. Akaike, "A New Look at the Statistical Model Identification," *IEEE Trans. Automatic Control*, Vol. AC-19, pp. 716–723, Dec. 1974.

APPENDIX 2.1
DERIVATION OF PERIODOGRAM STATISTICS

The statistics of the periodogram are derived in this appendix for the case of zero-mean white Gaussian noise. It is assumed that the periodogram is evaluated at $\omega_k = k2\pi/N$ for $k = 0, 1, \ldots, N/2$, where N is even. Let

$$A(\omega_k) = \frac{1}{\sqrt{N}} \sum_{n=0}^{N-1} x(n) \cos \omega_k n$$

$$B(\omega_k) = \frac{1}{\sqrt{N}} \sum_{n=0}^{N-1} x(n) \sin \omega_k n \tag{2.1.1}$$

Then

$$\hat{P}_{\text{PER}}(\omega_k) = A^2(\omega_k) + B^2(\omega_k) \tag{2.1.2}$$

Note that $B(\omega_0)$ and $B(\omega_{N/2})$ are identically zero. Since $x(n)$ is a Gaussian random process, $A(\omega_k)$ and $B(\omega_k)$ are jointly Gaussian. Furthermore, they are uncorrelated and hence independent, as we will now show.

$$E[A(\omega_k)] = \frac{1}{\sqrt{N}} \sum_{n=0}^{N-1} E[x(n)] \cos \omega_k n = 0$$

$$E[B(\omega_k)] = \frac{1}{\sqrt{N}} \sum_{n=0}^{N-1} E[x(n)] \sin \omega_k n = 0$$

$$E[A(\omega_k)B(\omega_k)] = \frac{1}{N} \sum_{n=0}^{N-1} \sum_{m=0}^{N-1} E[x(n)x(m)] \cos \omega_k n \sin \omega_k m$$

$$= \frac{1}{N} \sum_{n=0}^{N-1} \sum_{m=0}^{N-1} \sigma_x^2 \delta(n - m) \cos \omega_k n \sin \omega_k m$$

$$= \frac{\sigma_x^2}{N} \sum_{n=0}^{N-1} \sin \omega_k n \cos \omega_k n$$

$$= \frac{\sigma_x^2}{2N} \sum_{n=0}^{N-1} \sin 2\omega_k n$$

$$= \frac{\sigma_x^2}{2N} \mathscr{I}m\left\{ \sum_{n=0}^{N-1} \exp(j2\omega_k n) \right\}$$

But

$$\sum_{n=0}^{N-1} \exp(j2\omega_k n) = \frac{\sin N\omega_k}{\sin \omega_k} \exp[j(N - 1)\omega_k]$$

Hence,

$$E[A(\omega_k)B(\omega_k)] = \frac{\sigma_x^2}{2N} \sin(N-1)\omega_k \frac{\sin N\omega_k}{\sin \omega_k}$$

$$= \frac{\sigma_x^2}{2N} \sin[(N-1)2\pi k/N]\frac{\sin 2\pi k}{\sin 2\pi k/N}$$

$$= 0 \qquad \text{for } k = 1, \ldots, N/2 - 1$$

which proves that $A(\omega_k)$ and $B(\omega_k)$ are independent. Next we compute the variances of $A(\omega_k)$ and $B(\omega_k)$.

$$\text{VAR}[A(\omega_k)] = E[A^2(\omega_k)]$$

$$= \frac{1}{N}\sum_{n=0}^{N-1}\sum_{m=0}^{N-1} E[x(n)x(m)] \cos \omega_k n \cos \omega_k m$$

$$= \frac{\sigma_x^2}{N}\sum_{n=0}^{N-1} \cos^2 \omega_k n$$

$$= \begin{cases} \dfrac{\sigma_x^2}{2} & \text{for } k = 1, 2, \ldots, N/2 - 1 \\ \sigma_x^2 & \text{for } k = 0, N/2 \end{cases}$$

since $\sum_{n=0}^{N-1} \cos^2 \omega_k n = N/2$ for $k = 1, 2, \ldots, N/2 - 1$. Similarly,

$$\text{VAR}[B(\omega_k)] = \sigma_x^2/2, \qquad k = 1, 2, \ldots, N/2 - 1$$

In summary,

$$A(\omega_k) \sim N(0, \sigma_x^2/2), \qquad k = 1, 2, \ldots, N/2 - 1$$

$$\sim N(0, \sigma_x^2), \qquad k = 0, N/2 \qquad\qquad (2.1.3)$$

$$B(\omega_k) \sim N(0, \sigma_x^2/2), \qquad k = 1, 2, \ldots, N/2 - 1$$

$$B(\omega_0) = B(\omega_{N/2}) = 0$$

and $A(\omega_k)$ and $B(\omega_k)$ are independent. Hence,

$$\frac{\hat{P}_{\text{PER}}(\omega_k)}{\sigma_x^2/2} = \left[\frac{A(\omega_k)}{\sqrt{\sigma_x^2/2}}\right]^2 + \left[\frac{B(\omega_k)}{\sqrt{\sigma_x^2/2}}\right]^2, \qquad k = 1, 2, \ldots, N/2 - 1$$

$$\frac{\hat{P}_{\text{PER}}(\omega_k)}{\sigma_x^2} = \left[\frac{A(\omega_k)}{\sigma_x}\right]^2, \qquad k = 0, N/2 \qquad\qquad (2.1.4)$$

Since each random variable in brackets is $N(0, 1)$, it follows from Section 2.3.3 that

$$\frac{2\hat{P}_{\text{PER}}(\omega_k)}{\sigma_x^2} \sim \chi_2^2, \qquad k = 1, 2, \ldots, N/2 - 1$$

$$\frac{\hat{P}_{\text{PER}}(\omega_k)}{\sigma_x^2} \sim \chi_1^2, \qquad k = 0, N/2$$

Also, $P_x(\omega) = \sigma_x^2$, so

$$\frac{2\hat{P}_{\mathrm{PER}}(\omega_k)}{P_x(\omega)} \sim \chi_2^2, \qquad k = 1, 2, \ldots, N/2 - 1$$

$$\frac{\hat{P}_{\mathrm{PER}}(\omega_k)}{P_x(\omega)} \sim \chi_1^2 \qquad k = 0, N/2$$

ACKNOWLEDGMENTS

The author wishes to thank R. Gross, P. Lin, J. Mulhearn, and D. Sengupta of the University of Rhode Island for providing many of the computer-generated figures used throughout this chapter.

3

Multirate Processing of Digital Signals

Ronald E. Crochiere
Lawrence R. Rabiner
AT&T Bell Laboratories

3.0 INTRODUCTION

As more and more signals in the real world are represented, stored, and transmitted in digital formats, the importance of being able to process a signal digitally from its inception to its final destination grows. The general theory of digital signal processing has emerged in the past two decades and has grown to prominence in the engineering community, as evidenced by the publication of several key texts and reprint collections [1–9]. Almost all of the theory presented in those references deals with processing signals at a fixed sampling rate. However, in the past few years, there has been an increasing need for a deeper understanding of how to process digital signals in systems that require more than one sampling rate. An entire subfield of digital signal processing—multirate signal processing—has developed to meet this need [10].

This chapter presents the main ideas and concepts of multirate digital signal processing, with particular emphasis on digital techniques for changing the sampling rate of a signal. We begin our discussion with a thorough review of the Nyquist sampling theorem and its interpretations in terms of modulated signals. We then show how a continuous-time signal can be reconstructed from its digital samples. This discussion leads naturally to the topic of sampling rate conversion, in which we show that such conversion is essentially a digital "resampling" of the signal at the required rate. We show that we can readily implement such sampling rate conversion systems by paying careful attention to the structure for computation, to the digital filters that perform the antialiasing or anti-imaging functions, and to the use of cascade architectures, when appropriate. Finally, we conclude by discussing several applications of

Ronald E. Crochiere and Lawrence R. Rabiner are with AT&T Bell Laboratories, Murray Hill, NJ 07974.

sampling rate conversion to illustrate the power of the multirate signal processing techniques discussed in this chapter.

3.1 SAMPLING AND SIGNAL RECONSTRUCTION

3.1.1 The Sampling Theorem

Assume that we have a continuous-time signal $x_c(t)$, $-\infty < t < \infty$, for which we would like to obtain a discrete-time representation $x(n)$, $-\infty < n < \infty$. The signals $x(n)$ and $x_c(t)$ are related by sampling, that is,

$$x(n) = x_c(t)\big|_{t=q(n)} \tag{3.1}$$

where $q(n)$ is the "sampler" applied to $x_c(t)$. The most widely used sampler is the *uniform sampler,* in which

$$t = q(n) = nT \tag{3.2}$$

where T is the sampling period, and the quantity

$$F_s = 1/T \tag{3.3}$$

is the sampling rate. Figure 3.1 gives an illustration of uniform sampling. Part (a) of the figure depicts a typical waveform of the continuous-time signal $x_c(t)$. Parts (b) and (c) show the resulting digital signals (called $x(n)$) for two different choices of sampling period T. The most important consideration in sampling is determination of the proper

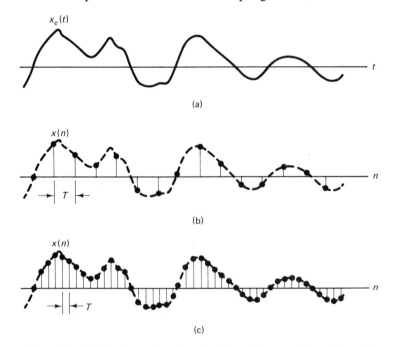

Figure 3.1   (a) Continuous-time signal and (b, c) two sampled versions of it.

value of T. We will discuss this point later in this section. First, however, we give a few simple interpretations of sampling.

The simplest interpretation of sampling is as a modulation process, as shown in Fig. 3.2(a). The continuous-time signal $x_c(t)$ (Fig. 3.2b) is first multiplied (modulated) by the sampling pulse train signal $s_c(t)$ (Fig. 3.2c), where $s_c(t)$ has the form

$$s_c(t) = \sum_{\ell=-\infty}^{\infty} \delta_c(t - \ell T) \tag{3.4}$$

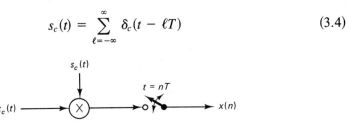

(a)

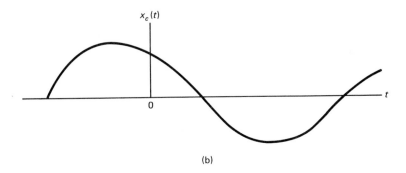

(b)

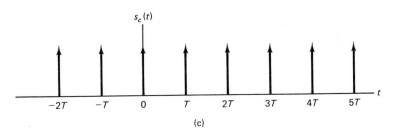

(c)

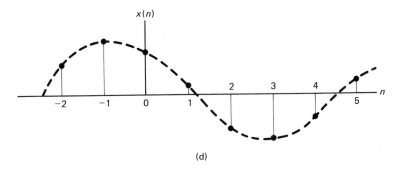

(d)

Figure 3.2 Periodic sampling of $x_c(t)$ via modulation to obtain $x(n)$.

and where $\delta_c(t)$ denotes an ideal unit impulse function. The resulting pulselike signal is measured for values of $t = nT$ via the switch, which closes for a brief instant once every T seconds and then remains open the rest of the time. The resulting digital signal $x(n)$ is shown in Fig. 3.2(d) and has the form

$$x(n) = \lim_{\epsilon \to 0} \int_{t=nT-\epsilon}^{nT+\epsilon} x_c(t)s_c(t) \, dt = x_c(nT) \tag{3.5}$$

The sampling network of Fig. 3.2(a) is often referred to as an analog-to-digital (A/D) converter since an analog signal goes in and a digital signal comes out.

An alternative interpretation of sampling can be obtained by examining the modulation system of Fig. 3.2 in the frequency domain. To do this we must rely on Fourier transform theory and some elementary properties of such transforms. We assume that $x_c(t)$ has a Fourier transform

$$X_c(\Omega) = \int_{-\infty}^{\infty} x_c(t)e^{-j\Omega t} \, dt \tag{3.6}$$

and $s_c(t)$, the sampling pulse train, has a Fourier transform

$$S_c(\Omega) = \int_{-\infty}^{\infty} s(t) \, e^{-j\Omega t} \, dt \tag{3.7}$$

Since $s_c(t)$ is a periodic impulse train (Eq. 3.4), $S_c(\Omega)$ has the form

$$S_c(\Omega) = \frac{2\pi}{T} \sum_{\ell=-\infty}^{\infty} \delta_c\left(\Omega - \frac{2\pi\ell}{T}\right) \tag{3.8}$$

i.e., a uniformly spaced impulse train in frequency with period $\Omega_F = 2\pi/T = 2\pi F_s$.

Since multiplication in the time domain is equivalent to convolution in the frequency domain,

$$X_c(\Omega) * S_c(\Omega) = \int_{-\infty}^{\infty} [x_c(t)s(t)]e^{-j\Omega t} \, dt \tag{3.9}$$

where $*$ denotes linear convolution.

Figure 3.3 shows typical plots of $X_c(\Omega)$ (part a), $S_c(\Omega)$ (part b), and the convolution $X_c(\Omega) * S_c(\Omega)$ (part c), where it is assumed that $X_c(\Omega)$ is bandlimited and its highest frequency component, $2\pi F_c$, is less than one-half of the sampling frequency $\Omega_F = 2\pi F_s$. From Fig. 3.3 we see that the process of pulse amplitude modulation (PAM) periodically repeats the spectrum $X_c(\Omega)$ at harmonics of the sampling frequency, due to the convolution of $X_c(\Omega)$ and $S_c(\Omega)$.

Because of the direct correspondence between the sequence $x(n)$ and the pulse amplitude modulated signal $x_c(t)s_c(t)$ (as seen in Eq. 3.5), it is clear that the information content and the spectral interpretations of the two signals are synonymous. This correspondence can be shown more formally by considering the (discrete-time) Fourier transform of the sequence $x(n)$, defined as

$$X(\omega) = \sum_{n=-\infty}^{\infty} x(n)e^{-j\omega n} \tag{3.10}$$

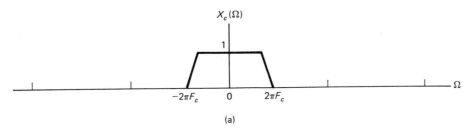

(a)

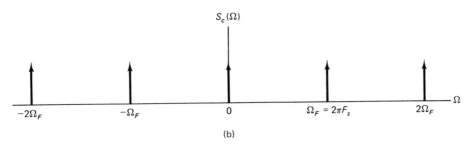

(b)

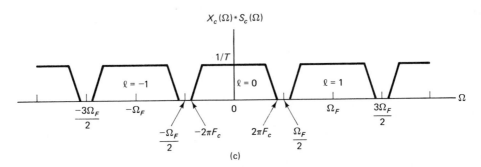

(c)

Figure 3.3 Spectra of signals obtained from periodic sampling via modulation.

where $\omega = \Omega T = \Omega/F_s$. Since $x_c(t)$ and $x(n)$ are related by Eq. (3.5), a relation can be derived between $X_c(\Omega)$ and $X(\omega)$ with the aid of Eqs. (3.6) and (3.10) as follows. The inverse Fourier transform of $X_c(\Omega)$ gives $x_c(t)$ as

$$x_c(t) = \frac{1}{2\pi} \int_{-\infty}^{\infty} X_c(\Omega) e^{j\Omega t} \, d\Omega \tag{3.11}$$

Evaluating Eq. (3.11) for $t = nT$, we get

$$x(n) = x_c(nT) = \frac{1}{2\pi} \int_{-\infty}^{\infty} X_c(\Omega) e^{j\Omega nT} \, d\Omega \tag{3.12}$$

The sequence $x(n)$ may also be obtained as the inverse (discrete-time) Fourier transform of $X(\omega)$:

$$x(n) = \frac{1}{2\pi} \int_{-\pi}^{\pi} X(\omega) e^{j\omega n} \, d\omega \tag{3.13}$$

Thus, combining Eqs. (3.12) and (3.13), we get

$$\frac{1}{2\pi} \int_{-\pi}^{\pi} X(\omega)e^{j\omega n}\, d\omega = \frac{1}{2\pi} \int_{-\infty}^{\infty} X_c(\Omega)e^{j\Omega nT}\, d\Omega \qquad (3.14)$$

By expressing the right-hand integral as a sum of integrals, each of width $2\pi/T$, we can put Eq. (3.14) in the form

$$\frac{1}{2\pi} \int_{-\pi}^{\pi} [X(\omega)]e^{j\omega n}\, d\omega = \frac{1}{2\pi} \int_{-\pi}^{\pi} \left[\frac{1}{T} \sum_{\ell=-\infty}^{\infty} X_c(\Omega + \ell\Omega_F)\right] e^{j\omega n}\, d\omega \qquad (3.15)$$

By equating terms within the brackets of Eq. (3.15), we get

$$X(\omega) = \frac{1}{T} \sum_{\ell=-\infty}^{\infty} X_c(\Omega + \ell\Omega_F) = \frac{1}{T} \sum_{\ell=-\infty}^{\infty} X_c\left[\frac{1}{T}(\omega + 2\pi\ell)\right] \qquad (3.16)$$

Equation (3.16) shows that the Fourier transform of the digital signal is the sum of frequency-shifted and scaled versions of the Fourier transform of the continuous signal.

We can now see quite clearly the effect of different choices of $\Omega_F = 2\pi/T$ on the resulting Fourier transform of the digital signal. These possibilities are illustrated in Fig. 3.4. Figure 3.4(a) shows the Fourier transform of the continuous signal and Fig. 3.4(b) shows the resulting digital Fourier transform when $\Omega_F > 4\pi F_c$. The individual terms for $\ell = 0$ and $\ell = \pm 1$ of Eq. (3.16) are shown in this figure. Figure 3.4(c) shows the resulting discrete-time Fourier transform when $\Omega_F = 4\pi F_c$. In this case the individual terms of Eq. (3.16) come right up to each other in frequency. This case is referred to as *critical sampling* of the signal. Finally, Fig. 3.4(d) shows the resulting discrete-time Fourier transform when $\Omega_F < 4\pi F_c$. In this case the individual terms in Eq. (3.16) overlap in frequency, and the resulting digital frequency response, in general, bears no simple, direct relationship to the continuous frequency response of Fig. 3.4(a). In this case we say that the digital signal is an *aliased* representation of the continuous signal.

The implications of the three cases of sampling discussed above are summarized in a simple and straightforward manner in the Nyquist sampling theorem:

If a continuous-time signal $x_c(t)$ has a bandlimited Fourier transform $X_c(\Omega)$ that satisfies the condition $|X_c(\Omega)| = 0$ for $\Omega \geq 2\pi F_c$, then $x_c(t)$ can be *uniquely* reconstructed, without error, from equally spaced samples $x(n) = x_c(nT)$, $-\infty < n < \infty$, if $F_s \geq 2F_c$, where $F_s = 1/T$.

The sampling theorem tells us that the digital signals of Fig. 3.4(b) and (c) are suitable for exact reconstruction of the continuous-time signal from which they were obtained; however, the aliased digital signal of Fig. 3.4(d) cannot be used to reconstruct the continuous signal from which it was obtained.

There are several points worth noting about the sampling theorem and its practical implementation. The first point concerns the bandlimited requirement on $x_c(t)$. In practice there are no truly bandlimited signals for which $|X_c(\Omega)| = 0$ for $\Omega \geq 2\pi F_c$. Thus we substitute the more reasonable bandlimited criterion $|X_c(\Omega)| \leq \delta$ for $\Omega \geq 2\pi F_c$, where δ is some suitably small constant. For example, if the largest

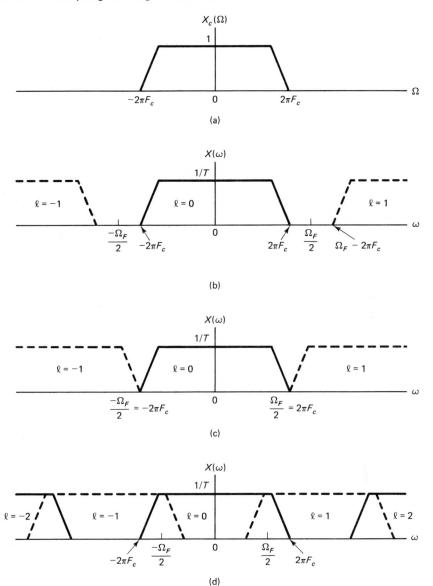

Figure 3.4 Spectral interpretations of the sampling theorem.

magnitude of $X_c(\Omega)$ is on the order of 1.0, values of δ from 0.0001 (80 dB down) to 0.01 (40 dB down) are generally used.

A second point of note in the practical use of the sampling theorem is that, in general, we are using a sampling rate F_s as close to the critical rate $2F_c$ as possible. In this manner the computation of the digital signal processing is kept as small as possible, consistent with the sampling requirements.

A final point about the sampling theorem concerns cases that violate the rule to keep the sampling rate as low as possible. For digital processing systems with nonlinearities it is often advisable to use sampling rates much greater than the minimum possible. The reasons for this are related to the nonlinearities that cause the frequency spectrum of the signal to be smeared across all frequencies. In a digital signal the frequency range is inherently limited to the band $|f| \leq F_c$. Thus the higher the sampling rate, the broader the range of frequencies in which to spread the energy of the signal due to the nonlinearity. An example of this type of effect is given in [14], which discusses a digital simulation of an ordinary telephone line. For this system a sampling rate of five times the minimum rate was used.

3.1.2 Signal Reconstruction from Samples

The sampling theorem gives us the set of necessary and sufficient conditions for being able to reconstruct a continuous-time signal from its samples. To actually do the reconstruction we have to use a system, called a digital-to-analog (D/A) converter, to transform $x(n)$ back to the sampled signal $x_c(t)s_c(t)$ and then pass the sampled signal through an ideal lowpass filter $h_I(t)$ to recover $x_c(t)$. This sequence of operations is depicted in Fig. 3.5.

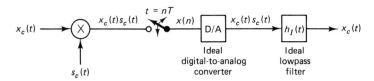

Figure 3.5 Sampling and reconstruction of a continuous-time signal.

To implement this process, an ideal digital-to-analog converter is required to get $x_c(t)s_c(t)$ from $x(n)$. Assuming that we do not worry about the reliability of such an ideal converter, the reconstruction formula for $x_c(t)$ is

$$x_c(t) = \int_{\tau=-\infty}^{\infty} x_c(\tau)s_c(\tau)h_I(t - \tau) \, d\tau \tag{3.17}$$

and applying Eqs. (3.4) and (3.5) gives

$$x_c(t) = \sum_{n=-\infty}^{\infty} x(n)h_I(t - nT) \tag{3.18}$$

If we use the reconstruction method, the relationship between the reconstructed signal $x_c(t)$ and the set of samples $x(n)$ is given by the convolutional formula

$$x_c(t) = \sum_{n=-\infty}^{\infty} x(n)h_I(t - nT) \tag{3.19}$$

The ideal lowpass filter $h_I(t)$ has the frequency-domain characteristics

$$|H_I(\Omega)| = \begin{cases} 1, & 0 \leq |\Omega| \leq \pi F_s = \pi/T \\ 0, & |\Omega| > \pi F_s = \pi/T \end{cases} \tag{3.20}$$

giving the ideal impulse response

$$h_I(t) = \frac{\sin(\pi t/T)}{(\pi t/T)}, \qquad -\infty < t < \infty \tag{3.21}$$

Combining Eqs. (3.19)–(3.21) gives the ideal reconstruction formula

$$x_c(t) = \sum_{n=-\infty}^{\infty} x(n) \left[\frac{\sin[\pi(t-nT)]/T}{\pi(t-nT)/T} \right] \tag{3.22}$$

Figure 3.6 illustrates the application of the reconstruction formula of Eq. (3.22). At each instant n_0 the digital sample $x(n_0)$ weights the ideal lowpass filter response. All such terms are then summed to give $x_c(t)$.

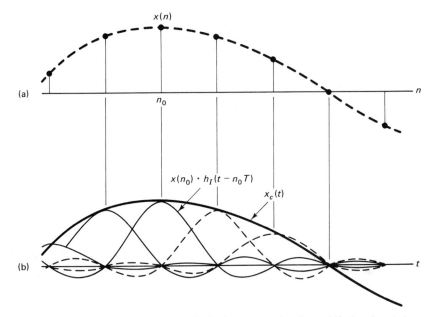

Figure 3.6 Illustration of a bandlimited reconstruction from shifted and scaled lowpass filter responses.

Unfortunately, the reconstruction formula of Eq. (3.22) cannot be implemented because the ideal lowpass filter is unrealizable. As such, in practical systems, the ideal lowpass filter is replaced with a real desampling filter giving a reasonable approximation to the desired response.

3.2 SAMPLING RATE CONVERSION

Once we have a signal in digital form, it is convenient to be able to process it digitally until it reaches its final destination and must ultimately be converted back to analog (continuous-time) form. When the sampling rate required in the digital processing remains constant, there is no problem in handling the digital signal. However, in many practical systems, the sampling rates at different points in the system are different. Hence we must be able to change freely the sampling rate of a digital signal for maximum signal processing efficiency.

There is one very simple and straightforward approach to changing the sampling rate of a digital signal. This approach, called the analog approach, merely reconstructs the continuous-time signal from the original set of samples and then resamples the signal at the new rate (assuming no additional antialiasing lowpass filtering is required). Thus, using Eq. (3.19), if the original sampling period is T and if the new sampling period is T', we form

$$y(m) = x_c(t)\big|_{t=mT'} = \sum_{n=-\infty}^{\infty} x(n)h_I(mT' - nT) \tag{3.23}$$

If we restrict ourselves to practical lowpass filters (with a finite-duration impulse response), then we can rewrite Eq. (3.23) to express $y(m)$ as the finite summation

$$y(m) = \sum_{n=N_1}^{N_2} x(n)h_I(mT' - nT) \tag{3.24}$$

In theory the analog approach to sampling rate conversion works well. In practice it suffers from one major problem, namely, that the ideal operations required to reconstruct the continuous-time signal from the original samples and to resample the signal at the new rate cannot be implemented exactly. When we resort to using practical D/A and A/D converters, we find that the resulting signal has additive noise (due to resampling), signal-dependent distortions (due to nonideal samplers), and frequency distortions (due to nonideal frequency responses of the filters). Although these real-world distortions can be minimized, they cannot be eliminated using the analog approach.

An alternative is the so-called direct digital approach to sampling rate conversion. The key to this approach is the realization that the processing of Eq. (3.24) is all that is required to change the sampling rate of the signal. Hence, if the computations of Eq. (3.24) could be directly implemented, that is, without going through D/A or A/D conversion, in theory a distortion-free sampling rate conversion could be achieved. We now show how this is achieved in practice. Before presenting the general equations for the digital approach to sampling rate conversion, we first discuss three special cases that essentially provide a complete understanding of the nature of the digital approach:

1. Decimation (sampling rate decrease) by an integer factor M
2. Interpolation (sampling rate increase) by an integer factor L
3. Sampling rate conversion by a ratio of integer factors M/L

3.2.1 Decimation by an Integer Factor *M*

Consider the process of reducing the sampling rate (decimation) of $x(n)$ by an integer factor M. If we denote the sampling period and rate of $x(n)$ by T and $F_s = 1/T$ and the sampling period and rate of the decimated signal $y(n)$ by T' and $F'_s = 1/T'$, then we have

$$\frac{T'}{T} = M = \frac{F_s}{F'_s} \tag{3.25}$$

Assume that $x(n)$ represents a fullband signal, i.e., its spectrum is nonzero for all frequencies in the range $|f| \le F_s/2$, except possibly at an isolated set of points. Based on the analog interpretation of sampling, we see that to lower the sampling rate and to avoid aliasing at this lower rate, it is necessary to filter the signal $x(n)$ with a *digital* lowpass filter that approximates the ideal characteristic

$$H_I(\omega) = \begin{cases} 1, & |\omega| < \dfrac{\pi}{M} \\ 0, & \text{otherwise} \end{cases} \tag{3.26}$$

The sampling rate reduction is then achieved by forming the sequence $y(m)$ by saving only every Mth sample of the filtered output. This process is illustrated in Fig. 3.7(a). If we denote the actual lowpass filter impulse response as $h(n)$, then we have

$$w(n) = \sum_{k=-\infty}^{\infty} h(k)x(n - k) \tag{3.27}$$

where $w(n)$ is the filtered output as seen in Fig. 3.7(a). The final output, $y(m)$, is then obtained as

$$y(m) = w(Mm) \tag{3.28}$$

as denoted by the operation of the second box in Fig. 3.7(a). The use of a down arrow followed by an integer is called a sampling rate compressor and corresponds to the resampling operation of Eq. (3.28).

Figure 3.7(b) shows typical spectra (magnitude of the discrete Fourier transforms) of the signals $x(n)$, $h(n)$, $w(n)$, and $y(m)$ for an M-to-1 reduction in sampling rate. Note that the frequencies $\omega = 2\pi fT$ and $\omega' = 2\pi f'T'$ are normalized with respect to the sampling frequencies F_s and F'_s.

When Eqs. (3.27) and (3.28) are combined, the relation between $y(m)$ and $x(n)$ is of the form

$$y(m) = \sum_{k=-\infty}^{\infty} h(k)x(Mm - k) \tag{3.29}$$

or, by a change of variables, it becomes

$$y(m) = \sum_{n=-\infty}^{\infty} h(Mm - n)x(n) \tag{3.30}$$

It is valuable to derive the relationship between the z-transforms of $y(m)$ and $x(n)$ so as to be able to study the nature of the errors in $y(m)$ caused by a practical (nonideal) lowpass filter. To obtain this relationship, we define the signal

$$w'(n) = \begin{cases} w(n), & n = 0, \pm M, \pm 2M, \ldots \\ 0, & \text{otherwise} \end{cases} \tag{3.31}$$

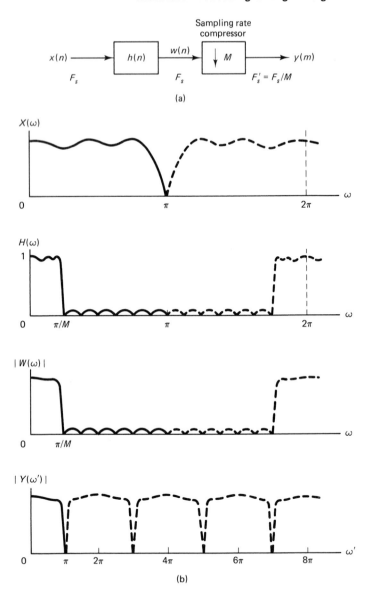

Figure 3.7   Block diagram and typical spectra for decimation by an integer factor M.

i.e., $w'(n) = w(n)$ at the sampling instants of $y(m)$ but is zero otherwise. A convenient and useful representation of $w'(n)$ is then

$$w'(n) = w(n)\left[\frac{1}{M}\sum_{\ell=0}^{M-1} e^{j(2\pi\ell n/M)}\right], \quad -\infty < n < \infty \tag{3.32}$$

where the term in brackets corresponds to a discrete Fourier series representation of a periodic impulse train with a period of M samples. (This pulse train is the digital equivalent of the analog pulse amplitude modulated sampling function of Section 3.1.) Thus we have

$$y(m) = w'(Mm) = w(Mm) \tag{3.33}$$

We now write the z-transform of $y(m)$ as

$$Y(z) = \sum_{m=-\infty}^{\infty} y(m)z^{-m} = \sum_{m=-\infty}^{\infty} w'(Mm)z^{-m} \tag{3.34}$$

and since $w'(n)$ is zero, except at integer multiples of M, Eq. (3.34) becomes (after some simple manipulations)

$$Y(z) = \sum_{m=-\infty}^{\infty} w'(m)z^{-m/M} = \frac{1}{M} \sum_{\ell=0}^{M-1} W(e^{-j2\pi\ell/M} z^{\ell/M}) \tag{3.35}$$

Since

$$W(z) = H(z)X(z) \tag{3.36}$$

we can express $Y(z)$ as

$$Y(z) = \frac{1}{M} \sum_{\ell=0}^{M-1} H(e^{-j2\pi\ell/M} z^{1/M})X(e^{-j2\pi\ell/M} z^{1/M}) \tag{3.37}$$

Evaluating $Y(z)$ on the unit circle, $z = e^{j\omega'}$, leads to

$$Y(\omega') = \frac{1}{M} \sum_{\ell=0}^{M-1} H((\omega' - 2\pi\ell)/M)X((\omega' - 2\pi\ell)/M) \tag{3.38}$$

Equation (3.38) expresses the Fourier transform of the output signal $y(m)$ in terms of the transform of the aliased components of the filtered input signal $x(n)$. By writing out the individual terms, we get

$$Y(\omega') = \frac{1}{M}[H(\omega'/M)X(\omega'/M) + H((\omega' - 2\pi)/M)X((\omega' - 2\pi)/M) + \cdots] \tag{3.39}$$

The purpose of the lowpass filter, $H(\omega)$, is to filter $x(n)$ sufficiently so that its spectral components above the frequency $\omega = \pi/M$ are negligible (see Fig. 3.7). Thus it serves as an antialiasing filter. If the lowpass filter sufficiently removes all energy of $x(n)$ above the frequency $\omega = \pi/M$ (i.e., terms of Eq. 3.38 with $\ell \neq 0$) and if the filter $H(\omega)$ closely approximates the ideal response of Eq. (3.26), then Eq. (3.39) becomes

$$Y(\omega) = \frac{1}{M}X(\omega'/M), \qquad |\omega'| < \pi \tag{3.40}$$

i.e., the desired resampled signal.

One of the most interesting properties of the decimation system can be seen by comparing Eq. (3.38) and Eq. (3.16). Both equations express the frequency-domain content of the digital signal in terms of sums of shifted components of the frequency contents of the input signal to the system. Hence the decimation system is truly a digital resampling of a signal with potential aliasing components. Much as with the sampling theorem, we must keep the level of the digital aliasing components low by applying a properly designed digital antialiasing lowpass filter.

3.2.2 Interpolation by an Integer Factor L

Consider the process of increasing the sampling rate (interpolation) of $x(n)$ by an integer factor L. If we again denote the sampling period and rate of $x(n)$ by T and $F_s = 1/T$, and the sampling period and rate of the interpolated signal $y(n)$ by T' and $F'_s = 1/T'$, then we have

$$\frac{T'}{T} = \frac{1}{L} = \frac{F_s}{F'_s} \tag{3.41}$$

This process of increasing the sampling rate (interpolation) of a signal $x(n)$ by the integer factor L implies that we must interpolate $L - 1$ new sample values between each pair of sample values of $x(n)$. The process is similar to that of digital-to-analog conversion in which all continuous-time values of a signal $x_c(t)$ must be interpolated from its sequence $x(n)$. (In this case only specific values must be determined.)

Figure 3.8 illustrates an example of interpolation by a factor $L = 3$. The input signal $x(n)$ is "filled in" with $L - 1$ zero-valued samples between each pair of samples of $x(n)$, giving the signal

$$w(m) = \begin{cases} x(m/L), & m = 0, \pm L, \pm 2L, \ldots \\ 0, & \text{otherwise} \end{cases} \tag{3.42}$$

This process is the digital equivalent to the digital-to-PAM conversion process discussed in Section 3.1.2, and it is illustrated by the first box in the block diagram of Fig. 3.8(a). As with the resampling operation, the block diagram symbol of an up arrow with an integer corresponds to increasing the sampling rate as given by Eq. (3.42), and it will be referred to as a *sampling rate expander*. The resulting signal $w(n)$ has the z-transform

$$W(z) = \sum_{m=-\infty}^{\infty} w(m)z^{-m} \tag{3.43a}$$

$$= \sum_{m=-\infty}^{\infty} x(m)z^{-mL} \tag{3.43b}$$

$$= X(z^L) \tag{3.43c}$$

Evaluating $W(z)$ on the unit circle, $z = e^{j\omega'}$, gives the result

$$W(\omega') = X(\omega'L) \tag{3.44}$$

which is the Fourier transform of the signal $w(m)$ expressed in terms of the spectrum of the input signal $x(n)$ (where $\omega' = 2\pi f T'$ and $\omega = 2\pi f T$).

As illustrated by the spectral interpretation in Fig. 3.8(c), the spectrum of $w(m)$ contains not only the baseband frequencies of interest (i.e., $-\pi/L$ to π/L) but also *images* of the baseband centered at harmonics of the original sampling frequency $\pm 2\pi/L, \pm 4\pi/L, \ldots$. To recover the baseband signal of interest and eliminate the

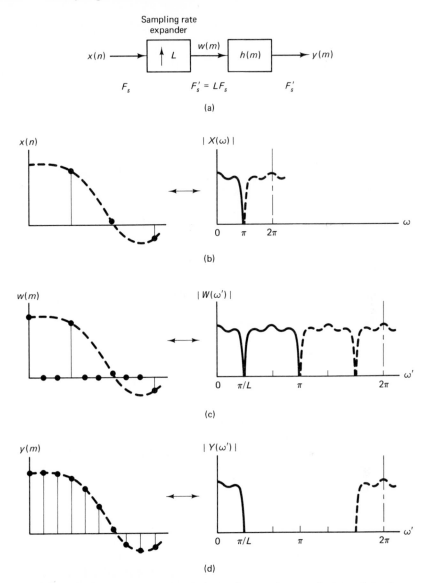

Figure 3.8 Block diagram and typical waveforms and spectra for interpolation by an integer factor L.

unwanted image components, it is necessary to filter the signal $w(m)$ with a digital lowpass (anti-imaging) filter that approximates the ideal characteristic

$$H_I(\omega') = \begin{cases} G, & |\omega'| \leq \dfrac{2\pi F_s T'}{2} = \dfrac{\pi}{L} \\ 0, & \text{otherwise} \end{cases} \qquad (3.45)$$

It will be shown that to ensure that the amplitude of $y(m)$ is correct, the gain of the filter, G, must be L in the passband.

Letting $H(\omega')$ denote the frequency response of an actual filter that approximates the characteristic in Eq. (3.45), we see that

$$Y(\omega') = H(\omega')X(\omega'L) \qquad (3.46)$$

and within the approximation of Eq. (3.45),

$$Y(\omega') \approx \begin{cases} GX(\omega'L), & |\omega'| \leq \dfrac{\pi}{L} \\ 0, & \text{otherwise} \end{cases} \qquad (3.47)$$

It is easy to see why we need a gain of G in $H_I(\omega')$, whereas for the decimation filter a gain of 1 is adequate. For the "ideal" sampling system (with no aliasing error) we have seen from Eq. (3.16) that we desire

$$X(\omega) = \frac{1}{T}X_c\left(\frac{\omega}{T}\right)$$

For the "ideal" decimator, we have shown in Eq. (3.40) that

$$Y(\omega') = \frac{1}{M}X(\omega'/M)$$

$$= \frac{1}{MT}X_c\left(\frac{\omega'}{MT}\right)$$

$$= \frac{1}{T'}X_c\left(\frac{\omega'}{T'}\right)$$

Thus the necessary scaling is taken care of directly in the decimation process, and a filter gain of 1 is suitable. For the "ideal" interpolator, however, we have

$$Y(\omega') = GX(\omega'L)$$

$$= \frac{G}{T}X_c\left(\frac{\omega'L}{T}\right)$$

$$= \frac{G}{L}\left(\frac{1}{T'}\right)X_c\left(\frac{\omega'}{T'}\right)$$

Clearly, a gain $G = L$ is required to meet the conditions of Eq. (3.16).

If $h(m)$ denotes the unit sample response of $H(\omega')$, then from Fig. 3.8, $y(m)$ can be expressed as

$$y(m) = \sum_{k=-\infty}^{\infty} h(m - k)w(k) \qquad (3.48)$$

Combining Eqs. (3.42) and (3.48) leads to the time-domain input-to-output relation of the interpolator

$$y(m) = \sum_{k=-\infty}^{\infty} h(m - k)x(k/L), \qquad k/L \text{ an integer}$$

$$= \sum_{r=-\infty}^{\infty} h(m - rL)x(r) \qquad (3.49)$$

An alternative formulation of this equation can be obtained by introducing the change of variables

$$r = \left\lfloor \frac{m}{L} \right\rfloor - n \tag{3.50}$$

where $\lfloor u \rfloor$ denotes the integer less than or equal to u. Then by noting that

$$m - \left\lfloor \frac{m}{L} \right\rfloor L = ((m))_L \tag{3.51}$$

where $((m))_L$ denotes m modulo L,

$$
\begin{aligned}
y(m) &= \sum_{n=-\infty}^{\infty} h\left(m - \left\lfloor \frac{m}{L} \right\rfloor L + nL\right) x\left(\left\lfloor \frac{m}{L} \right\rfloor - n\right) \\
&= \sum_{n=-\infty}^{\infty} h(nL + ((m))_L) x\left(\left\lfloor \frac{m}{L} \right\rfloor - n\right)
\end{aligned}
\tag{3.52}
$$

Equation (3.52) expresses the output $y(m)$ in terms of the input $x(n)$ and the filter coefficients $h(m)$.

3.2.3 Sampling Rate Conversion by a Rational Factor M/L

In Sections 3.2.1 and 3.2.2 we have considered the cases of decimation by an integer factor M and interpolation by an integer factor L. In this section we consider the general case of conversion by the ratio

$$\frac{T'}{T} = \frac{M}{L} = \frac{F_s}{F_s'} \tag{3.53}$$

This conversion can be achieved by a cascade of the two processes of integer conversions above by first increasing the sampling rate by L and then decreasing it by M. Figure 3.9(a) illustrates this process. It is important to recognize that the interpolation by L *must* precede the decimation process by M so that the width of the baseband of the intermediate signal $s(k)$ is greater than or equal to the width of the basebands of $x(n)$ or $y(m)$.

It can be seen from Fig. 3.9(a) that the two filters $h_1(k)$ and $h_2(k)$ are operating in cascade at the same sampling rate LF_s. Thus a more efficient implementation of the overall process can be achieved if the filters are combined into one composite lowpass filter, as shown in Fig. 3.9(b). Since this digital filter, $h(k)$, must serve the purposes of both the decimation and interpolation operations described in the preceding two sections, it is clear from Eqs. (3.26) and (3.45) that it must approximate the ideal digital lowpass characteristic

$$H_I(\omega'') = \begin{cases} L, & |\omega''| \le \min \left| \dfrac{\pi}{L}, \dfrac{\pi}{M} \right| \\ 0, & \text{otherwise} \end{cases} \tag{3.54}$$

where

$$\omega'' = 2\pi f T'' = 2\pi f \frac{T}{L} \tag{3.55}$$

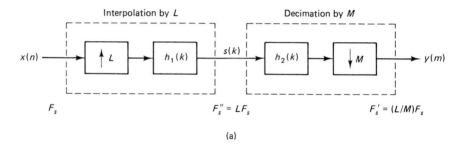

(a)

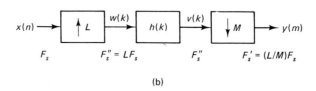

(b)

Figure 3.9   (a) Cascade of an integer interpolator and an integer decimator for achieving sampling rate changes by rational fractions; (b) a more efficient implementation of this process.

That is, the ideal cutoff frequency must be the minimum of the two cutoff frequency requirements for the decimator and interpolator, and the sampling rate of the filters is $F_s'' = LF_s$.

The time-domain input-to-output relation for the general conversion circuit of Fig. 3.9(b) can be derived by considering the integer interpolation and decimation relations derived in Sections 3.2.1 and 3.2.2; that is, from Eq. (3.49) it can be seen that $v(k)$ can be expressed as

$$v(k) = \sum_{r=-\infty}^{\infty} h(k - rL)x(r) \tag{3.56}$$

and from Eq. (3.28) $y(m)$ can be expressed in terms of $v(k)$ as

$$y(m) = v(Mm) \tag{3.57}$$

Combining Eqs. (3.56) and (3.57) gives the desired result

$$y(m) = \sum_{r=-\infty}^{\infty} h(Mm - rL)x(r) \tag{3.58}$$

Alternatively, by making the change of variables

$$r = \left\lfloor \frac{mM}{L} \right\rfloor - n \tag{3.59}$$

and using the relation

$$mM - \left\lfloor \frac{mM}{L} \right\rfloor L = ((mM))_L \tag{3.60}$$

we get

$$y(m) = \sum_{n=-\infty}^{\infty} h\left(Mm - \left\lfloor \frac{mM}{L} \right\rfloor L + nL\right) x\left(\left\lfloor \frac{mM}{L} \right\rfloor - n\right)$$

(3.61)

$$= \sum_{n=-\infty}^{\infty} h[nL + ((mM))_L] x\left(\left\lfloor \frac{mM}{L} \right\rfloor - n\right)$$

Similarly, by considering the transform relationships of the individual integer decimation and interpolation systems, the output spectrum $Y(\omega')$ can be determined in terms of the input spectrum $X(\omega)$ and the frequency response of the filter $H(\omega'')$. From Eq. (3.46) we see that $V(\omega'')$ can be expressed in terms of $X(\omega)$ and $H(\omega'')$ as

$$V(\omega'') = H(\omega'')X(\omega''L) \qquad (3.62)$$

and from Eq. (3.35) $Y(\omega')$ can be expressed in terms of $V(\omega'')$ as

$$Y(\omega') = \frac{1}{M} \sum_{\ell=0}^{M-1} V((\omega' - 2\pi\ell)/M)$$

$$= \frac{1}{M} \sum_{\ell=0}^{M-1} H((\omega' - 2\pi\ell)/M)X((\omega'L - 2\pi\ell)/M) \qquad (3.63)$$

When $H(\omega'')$ closely approximates the ideal characteristic of Eq. (3.54), we see that this expression reduces to

$$Y(\omega') \approx \begin{cases} \frac{L}{M}X(\omega'L/M), & \text{for } |\omega'| \leq \min\left[\pi, \pi\frac{M}{L}\right] \\ 0, & \text{otherwise} \end{cases} \qquad (3.64)$$

3.2.4 General Form of Digital Sampling Rate Conversion

It is possible to generalize the discussion of Sections 3.2.1–3.2.3 so as to give a canonic form of a digital system for sampling rate conversion. The form of this canonic system is illustrated in Fig. 3.10, in which an input signal $x(n)$, sampled at rate $F_s = 1/T$, is sent to a linear, time-varying, digital system with impulse response $g_m(n)$ to give the output signal $y(m)$, with new sampling rate $F_s' = 1/T'$. If we assume

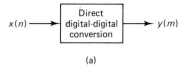

(a)

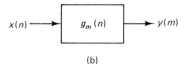

(b)

Figure 3.10 (a) Direct digital conversion of $x(n)$ to $y(m)$ and (b) a time-varying filter interpretation of the process.

that the ratio of sampling periods of $y(m)$ and $x(n)$ can be expressed as the rational fraction

$$\frac{T'}{T} = \frac{F_s}{F'_s} = \frac{M}{L} \tag{3.65}$$

with M and L integers, then the digital system response $g_m(n)$ becomes the response at output sample time m to an input at sample time $\lfloor mM/L \rfloor - n$, where $\lfloor u \rfloor$ again denotes the integer less than or equal to u.

Since the system is linear, each output sample $y(m)$ can be expressed as a linear combination of input samples. A general form [10] for this expression is

$$y(m) = \sum_{n=-\infty}^{\infty} g_m(n)x\left(\left\lfloor \frac{mM}{L} \right\rfloor - n\right) \tag{3.66}$$

where $g_m(n)$ is periodic in m with period L, i.e.,

$$g_m(n) = g_{m+rL}(n), \quad r = 0, \pm 1, \pm 2, \ldots \tag{3.67}$$

Thus the system $g_m(n)$ belongs to the class of linear, periodically time-varying systems.

Consider now several specific cases of sampling rate conversion systems. First consider the trivial case $T' = T$ or $L = M = 1$, in which case Eq. (3.66) reduces to the simple time-invariant digital convolution equation, i.e.,

$$y(m) = \sum_{n=-\infty}^{\infty} g(n)x(m - n) \tag{3.68}$$

since the period of $g_m(n)$ is 1 and the integer part of $m - n$ is the same as $m - n$.

Next consider the case of sampling rate reduction (decimation) by an integer factor M. In this case we get

$$y(m) = \sum_{n=-\infty}^{\infty} g_m(n)x(mM - n) \tag{3.69}$$

where $g_m(n) = g(n) = h(n)$ for all m and n, with $h(n)$ the lowpass filter impulse response of the system of Fig. 3.7. Although $g_m(n)$ is *not* a function of m for this case, it can readily be shown that the overall system of Eq. (3.69) is not time-invariant [10].

Next consider the case of sampling rate increase (interpolation) by an integer factor L. By comparing Eqs. (3.66) and (3.52), we see that the form of $g_m(n)$ is

$$g_m(n) = h[nL + ((m))_L], \qquad \text{for all } m \text{ and } n \tag{3.70}$$

and also that $g_m(n)$ is periodic in m with period L.

Finally, if we consider the general case of sampling rate conversion by the rational fraction M/L, then from Eq. (3.61) we get the result

$$g_m(n) = h[nL + ((mM))_L], \qquad \text{for all } m \text{ and } n \tag{3.71}$$

where $h(k)$ is the time-invariant unit sample response of the lowpass filter at the sampling rate LF_s.

3.3 PRACTICAL STRUCTURES FOR DECIMATORS
AND INTERPOLATORS

It is easy to understand the need for studying structures for realizing sampling rate conversion systems by examining the simple block diagram of Fig. 3.9(b), which can be used to convert the sampling rate of a signal by a factor L/M. As discussed in Section 3.2, the theoretical model for this system first increases the signal sampling rate by a factor L (by filling in $L - 1$ zero-valued samples between each pair of samples of $x(n)$ to give the signal $w(k)$), then filters $w(k)$ (to eliminate the images of $X(\omega)$) by a standard linear, time-invariant, lowpass filter $h(k)$ to give $v(k)$, and then compresses the sampling rate of $v(k)$ by a factor M (by retaining 1 of each M samples of $v(k)$). A direct implementation of this system is grossly inefficient since the lowpass filter $h(k)$ is operating at the high sampling rate on a signal for which $L - 1$ out of each L input values are zero, and the values of the filtered output are required only once each M samples. For this example, we can directly apply this knowledge in implementing the system of Fig. 3.9(b) in a more efficient manner, as will be discussed in this section.

3.3.1 Signal Flow Graphs

To precisely define the sets of operations necessary to implement these digital systems, we will strongly rely on the concepts of signal flow graph representation [10]. Signal flow graphs provide a graphical representation of the explicit set of equations that are used to implement such systems. Furthermore, manipulating the flow graphs in a pictorial way is equivalent to manipulating the mathematical equations.

Figure 3.11 illustrates an example of a signal flow graph of a direct-form finite impulse response (FIR) digital filter. The input branch applies the external signal $x(n)$ to the network, and the output of the network $y(n)$ is identified as one of the node

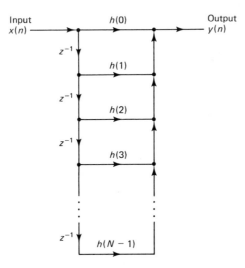

Figure 3.11 Direct-form structure for an FIR digital filter.

values. Branches define the signal operations in the structure such as delays, gains, and sampling rate expanders and compressors. Nodes define the connection points and summing points. The signal entering a branch is taken as the signal associated with the input node value of the branch. The node value of a branch is the sum of all branch signals entering the node.

From the signal flow graph in Fig. 3.11 we can immediately write down the network equation as

$$y(n) = x(n)h(0) + x(n - 1)h(1) + \cdots + x(n - N + 1)h(N - 1)$$

An important concept in the manipulation of signal flow graphs is the principle of commutation of branch operations. Two branch operations commute if the order of their cascade operation can be interchanged without affecting the input-to-output response of the cascaded system. Thus interchanging commutable branches in a network is one way of modifying the network without affecting the desired input-to-output network response. This operation will be used extensively in constructing efficient structures for decimation and interpolation, as we will see shortly.

Another important network concept on which we rely heavily is that of transposition and duality [10]. Basically, a dual system is one that performs a complementary operation to that of an original system, and it can be constructed from the original system through the process of transposition. We have already seen an example of dual systems, namely, the integer decimator and interpolator (Fig. 3.7a and Fig. 3.8a) for the case $M = L$.

Basically the transposition operation is one in which the direction of all branches in the network are reversed and the roles of the input and output of the network are interchanged. Furthermore, all branch operations are replaced by their transpose operations. In the case of linear time-invariant branch operations, such as gains and delays, these branch operations remain unchanged. Thus, for example, the transpose of the direct-form structure of Fig. 3.11 is the transposed direct-form structure shown in Fig. 3.12. Also it can be shown that for the case of linear time-invariant systems

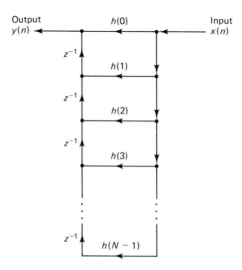

Figure 3.12 Transposed direct-form FIR filter structure.

the input-to-output system response of a system and its dual are identical (e.g., it can be verified that the networks of Fig. 3.11 and Fig. 3.12 have identical system functions).

For the time-varying systems this is not necessarily the case. For example, the transpose of a sampling rate compressor is a sampling rate expander, and the transpose of a sampling rate expander is a sampling rate compressor, as shown in Fig. 3.13. Clearly these systems do not have the same system response.

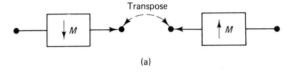

(a)

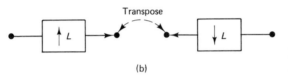

(b)

Figure 3.13   Transpositions of a sampling rate compressor and expander.

By extending the concepts of transposition rigorously, we can also show that the transposition of a network that performs a sampling rate conversion by the factor L/M is a network that performs a sampling rate conversion by the factor M/L. This is illustrated in Fig. 3.14.

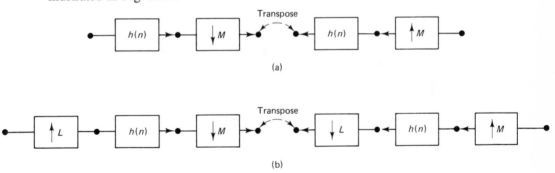

(a)

(b)

Figure 3.14   Transpositions of a decimator and a generalized L/M sampling rate changer.

3.3.2 Direct-Form FIR Structures for Integer Changes in Sampling Rates

Consider the model of an M-to-1 decimator as shown in Fig. 3.15(a). According to this model the filter $h(n)$ operates at the high sampling rate F_s, and $M - 1$ out of every M output samples of the filters are discarded by the M-to-1 sampling rate compressor. In particular, if we assume that the filter $h(n)$ is an N-point FIR filter realized with a direct-form structure, the network of Fig. 3.15(b) results. The multiplications by $h(0)$,

$h(1)$, . . . , $h(N − 1)$ and the associated summations in this network must be performed at the rate F_s.

A more efficient realization of the above structure can be achieved by noting that the branch operations of sampling rate compression and gain can be commuted. By performing a series of commutative operations on the network, we obtain the modified network of Fig. 3.15(c). The multiplications and additions associated with the coefficients $h(0)$ to $h(N − 1)$ now occur at the low sampling rate F_s/M and therefore the total computation rate in the system has been reduced by a factor M. For every M samples of $x(n)$ that are shifted into the structure (the cascade of delays), one output

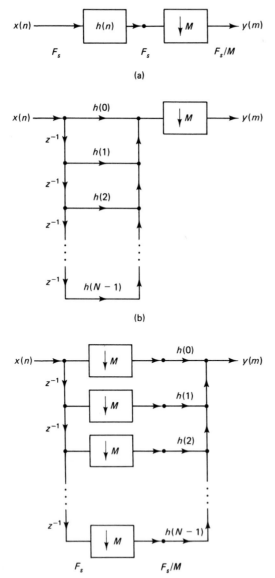

Figure 3.15 Generation of an efficient direct-form structure of an M-to-1 decimator.

sample $y(m)$ is computed. Thus the structure of Fig. 3.15(c) is seen to be a direct realization of Eq. (3.29).

An efficient structure for the 1-to-L integer interpolator, using an FIR filter, can be derived in a similar manner. We begin with the cascade model for the interpolator shown in Fig. 3.16(a). In this case however, if $h(m)$ is realized with the direct-form structure of Fig. 3.11 we are faced with the problem of commuting the 1-to-L sampling rate expander with a series of unit delays. One way around this problem is to realize $h(m)$ with the transposed direct-form FIR structure as shown in Fig. 3.12. The sampling rate expander can then be commuted into the network as shown by the series

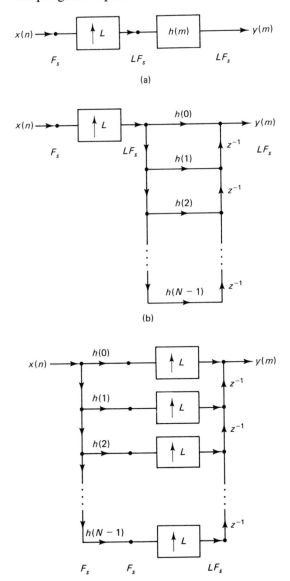

(a)

(b)

(c)

Figure 3.16 Steps in the generation of an efficient structure of a 1-to-L interpolator.

of operations in Fig. 3.16. Since the coefficients $h(0)$, $h(1)$, . . . , $h(N - 1)$ in Fig. 3.16(c) are now commuted to the low sampling rate side of the network, this structure requires a factor of L times less computation than the structure in Fig. 3.16(b).

An alternative way of deriving the structure of Fig. 3.16(c) is by a direct transposition of the network of Fig. 3.15(c) (letting $L = M$). This is a direct consequence of the fact that decimators and interpolators are duals. A further property of transposition is that for the resulting network, neither the number of multipliers nor the rate at which these multipliers operate will change [15]. Thus if we are given a network that is minimized with respect to its multiplication rate, then its transpose will also be minimized with respect to its multiplication rate.

3.3.3 Polyphase FIR Structures for Integer Decimators and Interpolators

A second general class of structures of interest in multirate digital systems is the polyphase networks. We will find it convenient to first derive this structure for the L-to-1 interpolator and then obtain the structure for the decimator by transposing the interpolator structure.

In Section 3.2 it was shown that a general form for the input-to-output time-domain relationship for the 1-to-L interpolator is

$$y(m) = \sum_{n=-\infty}^{\infty} g_m(n)x\left(\left\lfloor \frac{m}{L} \right\rfloor - n\right) \tag{3.72}$$

where

$$g_m(n) = h(nL + ((m))_L) \quad \text{for all } m \text{ and } n \tag{3.73}$$

is a periodically time-varying filter with period L. Thus to generate each output sample $y(m)$, $m = 0, 1, 2, \ldots, L - 1$, a different set of coefficients $g_m(n)$ is used. After L outputs are generated, the coefficient pattern repeats; thus $y(L)$ is generated using the same set of coefficients $g_0(n)$ as $y(0)$, $y(L + 1)$ uses the same set of coefficients $g_1(n)$ as $y(1)$, and so on.

Similarly the term $\lfloor m/L \rfloor$ in Eq. (3.72) increases by 1 for every L samples of $y(m)$. Thus for output samples $y(L)$, $y(L + 1)$, . . . , $y(2L - 1)$ the coefficients $g_m(n)$ are multiplied by samples $x(1 - n)$. In general, for output samples $y(rL)$, $y(rL + 1)$, . . . , $y(rL + L - 1)$, the coefficients $g_m(n)$ are multiplied by samples $x(r - n)$. Thus we see that $x(n)$ is updated at the low sampling rate F_s, whereas $y(m)$ is evaluated at the high sampling rate LF_s.

An implementation of the 1-to-L interpolator based on the computation of Eq. (3.72) is shown in Fig. 3.17(a). The way in which this structure operates is as follows. The partitioned subsets $g_0(n)$, $g_1(n)$, . . . , $g_{L-1}(n)$, of $h(m)$ can be identified with L separate linear time-invariant filters that operate at the low sampling rate F_s. To make this subtle notational distinction between the time-varying coefficients and the time-invariant filters, we will refer to the time-invariant filters respectively as $p_0(n)$, $p_1(n)$, . . . , $p_{L-1}(n)$. Thus

$$p_\rho(n) = g_\rho(n), \quad \text{for } \rho = 0, 1, 2, \ldots, L - 1 \text{ and all } n \tag{3.74}$$

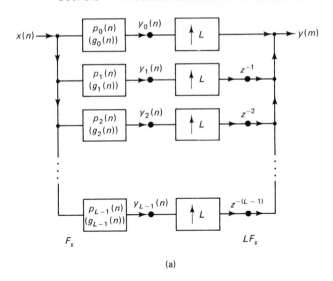

(a)

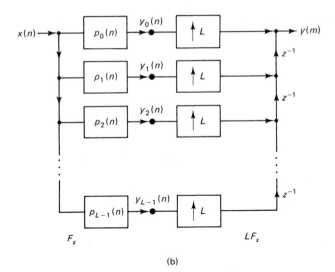

(b)

Figure 3.17 Polyphase structures for a 1-to-L interpolator.

These filters $p_\rho(n)$ will be referred to as the *polyphase filters*. Furthermore, by combining Eqs. (3.73) and (3.74) we see that

$$p_\rho(n) = h(nL + \rho), \qquad \text{for } \rho = 0, 1, 2, \ldots, L - 1 \text{ and all } n \qquad (3.75)$$

For each new input sample $x(n)$, there are L output samples (see Fig. 3.17). The output from the upper path $y_0(m)$ has nonzero values for $m = nL$, $n = 0, \pm 1, \pm 2,$ . . . , which correspond to system outputs $y(nL)$, $n = 0, \pm 1, \ldots$. The output from the next path $y_1(m)$ is nonzero for $m = nL + 1$, $n = 0, \pm 1, \pm 2, \ldots$ because of the delay of one sample at the high sampling rate. Thus $y_1(m)$ corresponds to the interpolation output samples $y(nL + 1)$, $n = 0, \pm 1, \ldots$. In general, the output of the

ρth path, $y_\rho(m)$ corresponds to the interpolation output samples $y(nL + \rho)$, $n = 0$, $\pm 1, \ldots$. For each input sample $x(n)$, each of the L branches of the polyphase network contributes one nonzero output that corresponds to one of the L outputs of the network. In the polyphase interpolation network of Fig. 3.17(a), the filtering is performed at the low sampling rate, and thus it is an efficient structure. A simple manipulation of the structure of Fig. 3.17(a) leads to the equivalent network of Fig. 3.17(b), in which all the delays are single sample delays.

The individual polyphase filters $\rho_\rho(n)$, $\rho = 0, 1, 2, \ldots, L - 1$ have a number of interesting properties. This is a consequence of the fact that the impulse responses $p_\rho(n)$, $\rho = 0, 1, 2, \ldots, L - 1$, correspond to decimated versions of the impulse response of the prototype filter $h(m)$ (decimated by a factor of L according to Eq. 3.73 or 3.75). Figure 3.18 illustrates this for the case $L = 3$ and for an FIR filter $h(m)$ with $N = 9$ taps. Part (a) shows the samples of $h(m)$ for $h(m)$ symmetric about $m = 4$. Thus $h(m)$ has a flat delay of 4 samples. The filter $\rho_0(n)$ has three samples corresponding to $h(0)$, $h(3)$, $h(6) = h(2)$. Since the point of symmetry of the envelope of $\rho_0(n)$ is $n = \frac{4}{3}$, it has a flat delay of $\frac{4}{3}$ samples. Similarly, $\rho_1(n)$ has samples $h(1)$, $h(4)$, $h(7) = h(1)$, and because its zero reference ($n = 0$) is offset by $\frac{1}{3}$ sample (with respect to $m = 0$) it has a flat delay of 1 sample. Thus different fractional sample delays and consequently different phase shifts are associated with the different filters $p_\rho(n)$, as seen in Fig. 3.18(b). These delays are compensated for by the delays that occur at the high sampling rate LF in the network (see Fig. 3.17). The fact that different phases are associated with different paths of the network is, of course, the reason for the term *polyphase network*.

A second property of the polyphase filters is shown in Fig. 3.19. The frequency response of the prototype filter $h(m)$ approximates the ideal lowpass characteristic $H_I(\omega)$ shown in Fig. 3.19(a). Since the polyphase filters $p_\rho(n)$ are decimated versions of $h(m)$ (decimated by L), the frequency response $0 \leq \omega \leq \pi/L$ of $H_I(\omega)$ scales to the range $0 \leq \omega' \leq \pi$ for $P_{\rho,I}(\omega')$ as seen in Fig. 3.19, where $P_{\rho,I}(\omega')$ is the ideal characteristic that the polyphase filter $p_\rho(n)$ approximates. Thus the polyphase filters approximate allpass functions and each value of ρ, $\rho = 0, 1, 2, \ldots, L - 1$, corresponds to a different phase shift.

The polyphase filters can be realized in a variety of ways. If the prototype filter $h(m)$ is an FIR filter of length N, then the filters $p_\rho(n)$ will be FIR filters of length N/L. In this case it is often convenient to choose N to be a multiple of L so that all of the polyphase filters are of equal length. These filters may be realized by any of the conventional methods for implementing FIR filters such as the direct-form structure or the methods based on fast convolution [10]. If a direct-form FIR structure is used for the polyphase filters, the polyphase structure of Fig. 3.17 will require the same multiplication rate as the direct-form interpolator structure of Fig. 3.16.

By transposing the structure of the polyphase 1-to-L interpolator of Fig. 3.17(b), we get the polyphase M-to-1 decimator structure of Fig. 3.20, where L is replaced by M. Again the filtering operations of the polyphase filters occur at the low sampling rate side of the network, and they can be implemented by any of the conventional structures discussed above.

In the preceding discussion for the 1-to-L interpolator, we have identified the coefficients of the polyphase filters $p_\rho(n)$ with the coefficient sets $g_m(n)$ of the time-

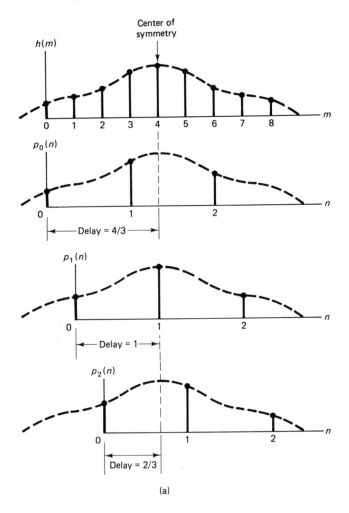

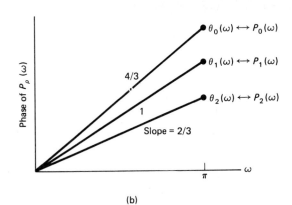

Figure 3.18 Illustration of the properties of polyphase network filters.

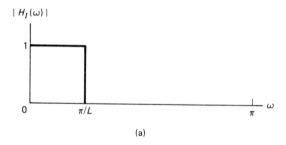

(a)

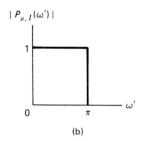

(b)

Figure 3.19 Ideal frequency response of the polyphase network filters.

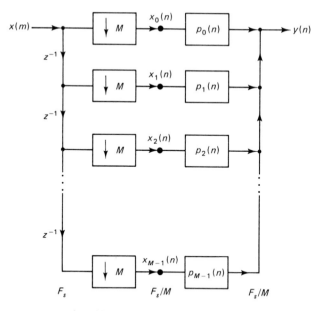

Figure 3.20 Polyphase structure for an M-to-1 decimator.

varying filter model. In the case of the M-to-1 decimator, however, this identification cannot be made directly. According to the time-varying filter model, discussed in Section 3.2, the coefficients $g_m(n)$ for the M-to-1 decimator are

$$g_m(n) = g(n) = h(n), \qquad \text{for all } n \text{ and } m \qquad (3.76)$$

Alternatively, according to the transpose network of Fig. 3.20, the coefficients of the M-to-1 polyphase decimator are

$$p_\rho(n) = h(nM + \rho), \qquad \text{for } \rho = 0, 1, 2, \ldots, M - 1, \text{ and all } n \qquad (3.77)$$

where ρ denotes the ρth polyphase filter. Thus the polyphase filters $p_\rho(n)$ for the M-to-1 decimator are equal to the time-varying coefficients $g_m(n)$ of the transpose (interpolator) of this decimator.

From a practical point of view it is often convenient to implement the polyphase structures in terms of a commutator model. By careful examination of the interpolator structure of Fig. 3.17 we can see that the outputs of each of the polyphase branches contributes samples of $y(m)$ for different time slots. Thus the 1-to-L sampling rate expander and delays can be replaced by a commutator, as shown in Fig. 3.21. The commutator rotates in a counterclockwise direction starting with the zeroth-polyphase branch at time $m = 0$.

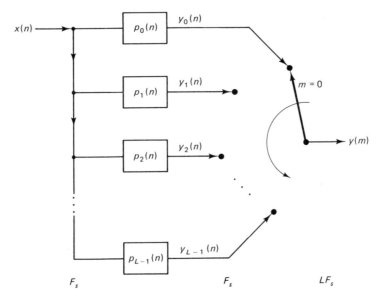

Figure 3.21 Counterclockwise commutator model for a 1-to-L interpolator.

A similar commutator model can be developed for the M-to-1 polyphase decimator by starting with the structure of Fig. 3.20 and replacing the delays and M-to-1 sampling rate compressors with a commutator. This leads to the structure of Fig. 3.22. Again the commutator rotates in a counterclockwise direction starting with the zeroth-polyphase branch at time $m = 0$.

3.3.4 FIR Structures with Time-Varying Coefficients for Interpolation/Decimation by a Factor of L/M

In the previous two sections we have considered implementations of decimators and interpolators using the direct-form and polyphase structures for the case of integer changes in the sampling rate. We obtained efficient realizations of these structures by commuting the filtering operations to occur at the low sampling rate. For the case of a network that realizes a change in sampling rate by a factor of L/M, it is difficult to

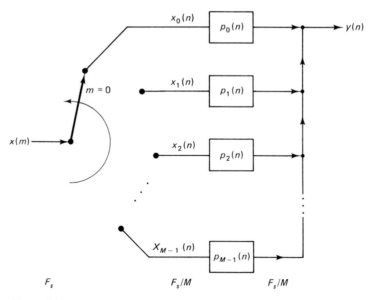

Figure 3.22   Counterclockwise commutator model for an M-to-1 decimator.

achieve such efficiencies. The difficulty is illustrated in Fig. 3.23. If we realize the 1-to-L interpolation part of the structure using the techniques described earlier, then we are faced with the problem of commuting the M-to-1 sampling rate compressor into the resulting network (Fig. 3.23a). If we realize the decimator part of the structure first, then the 1-to-L sampling rate expander must be commuted into the structure (Fig. 3.23b). In both cases difficulties arise and we are faced with a network that cannot be implemented efficiently simply using the techniques of commutation and trans-position.

Efficient structures exist for implementing a sampling rate converter with a ratio in sampling rates of L/M, and in this section we discuss one such class of FIR structures with time-varying coefficients [10]. This structure can be derived from the time-domain input-to-output relation of the network, as derived in Section 3.2, namely

$$y(m) = \sum_{n=-\infty}^{\infty} g_m(n)x\left(\left\lfloor \frac{mM}{L} \right\rfloor - n\right) \tag{3.78}$$

where

$$g_m(n) = h(nL + ((mM))_L), \qquad \text{for all } m \text{ and all } n \tag{3.79}$$

and $h(k)$ corresponds to the lowpass (or bandpass) FIR prototype filter. It will be convenient for our discussion to assume that the length of the filter $h(k)$ is a multiple of L, i.e.,

$$N = QL \tag{3.80}$$

where Q is an integer. Then all of the coefficient sets $g_m(n)$, $m = 0, 1, 2, \ldots, L - 1$ contain *exactly* Q coefficients. Furthermore, $g_m(n)$ is periodic in m with period L, i.e.,

$$g_m(n) = g_{m+rL}(n), \qquad r = 0, \pm 1, \pm 2, \ldots \tag{3.81}$$

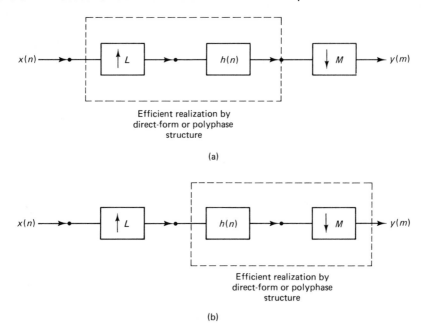

Efficient realization by
direct-form or polyphase
structure

(a)

Efficient realization by
direct-form or polyphase
structure

(b)

Figure 3.23   Possible realizations of an L/M sampling rate converter.

Therefore, Eq. (3.78) can be expressed as

$$y(m) = \sum_{n=0}^{Q-1} g_{((m))_L}(n) x\left(\left\lfloor \frac{mM}{L} \right\rfloor - n\right) \tag{3.82}$$

Equation (3.82) shows that the computation of an output sample $y(m)$ is obtained as a weighted sum of Q sequential samples of $x(n)$ starting at the $x(\lfloor mM/L \rfloor)$ sample and going backward in n sequentially. The weighting coefficients are periodically time-varying, so the $((m))_L$ coefficient set $g_{((m))_L}(n)$, $n = 0, 1, 2, \ldots, Q - 1$, is used for the mth output sample. Figure 3.24 illustrates this timing relationship for the $n = 0$

$y(m)$	$x\left(\left\lfloor \dfrac{mM}{L} \right\rfloor\right)$	$g_{((m))_L}(0)$
m	$\left\lfloor \dfrac{2m}{3} \right\rfloor$	$((m))_L$
0	0	0
1	0	1
2	1	2
3	2	0
4	2	1
5	3	2
6	4	0

(a)

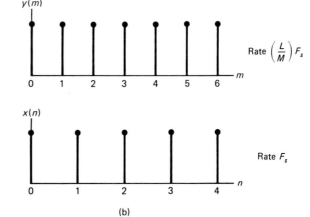

(b)

Figure 3.24   Timing relationships between $y(m)$ and $x(n)$ for the case $M = 2$, $L = 3$.

term in Eq. (3.82) and for the case $M = 2$ and $L = 3$. Figure 3.24(a) shows the index values of $y(m)$, $x(\lfloor mM/L \rfloor)$, and $g_{((m))_L}(0)$ for $m = 0, 1, 2, \ldots, 6$. Figure 3.24(b) illustrates the relative timing positions of the signals $y(m)$ and $x(n)$ drawn on an absolute time scale. By comparison of parts (a) and (b) we can see that the value $x(\lfloor mM/L \rfloor)$ always represents the most recent available sample of $x(n)$, i.e., $y(0)$ and $y(1)$ are computed on the basis of $x(0 - n)$. For $y(2)$ the most recent available value of $x(n)$ is $x(1)$, for $y(3)$ it is $x(2)$, and so on.

Based on Eq. (3.82) and the preceding description of how the input, output, and coefficients enter into the computation, the structure of Fig. 3.25 is suggested for realizing an L/M sampling rate converter. The structure consists of the following:

1. A Q sample "shift register" operating at the input sampling rate F_s, which stores sequential samples of the input signal
2. A direct-form FIR structure with time-varying coefficients $(g_{((m))_L}(n), n = 0, 1, 2, \ldots, Q - 1)$ that operates at the output sampling rate $(L/M)F_s$
3. A series of digital "hold-and-sample" boxes that couple the two sampling rates. The input side of the box "holds" the most recent input value until the next input value comes along; the output side of the box "samples" the input values at times $n = mM/L$. For times when mM/L is an integer (i.e., input and output sampling times are the same), the input changes first and the output samples the changed input.

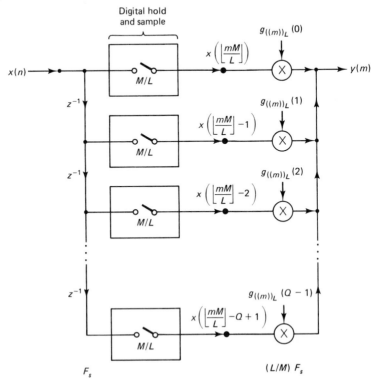

Figure 3.25   Efficient structure for realizing an L/M sampling rate converter.

It should be clear that the structure of Fig. 3.25 is an efficient one for implementing an (L/M) sampling rate converter since the filtering operations are all performed at the output sampling rate with the minimum required number of coefficients used to generate each output.

Figure 3.26 shows a diagram of a program configuration to implement this structure in a block-by-block manner. The program takes in a block of M samples of the input signal, denoted as $x(n')$, $n' = 0, 1, 2, \ldots, M - 1$, and computes a block of L output samples $y(m')$ $m' = 0, 1, 2, \ldots, L - 1$. For each output sample time m', $m' = 0, 1, 2, \ldots, L - 1$, the Q samples from the state-variable buffer are multiplied respectively with Q coefficients from one of the coefficient sets $g_{m'}(n')$ and the products are accumulated to give the output $y(m')$. Each time the quantity $\lfloor m'M/L \rfloor$ increases by 1, one sample from the input buffer is shifted into the state-variable buffer. (This information can be stored in a control array.) Thus after L output values are computed, M input samples have been shifted into the state-variable buffer and the process can be repeated for the next block of data. In the course of processing one block of data (M input samples and L output samples), the state-variable buffer is sequentially addressed L times and the coefficient storage buffer is sequentially addressed once. A program that performs this computation can be found in [16].

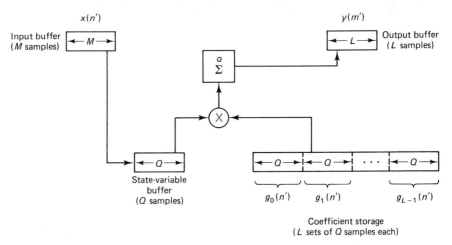

Figure 3.26   Block diagram of a program structure to implement the signal flow graph of Fig. 3.25 in a block-by-block manner.

3.3.5 Comparison of Structures

In Section 3.3 so far, we have discussed three principal classes of FIR structures for decimators and interpolators. In Section 3.5 we will discuss multistage cascades of these structures and show how such cascading can lead to additional gains in computational efficiency when conversion ratios are large. A natural question to ask at this point is which of these methods is most efficient. The answer, unfortunately, is nontrivial and is highly dependent on the application being considered. Some insight and direction, however, can be provided by observing a few general properties of the classes of structures discussed here.

The direct-form structures have the advantage that they can be easily modified to exploit symmetry in the system function to gain an additional reduction in computation by a factor of approximately two. The polyphase structures have the advantage that the filters $p_\rho(n)$ can be easily realized with efficient techniques such as the fast convolution methods based on the FFT. As such, this structure has been found useful for filter banks [10]. The structures with time-varying coefficients are particularly useful when considering conversions by factors of L/M.

3.4 DESIGN OF FIR FILTERS FOR DECIMATION AND INTERPOLATION

In the previous discussion we have assumed that the filter $h(k)$ approximates some ideal lowpass (or bandpass) characteristic. Consequently the effectiveness of these systems is directly related to the type and quality of design of this digital filter. The purpose of this section is to review digital filter design techniques that are especially applicable to the design of the digital filter in sampling rate-changing systems.

The filter design problem is essentially one of determining suitable values of $h(k)$ to meet given performance specifications on the filter. Such performance specifications can be made on the time response $h(k)$ or the frequency response of the filter $H(\omega)$ defined as

$$H(\omega) = \sum_{k=-\infty}^{\infty} h(k)e^{-j\omega k} = H(z)\big|_{z=e^{j\omega}} \tag{3.83}$$

Before proceeding to a discussion of filter design techniques for decimators and interpolators, it is important to consider the ideal frequency domain and the time-domain criteria that specify such designs. It is also important to consider, in more detail, the representation of such filters in terms of a single prototype filter or as a set of polyphase filters. Although both representations are equivalent, it is sometimes easier to view filter design criteria in terms of one representation or the other.

3.4.1 Relationship Between the Prototype Filter and Its Polyphase Representation

As discussed in Section 3.3, the coefficients, or impulse responses, of the polyphase filters correspond to sampled (and delayed) versions of the impulse response of the prototype filter. For a 1-to-L interpolator there are L polyphase filters and they are defined as (see Fig. 3.18)

$$p_\rho(n) = h(\rho + nL), \qquad \rho = 0, 1, 2, \ldots, L - 1, \text{ and all } n \tag{3.84}$$

Similarly, for an M-to-1 decimator there are M polyphase filters in the polyphase structure and they are defined as

$$p_\rho(n) = h(\rho + nM), \qquad \rho = 0, 1, 2, \ldots, M - 1, \text{ and all } n \tag{3.85}$$

Taken as a set, the samples $p_\rho(n)$ ($\rho = 0, 1, \ldots, L - 1$, for an interpolator or

$\rho = 0, 1, \ldots, M - 1$, for a decimator) represent all of the samples of $h(k)$. Since the development of the filter specifications is identical for both cases (1-to-L interpolators and M-to-1 decimators) we will consider only the case of interpolators. The results for decimators can then simply be obtained by replacing L by M in the appropriate equations.

The samples $h(k)$ can be recovered from $p_\rho(n)$ by sampling rate expanding the sequences $p_\rho(n)$ by a factor L. Each expanded set is then delayed by ρ samples and the L sets are then summed to give $h(k)$ (the reverse operation to that of Fig. 3.18). If we let $\hat{p}_\rho(k)$ represent the sampling rate expanded set

$$\hat{p}_\rho(k) = \begin{cases} p_\rho(k/L), & k = 0, \pm L, \pm 2L, \ldots \\ 0, & \text{otherwise} \end{cases} \tag{3.86}$$

then $h(k)$ can be reconstructed from $\hat{p}_\rho(k)$ via the summation

$$h(k) = \sum_{\rho=0}^{L-1} \hat{p}_\rho(k - \rho) \tag{3.87}$$

The z-transform $H(z)$ of the prototype filter can similarly be expressed in terms of the z-transforms of the polyphase filters $P_\rho(z)$. It can be shown that

$$H(z) = \sum_{\rho=0}^{L-1} z^{-\rho} P_\rho(z^L) \tag{3.88}$$

Finally, the z-transform $P_\rho(z)$ can be expressed in terms of $H(z)$ according to the following derivation. If we define a sampling function $\delta_\rho(k)$ such that

$$\delta_\rho(k) = \begin{cases} 1, & k = \rho, \rho \pm L, \rho \pm 2L, \ldots \\ 0, & \text{otherwise} \end{cases} \tag{3.89}$$

$$= \frac{1}{L} \sum_{\ell=0}^{L-1} e^{j2\pi\ell(k-\rho)/L} \tag{3.90}$$

then the sampling rate expanded sequences $\hat{p}_\rho(k)$ in Eq. (3.86) can be expressed as

$$\hat{p}_\rho(k) = \delta_\rho(k)h(k) = h(k)\frac{1}{L}\sum_{\ell=0}^{L-1} e^{j2\pi\ell(k-\rho)/L} \tag{3.91}$$

The z-transform $P_\rho(z)$ can then be expressed in the form

$$P_\rho(z) = \sum_{n=-\infty}^{\infty} p_\rho(n)z^{-n} = \sum_{n=-\infty}^{\infty} \hat{p}_\rho(\rho + nL)z^{-n} \tag{3.92}$$

and by the substitution of variables $k = \rho + nL$,

$$P_\rho(z) = \sum_{k=-\infty}^{\infty} \hat{p}_\rho(k)z^{-(k-\rho)/L} \tag{3.93}$$

Combining Eqs. (3.91) and (3.93), we get

$$P_\rho(z) = \frac{1}{L}\sum_{k=-\infty}^{\infty}\sum_{\ell=0}^{L-1} h(k)e^{j2\pi\ell(k-\rho)/L}z^{-(k-\rho)/L} \tag{3.94}$$

Letting $z = e^{j\omega}$ and rearranging terms gives

$$P_\rho(\omega) = \frac{1}{L} \sum_{\ell=0}^{L-1} e^{j(\omega-2\pi\ell)\rho/L} \sum_{k=-\infty}^{\infty} h(k)e^{-j(\omega-2\pi\ell)k/L}$$

$$= \frac{1}{L} \sum_{\ell=0}^{L-1} e^{j(\omega-2\pi\ell)\rho/L} H(\omega - 2\pi\ell)/L, \qquad \rho = 0, 1, 2, \ldots, L-1$$

(3.95)

Equation (3.95) shows the relationships of the Fourier transforms of the polyphase filters to the Fourier transform of the prototype filter.

3.4.2 Ideal Frequency-Domain Characteristics for Interpolation and Decimation Filters

In the previous sections we have assumed that the filter $h(k)$ approximates some ideal lowpass (or bandpass) characteristic. We will elaborate on these "ideal" characteristics in somewhat more detail in the next two sections. In practice it is also necessary to specify a performance criterion to measure (in a consistent manner) how closely an actual filter design approximates this ideal characteristic. Since different design techniques are often based on different criteria, we will consider these criteria as they arise.

Recall from the discussion in Section 3.2 that the interpolator filter $h(k)$ must approximate the ideal lowpass characteristic defined as

$$H_I(\omega') = \begin{cases} L, & |\omega'| < \pi/L \\ 0, & \text{otherwise} \end{cases}$$

(3.96)

where the subscript I refers to the "ideal" characteristic.

By combining Eqs. (3.95) and (3.96) it is possible to derive the equivalent ideal characteristics $P_{\rho,I}(\omega)$ that are implied in the polyphase filters. Note that the frequency variable ω' refers to $h(m)$ whereas the frequency variable $\omega = \omega'L$ refers to the polyphase filters $p_\rho(n)$. Because of the constraint imposed by Eq. (3.96), only the $\ell = 0$ term in Eq. (3.95) is nonzero, and the equation simplifies to the form

$$P_{\rho,I}(\omega) = \frac{1}{L} e^{j\omega\rho/L} H_I(\omega/L)$$

(3.97)

$$= e^{j\omega\rho/L}, \qquad \rho = 0, 1, 2, \ldots, L-1$$

Equation (3.97) shows that the "ideal" polyphase filters $P_{\rho,I}(n)$ should approximate allpass filters with linear phase shifts corresponding to fractional advances of ρ/L samples ($\rho = 0, 1, 2, \ldots, L-1$) (ignoring any fixed delays that must be introduced in practical implementations of such filters).

In some cases it is known that the spectrum of $x(n)$ does not occupy its full bandwidth. This property can be used to advantage in the filter design, and we will see examples of this in the next section on cascaded (multistage) implementations of sampling rate changing systems. If we define ω_c as the highest frequency of interest in $X(\omega)$, i.e.,

$$|X(\omega)| < \epsilon, \qquad \text{for } \pi > |\omega| > \omega_c$$

(3.98)

where ϵ is a small quantity (relative to the peak of $|X(\omega)|$), as shown in Fig. 3.27 (for $L = 5$). In this case, the ideal interpolator filter has to remove only the $(L - 1)$ repetitions of the band of $X(\omega)$ where $|X(\omega)| > \epsilon$. Thus in the frequency domain, the ideal interpolator filter satisfies the constraints

$$H_I(\omega') = \begin{cases} L, & 0 \le |\omega'| \le \omega_c/L \\ 0, & (2\pi r - \omega_c)/L \le |\omega'| \le (2\pi r + \omega_c)/L, \ r = 1, 2, \ldots, L - 1 \end{cases}$$

$$\text{(3.99)}$$

as illustrated in Fig. 3.27(c). The bands from $(2\pi r + \omega_c)/L$ to $[2\pi(r + 1) - \omega_c]/L$, $r = 0, 1, \ldots$, are "don't care" (ϕ) bands in which the filter frequency response is essentially unconstrained. (In practice, however, $|H(\omega')|$ should not be very large in these ϕ bands, e.g., not larger than L, to avoid amplification of any noise (or tails of $X(\omega)$) that may exist in these bands.) We will see later how these ϕ bands can have a significant effect on the filter design problem. Figure 3.27(d) shows the response of the ideal polyphase filter that is converted from an all-pass to a lowpass filter with cutoff frequency ω_c. Of course, the phase response of each polyphase filter is unaltered by the "don't care" bands.

As discussed in Section 3.2 for a decimator, the filter $H(\omega)$ should approximate the ideal lowpass characteristic

$$H_I(\omega) = \begin{cases} 1, & 0 \le \omega' \le \pi/M \\ 0, & \text{otherwise} \end{cases} \qquad \text{(3.100)}$$

Alternatively, the polyphase filters should approximate the ideal all-pass characteristics

$$P_{\rho,I}(\omega) = \frac{1}{M} e^{j\omega\rho/M}, \qquad \rho = 0, 1, 2, \ldots, M - 1 \qquad \text{(3.101)}$$

If we are interested only in preventing aliasing in a band from 0 to ω_c, where $\omega_c < \pi/M$, and we are willing to tolerate aliased components for frequencies above ω_c, then we again have a situation where "don't care" bands are permitted in the filter design. The "don't care" regions are the same as those illustrated in Fig. 3.27(c) (with L replaced by M). In fact, all of the frequency-domain constraints that apply to the design of interpolation filters also apply to the design of decimation filters, a consequence of the property that they are transpose systems.

3.4.3 Time-Domain Properties of Ideal Interpolation and Decimation Filters

If we view the interpolation filter design problem in the time-domain, we obtain an alternative picture of the "ideal" interpolation filter. By taking the inverse transform of the ideal filter characteristic defined by Eq. (3.96) we get the well-known $\sin(x)/x$ characteristic

$$h_I(k) = \frac{\sin(\pi k/L)}{(\pi k/L)}, \qquad k = 0, \pm 1, \pm 2, \ldots \qquad \text{(3.102)}$$

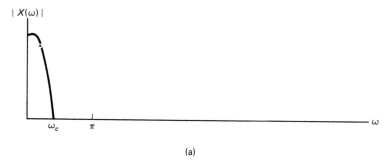

(a)

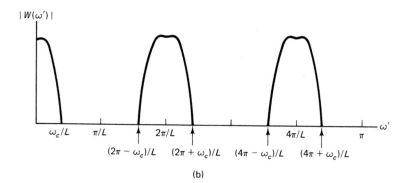

(b)

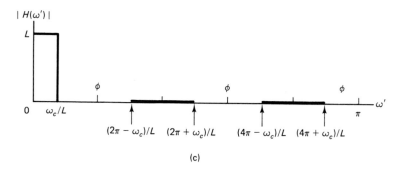

(c)

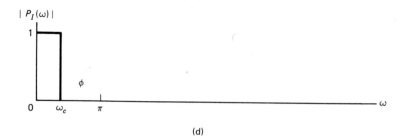

(d)

Figure 3.27   Illustration of ϕ bands in the specification of an interpolation filter ($L = 5$).

In a similar manner we can determine the ideal time responses of the polyphase filters, either by taking the inverse transform of Eq. (3.101) or by sampling the above time response $h_I(k)$ according to Eq. (3.84). The net result is that the ideal time responses of the polyphase filters are

$$p_{\rho,I}(n) = \frac{\sin[\pi(n + \rho/L)]}{\pi(n + \rho/L)}, \qquad \rho = 0, 1, 2, \ldots, L - 1, \quad \text{and all } n \qquad (3.103)$$

A number of interesting observations can be made about the above ideal time responses. First we see that they constrain every Lth value of $h_I(k)$ such that

$$h_I(k) = \begin{cases} 1, & k = 0 \\ 0, & k = rL, r = \pm 1, \pm 2, \ldots \end{cases} \qquad (3.104)$$

Alternatively, this implies the constraint that the zeroth-polyphase filter have an impulse response that is a unit pulse, i.e.,

$$p_{0,I}(n) = \delta(n), \qquad \text{for all } n \qquad (3.105)$$

In terms of the polyphase structure of Fig. 3.17 and its signal processing interpretation in Fig. 3.18, the above constraint is easy to visualize. It simply implies that the output $y_0(m)$ of the zeroth polyphase branch is identical to the input $x(n)$ filled in with $L - 1$ zeros, i.e., these sample values are already known. The remaining $L - 1$ samples between these values must be interpolated by the polyphase filters $p_\rho(m), \rho = 1, 2, \ldots, L - 1$. Since these filters are theoretically infinite in duration, they must be approximated, in practice, with finite-duration filters. Thus the interpolation "error" between the outputs of a practical system and an ideal system can be zero for $m = 0, \pm L, \pm 2L, \ldots$. However "in between" these samples, the error will always be nonzero.

By choosing a design that does not specifically satisfy the constraints of Eq. (3.104) or (3.105), we can make a tradeoff between errors that occur at sample times $m = 0, \pm L, \pm 2L, \ldots$, and errors that occur between these samples.

Another "time-domain" property that can be observed is that the ideal filter $h_I(k)$ is symmetric about zero, i.e.,

$$h_I(k) = h_I(-k) \qquad (3.106)$$

(Alternatively, for practical systems it may be symmetric about some fixed nonzero delay.) This symmetry does not necessarily extend directly to the polyphase filters since they correspond to sample values of $h_I(k)$ offset by some fraction of a sample. Their envelopes, however, are symmetrical (see Fig. 3.18).

The ideal time responses for $h_I(k)$ and $p_{\rho,I}(n)$ for decimators are the same as those of Eqs. (3.102) and (3.103), respectively, with L replaced by M.

3.4.4 Filter Designs Based on Conventional Techniques

Many filter design techniques can be applied to the above ideal filter characteristics to achieve practical designs. They may be applied for a variety of reasons depending on the nature of the application and on the degree of accuracy necessary in meeting desired error criteria. For example, window designs offer a simple, classical design procedure, and they have the property that they preserve the zero-crossing pattern of $h_i(k)$ in the actual design. They are limited, however, in the ability to control cutoff frequencies and stopband errors.

Equiripple filters have the advantage that highly efficient optimization techniques are available to design such filters with full control over the choice of error criteria and cutoff frequencies. They generally lead to very efficient designs, i.e., designs with a minimum filter length for a given tolerance specification. They can also be modified to accommodate the multiband design criteria illustrated in Fig. 3.27.

A third class of designs can be derived based on classical linear and Lagrange interpolation techniques. They are of interest from a historical point of view but are not widely used in modern digital signal processing.

Halfband filters are another class of designs of particular interest for interpolation or decimation by a factor of two. They are obtained by specifying the passband and stopband cutoff frequencies, ω_p, and ω_s, to be symmetrical about $\omega = \pi/2$, that is,

$$\omega_s = \pi - \omega_p \tag{3.107}$$

and the passband and stopband error tolerances δ_p and δ_s to be equal. This leads to filter designs with the symmetric property that

$$H(\omega) = 1 - H(\pi - \omega) \tag{3.108}$$

and with coefficient constraints

$$h(k) = \begin{cases} 1, & k = 0 \\ 0, & k = \pm 2, \pm 4, \ldots \end{cases} \tag{3.109}$$

The condition specified in Eq. (3.109) satisfies the zero-crossing criterion of ideal filters and results in efficient designs in that every other coefficient is zero and need not be computed in a practical implementation.

Comb filters are another class of designs that are used, particularly for multistage designs (to be discussed in Section 3.5). They are characterized by the impulse response

$$h(k) = \begin{cases} 1, & 0 \leq n \leq N - 1 \\ 0, & \text{elsewhere} \end{cases} \tag{3.110}$$

where N is the length of the filter. N is usually chosen such that $N = M$ for decimation by M, or $N = L$ for interpolation by L. The frequency response of this class of filters can be shown to be

$$H(\omega) = \left| \frac{\sin(\omega N/2)}{\sin(\omega/2)} \right| \tag{3.111}$$

Under restricted conditions, comb filters can be applied to meet multiband requirements of the type illustrated in Fig. 3.27. When applicable, they are very useful in practical implementations since all coefficients are 1.0 and they can be efficiently implemented by simple sum-and-dump or sample-and-hold procedures.

Finally, a major class of filter designs that may be applied to decimation and interpolation designs are the infinite impulse response (IIR) designs. They can be used when linear phase response is not required in the application. Numerous classical design techniques are available including well-known Butterworth, Bessel, Chebyshev, and elliptic designs.

3.4.5 Minimum Mean-Square Error Design of FIR Interpolator: Deterministic Signals

Thus far we have considered the design of decimation and interpolation filters based on conventional techniques applied to approximating the ideal choice of $h(k)$. In this section we consider an alternative point of view of filter design based on a minimum mean-square error criterion. In this approach the error criterion to be minimized is a function of the difference between the actual interpolated *signal* and its ideal value rather than a direct specification on the filter itself. We see in this section and in following sections that such an approach leads to a number of filter design techniques [17] that are capable of accounting directly for the spectrum of the signal being interpolated.

Figure 3.28(a) depicts the basic theoretical framework used for defining the above interpolator error criterion. We wish to design the FIR filter $h(m)$ such that it can be used to interpolate the signal $x(n)$ by a factor of L with minimum interpolation error. To define this error we need to compare the output of this actual interpolator with that of an ideal (infinite-duration) interpolator $h_l(m)$ whose characteristics were derived in Sections 3.4.2 and 3.4.3. This signal error is defined as

$$\Delta y(m) = y(m) - y_l(m) \tag{3.112}$$

where $y(m)$ is the output of the actual interpolator and $y_l(m)$ is the ideal output. We will consider interpolator designs that minimize the mean-square value of $\Delta y(m)$, defined as

$$E^2 = \| \Delta y(m) \|^2 = \lim_{K \to \infty} \frac{1}{2K + 1} \sum_{m=-K}^{K} \Delta y(m)^2 \tag{3.113a}$$

$$= \frac{1}{2\pi} \int_{-\pi}^{\pi} |\Delta Y(\omega')|^2 \, d\omega' \tag{3.113b}$$

Alternative approaches lead to designs that minimize the maximum value of $|\Delta Y(\omega')|$ over a prescribed frequency range or to designs that minimize the maximum value of $|\Delta y((m))|$ in the time domain [10].

The above design problems are greatly simplified by considering them in the framework of the polyphase structures as illustrated in Fig. 3.17. Here we see that the signal $y(m)$ is actually composed of interleaved samples of the signals $u_p(n)$, $\rho = 0$, 1, 2, . . . , $L - 1$, as shown by Fig. 3.28(b), where $u_p(n)$ is the output of the ρth-polyphase filter. Thus the errors introduced by each polyphase branch are orthogonal to each other (since they do not coincide in time), and we can define the error in the ρth branch as the error between the actual output and the output of the ρth branch of an ideal polyphase interpolator as shown in Fig. 3.28(b), i.e.,

$$\Delta u_\rho(n) = u_\rho(n) - u_{\rho,l}(n) \tag{3.114}$$

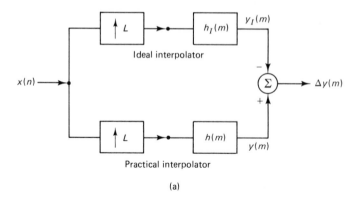

(a)

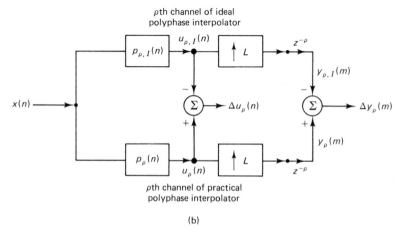

(b)

Figure 3.28 Framework for defining error criteria for interpolation filters.

Because of this orthogonality property we can *separately* and *independently* design each of the polyphase filters for minimum error and arrive at an overall interpolator design that minimizes the error $\|\Delta y(m)\|$. Thus a large (multirate) filter design problem can be broken down into L smaller (time-invariant) filter design problems.

In the case of the mean-square error criterion it can be seen that

$$E^2 = \|\Delta y(m)\|^2 = \frac{1}{L} \sum_{\rho=0}^{L-1} E_\rho^2 \tag{3.115}$$

where

$$E_\rho^2 = \|\Delta u_\rho(n)\|^2 \tag{3.116}$$

To minimize E^2 we then need to design L independent polyphase filters $p_\rho(n)$, $\rho = 0$, 1, 2, ..., $L - 1$, which independently minimize the respective mean-square errors E_ρ^2.

To analytically set up the filter design problem, we can note that the ideal polyphase filter response is

$$P_{\rho,I}(\omega) = e^{j\omega\rho/L} \tag{3.117}$$

which then leads to the form

$$E_\rho^2 = \| \Delta u_\rho(n) \|^2$$

$$= \frac{1}{2\pi} \int_{-\pi}^{\pi} |P_\rho(\omega) - e^{j\omega\rho/L}|^2 |X(\omega)|^2 \, d\omega \tag{3.118}$$

Equation (3.118) reveals that in the minimum-mean-square error design, we are in fact attempting to design a polyphase filter such that the integral of the squared difference between its frequency response $P_\rho(\omega)$ and a linear (fractional sample) phase delay $e^{j\omega\rho/L}$, weighted by the spectrum of the input signal $|X(\omega)|^2$, is minimized. Note also that the integral from $-\pi$ to π in Eq. (3.118) is taken over the frequency range of the input signal of the interpolator, not the output signal.

In practice this error criterion is often modified slightly by specifying that $X(\omega)$ is bandlimited to the range $0 \leq \omega \leq \alpha\pi$, where $0 < \alpha < 1$, i.e.,

$$|X(\omega)| = 0, \qquad \text{for } |\omega| \geq \alpha\pi \tag{3.119}$$

Then Eq. (3.118) can be expressed as

$$E_{\rho,\alpha}^2 = \frac{1}{2\pi} \int_{-\alpha\pi}^{\alpha\pi} |P_\rho(\omega) - e^{j\omega\rho/L}|^2 |X(\omega)|^2 \, d\omega \tag{3.120}$$

where the subscript α is used to distinguish this norm from the one in Eq. (3.118). Alternatively we can consider the modification in Eq. (3.120) as a means of specifying that we want the design of $P_\rho(\omega)$ to be minimized only over the frequency range $0 \leq \omega \leq \alpha\pi$ and that the range $\alpha\pi \leq \omega \leq \pi$ is allowed to be a transition region. Then α can be used as a parameter in the filter design procedure.

The solution to the minimization problem of Eq. (3.120) involves expressing the norm $E_{\rho,\alpha}^2$ directly in terms of the filter coefficients $p_\rho(n)$. Then, since the problem is formulated in a classical mean-square sense, it can be seen that $E_{\rho,\alpha}^2$ is a quadratic function of the coefficients $p_\rho(n)$ and thus it has a single, unique minimum for some optimum choice of coefficients. At this minimum point, the derivative of $E_{\rho,\alpha}^2$ with respect to all the coefficients $p_\rho(n)$ is zero. Thus the second step in the solution is to take the derivative of $E_{\rho,\alpha}^2$ with respect to the coefficients $p_\rho(n)$ and set it equal to zero. This leads to a set of linear equations in terms of the coefficients $p_\rho(n)$, and the solution to this set of equations gives the optimum choice of coefficients that minimize $E_{\rho,\alpha}^2$. This minimization problem is solved for each value of ρ, $\rho = 0, 1, 2, \ldots, L-1$, and each solution provides the optimum solution for one of the polyphase filters. Finally, these optimum polyphase filters can be combined as in Eqs. (3.86) to (3.88) to obtain the optimum prototype filter $h(m)$ that minimizes the overall norm. The details for this approach can be found in [17]. Also, reference [17] contains a computer program that designs interpolation filters according to the above techniques and that greatly simplifies the task of designing these filters.

The minimum mean-square error interpolators designed using the procedure described have a number of interesting properties.

1. The resulting filters have the same symmetry properties as the ideal filters in Eq. (3.106).
2. The minimum error $\min E^2_{\rho,\alpha}$ for the polyphase filters also satisfies the symmetry condition

$$\min E^2_{\rho,\alpha} = \min E^2_{L-\rho,\alpha} \qquad (3.121)$$

This error increases monotonically as ρ increases (starting with $E^2_{0,\alpha} = 0$) until $\rho = L/2$, at which point it decreases monotonically according to Eq. (3.121). Thus the greatest error occurs in interpolating sample values that are halfway between two given samples. This normalized error is closely approximated by the sine-squared function, i.e.,

$$\frac{\min E^2_{\rho,\alpha}}{\min E^2_{L/2,\alpha}} \approx \sin^2\left(\frac{\rho\pi}{L}\right) \qquad (3.122)$$

3. If an interpolator is designed for a given signal with a large value of L, all interpolators whose lengths are fractions of L are obtained by simply sampling the original filter; i.e., if we design an interpolator for $L = 100$, then for the same parameters α and R we can derive from this filter the optimum mean-square error interpolators for $L = 50$, 25, 20, 10, 5, and 2 by taking appropriate samples (or appropriate polyphase filters).

Figure 3.29 shows an example of the impulse response and frequency response for a minimum mean-square error interpolation filter with parameter values $\alpha = 0.5$, $N = 49$, $R = (N - 1)/2L = 3$, $L = 8$, and assuming that $|X(\omega)| = 1$.

3.5 MULTISTAGE IMPLEMENTATIONS OF SAMPLING RATE CONVERSION

The concept of using a series of stages to implement a sampling rate conversion system can be extended to the case of simple interpolators and decimators [10], as shown in Fig. 3.30 and 3.31. Consider first a system for interpolating a signal by a factor of L as shown in Fig. 3.30(a). We denote the original sampling frequency of the input signal $x(n)$ as F_0, and the interpolated signal $y(m)$ has a sampling rate of LF_0. If the interpolation rate L can be factored into the product

$$L = \prod_{i=1}^{I} L_i \qquad (3.123)$$

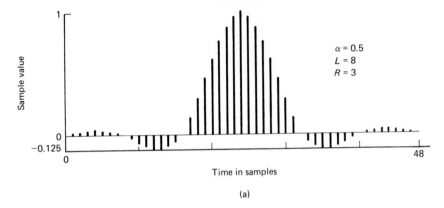

(a)

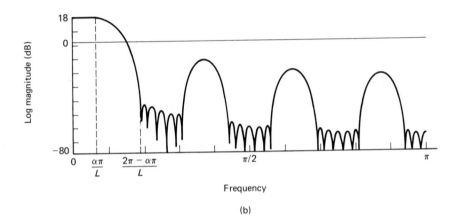

(b)

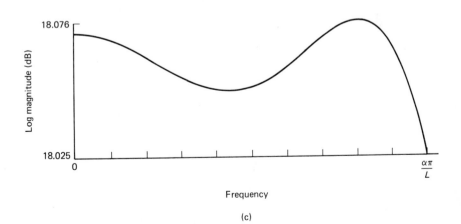

(c)

Figure 3.29 The impulse and frequency responses of a minimum mean-square error interpolation filter with $\alpha = 0.5$, $R = 3$, and $L = 8$.

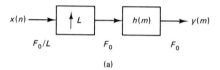

(a)

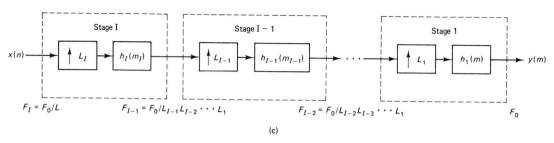

(b)

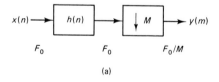

(c)

Figure 3.30　Steps in constructing a multistage interpolator for interpolation by a factor L.

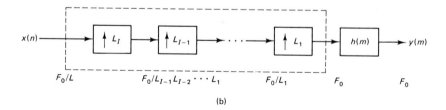

(a)

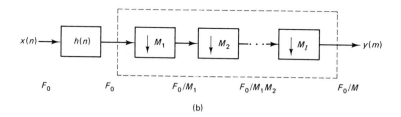

(b)

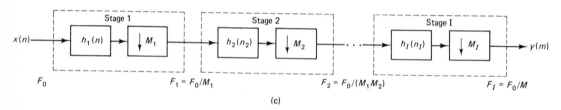

(c)

Figure 3.31　Steps in constructing a multistage decimator for decimation by a factor M.

where each L_i is an integer, then we can express this network in the form shown in Fig. 3.30(b). This structure, by itself, does not provide any inherent advantage over the structure of Fig. 3.30(a). However, if we modify the structure by introducing a lowpass filter between *each pair* of the sampling rate increasing boxes, we produce the structure of Fig. 3.30(c). This structure has the property that the sampling rate increase occurs in a series of I stages, where each stage (shown within dashed boxes) is an *independent* interpolation stage.

Similarly, for an M-to-1 decimator, if the overall decimation rate M can be factored into the product

$$M = \prod_{j=1}^{J} M_j \qquad (3.124)$$

then the general single-stage decimator structure of Fig. 3.31(a) can be converted into the multistage structure of Fig. 3.31(b). Again, each of the stages within the structure of Fig. 3.31(b) is an *independent* decimation stage.

Perhaps the most obvious question that arises from the preceding discussion is why we should consider such multistage structures. At first glance it would appear as if we are greatly increasing the overall computation (since we have inserted filters between each pair of stages) of the structure. This, however, is precisely the opposite of what occurs in practice. The reasons for considering multistage structures, of the types shown in Figs. 3.30(c) and 3.31(b), are as follows:

1. Significantly reduced computation to implement the system
2. Reduced storage in the system
3. Simplified filter design problem
4. Reduced finite word length effects, i.e., roundoff noise and coefficient sensitivity, in the implementations of the digital filters

These structures, however, are not without some drawbacks:

1. Increased control structure required to implement a multistage process
2. Difficulty in choosing the appropriate values of I (or J) of Eq. (3.123) and the best factors L_i (or M_j)

It is the purpose of this section to briefly show why and how a multistage implementation of a sampling rate conversion system can be (and generally is) more efficient than the standard single-stage structure for the following cases:

$$\begin{aligned}
&\textit{Case 1:} \quad L \gg 1 \quad (M = 1) \\
&\textit{Case 2:} \quad M \gg 1 \quad (L = 1) \\
&\textit{Case 3:} \quad L/M \approx 1 \quad \text{but } L \gg 1, M \gg 1
\end{aligned}$$

Cases 1 and 2 are high-order interpolation and decimation systems, and Case 3 occurs when a slight change in sampling rate is required (e.g., $L/M = 80/69$).

3.5.1 Computational Efficiency of a Two-Stage Structure: A Design Example

Since the motivation for considering multistage implementations of sampling rate conversion systems is the potential reduction in computation, it is worthwhile to present a simple design example that illustrates the manner in which the computational efficiency is achieved.

The design example is one in which a signal $x(n)$ with a sampling rate of 10,000 Hz is to be decimated by a factor of $M = 100$ to give the signal $y(m)$ at a 100-Hz rate. Figure 3.32(a) shows the standard single-stage decimation network that implements the desired process. It is assumed that the passband of the signal is from 0 to 45 Hz and that the band from 45 to 50 Hz is a transition band. Hence the specifications of the required lowpass filter are as shown in Fig. 3.32(b). We assume, for simplicity, that the design formula [10]

$$N \approx \frac{D(\delta_p, \delta_s)}{(\Delta F/F_s)} \tag{3.125}$$

can be used to give the order N of a symmetric FIR filter with maximum passband ripple δ_p, maximum stopband ripple δ_s, transition width ΔF, and sampling frequency F_s. For the lowpass filter of Fig. 3.32(b), we have

$$\Delta F = 50 - 45 = 5 \text{ Hz}$$

$$F_s = 10,000 \text{ Hz}$$

$$\delta_p = 0.01$$

$$\delta_s = 0.001$$

$$D(\delta_p, \delta_s) = 2.54$$

giving, from Eq. (3.125), $N \approx 5080$. The overall computation in multiplications per second (MPS) necessary to implement this system is

$$R = \frac{NF}{2M} = \frac{(5080)10,000}{2(100)} = 250,000 \text{ MPS}$$

i.e., a total of 250,000 MPS at the 10,000-Hz rate is required to implement the system of Fig. 3.32(a) (assuming the use of symmetry of $h(n)$).

Consider now the two-stage implementation shown in Fig. 3.32(c). The first stage decimates the signal by a factor of 50, and the second stage decimates the (already decimated) signal by a factor of 2, giving a total decimation factor of 100. The resulting filter specifications are illustrated in Fig. 3.32(d). For the first stage the passband is from 0 to 45 Hz, but the transition band extends from 45 to 150 Hz. Since the sampling rate at the output of the first stage is 200 Hz, the residual signal energy from 100 to 150 Hz gets aliased back into the range 50 to 100 Hz after decimation by the factor of 50. This aliased signal then gets removed in the second stage. For the second stage the passband extends from 0 to 45 Hz, and the transition band extends

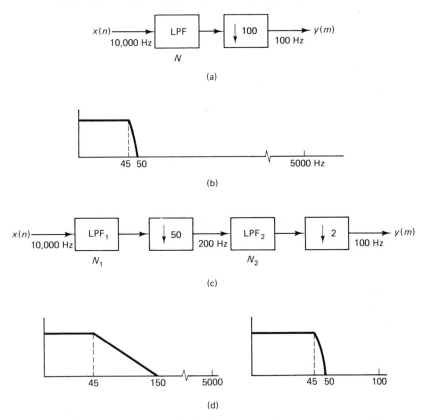

Figure 3.32 Simple example of (a) a one-stage and (c) a two-stage network for decimation by a factor of 100:1 and their respective filter design requirements (b and d).

from 45 to 50 Hz with a sampling rate of 200 Hz. One other change in the filter specifications occurs because we are using a two-stage filtering operation. The pass-band ripple specification of the two-stage structure is reduced to $\delta_p/2$ (since each stage can theoretically add passband ripple to each preceding stage). The stopband ripple specification does not change since the cascade of two lowpass filters reduces only the stopband ripples. Hence the $D(\delta_p, \delta_s)$ function in the filter design equation becomes $D(\delta_p/2, \delta_s)$ for the filters in the two-stage implementation. Since $D(\delta_p, \delta_s)$ is relatively insensitive to factors of 2, only slight changes occur (from 2.54 to 2.76) due to this factor. For the specific example of Fig. 3.32(c) we get (for the first stage)

$$N_1 = \frac{2.76}{(150 - 45)/10,000} = 263$$

$$R_1 = \frac{N_1 F}{2(M_1)} = \frac{263(10,000)}{(2)(50)}$$

$$= 26,300 \text{ MPS}$$

For the second stage we get

$$N_2 = \frac{2.76}{(5/200)} = 110.4$$

$$R_2 = \frac{110.4(200)}{(2)(2)} = 5500 \text{ MPS}$$

The total computation for the two-stage implementation is $R_1 + R_2 = 26,300 + 5500 = 31,800$ MPS. Thus a reduction in computation of almost 8 to 1 is achieved in the two-stage decimator over a single-stage decimation for this design example.

It is easy to see where the reduction in computation comes from for the multistage decimator structure by examining Eq. (3.125). We see that the required filter orders are directly proportional to $D(\delta_p, \delta_s)$ and F and inversely proportional to ΔF, the filter transition width. For the early stages of a multistage decimator, although the sampling rates are large, equivalently the transition widths are very large, thereby leading to relatively small values of filter length N. For the last stages of a multistage decimator, the transition width becomes small but so does the sampling rate, and the combination again leads to relatively small values of required filter lengths. We see from the preceding analysis that computation is kept low in each stage of the overall multistage structure.

The simple example presented above is by no means a complete picture of the capabilities and sophistication that can be found in multistage structures for sampling rate conversion. It is merely intended to show why such structures are of fundamental importance for many practical systems in which sampling rate conversion is required.

3.5.2 Design Considerations for Multistage Decimators and Interpolators

A large number of parameters and tradeoffs are involved in the design of multistage decimator and interpolator systems, including the following:

1. The number of stages to realize an overall decimation (or interpolation) factor M (or L) most efficiently
2. The choice of decimation ratios that are appropriate for each stage
3. The types of digital filters used in each stage
4. The structure used to implement filters in each stage
5. The required filter order in each stage
6. The resulting amount of computation, storage, and processing delay incurred in each stage and in the overall structure

As in most signal processing design problems, a number of factors influence each choice, and it is not a simple matter to make any one choice over all others. However, three general approaches and design philosophies have been applied to these designs. Each approach has slightly different advantages and disadvantages, and often a mix of these strategies is applied in practical applications to obtain the most effective design.

The first general approach is based on the formulation of the design problem in terms of a mathematically defined optimization problem with decimation (or interpolation) factors in each stage treated as continuous variables. The objective function to be minimized is then expressed as an analytical measure of the efficiency of the design. The design procedure is then performed by finding the most efficient solution (i.e., the best choice of decimation or interpolation factors for each stage) for each value of I (the number of stages) and then selecting that choice of I and its solution that give the best the overall solution [10].

A second approach is based on the use of halfband filter designs with a 2-to-1 decimation (or interpolation) factor for each stage. These filters have the advantage that approximately half of the filter coefficients are zero value (because of the halfband designs) and need not be implemented. This design procedure works best when the overall decimation or interpolation ratio is a power of two.

A third design approach combines the use of simple comb filters, where possible, in the initial stages of multistage decimators (or the final stages of multistage interpolators), followed by halfband filters and other special classes of filter designs. The idea here is to use a large number of stages to implement a large change in sampling rates and to use extremely simple linear phase FIR filters when possible.

In the preceding discussion we have very briefly outlined the issues and approaches used to design multistage decimation and interpolation systems. In addition to issues of overall computation rate, the cost of the control structure associated with multiple stages must be considered in the implementation of the design. As in most real-world problems there is no simple or universal answer about what design approach is best, and in practice a combination of these techniques often yields the most appropriate tradeoffs.

3.6 SIGNAL PROCESSING OPERATIONS BASED ON DECIMATION AND INTERPOLATION CONCEPTS

In the previous sections we discussed the basic concepts of sampling rate conversion and the issues involved in the efficient design of decimators and interpolators. In this section we show that these concepts not only are useful in designing efficient integer or ratio-of-integer sampling rate conversion systems but also can be applied more generally to the efficient design of a broad range of signal processing operations.

3.6.1 Sampling Rate Conversion Between Systems with Incommensurate Sampling Rates

So far we have considered multirate systems in which the various sampling rates within the system are related by exact integer or rational fraction ratios. That is, the sampling rates within the process are all generated from some common (higher-rate) clock. In practice, however, it is sometimes desired to interface digital systems that are controlled by *independent* (incommensurate) clocks. For example, it may be desired to exchange signals between two different systems, both of which are intended to be sampled at a rate F_s. However, due to practical limitations on the accuracy of the clocks, the first system may be sampled at an actual rate of $F_s + \epsilon_1$ and the second system may be sampled at an actual rate of $F_s + \epsilon_2$ where ϵ_1 and ϵ_2 represent slowly varying components of drift as a function of time. If we simply exchange digital signals between these two systems (e.g., by means of a sample-and-hold process) samples may be either lost or repeated in the exchange due to the relative "slippage" of the two clocks. If the signals are highly uncorrelated from sample to sample (i.e., sampled at their Nyquist rates), this process can introduce large spikes or errors into the signals. The rate of occurrence of these errors is directly related to the amount of sampling rate slippage, that is, to the ratio $(\epsilon_1 - \epsilon_2)/F_s$.

One way to avoid the above problem is to interface the two digital systems through an analog connection, that is, to convert the signals to analog signals and then resample them with the new clock. In principle this process provides an error-free interface. In practice, however, it is limited by the practical capabilities and expense of the A/D and D/A conversion process as well as the dynamic range of the analog connection.

A more attractive all-digital approach to this problem can be accomplished by applying the multirate techniques discussed above to, in effect, duplicate the analog process in digital form. Figure 3.33 shows an example of a system for transferring a digital signal $x(n)$ from system 1 (with sampling rate $F_s + \epsilon_1$) to system 2 (with sampling rate $F_s + \epsilon_2$). The signal $x(n)$ is first interpolated by a 1-to-L interpolator (where $L \gg 1$) to produce the highly oversampled signal $y(m)$ (at sampling rate $L(F_s + \epsilon_1)$). This signal is then converted to a signal $\hat{y}(m)$ (at the sampling rate $L(F_s + \epsilon_2)$) through a digital sample-and-hold procedure that interfaces the two incommensurate sampling rates. It is then decimated by an L-to-1 decimator to produce the signal $\hat{x}(n)$ for input to system 2. The interface between the oversampled digital signals $y(m)$ and $\hat{y}(m)$ plays the same role as that of an analog interface and as $L \to \infty$ it is equivalent. The advantage is that the system is all-digital and the accuracy of the conversion can be designed with any degree of desired precision.

Although samples may be repeated or dropped in the sample-and-hold conversion process between $y(m)$ and $\hat{y}(m)$, it can be shown that the effects of these errors become small as L becomes large. This can be seen by considering the sample-to-sample difference of the signal $y(m)$ (or $\hat{y}(m)$) relative to its actual value as a function of L [10]. The sampling rate conversion can be efficiently realized using the structures and filter designs in Sections 3.3–3.5. In particular, for large values of L, the multi-

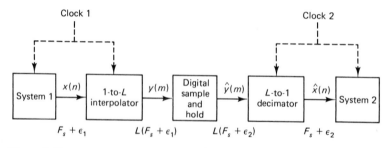

Figure 3.33   A multirate approach to sampling rate conversion between systems with incommensurate sampling rates.

stage designs are appropriate. Also it should be noted that although the system in Fig. 3.33 is described in terms of a conversion process between two "equivalent" but incommensurate sampling rates $F_s + \epsilon_1$ and $F_s + \epsilon_2$, it can be readily extended to a process of conversion between any two incommensurate sampling rates by choosing different ratios for the interpolator and decimator. It is necessary only that the signals $y(m)$ and $\hat{y}(m)$ have "equivalent" rates to minimize the amount of sample repeating or dropping in the sample-and-hold process.

3.6.2 Design of Fractional Sample Phase Shifters Based on Multirate Concepts

Many signal processing applications require a network that essentially delays the input signal by a fixed number of samples. When the desired delay is an integer number of samples, at the current sampling rate, such a network is trivially realized as a cascade of unit delays. However, when delays of a fraction of a sample are required, the processing required to achieve such a delay is considerably more difficult. We show here how multirate signal processing concepts can be used to greatly simplify the process required to design noninteger delay as long as the desired delay is a rational fraction of a sample.

Consider the ideal delay network $h_{AP}(n)$ of Fig. 3.34(a). The desired allpass network processes $x(n)$ to give the output $y(n)$ so that the relationship between their Fourier transforms is

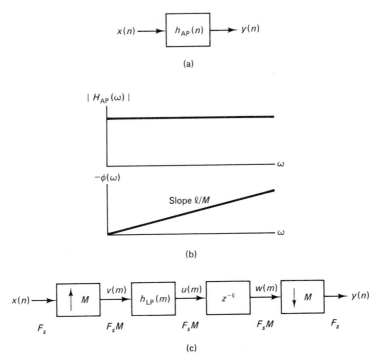

Figure 3.34   A multirate structure for realizing a fixed delay of ℓ/M samples.

$$Y(\omega) = e^{-j\omega\ell/M} X(\omega) \tag{3.126}$$

where ℓ and M are any integers. In the time domain this amounts to a delay of the envelope of the signal $x(n)$ by a fraction of a sample ℓ/M. The magnitude and phase responses of $h_{AP}(n)$, shown in Fig. 3.34, are of the form

$$H_{AP}(\omega) = |H_{AP}(\omega)| e^{j\phi(\omega)} \tag{3.127a}$$

where

$$|H_{AP}(\omega)| = 1 \tag{3.127b}$$

$$\phi(\omega) = -\frac{\ell\omega}{M} \tag{3.127c}$$

and where $\phi(\omega)$ denotes the phase of $H_{AP}(\omega)$. (At this point the observant reader should notice the similarity between the desired response and that of a polyphase filter. We return to this equivalence later in this section.) It should be clear that for arbitrary values of ℓ and M, the desired frequency response of Eqs. (3.127) cannot be achieved exactly by an FIR or an IIR filter. An FIR filter (all zeros) cannot achieve an exact allpass magnitude response, and an IIR filter (poles and zeros) cannot achieve an exact linear phase response. Thus the desired noninteger delay network $h_{AP}(n)$ cannot be realized exactly but can only be approximated through a design procedure.

Using multirate principles, this design problem can be clearly defined as illustrated in Fig. 3.34(c). The key to this procedure is the realization that a delay of ℓ/M samples at rate F_s is equivalent to a delay of ℓ samples (i.e., an integer delay) at rate $F_s M$. Hence the structure in Fig. 3.34(c) first raises the sampling rate of the signal to $F_s M$, filters the signal with a lowpass filter $h_{LP}(m)$ to eliminate images of $x(n)$, delays the signal by ℓ samples, and then decimates it back to the original sampling rate.

A simple analysis of the structure of Fig. 3.34(c) gives

$$V(\omega) = X(\omega M) \tag{3.128}$$

$$U(\omega) = H_{LP}(\omega)X(\omega M) \tag{3.129}$$

$$W(\omega) = U(\omega)e^{-j\omega\ell}$$

$$= H_{LP}(\omega)X(\omega M)e^{-j\omega\ell} \tag{3.130}$$

$$Y(\omega) = \frac{1}{M}\sum_{r=0}^{M-1} W((-2\pi r/M) + (\omega/M)) \tag{3.131}$$

We assume that $H_{LP}(\omega)$ sufficiently attenuates the images of $X(\omega)$ so that they are negligible in Eq. (3.131), thereby giving only the $r = 0$ term, that is,

$$Y(\omega) \approx \frac{1}{M}W(\omega/M)$$

$$\approx \frac{1}{M}H_{LP}(\omega/M)e^{-j\omega\ell/M}X(\omega) \tag{3.132}$$

We further assume that $H_{LP}(\omega)$ is an FIR filter with exactly linear phase, whose delay (at the high rate) is $(N-1)/2$ samples, and this value is chosen to be an integer delay at the low rate, that is,

$$\frac{N-1}{2} = IM \qquad (3.133)$$

or

$$N = 2IM + 1 \qquad (3.134)$$

We also require $H_{LP}(\omega)$ to have a magnitude response essentially equal to M (to within a small tolerance) in the passband, thereby giving for $Y(\omega)$

$$Y(\omega) \approx e^{-j\omega l} e^{-j\omega \ell/M} X(\omega) \qquad (3.135)$$

or, as an equivalent network,

$$\frac{Y(\omega)}{X(\omega)} \approx e^{-j\omega l} e^{-j\omega \ell/M} \qquad (3.136)$$

Thus the structure of Fig. 3.34(c) is essentially an allpass network with a fixed integer delay of I samples and a variable, noninteger delay of ℓ/M samples.

One efficient implementation of the multirate allpass filter of Fig. 3.34(c) is given in Fig. 3.35. A polyphase structure is used to realize the sampling rate increase and lowpass filtering based on the counterclockwise commutator structure where

$$p_p(n) = h_{LP}(nM + \rho), \qquad 0 \le \rho \le M - 1 \qquad (3.137)$$

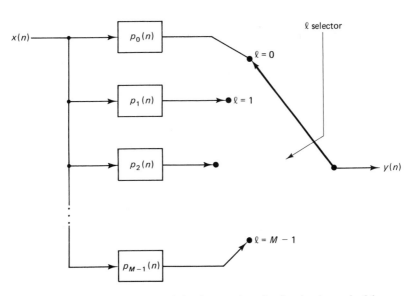

Figure 3.35 A polyphase network implementation of a fractional sample delay network.

The delay of ℓ samples is implemented as a new initial position of the commutator corresponding to the $m = 0$ sample. Finally, the decimation by M is implemented as a fixed arm position of the commutator, since it is back at the original position each M samples.

Thus for a single *fixed* delay of ℓ/M samples, only one branch (corresponding to the $\rho = \ell$th polyphase filter branch) is required, and for a network that requires all possible values of ℓ (from 0 to $M - 1$) the entire network of Fig. 3.35 is required, i.e., it represents a selectable choice of M different fractional delays.

A key point made in this section is that the design of an N-tap time-invariant filter with a noninteger delay (and flat magnitude response), such as $h_{AP}(n)$ in Fig. 3.34(a), can be readily transformed to that of an NM-tap lowpass filter design. This transformation is accomplished by means of an appropriate multirate interpretation of the problem and an application of the concept of polyphase filters.

3.6.3 Multirate Implementation of Lowpass Filters

In Section 3.2 we introduced the idea of cascading a 1-to-L interpolator with an M-to-1 decimator and showed that the resulting structure implemented a sampling rate conversion by a factor of L/M. Consider now cascading an M-to-1 decimator with a 1-to-M interpolator as shown in Fig. 3.36(a). Intuitively, we see that the overall system relating the output $y(n)$ to the input $x(n)$ acts like a lowpass filter (due to $h_1(n)$ and $h_2(n)$) with aliasing (due to the decimation) and imaging (due to the interpolation). When these components are appropriately removed by filters $h_1(n)$ and $h_2(n)$ it can be shown that the overall system acts like a well-behaved lowpass digital filter [10].

This same line of reasoning can be applied to the multirate lowpass system of Fig. 3.36(b). It can be shown that as the bandwidth of the resulting lowpass filter (relative to the sampling rate) becomes small, the benefits of the multistage approach become large. Since the design is identical to that of the cascaded multistage decimators and interpolators, the same methodologies discussed in Section 3.5 apply to the efficient design of narrowband lowpass digital filters. This approach leads to designs that can be significantly more efficient and also less sensitive to effects of

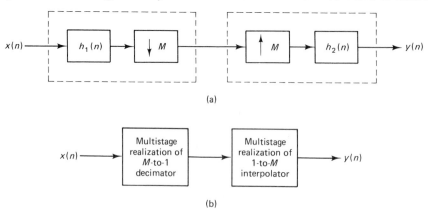

Figure 3.36 Block diagram of a multistage realization of a multirate lowpass filter structure.

coefficient quantization and roundoff noise than classical implementations. To illustrate the above concepts more explicitly we conclude this section with a design example that illustrates the gains in efficiency attainable using a multirate, multistage implementation of a narrowband lowpass filter over a standard, direct-form implementation.

Consider the design of a narrowband lowpass filter with specifications $F_s = 1.0$, $f_p = \omega_p/2\pi = 0.00475$, $f_s = \omega_s/2\pi = 0.005$, $\delta_p = 0.001$, and $\delta_s = 0.0001$. Clearly, this filter is a very narrowband lowpass filter with very tight ripple specifications.

For the single-rate, standard, direct-form FIR implementation, the *estimate* of filter order, based on the standard equiripple design formula [2] is $N_0 = 15{,}590$; that is, an extremely high-order FIR filter is required to meet these filter specifications. Such a filter would never be designed in practice, and even if it could be designed, the implementation would yield excessive roundoff noise and therefore would not be useful. However, for theoretical purposes, we postulate such a filter and compute its multiplication rate (using symmetry of the impulse response) as

$$R_0 = \frac{N_0}{2} = 7795 \text{ MPS}$$

For a multirate implementation, a decimation ratio of $M = 100 = 0.5/0.005$ can be used. For a one-stage implementation of the decimator and interpolator as shown in Fig. 3.36(a), the (estimated) required filter order for $h_1(n)$ and $h_2(n)$ is $N_1 = 16{,}466$. Again, we could never really design such a high-order filter, but, for theoretical purposes, we can compute its multiplication rate (again employing symmetry in both the decimator and interpolator) to give

$$R_1 = \frac{N_1}{2(100)}(2) = 165 \text{ MPS}$$

resulting in a potential savings (of multiplications per second) of about 47.2 to 1 over the direct-form implementation.

For a two-stage implementation of the decimator and interpolator (i.e., two stages of decimation followed by two stages of interpolation), a reasonable set of ratios for decimation is $M_1 = L_1 = 50, M_2 = L_2 = 2$, resulting in filter orders of $N_1 = 423$ and $N_2 = 347$. The total multiplication rate for the two-stage structure is

$$R_2 = 2\left[\frac{N_1}{2(50)} + \frac{N_2}{2(100)}\right] = 11.9 \text{ MPS}$$

resulting in a potential savings of about 655 to 1 over the direct-form structure.

Finally, if we use a three-stage implementation of the decimator and interpolator, a reasonable set of ratios is $M_1 = L_1 = 10, M_2 = L_2 = 5, M_3 = L_3 = 2$, resulting in filter orders of $N_1 = 50, N_2 = 44$, and $N_3 = 356$. The total multiplication rate is then

$$R_3 = 2\left[\frac{N_1}{2(10)} + \frac{N_2}{2(50)} + \frac{N_3}{2(100)}\right] = 9.4 \text{ MPS}$$

resulting in a savings of 829 to 1 over the direct-form implementation.

By way of comparison, an elliptic filter meeting the given filter design specifications is of 14th order and, in a cascade realization, requires 22 MPS. This shows that for the given design example, the three-stage FIR design is about 2.3 times more efficient than a single-stage, fixed-rate IIR filter (and it has linear phase). However, it requires a significantly larger amount of storage for coefficients and data than does the IIR design.

A summary of these results (for the FIR filters) is presented in Table 3.1. The key point to note is the spectacular gains in efficiency that are readily achievable by a multistage, multirate implementation over a direct FIR filter implementation. Furthermore, we see that, for the three-stage structure, the resulting FIR designs can be readily achieved and implemented without problem due to excessive length of the impulse response or the excessive roundoff noise in the implementation. Thus we have achieved both an efficient implementation *and* an efficient method of designing very long, very narrow-bandwidth lowpass filters with tight ripple tolerances.

TABLE 3.1 COMPARISONS OF FILTER CHARACTERISTICS FOR SEVERAL MULTISTAGE IMPLEMENTATIONS OF A LOWPASS FILTER WITH SPECIFICATIONS $F_s = 1.0$, $f_p = 0.00475$, $f_s = 0.005$, $\delta_p = 0.001$, $\delta_s = 0.0001$

	Direct Form	One-Stage	Two-Stage	Three-Stage
Decimation rates	—	100	50 2	10 5 2
Filter lengths	15,590	16,466	423 347	50 44 356
MPS	7795	165	11.9	9.4
Rate reduction (MPS)	1	47.2	655	829
Total storage for filter coefficients	7795	8233	385	225

3.6.4 Sampling Rate Conversion Applied to Bandpass Signals

Until now we have assumed that the signals with which we are dealing are lowpass signals and that the filters required for decimation and interpolation are therefore lowpass filters that preserve the baseband signals of interest. In many practical systems, however, it is often necessary to deal with bandpass signals as well as lowpass signals. In this section we show how the concepts of decimation and interpolation can be applied to systems in which bandpass signals are present.

Figure 3.37(a) shows an example of the discrete-time Fourier transform of a digital bandpass signal $S(2\pi fT)$ that contains spectral components only in the frequency range $f_\ell < |f| < f_\ell + f_\Delta$. If we apply directly the concepts of lowpass sam-

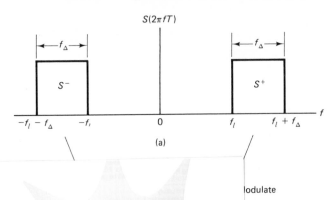

$S(2\pi fT)$

S^- S^+

$-f_l - f_\Delta$ $-f_l$ 0 f_l $f_l + f_\Delta$ f

(a)

Figure 3.37 Bandpass signal and its lowpass translated representation.

rate, F_w, necessary to represent this signal must be [ ]y component in $S(2\pi fT)$, that is, $F_w \geq 2(f_\ell + f_\Delta)$. [ ]mponent of $S(2\pi fT)$ associated with $f > 0$ and S^- [ ]) associated with $f < 0$, as seen in Fig. 3.37. Then, [ ]g) S^+ to the band 0 to f_Δ and S^- to the band $-f_\Delta$ to [ ]ve see that a new signal $S_\gamma(2\pi fT)$ can be generated [ ]in the sense that $S(2\pi fT)$ can uniquely be recon- [ ]erse process of bandpass translation. (Actually, we [ ]ideband" modulated version of $S_\gamma(2\pi fT)$.) By ap- [ ]oling to $S_\gamma(2\pi fT)$, however, we can see that the [ ]epresent this signal is now $F_\Delta \geq 2f_\Delta$, which can be [ ]pecified above (if $f_1 \gg f_\Delta$). Thus we see that by an [ ]ation followed by lowpass sampling, any real band- pass signal with (positive-frequency) bandwidth f_Δ can be uniquely sampled at a rate $F_\Delta \geq 2f_\Delta$ (i.e., such that the original bandpass signal can be uniquely reconstructed from the sampled representation).

Perhaps the simplest and most direct approach to decimating or interpolating digital bandpass signals, sometimes referred to as *integer-band sampling,* is to take advantage of the inherent frequency translating (i.e., aliasing or imaging) properties of decimation and interpolation. As discussed in Section 3.1, sampling and sampling rate conversion can be viewed as a modulation process in which the spectrum of the digital signal contains periodic repetitions of the baseband signal (images) spaced at harmonics of the sampling frequency. This property can be used to advantage when dealing with bandpass signals by associating the bandpass signal with one of these images instead of with the baseband.

Figure 3.38(a) illustrates an example of this process for the use of decimation

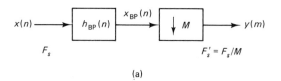

(a)

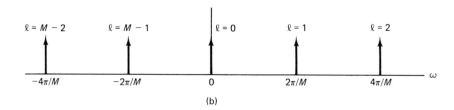

(b)

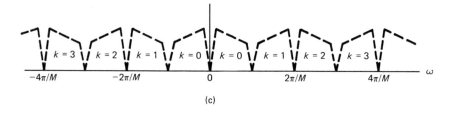

(c)

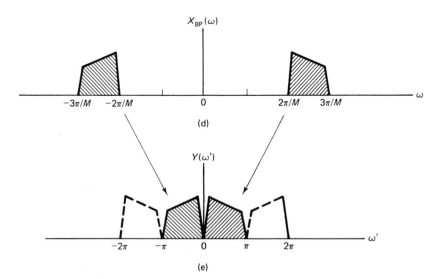

Figure 3.38 Integer-band decimation and a spectral interpretation for the $k = 2$ band.

by the integer factor M. The input signal $x(n)$ is first filtered by the bandpass filter $h_{BP}(n)$ to isolate the frequency band of interest. The resulting bandpass signal, $x_{BP}(n)$, is then directly recduced in sampling rate by an M-sample compressor giving the final output, $y(m)$. We see that this system is identical to that of the integer lowpass decimator discussed in Section 3.2.1, with the exception that the filter is a bandpass filter rather than a lowpass filter. Thus the output signal $Y(\omega')$ can be expressed as

$$Y(\omega') = \frac{1}{M} \sum_{\ell=0}^{M-1} H_{BP}((\omega' - 2\pi\ell)/M)X((\omega' - 2\pi\ell)/M) \qquad (3.138)$$

From Eq. (3.138) we see that $Y(\omega')$ is composed of M aliased components of $X(\omega')H_{BP}(\omega')$ modulated by factors of $2\pi\ell/M$. The function of the filter $H_{BP}(\omega)$ is to remove (attenuate) all aliasing components except those associated with the desired band of interest. Since the modulation is restricted to values of $2\pi\ell/M$, we can see that only specific frequency bands are allowed by this method. As a consequence, the choice of the filter $H_{BP}(\omega)$ is restricted to approximate one of the M ideal characteristics

$$H_{BP,I}(\omega) = \begin{cases} 1, & k\dfrac{\pi}{M} < |\omega| < (k+1)\dfrac{\pi}{M} \\ 0, & \text{otherwise} \end{cases} \qquad (3.139)$$

where $k = 0, 1, 2, \ldots, M - 1$; that is $H_{BP}(\omega)$ is restricted to bands $\omega = k\pi/M$ to $\omega = (k+1)\pi/M$, where π/M is the bandwidth.

Figures 3.38(b)–(e) illustrate this approach. Figure 3.38(b) shows the M possible modulating frequencies that are a consequence of the M-to-1 sampling rate reduction; that is, the digital sampling function (a periodic train of unit samples spaced M samples apart) has spectral components spaced $2\pi\ell/M$ apart. Figure 3.38(c) shows the "sidebands" that are associated with these spectral components, which correspond to the M choices of bands as defined by Eq. (3.139). They correspond to the bands that are aliased into the baseband of the output signal $Y(\omega')$ according to Eq. (3.139).

Figure 3.38(d) illustrates an example in which the $k = 2$ band is used, such that $X_{BP}(\omega)$ is bandlimited to the range $2\pi/M < |\omega| < 3\pi/M$. Since the process of sampling rate compression by M to 1 corresponds to a convolution of the spectra of $X_{BP}(\omega)$ (Fig. 3.38d) and the sampling function (Fig. 3.38b), this band is lowpass translated to the baseband of $Y(\omega')$ as seen in Fig. 3.38(e). Thus the processes of modulation and sampling rate reduction are achieved simultaneously by the M-to-1 compressor.

Figure 3.39 illustrates a similar example for the $k = 3$ band such that $X_{BP}(\omega)$ is bandlimited to the band $3\pi/M < |\omega| < 4\pi/M$. In this case it is seen that the spectrum is inverted in the process of lowpass translation. If the noninverted representation of $y(m)$ is desired, it can easily be achieved by modulating $y(m)$ by $(-1)^m$ (i.e., $\bar{y}(m) = (-1)^m y(m)$), which corresponds to inverting the signs of odd samples of $y(m)$. In general, bands associated with even values of k are directly lowpass translated to the baseband of $Y(\omega')$, whereas bands associated with odd values of k are translated and inverted. This is a consequence of the fact that even-numbered bands (k even) correspond to "upper sidebands" of the modulation frequencies $2\pi\ell/M$, whereas odd-numbered bands (k odd) correspond to "lower sidebands" of the modulation frequencies.

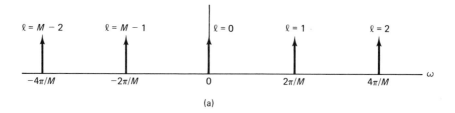

(a)

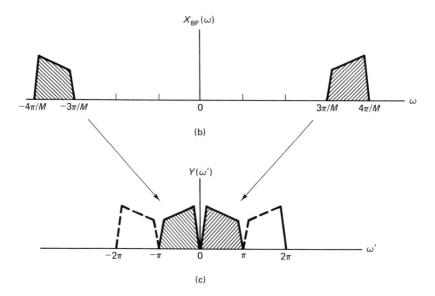

(b)

(c)

Figure 3.39 Spectral interpretation of integer-band decimation for the band $k = 3$.

The process of integer-band interpolation is the transpose of that of integer-band decimation; that is, it performs the reconstruction (interpolation) of a bandpass signal from its integer-band decimated representation. Figure 3.40(a) illustrates this process. The input signal, $x(n)$, is sampling rate expanded by L to produce the signal $w(m)$. The spectrum of $w(m)$ can be expressed as

$$W(\omega') = X(\omega'L) \tag{3.140}$$

and it corresponds to periodically repeated images of the baseband of $X(\omega)$ centered at the harmonics $\omega' = 2\pi\ell/L$. A bandpass filter $h_{BP}(m)$ is then used to select the appropriate image of this signal. We can see that to obtain the kth image, the bandpass filter must approximate the ideal characteristic

$$H_{BP,I}(\omega') = \begin{cases} L, & k\dfrac{\pi}{L} < |\omega'| < (k + 1)\dfrac{\pi}{L} \\ 0, & \text{otherwise} \end{cases} \tag{3.141}$$

where $k = 0, 1, 2, \ldots, L - 1$. Figure 3.40(d) shows an example of the output spectrum of the bandpass signal $Y(\omega')$ for the $k = 2$ band, and Fig. 3.40(e) illustrates

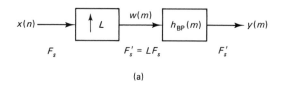

(a)

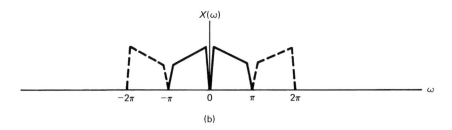

(b)

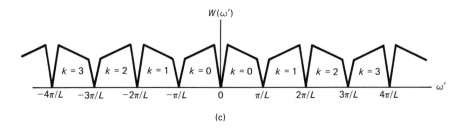

(c)

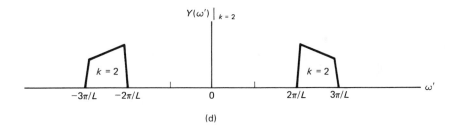

(d)

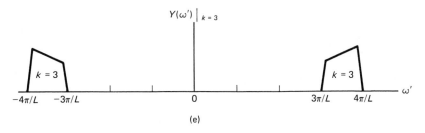

(e)

Figure 3.40   Spectral interpretation of integer-band interpolation of bandpass signals.

an example for the $k = 3$ band. As in the case of integer-band decimation, we also see that the spectrum of the resulting bandpass signal is inverted for odd values of k. If this inversion is not desired, the input signal $x(n)$ can first be modulated by $(-1)^n$, which inverts the spectrum of the baseband and consequently the bandpass signal.

From the discussion in Section 3.2.2 we see that the integer-band interpolator is identical to that of the lowpass interpolator with the exception that the filter may be a bandpass filter. Thus the spectrum of the output signal can be expressed as

$$Y(\omega') = H_{\mathrm{BP}}(\omega')X(\omega'L) \tag{3.142}$$

or

$$Y(\omega') \approx \begin{cases} LX(\omega'L), & k\dfrac{\pi}{L} < |\omega'| < (k+1)\dfrac{\pi}{L} \\ 0, & \text{otherwise} \end{cases} \tag{3.143}$$

The processes of integer-band decimation and interpolation can also be used for translating or modulating bandpass signals from one integer band to another. For example, a cascade of an integer-band decimator with the bandpass characteristic for $k = 2$ and an integer-band interpolator (with $L = M$) with a bandpass characteristic for $k = 3$ results in a system that translates the band from $2\pi/M$ to $3\pi/M$ to a band from $3\pi/M$ to $4\pi/M$.

3.6.5 Decimation and Interpolation Applied to Filter Bank Implementations

Decimation and interpolation concepts often arise in digital filter banks and spectrum analyzers. In this section we give one example of how the concepts discussed in this chapter can be used to achieve highly efficient implementations of filter banks. Figure 3.41(a) illustrates the basic framework for a K-channel filter bank analyzer, and Fig. 3.41(b) shows a similar framework for a K-channel synthesizer. In the analyzer the

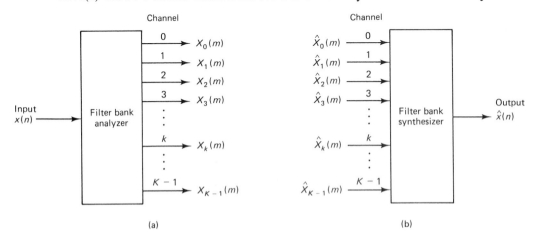

Figure 3.41 Basic framework for (a) a K-channel filter bank analyzer and (b) a K-channel filter bank synthesizer.

input signal $x(n)$ is divided into a set of K spectral components or channel signals denoted as $X_k(m)$, $k = 0, 1, \ldots, K - 1$. In the synthesizer a similar set of spectral components $\hat{X}_k(m)$ can be recombined to form a single output signal $\hat{x}(n)$. In practice the signals $X_k(m)$ and $\hat{X}_k(m)$ are often reduced in sampling rate for efficiency and therefore these systems are inherently multirate systems.

An important class of filter banks are those based on the framework of the discrete Fourier transform (DFT). Perhaps the simplest interpretation of the DFT filter bank, shown in Fig. 3.42(a), is that in which each channel, k, is separately bandpass modulated by a complex modulation signal $e^{-j\omega_k n}$, lowpass filtered by a filter $h(n)$, and then reduced in sampling rate by a factor M. Signal paths with double lines denote complex signals. Figure 3.42(b) shows the equivalent transpose operation for channel k in the synthesizer. In this framework the filter $h(n)$ is often referred to as the *analysis filter* and $f(n)$ is often referred to as the *synthesis filter*.

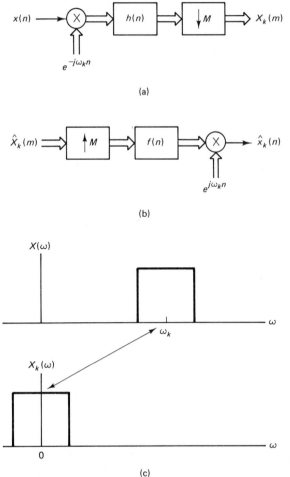

(a)

(b)

(c)

Figure 3.42 (a) Single channel of a DFT filter bank analyzer; (b) single channel of a DFT filter bank synthesizer; (c) a spectral interpretation.

In the DFT filter bank model the center frequencies of the channels are further defined to be uniformly spaced so that the modulation frequencies are

$$\omega_k = \frac{2\pi k}{K}, \qquad k = 0, 1, 2, \ldots, K - 1 \tag{3.144}$$

It is convenient to define

$$W_K = e^{j(2\pi/K)} \tag{3.145}$$

and the complex modulation function as

$$e^{-j\omega_k n} = e^{-j(2\pi kn/K)} = W_K^{-kn} \tag{3.146}$$

The channel signals can then be expressed as

$$X_k(m) = \sum_{n=-\infty}^{\infty} h(mM - n)x(n)W_K^{-kn}, \qquad k = 0, 1, 2, \ldots, K - 1 \tag{3.147}$$

Similarly, in the synthesizer the reconstructed channel signals $\hat{x}_k(n)$ can be expressed as

$$\hat{x}_k(n) = W_K^{kn} \sum_{m=-\infty}^{\infty} \hat{X}_k(m)f(n - mM), \qquad k = 0, 1, 2, \ldots, K - 1 \tag{3.148}$$

and the final output, $\hat{x}(n)$, is defined as the sum of the channel signals, i.e.,

$$\hat{x}(n) = \frac{1}{K} \sum_{k=0}^{K-1} \hat{x}_k(n) \tag{3.149}$$

where the scale factor of $1/K$ is inserted for convenience. Applying Eq. (3.148) to Eq. (3.149) and interchanging sums then gives the resulting overall expression for the DFT filter bank synthesizer in the form

$$\hat{x}(n) = \sum_{m=-\infty}^{\infty} f(n - mM)\frac{1}{K} \sum_{k=0}^{K-1} \hat{X}_k(m)W_K^{kn} \tag{3.150}$$

The properties of the analyzer and synthesizer are determined by the choice of the number of bands K, the decimation ratio M, and the designs of the analysis filter $h(n)$ and the synthesis filter $f(n)$.

For the purposes of this discussion we will restrict and simplify our example to an important subset of DFT filter banks called critically sampled filter banks, where the decimation ratio M is equal to the number of bands K. (For a discussion of the more general case, see [10].) For the analyzer we will further define a set of polyphase analysis filters of the form

$$\bar{p}_\rho(m) = h(mM - \rho), \qquad \rho = 0, 1, 2, \ldots, M - 1 \tag{3.151}$$

and a set of decimated input signals of the form

$$x_\rho(m) = x(mM + \rho), \qquad \rho = 0, 1, 2, \ldots, M - 1 \tag{3.152}$$

By carefully applying these definitions and the condition $M = K$ to Eq. (3.147), we can derive an expression for the filter bank analyzer in the form

$$X_k(m) = \sum_{\rho=0}^{M-1} \sum_{r=-\infty}^{\infty} \overline{p}_\rho(r) W_M^{-k\rho} x_\rho(m-r)$$

$$= \sum_{\rho=0}^{M-1} W_M^{-k\rho} [\overline{p}_\rho(m) * x_\rho(m)]$$

(3.153)

where $*$ denotes discrete convolution [10]. This form suggests the structure of Fig. 3.43(a) for a single filter bank channel k.

Equation (3.153) further suggests that $X_k(m)$ is of the form of a discrete Fourier transform of the convolved outputs $\overline{p}_\rho(m) * x_\rho(m)$ of the polyphase branches. In fact, these output signals are *independent* of the filter channel number k. Therefore, it is clear from Eq. (3.153) or Fig. 3.43(a) that computations involving the polyphase

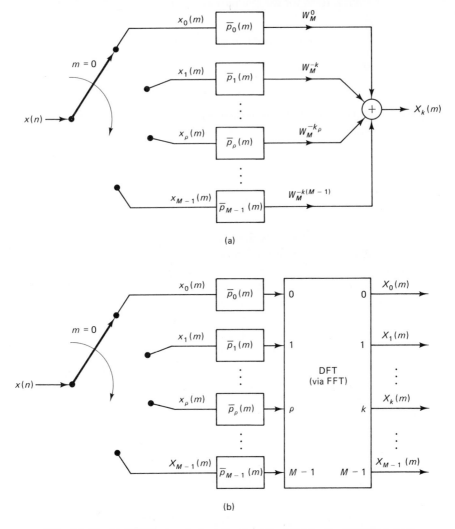

Figure 3.43 (a) Polyphase structure for the kth channel of a DFT filter bank analyzer; (b) the total polyphase DFT filter bank structure with an FFT modulator.

filters $\bar{p}_\rho(m)$ can be shared among all the filter bank channels, saving a factor of M in the total computation. An additional factor of M is gained from the polyphase form of the structure, as discussed in Section 3.3.3. Finally, we can recognize that the DFT in Eq. (3.153) can be performed with $M \log_2 M$ efficiency (as opposed to M^2 efficiency) by using the fast Fourier transform (FFT) algorithm. This then leads to the highly efficient filter bank structure shown in Fig. 3.43(b). In this implementation the input signal $x(n)$ is divided into a decimated set of branch signals, each of which is filtered by a separate polyphase filter (a decimated set of the coefficients $h(n)$). After one sweep of the commutator, the filtered outputs are transformed by the FFT to produce the desired filter bank signals. The overall structure is therefore very simple and elegant to implement.

A similar form for the synthesizer can be derived by defining the polyphase synthesis filters

$$q_\rho(m) = f(mM + \rho), \qquad \rho = 0, 1, 2, \ldots, M - 1 \qquad (3.154)$$

and the branch signals

$$\hat{x}_\rho(r) = \hat{x}(rM + \rho), \qquad \rho = 0, 1, 2, \ldots, M - 1 \qquad (3.155)$$

which define subsets of the output signal $\hat{x}(n)$. Then, from Eqs. (3.150), (3.154), (3.155) and the condition $K = M$, we can derive the ρth branch signal in the form

$$\hat{x}_\rho(r) = \frac{1}{M} \sum_{k=0}^{M-1} \sum_{m=-\infty}^{\infty} \hat{X}_k(m) q_\rho(r - m) W_M^{k\rho}$$

$$= \sum_{m=-\infty}^{\infty} q_\rho(r - m) \left[\frac{1}{M} \sum_{k=0}^{M-1} \hat{X}_k(m) W_M^{k\rho} \right] \qquad (3.156)$$

This form suggests the polyphase filter bank synthesis structure shown in Fig. 3.44 where, again, an efficient implementation is achieved by combining the advantages of the FFT and polyphase structures. It can be further observed that the structures of Figs. 3.43(b) and 3.44 are transpose structures.

An additional subtle but critical issue that is revealed in the preceding derivation

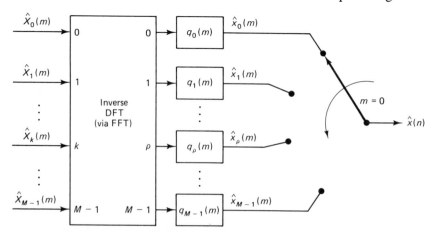

Figure 3.44 Synthesis structure for the polyphase DFT filter bank.

is that the transpose of a clockwise commutator is a counterclockwise commutator. This accounts for the fact that forms of the polyphase filters in Eqs. (3.151) and (3.154) must be defined with different signs to reflect the direction of commutation implied by the polyphase structure [10].

This concludes our discussion of filter bank implementations. The example derived above represents one of a class of filter bank structures that can be derived to take full advantage of efficient decimation and interpolation structures combined with the efficiency of fast transform algorithms. The issues are further complicated when we consider back-to-back arrangements of filter banks in which the processes of both analysis and synthesis must be performed. The interested reader is referred to Chapter 7 in [10]. For a discussion of the related topic of short-time Fourier transform, see Chapter 6 of this book.

3.7 CONCLUSIONS AND PRACTICAL EXAMPLES OF MULTIRATE DIGITAL SYSTEMS

In this chapter we have described the basic concepts of sampling rate conversion and have shown how these concepts can be efficiently utilized and extended to a broad range of signal processing operations. We conclude this chapter by showing specific examples of how these concepts have been utilized in practical applications and systems.

3.7.1 Sampling Rate Conversion in Digital Audio Systems

An important practical application of multirate digital signal processing and sampling rate conversion is in the field of professional digital audio. A variety of different types of digital processing systems have emerged for storage, transmission, and processing of audio program material. For a number of reasons such systems may have different sampling rates depending on whether they are used for broadcasting, digital storage, consumer products, or other professional applications. Also in digital processing of audio material, signals may be submitted to different types of digital rate control for varying the speed of the program material. This process can inherently vary the sampling frequency of the digital signal.

In practice it is often desired to convert audio program material from one digital format to another. One way to achieve this format conversion is to convert the audio signal back to analog form and digitize it in the new format. This process inherently introduces noise at each stage of conversion because of the limited dynamic range of the analog circuitry associated with the D/A and A/D conversion processes. Furthermore, this noise accumulates at each new interface.

An alternative and more attractive approach is to convert directly between the two digital formats by a process of waveform interpolation. This process is depicted in Fig. 3.45 and it is seen to be basically a sampling rate conversion problem. Since the accuracy of this sampling rate conversion can be maintained with any desired degree of precision (by controlling the wordlengths and the interpolator designs), essentially a noise free interconnection between the two systems can be achieved.

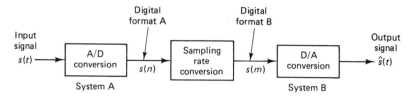

Figure 3.45 Example of a digital-to-digital translation between two audio signal formats.

3.7.2 Conversion Between Delta Modulation and PCM Signal Coding Formats

A second application of sampling rate conversion is in the area of digital communications. In communication networks a variety of different coding formats may be used in different parts of the network to achieve flexibility and efficiency. Conversion between these coding formats often involves a conversion of the basic sampling rate.

By way of example, delta modulation (DM) is sometimes used in A/D conversion or in voice terminals because of its simplicity and low cost. DM is basically a 1-bit/sample coding technique in which only the sign of the sample-to-sample difference of a highly oversampled signal is encoded. This approach eliminates the need for expensive antialiasing filters and allows the signal to be manipulated in a simple unframed serial bit stream format.

Alternatively, in long-distance transmission or in signal processing operations, such as digital filtering, it is generally desired to have the signal in a pulse code modulation (PCM) format. Thus it is necessary to convert between the high-sampling-rate, single-bit format of DM and the low-sampling-rate, multiple-bit format of PCM. Figure 3.46(a) shows an example of this process for the DM-to-PCM conversion, and Fig. 3.46(b) shows the reverse process of converting from a PCM to a DM format. When used as a technique for A/D conversion, this approach can combine the advantages of both the DM and the PCM signal formats.

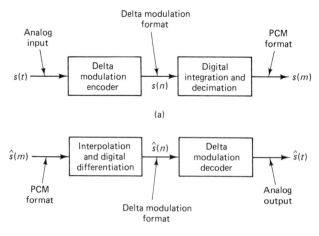

Figure 3.46 Illustration of (a) a DM-to-PCM conversion and (b) a PCM-to-DM conversion.

3.7.3 Digital Time-Division Multiplexing (TDM) to Frequency-Division Multiplexing (FDM) Translation

A third example of multirate digital systems is the translation of signals in a telephone system between time-division-multiplexed (TDM) and frequency-division-multiplexed (FDM) formats. The FDM format is often used for long-distance transmission, whereas the TDM format is more convenient for digital switching.

Figure 3.47 illustrates the basic process of translating a series of 12 TDM digital speech signals, $s_1(n)$, $s_2(n)$, . . . , $s_{12}(n)$, to a single FDM signal $r(m)$, and Fig. 3.47(b) illustrates the reverse (FDM-to-TDM) translation process. The sampling rate of the TDM speech signals is 8 kHz, whereas the sampling rate of the FDM signal is much higher to accommodate the increased total bandwidth. In each channel of the TDM-to-FDM translator the sampling rate is effectively increased (by interpolation) to the higher FDM sampling rate. The signal is then modulated by single-sideband modulation techniques to its appropriate frequency-band location in the range 56 to

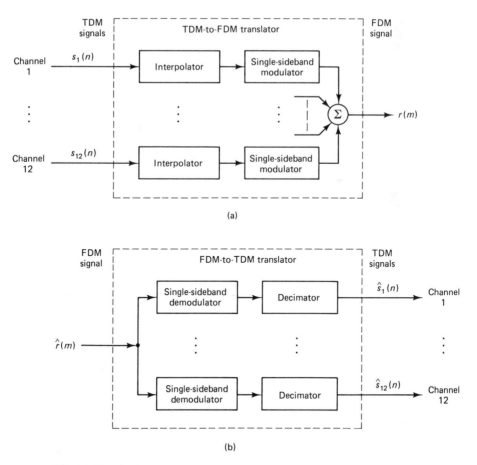

(a)

(b)

Figure 3.47   Illustration of (a) a TDM-to-FDM translator and (b) an FDM-to-TDM translator.

112 kHz, as illustrated in Fig. 3.48. The interpolated and modulated channel signals are then digitally summed to give the desired FDM signal. In the FDM-to-TDM translator, the reverse process takes place. As seen in Fig. 3.47 the process of translation between TDM and FDM formats involves sampling rate conversion, and therefore these systems are inherently multirate systems.

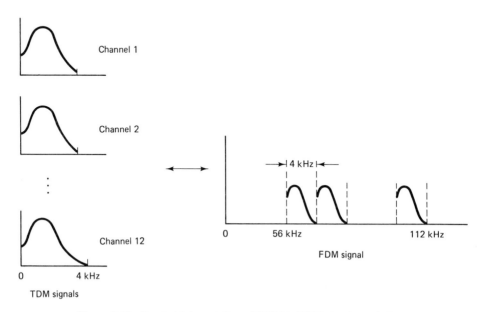

Figure 3.48 Spectral interpretation of TDM-to-FDM signal translation.

3.7.4 Subband Coding of Speech Signals

A fourth example of a practical multirate digital system is that of subband coding. Subband coding is a technique that is used to efficiently encode speech signals at low bit rates by taking advantage of the time-varying properties of the speech spectrum as well as some well-known properties of speech perception. It is based on the principle of decomposing the speech signal into a set of subband signals and separately encoding each signal with an adaptive PCM quantizer. By carefully selecting the number of bits per sample used to quantize each subband, according to perceptual criteria, one can achieve an efficient encoding of the speech.

Figure 3.49(a) shows a block diagram of a subband coder-decoder, and Fig. 3.49(b) shows a typical five-band filter bank arrangement for subband coding. A key element in this design is the implementation of the filter bank analysis and synthesis systems. Each subband in the filter bank analyzer is effectively obtained by a process of bandpass filtering, modulation to zero frequency, and decimation in a manner similar to that of the FDM-to-TDM translator. In the receiver the reverse process takes place, with the filter bank synthesizer reconstructing an output signal from the decoded subband signals. The process is similar to that of the TDM-to-FDM translator.

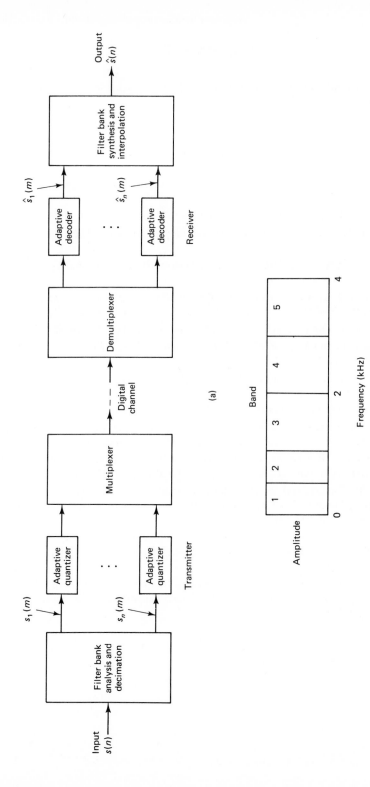

Figure 3.49 (a) Illustration of an N-band subband coder; (b) an example of the frequency bands for a 5-band subband coder design.

REFERENCES

1. A. V. Oppenheim and R. W. Schafer, *Digital Signal Processing,* Prentice-Hall, Englewood Cliffs, NJ, 1975.

2. L. R. Rabiner and B. Gold, *Theory and Application of Digital Signal Processing,* Prentice-Hall, Englewood Cliffs, NJ, 1975.

3. A. V. Oppenheim, Ed., *Applications of Digital Signal Processing,* Prentice-Hall, Englewood Cliffs, NJ, 1978.

4. L. R. Rabiner and R. W. Schafer, *Digital Processing of Speech Signals,* Prentice-Hall, Englewood Cliffs, NJ, 1978.

5. A. Peled and B. Liu, *Digital Signal Processing,* Wiley, New York, 1976.

6. L. R. Rabiner and C. M. Rader, Eds., *Digital Signal Processing,* IEEE Press, New York, 1972.

7. Digital Signal Processing Committee, Eds., *Selected Papers in Digital Signal Processing II,* IEEE Press, New York, 1976.

8. H. D. Helms, J. F. Kaiser, and L. R. Rabiner, Eds., *Literature in Digital Signal Processing: Author and Permuted Title Index,* IEEE Press, New York, 1975.

9. Special Issue on Digital Signal Processing, *Proc. IEEE,* Vol. 63, April 1975.

10. R. E. Crochiere and L. R. Rabiner, *Multirate Digital Signal Processing,* Prentice-Hall, Englewood Cliffs, NJ, 1983.

11. R. W. Schafer and L. R. Rabiner, "A Digital Signal Processing Approach to Interpolation," *Proc. IEEE,* Vol. 61, pp. 692–702, June 1973.

12. M. Bellanger, *Traitement Numérique du Signal,* Masson, Paris, 1981.

13. R. E. Crochiere and L. R. Rabiner, "Interpolation and Decimation of Digital Signals: A Tutorial Review," *Proc. IEEE,* Vol. 69, pp. 300–331, March 1981.

14. C. E. Schmidt, L. R. Rabiner, and D. A. Berkley, "A Digital Simulation of the Telephone System," *Bell Syst. Tech. J.,* Vol. 58, pp. 839–855, April 1979.

15. T. A. C. M. Claasen and W. F. G. Mecklenbrauker, "On the Transposition of Linear Time-Varying Discrete-Time Networks and Its Application to Multirate Digital Systems," *Philips J. Res.,* Vol. 23, pp. 78–102, 1978.

16. R. E. Crochiere, "A General Program to Perform Sampling Rate Conversion of Data by Rational Ratios," in *Programs for Digital Signal Processing,* IEEE Press, New York, 1979, pp. 8.2-1–8.2-7.

17. G. Oetken, T. W. Parks, and H. W. Schuessler, "A Computer Program for Digital Interpolator Design," in *Programs for Digital Signal Processing,* IEEE Press, New York, 1979, pp. 8.1-1–8.1-6.

4

Efficient Fourier Transform and Convolution Algorithms

C. S. Burrus
Rice University

4.0 INTRODUCTION

This chapter focuses on the discrete Fourier transform (DFT) and discrete convolution and the fast algorithms to calculate them. These topics have been at the center of digital signal processing since the beginning, and new results in hardware, theory, and applications continue to keep them important and exciting.

We develop three approaches to formulating efficient DFT algorithms. The first two approaches break a DFT into multiple shorter algorithms by using an index map (Section 4.1) and by polynomial reduction (Section 4.2). Section 4.3 develops the third method, which converts a prime-length DFT into cyclic convolution.

The very important computational complexity theorems of Winograd are stated and briefly discussed in Section 4.4. The specific details and evaluations of the Cooley-Tukey fast Fourier transform (FFT) and split-radix FFT are given in Section 4.5, and the prime factor algorithm (PFA) and Winograd Fourier transform algorithm (WFTA) are covered in Section 4.6. Short discussions of real-data algorithms and high-speed convolution are given in Sections 4.7 and 4.8, for both their own importance and their theoretical connection to the DFT. A basic knowledge of the Cooley-Tukey FFT at the level described in Oppenheim and Schafer [1] is assumed.

It is hard to overemphasize the importance of the DFT, convolution, and fast algorithms. Recent discoveries [2] show the history of the FFT goes back as far as Gauss and a compilation of references on these topics resulted in over 2400 entries [3].

C. S. Burrus is with the Department of Electrical and Computer Engineering, Rice University, Houston, TX 77251-1892.

New theoretical results still are appearing, advances in computers and hardware continually restate the basic questions, and new applications open new areas for research. It is hoped that this chapter will provide the background, references, and incentive to encourage further research and results in this area as well as provide tools for practical applications.

4.1 MULTIDIMENSIONAL INDEX MAPPING

A powerful approach to the development of efficient algorithms is to break a large problem into multiple small ones. A method for doing this with both the DFT and convolution uses a linear change of index variables to map the original one-dimensional problem into a multidimensional problem. This approach provides a unified derivation of the FFT, PFA, and WFTA.

The basic definition of the DFT is

$$X(k) = \sum_{n=0}^{N-1} x(n) W_N^{nk} \tag{4.1}$$

where

$$W_N = e^{-j2\pi/N}$$

If N values of the transform are calculated from N values of the data, $x(n)$, we easily see that N^2 multiplications and approximately that same number of additions are required. One method for reducing this required arithmetic is to use an index mapping (a change of variables) to change the one-dimensional DFT into a two- or higher-dimensional DFT. This is one of the ideas behind the very efficient Cooley-Tukey [4] and Winograd [5] algorithms. The purpose of index mapping is to change a large problem into several easier ones [6], sometimes called the "divide and conquer" approach [7].

For a length-N sequence, the time index takes on the values

$$n = 0, 1, 2, \ldots, N - 1 \tag{4.2}$$

When the length is not prime, N can be factored as $N = N_1 N_2$ and two new independent variables can be defined over the ranges

$$n_1 = 0, 1, 2, \ldots, N_1 - 1, \qquad n_2 = 0, 1, 2, \ldots, N_2 - 1 \tag{4.3}$$

A linear change of variables is defined that maps n_1 and n_2 to n and is expressed by

$$n = ((K_1 n_2 + K_2 n_2))_N \tag{4.4}$$

where the notation $((x))_N$ means the residue of x modulo N. This map defines a relation between all possible combinations of n_1 and n_2 in Eq. (4.3) and the values for n. The question as to whether all of the n in Eq. (4.2) are represented, i.e., whether the map is one-to-one (unique), has been answered in [6], which shows that certain K_i always exist such that the map in Eq. (4.4) is one-to-one. Two cases must be considered.

Case 1. N_1 and N_2 are relatively prime, i.e., $(N_1, N_2) = 1$.
The integer map of Eq. (4.4) is one-to-one if and only if

$$(K_1 = aN_2) \quad \text{and/or} \quad (K_2 = bN_1) \quad \text{and} \quad (K_1, N_1) = (K_2, N_2) = 1 \qquad (4.5)$$

Case 2. N_1 and N_2 are not relatively prime, i.e., $(N_1, N_2) > 1$.
The integer map of Eq. (4.4) is one-to-one if and only if

$$(K_1 = aN_2) \quad \text{and} \quad (K_2 \neq bN_1) \quad \text{and} \quad (a, N_1) = (K_2, N_2) = 1$$

or   (4.6)

$$(K_1 \neq aN_2) \quad \text{and} \quad (K_2 = bN_1) \quad \text{and} \quad (K_1, N_1) = (b, N_2) = 1$$

The notation (n, m) is used for the greatest common divisor of n and m.

Consult [6] for the details of these cases and examples. Two classes of index maps are defined from these conditions.

Type 1 index map.   The map of Eq. (4.4) is called a type 1 map when

$$K_1 = aN_2 \quad \text{and} \quad K_2 = bN_1 \qquad (4.7)$$

Type 2 index map.   The map of Eq. (4.4) is called a type 2 map when

$$K_1 = aN_2 \quad \text{or} \quad K_2 = bN_1, \quad \text{but not both} \qquad (4.8)$$

Type 1 can be used only if the factors are relatively prime, but type 2 can be used whether or not the factors are relatively prime. Good [8], Thomas [2], and Winograd [5] all used the type 1 map in their DFT algorithms. Cooley and Tukey [4] used type 2 in their algorithms, both for a fixed radix $(N = R^M)$ and a mixed radix [9].

The frequency index is defined by a map similar to Eq. (4.4) as

$$k = ((K_3 k_1 + K_4 k_2))_N \qquad (4.9)$$

where the same conditions, Eqs. (4.5) and (4.6), are used to determine the uniqueness of this map in terms of K_3 and K_4.

Two-dimensional arrays for the input data and its DFT are defined using these index maps to give

$$\hat{x}(n_1, n_2) = x((K_1 n_1 + K_2 n_2))_N \quad \text{and} \quad \hat{X}(k_1, k_2) = X((K_3 k_1 + K_4 k_2))_N$$

$$(4.10)$$

In some of the following equations, the residue reduction notation is omitted for clarity. These changes of variables applied to the definition of the DFT in Eq. (4.1) give

$$\hat{X} = \sum_{n_2=0}^{N_2-1} \sum_{n_1=0}^{N_1-1} \hat{x} W^{K_1 K_3 n_1 k_1} W^{K_1 K_4 n_1 k_2} W^{K_2 K_3 n_2 k_1} W^{K_2 K_4 n_2 k_2} \qquad (4.11)$$

The amount of arithmetic required to calculate Eq. (4.11) is the same as in the direct calculation of Eq. (4.1). However, because of the special nature of the DFT, the constants K_i can be chosen in such a way that the calculations are "uncoupled" and the arithmetic is reduced. The requirements for this are

$$((K_1 K_4))_N = 0 \quad \text{and/or} \quad ((K_2 K_3))_N = 0 \qquad (4.12)$$

When this condition and those for uniqueness in Eq. (4.4) are applied, we find that the K_i may always be chosen such that one of the terms in Eq. (4.12) is zero. If the N_i are relatively prime, it is always possible to make both terms zero. If the N_i are not relatively prime, *only* one of the terms can be set to zero. When they are relatively prime, there is a choice: it is possible to either set one or both to zero. This in turn causes one or both of the center two W terms in Eq. (4.11) to become unity.

An example of the Cooley-Tukey radix-4 FFT for a length-16 DFT uses the type 2 map:

$$n = 4n_1 + n_2, \quad k = k_1 + 4k_2 \qquad (4.13)$$

The residue reduction in Eq. (4.4) is not needed here since n does not exceed N as n_1 and n_2 take on their values. Since in this example the factors of N have a common factor, only one of the conditions in Eq. (4.12) can hold, and Eq. (4.11) therefore becomes

$$\hat{X}(k_1, k_2) = \sum_{n_2=0}^{3} \sum_{n_1=0}^{3} \hat{x}(n_1, n_2) W_4^{n_1 k_1} W_{16}^{n_2 k_1} W_4^{n_2 k_2} \qquad (4.14)$$

This has the form of a two-dimensional DFT with an extra term W_{16}, called a *twiddle factor* [1,10,11]. The inner sum over n_1 represents four length-4 DFTs, the W_{16} term represents 16 complex multiplications, and the outer sum over n_2 represents another four length-4 DFTs. This choice of the K_i "uncouples" the calculations since the first sum over n_1 for $n_2 = 0$ calculates the DFT of the first row of the data array $x(n_1, n_2)$, and those data values are never needed in the succeeding row calculations. The row calculations are independent, and examination of the outer sum shows that the column calculations are likewise independent. This is illustrated in Fig. 4.1.

The left-hand 4-by-4 array in Fig. 4.1 is the mapped input data, the center array has the rows transformed, and the right-hand array is the DFT array. The row DFTs and the column DFTs are independent of each other. The twiddle factors that are the

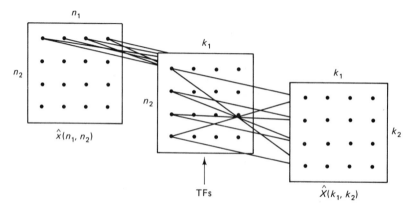

Figure 4.1   Uncoupling of the row and column calculations.

center W in Eq. (4.14) are the multiplications that take place on the center array of Fig. 4.1.

This uncoupling feature reduces the amount of arithmetic required and allows the results of each row DFT to be written back over the input data locations since that input row will not be needed again. This is called *in-place* calculation, and it results in a large memory requirement savings.

An example of the type 2 map used when the factors of N are relatively prime is given for $N = 15$ as

$$n = 5n_1 + n_2, \qquad k = k_1 + 3k_2 \qquad\qquad (4.15)$$

The residue reduction again is not explicitly needed. Although the factors 3 and 5 are relatively prime, use of the type 2 map sets only one of the terms in Eq. (4.12) to zero. The DFT in Eq. (4.11) becomes

$$\hat{X}(k_1, k_2) = \sum_{n_2=0}^{4} \sum_{n_1=0}^{2} \hat{x}(n_1, n_2) W_3^{n_1 k_1} W_{15}^{n_2 k_1} W_5^{n_2 k_2} \qquad\qquad (4.16)$$

which has the same form as Eq. (4.14), including the existence of the twiddle factors. Here the inner sum is five length-3 DFTs.

The type 1 map is illustrated next on the same length-15 example. This time the situation of Eq. (4.5) with the "and" condition is used in Eq. (4.7) to give an index map of

$$n = ((5n_1 + 3n_2))_{15}, \qquad k = ((10k_1 + 6k_2))_{15} \qquad\qquad (4.17)$$

The residue reduction is now necessary. Since the factors of N are relatively prime and the type 1 map is being used, both terms in Eq. (4.12) are zero, and Eq. (4.11) becomes

$$\hat{X}(k_1, k_2) = \sum_{n_2=0}^{4} \sum_{n_1=0}^{2} \hat{x}(n_1, n_2) W_3^{n_1 k_1} W_5^{n_2 k_2} \qquad\qquad (4.18)$$

which is similar to Eq. (4.16) except that now the type 1 map gives a pure two-dimensional DFT calculation with no twiddle factors, and the sums can be done in either order.

The purpose of index mapping is to improve the arithmetic efficiency. For example, a direct calculation of a length-16 DFT requires 256 real multiplications, and an uncoupled version requires 144. A direct calculation of a length-15 DFT requires 225 multiplications, but with a type 2 map only 135 and with a type 1 map, 120.

Algorithms of practical interest use short DFTs that require fewer than N^2 multiplications. For example, length-4 DFTs require no multiplications and, therefore, for the length-16 DFT, only the twiddle factors must be calculated. That calculation uses 16 multiplications, many fewer than the 256 or 144 required for the direct or uncoupled calculation.

The concept of using an index map can also be applied to convolution to convert a length $N = N_1 N_2$ one-dimensional cyclic convolution into an N_1-by-N_2 two-dimensional cyclic convolution [6,12]. There is no savings of arithmetic from the mapping alone as there is with the DFT, but savings can be obtained by using special short algorithms along each dimension. This is discussed in Section 4.8.

4.1.1 In-Place Calculation of the DFT and Scrambling

Because use of both the type 1 and type 2 index maps uncouples the calculations of the rows and columns of the data array, the results of each short-length N_i DFT can be written back over the data as they will not be needed again after that particular row or column is transformed. This is easily seen from Fig. 4.1, where the DFT of the first row of $\hat{x}(n_1, n_2)$ can be put back over the data rather than written into a new array. After all the calculations are finished, the total DFT is in the array of the original data. This gives a significant memory savings over using a separate array for the output.

Unfortunately, the use of in-place calculations results in the sequence order of the DFT being permuted or scrambled because the data are indexed according to the input map Eq. (4.4) and the results are put into the same locations rather than the locations dictated by the output map Eq. (4.9). For example, with a length-8 radix-2 FFT, the input index map is

$$n = 4n_1 + 2n_2 + n_3 \qquad (4.19)$$

which to satisfy Eq. (4.12) requires an output map of

$$k = k_1 + 2k_2 + 4k_3$$

The in-place calculations will place the DFT results in the locations of the input map, and these should be reordered or unscrambled into the locations given by the output map. Examination of these two maps shows the scrambled output to be in a "bit-reversed" order.

For certain applications, this scrambled output order is not important, but for many applications the order must be unscrambled before the DFT can be considered complete. Because the radix of the radix-2 FFT is the same as the base of the binary number representation, the correct address for any term is found by reversing the binary bits. The part of most FFT programs that does this reordering is called a bit-reversed counter. Examples of various unscramblers are found in [10,11].

The development here uses the input map, and the resulting algorithm is called *decimation-in-frequency*. If the output rather than the input map is used to derive the FFT algorithm so the correct output order is obtained, the input order must be scrambled so that its values are in locations specified by the output map rather than the input map. The resulting algorithm is called *decimation-in-time*. The scrambling is the same bit-reversed counting as before, but it precedes the FFT algorithm in this case. The same process of a post-unscrambler or prescrambler occurs for the in-place calculations with the type 1 maps. Details can be found in [11,13]. It is possible to do the unscrambling while calculating the FFT and to avoid a separate unscrambler. This is done for the Cooley-Tukey FFT in [14] and for the PFA in [11,13,15].

If a radix-2 FFT is used, the unscrambler is a bit-reversed counter. If a radix-4 FFT is used, the unscrambler is a base-4 reversed counter, and similarly for radix-8 and others. However, if for the radix-4 FFT the short length-4 DFTs have their outputs in bit-reversed order, the output of the total radix-4 FFT will be in bit-reversed order, not base-4 reversed order. This means that any radix-2^n FFT can use the same radix-2 bit-reversed counter as an unscrambler [16].

4.1.2 Efficiencies Resulting from Index Mapping with the DFT

In this section we examine the reductions in arithmetic in the DFT that result from the index mapping alone. In practical algorithms, several methods are always combined, but it is helpful in understanding the effects of a particular method to study it alone.

The most general form of an uncoupled two-dimensional DFT is given by

$$X(k_1, k_2) = \sum_{n_2=0}^{N_2-1} \left[\sum_{n_1=0}^{N_1-1} x(n_1, n_2) f_i(n_1, n_2, k_1) \right] f_2(n_2, k_1, k_2) \qquad (4.20)$$

where the inner sum calculates N_2 length-N_1 DFTs and, if for a type 2 map, the effects of the twiddle factors. If the number of arithmetic operations for a length-N DFT is denoted by $F(N)$, the number of operations for this inner sum is $F = N_2 F(N_1)$. The outer sum that gives N_1 length-N_2 DFTs requires $N_1 F(N_2)$ operations. The total number of arithmetic operations is then

$$F = N_2 F(N_1) + N_1 F(N_2) \qquad (4.21)$$

The first question to be considered is For a fixed length N, what is the optimal relation of N_1 and N_2 in the sense of minimizing the required amount of arithmetic? To answer this question, N_1 and N_2 are temporarily assumed to be real variables rather than integers. If the short length-N_i DFTs in Eq. (4.20) and any twiddle factor multiplications are assumed to require N_i^2 operations, i.e., $F(N_i) = N_i^2$, Eq. (4.21) becomes

$$\begin{aligned} F &= N_2 N_1^2 + N_1 N_2^2 \\ &= N(N_1 + N_2) \qquad (4.22) \\ &= N(N_1 + NN_1^{-1}) \end{aligned}$$

To find the minimum of F over N_1, the derivative of F with respect to N_1 is set to zero (temporarily assuming the variables to be continuous), and the result requires $N_1 = N_2$:

$$\frac{dF}{dN_1} = 0 \implies N_1 = N_2 \qquad (4.23)$$

This result is also easily seen from the symmetry of N_1 and N_2 in $N = N_1 N_2$. If a more general model of the arithmetic complexity of the short DFTs is used, the same result is obtained, but a closer examination must be made to ensure that $N_1 = N_2$ is a global minimum.

If only the effects of the index mapping are to be considered, then the $F(N) = N^2$ model is used, and Eq. (4.23) states that the two factors should be equal. If there are M factors, a similar reasoning shows that all M factors should be equal. For the sequence of length

$$N = R^M \qquad (4.24)$$

there are now M length-R DFTs and, since the factors are all equal, the index map must be type 2. This means there must be twiddle factors.

To simplify the analysis, we will consider only the number of multiplications.

If the number of multiplications for a length-R DFT is $F(R)$, then the formula for operation counts in Eq. (4.22) generalizes to

$$F = N \sum_{i=1}^{M} F(N_i)/N_i$$

$$= NM\frac{F(R)}{R} \qquad \text{for } N_i = R \tag{4.25}$$

$$= N \log_R(N)\frac{F(R)}{R}$$

$$= (N \ln N)\left[\frac{F(R)}{R \ln R}\right] \tag{4.26}$$

This is a very important formula derived by Cooley and Tukey in their well-known paper [4] on the FFT. It states that for a given R, called the radix, the number of multiplications (and additions) is proportional to $N \ln N$. It also shows the relation to the value of the radix R.

To get some idea of the "best" radix, we assume that the number of multiplications to compute a length-R DFT is $F(R) = R^x$. If this is used with Eq. (4.26), the optimal R can be found:

$$\frac{dF}{dR} = 0 \quad \Rightarrow \quad R = e^{1/(x-1)} \tag{4.27}$$

For $x = 2$, $R = e$, with the closest integer being 3.

The result of this analysis states that if no arithmetic-saving methods other than index mapping are used and if the length-R DFTs plus twiddle factors require $F = R^2$ multiplications, the optimal algorithm requires

$$F = 3N \log_3 N$$

multiplications for a length-$N = 3^M$ DFT. Compare this with N^2 for a direct calculation, and the improvement is obvious.

While this is an interesting result from the analysis of the effects of index mapping alone, in practice index mapping is almost always used in conjunction with special algorithms for the short length-N_i DFTs in Eq. (4.11). For example, if $R = 2$ or 4, there are no multiplications required for the short DFTs. Only the twiddle factors require multiplications. Winograd (see Section 4.4) has derived some algorithms for short DFTs that require $O(N)$ multiplications. This means that $F(N_i) = KN_i$, and the operation count F in Eq. (4.21) is independent of N_i. Therefore, the derivative of F is zero for all N_i. Obviously, these cases must be examined.

4.2 POLYNOMIAL DESCRIPTION OF SIGNALS

Polynomials are important in digital signal processing because calculating the DFT can be viewed as a polynomial evaluation problem and convolution can be viewed as polynomial multiplication [7,17]. Indeed, this is the basis for the important results of

Winograd discussed in Section 4.4. A length-N signal $x(n)$ will be represented by an $(N - 1)$-degree polynomial $\tilde{X}(s)$ defined by

$$\tilde{X}(s) = \sum_{n=0}^{N-1} x(n)s^n \qquad (4.28)$$

The polynomial $\tilde{X}(s)$ is a single entity, with the coefficients being the values of $x(n)$. This representation is somewhat similar to the use of matrix or vector notation to efficiently represent signals that allows use of new mathematical tools.

The convolution of two finite-length sequences, $x(n)$ and $h(n)$, gives an output sequence defined by

$$y(n) = \sum_{k=0}^{N-1} x(k)h(n - k), \qquad n = 0, 1, 2, \ldots, 2N - 1 \qquad (4.29)$$

where $h(k) = 0$ for $k < 0$. This is exactly the same operation as calculating the coefficients when multiplying two polynomials. Equation (4.29) is the same as

$$\tilde{Y}(s) = \tilde{X}(s)\tilde{H}(s) \qquad (4.30)$$

In fact, convolution of number sequences, multiplication of polynomials, and multiplication of integers (except for the carry operation) are all the same operations. To obtain cyclic convolution, where the indices in Eq. (4.29) are all evaluated modulo N, the polynomial multiplication in Eq. (4.30) is done modulo the polynomial $P(s) = s^N - 1$. This is seen by noting that $N = 0 \bmod N$; therefore, $s^N = 1$ and the polynomial modulus is $s^N - 1$.

4.2.1 Polynomial Reduction and the Chinese Remainder Theorem

Residue reduction of one polynomial modulo another is defined similarly to residue reduction for integers. A polynomial $F(s)$ has a residue polynomial $R(s)$ modulo $P(s)$ if, for a given $F(s)$ and $P(s)$, a $Q(S)$ and $R(s)$ exist such that

$$F(s) = Q(s)P(s) + R(s) \qquad (4.31)$$

with

$$\text{degree } \{R(s)\} < \text{degree } \{P(s)\}$$

The notation that we will use is

$$R(s) = ((F(s)))_{P(s)}$$

For example,

$$(s + 1) = ((s4 + s3 - s - 1))_{(s^2 - 1)} \qquad (4.32)$$

The concepts of factoring a polynomial and of primeness are an extension of these ideas for integers [18]. For a given allowed set of coefficients (values of $x(n)$), any polynomial has a unique factored representation

$$F(s) = \prod_{i=1}^{M} F_i(s)^{k_i}$$

where the $F_i(s)$ are relatively prime.

A very useful operation is an extension of the integer Chinese remainder theorem, which says that if the modulus polynomial can be factored into relatively prime factors

$$P(s) = P_1(s)P_2(s) \tag{4.33}$$

then there exist two polynomials, $K_1(s)$ and $K_2(s)$, such that any polynomial $F(s)$ can be recovered from its residues by

$$F(s) = K_1(s)F_1(s) + K_2(s)F_2(s) \bmod P(s) \tag{4.34}$$

where F_1 and F_2 are the residues given by

$$F_1(s) = ((F(s)))_{P_1(s)} \quad \text{and} \quad F_2(s) = ((F(s)))_{P_2(s)}$$

if the order of $F(s)$ is less than that of $P(s)$. This generalizes to any number of relatively prime factors of $P(s)$ and can be viewed as a means of representing $F(s)$ by several lower-degree polynomials $F_i(s)$.

This decomposition of $F(s)$ into lower-degree polynomials is the process used to break a DFT or convolution into several simple problems that are solved and then recombined using the Chinese remainder theorem of Eq. (4.34). This is another form of the "divide and conquer" approach similar to the index mappings in the preceding section.

One useful property of the Chinese remainder theorem is for convolution. If cyclic convolution of $x(n)$ and $h(n)$ is expressed in terms of polynomials by

$$\tilde{Y}(s) = \tilde{H}(s)\tilde{X}(s) \bmod P(s)$$

where $P(s) = s^N - 1$, and if $P(s)$ is factored into two relatively prime factors $P = P_1P_2$, using residue reduction of $H(s)$ and $X(s)$ modulo P_1 and P_2, the lower-degree residue polynomials can be multiplied and the results recombined with the Chinese remainder theorem. This is done by

$$Y(s) = ((K_1\tilde{H}_1\tilde{X}_1 + K_2\tilde{H}_2\tilde{X}_2))_P \tag{4.35}$$

where

$$\tilde{H}_1 = ((\tilde{H}))_{P_1} \qquad \tilde{X}_1 = ((\tilde{X}))_{P_1} \qquad \tilde{H}_2 = ((\tilde{H}))_{P_2} \qquad \tilde{X}_2 = ((\tilde{X}))_{P_2}$$

and K_1 and K_2 are the Chinese remainder theorem coefficient polynomials from Eq. (4.34). This allows two shorter convolutions to replace one longer one.

Another property of residue reduction that is useful in DFT calculation is polynomial evaluation. To evaluate $F(s)$ at $s = x$, $F(s)$ is reduced modulo $s - x$:

$$F(x) = ((F(s)))_{(s-x)} \tag{4.36}$$

This is easily seen from the definition in Eq. (4.31):

$$F(s) = Q(s)(s - x) + R(s)$$

Evaluating $s = x$ gives $R(s) = F(x)$, which is a constant. For the DFT this becomes

$$X(k) = ((\tilde{X}(s)))_{(s-W^k)} \tag{4.37}$$

Details of the polynomial algebra useful in digital signal processing can be found in [7,17,19].

4.2.2 The DFT as a Polynomial Evaluation

The z-transform of a number sequence $x(n)$ is defined as

$$\tilde{X}(z) = \sum_{n=0}^{\infty} x(n)z^{-n} \tag{4.38}$$

which is the same as the polynomial description in Eq. (4.28) but with a negative exponent. For a finite length-N sequence, Eq. (4.38) becomes

$$\tilde{X}(z) = \sum_{n=0}^{N-1} x(n)z^{-n}$$

$$= x(0) + x(1)z^{-1} + x(2)z^{-2} + \cdots + x(N-1)z^{-N+1} \tag{4.39}$$

This $(N-1)$-order polynomial takes on the values of the DFT of $x(n)$ when evaluated at

$$z = e^{j2\pi k/N}$$

which gives

$$X(k) = \tilde{X}(z) \bigg|_{z=e^{j2\pi k/N}} = \sum_{n=0}^{N-1} x(n)e^{-j2\pi nk/N} \tag{4.40}$$

In terms of the positive-exponent polynomial from Eq. (4.28), the DFT is

$$X(k) = \tilde{X}(s) \bigg|_{s=W^k}$$

where

$$W = e^{-j2\pi/N} \tag{4.41}$$

is an Nth root of unity. The N values of the DFT are found from $X(s)$ evaluated at the N Nth roots of unity, which are equally spaced around the unit circle in the complex s-plane.

One method of evaluating $X(z)$ is the so-called Horner's rule or nested evaluation. When expressed as a recursive calculation, Horner's rule becomes the Goertzel algorithm, which has some computational advantages, especially when only a few values of the DFT are needed. The details and programs can be found in [1,11].

Another method for evaluating $\tilde{X}(s)$ is the residue reduction modulo $(s - W^k)$ as shown in Eq. (4.37). Each evaluation requires N multiplications and, therefore, N^2 multiplications for the N values of $X(k)$:

$$X(k) = ((\tilde{X}(s)))_{(s-W^k)} \tag{4.42}$$

A considerable reduction in required arithmetic can be achieved if some operations can be shared between the reductions for different values of k. This is done by carrying out the residue reduction in stages that can be shared rather than done in one step for each k in Eq. (4.42).

The N values of the DFT are values of $X(s)$ evaluated at s equal to the N roots of the polynomial $P(s) = s^N - 1$, which are W^k. First, assuming N is even, we factor $P(s)$ as

$$P(s) = (s^N - 1) = P_1(s)P_2(s) = (s^{N/2} - 1)(s^{N/2} + 1)$$

$X(s)$ is reduced modulo these two factors to give two residue polynomials, $X_1(s)$ and $X_2(s)$. This process is repeated by factoring P_1 and further reducing X_1 and then factoring P_2 and reducing X_2. This is continued until the factors are of first degree, giving the desired DFT values as in Eq. (4.42). This is illustrated for a length-8 DFT. The polynomial whose roots are W^k factors as

$$P(s) = s^8 - 1$$
$$= [s^4 - 1][s^4 + 1]$$
$$= [(s^2 - 1)(s^2 + 1)][(s^2 - j)(s^2 + j)]$$
$$= [(s - 1)(s + 1)(s - j)(s + j)][(s - a)(s + a)(s - ja)(s + ja)]$$

where $a^2 = j$. Reducing $\tilde{X}(s)$ by the first factoring gives two third-degree polynomials

$$\tilde{X}(s) = x_0 + x_1 s + x_2 s^2 + \cdots + x_7 s^7$$

which give the residue polynomials

$$\tilde{X}_1(s) = ((\tilde{X}(s)))_{(s^4 - 1)} = (x_0 + x_4) + (x_1 + x_5)s + (x_2 + x_6)s^2 + (x_3 + x_7)s^3$$
$$\tilde{X}_2(s) = ((\tilde{X}(s)))_{(s^4 + 1)} = (x_0 - x_4) + (x_1 - x_5)s + (x_2 - x_6)s^2 + (x_3 - x_7)s^3$$

$$(4.43)$$

Two more levels of reduction are carried out to finally give the DFT. Close examination shows the resulting algorithm to be the decimation-in-frequency radix-2 Cooley-Tukey FFT [1,11]. Martens [20] has used this approach to derive a very efficient DFT algorithm.

4.3 RADER'S CONVERSION OF THE DFT INTO CONVOLUTION

In this section we develop a method quite different from the index mapping or polynomial evaluation. Rather than dealing with the DFT directly, we convert it into a cyclic convolution that must then be carried out by some efficient means. We cover those means later, but here we explain the conversion. This method requires use of some number theory, which can be found in an accessible form in [7,17,19] and is easy enough to verify on one's own. A good general reference on number theory is [18].

The DFT and cyclic convolution are defined by

$$X(k) = \sum_{n=0}^{N-1} x(n) \, W^{nk} \tag{4.44}$$

$$y(k) = \sum_{n=0}^{N-1} x(n) \, h(k - n) \tag{4.45}$$

For both, the indices are evaluated modulo N. To convert the DFT in Eq. (4.44) into the cyclic convolution of Eq. (4.45), the nk product must be changed to the $k - n$ difference. With real numbers, this can be done with logarithms, but it is more complicated when working in a finite set of integers modulo N. From number theory [7,17–19], it can be shown that if the modulus is a prime number, a base (called a *primitive root*) exists such that a form of integer logarithm can be defined. This is stated in the following way. If N is a prime number, a number r called a primitive root exists such that the integer equation

$$n = ((r^m))_N \tag{4.46}$$

creates a unique one-to-one map of the $N - 1$ member set $m = \{0, \ldots, N - 2\}$ and the $N - 1$-member set $n = \{1, \ldots, N - 1\}$. This is because the multiplicative group of integers modulo a prime p is isomorphic to the additive group of integers modulo $(p - 1)$ and is illustrated for $N = 5$ in Table 4.1.

TABLE 4.1   TABLE OF THE INTEGERS r^m modulo 5

r	0	1	2	3	4	5	6	7
1	1	1	1	1	1	1	1	1
2	1	2	4	3	1	2	4	3
3	1	3	4	2	1	3	4	2
4	1	4	1	4	1	4	1	4
5	*	0	0	0	*	0	0	0
6	1	1	1	1	1	1	1	1

* Not defined

From Table 4.1, it is easy to see that there are two primitive roots, 2 and 3, and Eq. (4.46) defines a permutation of the integers n from the integers m (except for zero). Equation (4.46) and a primitive root (usually chosen to be the smallest of those that exist) can be used to convert the DFT in Eq. (4.44) to the convolution in Eq. (4.45). Since Eq. (4.46) cannot give a zero, a new length-$(N - 1)$ data sequence is defined from $x(n)$ by removing the term with index zero. Let

$$n = r^{-m} \qquad \text{and} \qquad k = r^s$$

where the term with the negative exponent is defined as the integer that satisfies

$$((r^{-m}r^m))_N = 1$$

If N is a prime number, r^{-m} always exists. For example, $((2^{-1}))_5 = 3$. Equation (4.44) now becomes

$$X(r^s) = \sum_{m=0}^{N-2} x(r^{-m})W^{r^{-m}r^s} + x(0), \qquad s = 0, 1, \ldots, N - 2 \tag{4.47}$$

and

$$X(0) = \sum_{n=0}^{N-1} x(n)$$

New functions are defined, that are simply a permutation in the order of the original functions, as

$$x'(m) = x(r^{-m}), \qquad X'(s) = X(r^s), \qquad W'(n) = W^{r^n} \qquad (4.48)$$

Equation (4.47) then becomes

$$X'(s) = \sum_{m=0}^{N-2} x'(m)W'(s - m) + x(0) \qquad (4.49)$$

which is a cyclic convolution of length $N - 1$ (plus $x(0)$) and is denoted as

$$X'(k) = x'(k) * W'(k) + x(0) \qquad (4.50)$$

Applying this change of variables (use of logarithms) to the DFT can best be illustrated from the matrix formulation of the DFT. Equation (4.44) is written for a length-5 DFT as

$$\begin{bmatrix} X(0) \\ X(1) \\ X(2) \\ X(3) \\ X(4) \end{bmatrix} = \begin{bmatrix} 0 & 0 & 0 & 0 & 0 \\ 0 & 1 & 2 & 3 & 4 \\ 0 & 2 & 4 & 1 & 3 \\ 0 & 3 & 1 & 4 & 2 \\ 0 & 4 & 3 & 2 & 1 \end{bmatrix} \begin{bmatrix} x(0) \\ x(1) \\ x(2) \\ x(3) \\ x(4) \end{bmatrix} \qquad (4.51)$$

where the square matrix should contain the terms of W^{nk}. For clarity, only the exponents nk are shown. Separating the $X(0)$ term, applying the mapping of Eq. (4.48), and using the primitive roots $r = 2$ (and $r^{-1} = 3$) gives

$$\begin{bmatrix} X(1) \\ X(2) \\ X(4) \\ X(3) \end{bmatrix} = \begin{bmatrix} 1 & 3 & 4 & 2 \\ 2 & 1 & 3 & 4 \\ 4 & 2 & 1 & 3 \\ 3 & 4 & 2 & 1 \end{bmatrix} \begin{bmatrix} x(1) \\ x(3) \\ x(4) \\ x(2) \end{bmatrix} + \begin{bmatrix} x(0) \\ x(0) \\ x(0) \\ x(0) \end{bmatrix} \qquad (4.52)$$

and

$$X(0) = x(0) + x(1) + x(2) + x(3) + x(4)$$

which can be seen to be a reordering of the structure in Eq. (4.51). This is in the form of cyclic convolution as indicated in Eq. (4.49). Rader first showed this in 1968 [19], stating that a prime length-N DFT could be converted into a length-$(N - 1)$ cyclic convolution of a permutation of the data with a permutation of the W's. He also stated that a slightly more complicated version of the same idea would work for a DFT with a length equal to an odd prime to a power. The details of that theory can be found in [19,21].

    Until 1976, this conversion approach received little attention because it seemed to offer few advantages. It has specialized applications in calculating the DFT if the cyclic convolution is done by distributed arithmetic table look-up [46] or by use of number theoretic transforms [7,17,19]. It and the Goertzel algorithm [1,11] are efficient when only a few DFT values need to be calculated. It may also have advantages when used with pipelined or vector hardware designed for fast inner products. One example is the TMS 320 signal processing microprocessor, which is pipelined for inner products [23]. The general use of this scheme emerged when new, fast cyclic convolution algorithms were developed by Winograd [5].

4.4 WINOGRAD'S SHORT DFT ALGORITHMS

In 1976, Winograd [5] presented a new DFT algorithm that had significantly fewer multiplications than the Cooley-Tukey FFT, which had been published 11 years earlier. This new Winograd Fourier transform algorithm (WFTA) is based on the type 1 index map from Section 4.1, with each of the short DFTs having lengths that are relatively prime and each calculated by very efficient special algorithms. We develop these short algorithms in this section. The algorithms use the index permutation of Rader described in the previous section to convert the prime length DFTs into cyclic convolutions. Winograd developed a method for calculating digital convolution with the minimum number of multiplications. These optimal algorithms are based on the polynomial residue reduction techniques of Section 4.2.1 to break the convolution into multiple short convolutions [7,17,19,24–27].

The operation of discrete convolution defined by

$$y(n) = \sum_k h(n - k)x(k) \tag{4.53}$$

is called a bilinear operation because for a fixed $h(n)$, $y(n)$ is a linear function of $x(n)$ and for a fixed $x(n)$ it is a linear function of $h(n)$. The operation of cyclic convolution is the same but with all indices evaluated modulo N.

Recall from Eq. (4.30) that length-N cyclic convolution of $x(n)$ and $h(n)$ can be represented by polynomial multiplication

$$\tilde{Y}(s) = \tilde{X}(s)\tilde{H}(s) \bmod(s^N - 1) \tag{4.54}$$

This bilinear operation of Eqs. (4.53) and (4.54) can also be expressed in terms of linear matrix operators and a simpler bilinear operator denoted by $\circ$, which may be only a simple element-by-element multiplication of the two vectors [19,26,28]. This matrix formulation is

$$\mathbf{Y} = C[A\mathbf{X} \circ B\mathbf{H}] \tag{4.55}$$

where $\mathbf{X}$, $\mathbf{H}$, and $\mathbf{Y}$ are length-N vectors with elements of $x(n)$, $h(n)$, and $y(n)$, respectively. The matrices $\mathbf{A}$ and $\mathbf{B}$ have dimension $M \times N$, and $\mathbf{C}$ is $N \times M$ with $M \geq N$. The elements of $\mathbf{A}$, $\mathbf{B}$, and $\mathbf{C}$ are constrained to be simple, typically small integers or rational numbers. These matrix operators do the equivalent of the residue reduction on the polynomials in Eq. (4.54).

To derive a useful algorithm of the form of Eq. (4.55) to calculate Eq. (4.53), consider the polynomial formulation of Eq. (4.54) again. To use the residue reduction scheme, the modulus is factored into relatively prime factors. Fortunately, the factoring of this particular polynomial, $s^N - 1$, has been extensively studied and has been found to have considerable structure. When factored over the rationals, which means that the only coefficients allowed are rational numbers, the factors are called *cyclotomic polynomials* [7,17,19]. The most interesting property for our purposes is that most of the coefficients of cyclotomic polynomials are zero and the others are ± 1 for orders up to over 100. This means the residue reduction will generally not require multiplications.

The operations of reducing $X(s)$ and $H(s)$ in Eq. (4.54) are carried out by the matrices $\mathbf{A}$ and $\mathbf{B}$ in Eq. (4.55). The convolution of the residue polynomials is carried out by the $\circ$ operator, and the recombination by the CRT is done by the $\mathbf{C}$ matrix. More details are in [7,17,19,26,28], but the important fact is that the $\mathbf{A}$ and $\mathbf{B}$ matrices usually contain only zero and ± 1 entries and the $\mathbf{C}$ matrix contains only rational numbers. The only general multiplications are those represented by $\circ$. Indeed, in the theoretical results from computational complexity theory, these real or complex multiplications are usually the only ones counted. In practical algorithms, the rational multiplications represented by $\mathbf{C}$ could be a limiting factor.

The $h(n)$ terms are fixed for a digital filter, or they represent the W terms from Eq. (4.49) if the convolution is being used to calculate a DFT. Because of this, $\mathbf{d} = \mathbf{BH}$ in Eq. (4.55) can be precalculated and only the A and C operators represent the mathematics done at execution of the algorithm. To exploit this feature, it was shown [25,26] that the properties of Eq. (4.55) allow the exchange of the more complicated operator C with the simpler operator B. Specifically, this is given by

$$\mathbf{Y} = C[A\mathbf{X} \circ B\mathbf{H}], \qquad \mathbf{Y}' = B^T[A\mathbf{X} \circ C^T\mathbf{H}'] \qquad (4.56)$$

where $\mathbf{H}'$ has the same elements as $\mathbf{H}$, but in a permuted order, and likewise $\mathbf{Y}'$ and $\mathbf{Y}$. This very important property allows precomputing the more complicated $C^T\mathbf{H}'$ in Eq. (4.56) rather than $B\mathbf{H}$ as in Eq. (4.55).

Because $B\mathbf{H}$ or $C^T\mathbf{H}'$ can be precomputed, the bilinear form of Eqs. (4.55) and (4.56) can be written as a linear form. If an $M \times M$ diagonal matrix $\mathbf{D}$ is formed from $\mathbf{d} = C^T\mathbf{H}$, or in the case of Eq. (4.55) $\mathbf{d} = B\mathbf{H}$, assuming a commutative property for $\circ$, Eq. (4.56) becomes

$$\mathbf{Y}' = B^T DA\mathbf{X} \qquad (4.57)$$

and Eq. (4.55) becomes

$$\mathbf{Y} = CDA\mathbf{X}$$

In most cases there is no reason not to use the same reduction operations on $\mathbf{X}$ and $\mathbf{H}$; therefore, B can be the same as A, and Eq. (4.57) then becomes

$$\mathbf{Y}' = A^T DA\mathbf{X} \qquad (4.58)$$

To illustrate how the residue reduction is carried out and how the $\mathbf{A}$ matrix is obtained, the length-5 DFT algorithm started in Eq. (4.51) will be continued. The DFT is first converted to a length-4 cyclic convolution by the index permutation from Eq. (4.46) to give the cyclic convolution in Eq. (4.52). To avoid confusion from the permuted order of the data $x(n)$ in Eq. (4.52), the cyclic convolution will first be developed without the permutation, using the polynomial $U(s)$:

$$U(s) = x(1) + x(3)s + x(4)s^2 + x(2)s^3$$
$$= u_0 + u_1 s + u_2 s^2 + u_3 s^3 \qquad (4.59)$$

Then the results will be converted back to the permuted $x(n)$. The length-4 cyclic convolution in terms of polynomials is

$$Y(s) = U(s)H(s) \bmod(s^4 - 1) \qquad (4.60)$$

and the modulus factors into three cyclotomic polynomials

$$s^4 - 1 = (s^2 - 1)(s^2 + 1)$$
$$= (s - 1)(s + 1)(s^2 + 1) \tag{4.61}$$
$$= P_1 P_2 P_3$$

Both $U(s)$ and $H(s)$ are reduced modulo these three polynomials. The reduction modulo P_1 and P_2 is done in two stages. First it is done modulo $(s^2 - 1)$, and then that residue is further reduced modulo $(s - 1)$ and $(s + 1)$.

$$U(s) = u_0 + u_1 s + u_2 s^2 + u_3 s^3 \tag{4.62}$$

$$U'(s) = ((U(s)))_{(s^2-1)}$$
$$= (u_0 + u_2) + (u_1 + u_3)s \tag{4.63}$$

$$U_1(s) = ((U'(s)))_{P_1} = (u_0 + u_1 + u_2 + u_3) \tag{4.64}$$

$$U_2(s) = ((U'(s)))_{P_2} = (u_0 - u_1 + u_2 - u_3) \tag{4.65}$$

$$U_3(s) = ((U(s)))_{P_3} = (u_0 - u_2) + (u_1 - u_3)s \tag{4.66}$$

The reduction in Eq. (4.63) of the data polynomial in Eq. (4.62) can be denoted by a matrix operation on a vector that has the data as entries

$$\begin{bmatrix} 1 & 0 & 1 & 0 \\ 0 & 1 & 0 & 1 \end{bmatrix} \begin{bmatrix} u_0 \\ u_1 \\ u_2 \\ u_3 \end{bmatrix} = \begin{bmatrix} u_0 + u_2 \\ u_1 + u_3 \end{bmatrix} \tag{4.67}$$

and the reduction in Eq. (4.66) is

$$\begin{bmatrix} 1 & 0 & -1 & 0 \\ 0 & 1 & 0 & -1 \end{bmatrix} \begin{bmatrix} u_0 \\ u_1 \\ u_2 \\ u_3 \end{bmatrix} = \begin{bmatrix} u_0 - u_2 \\ u_1 - u_3 \end{bmatrix} \tag{4.68}$$

Combining Eqs. (4.67) and (4.68) gives one operator:

$$\begin{bmatrix} 1 & 0 & 1 & 0 \\ 0 & 1 & 0 & 1 \\ 1 & 0 & -1 & 0 \\ 0 & 1 & 0 & -1 \end{bmatrix} \begin{bmatrix} u_0 \\ u_1 \\ u_2 \\ u_3 \end{bmatrix} = \begin{bmatrix} u_0 + u_2 \\ u_1 + u_3 \\ u_0 - u_2 \\ u_1 - u_3 \end{bmatrix} = \begin{bmatrix} w_0 \\ w_1 \\ v_0 \\ v_1 \end{bmatrix} \tag{4.69}$$

Further reduction of $v_0 + v_1 s$ is not possible because $P_3 = s^2 + 1$ cannot be factored over the rationals. However, $s^2 - 1$ can be factored into $P_1 P_2 = (s - 1)(s + 1)$ and, therefore, $w_0 + w_1 s$ can be further reduced, as was done in Eqs. (4.64) and (4.65) by

$$r_0 = \begin{bmatrix} 1 & 1 \end{bmatrix} \begin{bmatrix} w_0 \\ w_1 \end{bmatrix} = w_0 + w_1 = u_0 + u_2 + u_1 + u_3 \tag{4.70}$$

$$r_1 = \begin{bmatrix} 1 & -1 \end{bmatrix} \begin{bmatrix} w_0 \\ w_1 \end{bmatrix} = w_0 - w_1 = u_0 + u_2 - u_1 - u_3 \tag{4.71}$$

Combining Eqs. (4.69), (4.70), and (4.71) gives

$$
\begin{bmatrix} 1 & 1 & 0 & 0 \\ 1 & -1 & 0 & 0 \\ 0 & 0 & 1 & 0 \\ 0 & 0 & 0 & 1 \end{bmatrix}
\begin{bmatrix} 1 & 0 & 1 & 0 \\ 0 & 1 & 0 & 1 \\ 1 & 0 & -1 & 0 \\ 0 & 1 & 0 & -1 \end{bmatrix}
\begin{bmatrix} u_0 \\ u_1 \\ u_2 \\ u_3 \end{bmatrix}
=
\begin{bmatrix} r_0 \\ r_1 \\ v_0 \\ v_1 \end{bmatrix}
\tag{4.72}
$$

The same reduction is done to $H(s)$ and then the convolution of Eq. (4.60) is done by multiplying each residue polynomial of $X(s)$ and $H(s)$ modulo each corresponding cyclotomic factor of $P(s)$ and, finally, a recombination is done using the polynomial Chinese remainder theorem as in Eqs. (4.34) and (4.35):

$$
Y(s) = K_1(s)U_1(s)H_1(s) + K_2(s)U_2(s)H_2(s) + K_3(s)U_3(s)H_3(s) \bmod (s^4 - 1)
\tag{4.73}
$$

where $U_1(s) = r_1$ and $U_2(s) = r_2$ are constants and $U_3(s) = v_0 + v_1 s$ is a first-degree polynomial. U_1 times H_1 and U_2 times H_2 are easy, but multiplying U_3 times H_3 modulo $(s^2 + 1)$ is more difficult.

The multiplication of $U_3(s)$ times $H_3(s)$ can be done by the Toom-Cook algorithm [7,17,19], which can be viewed as Lagrange interpolation or polynomial multiplication modulo a special polynomial with three arbitrary coefficients. To simplify the arithmetic, the constants are chosen to be ± 1 and 0. The details of this algorithm can be found in [7,17,19]. For this example it can be verified that

$$
(((v_0 + v_1 s)(h_0 + h_1 s)))_{s^2 + 1} = (v_0 h_0 - v_1 h_1) + (v_0 h_1 + v_1 h_0)s
$$

which by the Toom-Cook algorithm or by inspection is

$$
\begin{bmatrix} 1 & -1 & 0 \\ -1 & -1 & 1 \end{bmatrix}
\left[
\begin{bmatrix} 1 & 0 \\ 0 & 1 \\ 1 & 1 \end{bmatrix}
\begin{bmatrix} v_0 \\ v_1 \end{bmatrix}
\circ
\begin{bmatrix} 1 & 0 \\ 0 & 1 \\ 1 & 1 \end{bmatrix}
\begin{bmatrix} h_0 \\ h_1 \end{bmatrix}
\right]
=
\begin{bmatrix} y_0 \\ y_1 \end{bmatrix}
\tag{4.74}
$$

where $\circ$ signifies point-by-point multiplication. The total **A** matrix in Eq. (4.55) is a combination of Eqs. (4.72) and (4.74) giving

$$
A X = A_1 A_2 A_3 X =
$$

$$
\begin{bmatrix} 1 & 0 & 0 & 0 \\ 0 & 1 & 0 & 0 \\ 0 & 0 & 1 & 0 \\ 0 & 0 & 0 & 1 \\ 0 & 0 & 1 & 1 \end{bmatrix}
\begin{bmatrix} 1 & 1 & 0 & 0 \\ 1 & -1 & 0 & 0 \\ 0 & 0 & 1 & 0 \\ 0 & 0 & 0 & 1 \end{bmatrix}
\begin{bmatrix} 1 & 0 & 1 & 0 \\ 0 & 1 & 0 & 1 \\ 1 & 0 & -1 & 0 \\ 0 & 1 & 0 & -1 \end{bmatrix}
\begin{bmatrix} u_0 \\ u_1 \\ u_2 \\ u_3 \end{bmatrix}
\tag{4.75}
$$

where the matrix A_3 gives the residue reduction $s^2 - 1$ and $s^2 + 1$, the upper left-hand part of A_2 gives the reduction modulo $s - 1$ and $s + 1$, and the lower right-hand part of A_1 carries out the Toom-Cook algorithm modulo $s^2 + 1$ with the multiplication in Eq. (4.56). Notice that to calculate Eq. (4.75) in the three stages, seven additions are required. Also notice that A_1 is not square. It is this "expansion" that causes more than N multiplications to be required in $\circ$ in Eq. (4.56) or D in Eq. (4.57). This staged reduction will derive the A operator for Eq. (4.56).

The method described above is very straightforward for the shorter DFT lengths. For $N = 3$, both of the residue polynomials are constants, and the multiplication given by ∘ in Eq. (4.55) is trivial. For $N = 5$, which is the example used here, one first-degree polynomial multiplication is required, but the Toom-Cook algorithm uses simple constants and therefore works well, as indicated in Eq. (4.74). For $N = 7$, there are two first-degree residue polynomials, which can each be multiplied by the same techniques used in the $N = 5$ example. Unfortunately, for any longer lengths, the residue polynomials have an order of three or greater, which causes the Toom-Cook algorithm to require constants of ± 2 and worse. For that reason, the Toom-Cook method is not used, and other techniques such as index mapping are used that require more than the minimum number of multiplications but do not require an excessive number of additions. The resulting algorithms still have the structure of Eq. (4.58). Blahut [7] and Nussbaumer [17] have a good collection of algorithms for polynomial multiplication that can be used with the techniques discussed here to construct a wide variety of DFT algorithms.

The constants in the diagonal matrix **D** can be found from the Chinese remainder theorem matrix **C** in Eq. (4.56) using $\mathbf{d} = C^T \mathbf{H}'$ for the diagonal terms in **D**. As mentioned above, for the smaller prime lengths of 3, 5, and 7, this works well, but for longer lengths the Chinese remainder theorem becomes very complicated. An alternative method for finding **D** uses the fact that since the linear form of Eq. (4.57) or (4.58) calculates the DFT, it is possible to calculate a known DFT of a given $x(n)$ from the definition of the DFT in Eq. (4.1) and, given the **A** matrix in Eq. (4.58), solve for **D** by solving a set of simultaneous equations. The details of this procedure are described in [26].

A modification of this approach also works for a length that is an odd prime raised to some power: $N = P^M$. This is a bit more complicated [19,24] but has been done for lengths of 9 and 25. For longer lengths, the conventional Cooley-Tukey type 2 index map algorithm seems to be more efficient. For powers of two, there is no primitive root and, therefore, no simple conversion of the DFT into convolution. It is possible to use two generators [17,19,25] to make the conversion, and a set of length-4, -8, and -16 DFT algorithms of the form in Eq. (4.58) are given in [19].

Table 4.2 presents an operation count of several short DFT algorithms. These are practical algorithms that can be used alone or in conjunction with the index mapping to give longer DFTs, as shown in Section 4.6. Most are optimized in having either the theoretical minimum number of multiplications or the minimum number of multiplications without requiring a very large number of additions. Some allow other reasonable tradeoffs between numbers of multiplications and additions. There are two lists of the number of multiplications. The first is the number of actual floating-point multiplications that must be done for that length DFT. Some of these (one or two in most cases) will be by rational constants and the others will be by irrational constants. The second list is the total number of multiplications given in the diagonal matrix **D** in Eq. (4.58). At least one of these will be unity (the one associated with $X(0)$) and in some cases several will be unity (for $N = 2^M$). The second list is important in programming the WFTA in Section 4.6.2.

The number of real multiplications required for the DFT of real data is exactly

TABLE 4.2   NUMBER OF REAL MULTIPLICATIONS
AND ADDITIONS FOR A LENGTH-N DFT OF
COMPLEX DATA

Length N	Multiplications		Additions
	Nonunity	Total	
2	0	4	4
3	4	6	12
4	0	8	16
5	10	12	34
7	16	18	72
8	4	16	52
9	20	22	84
11	40	42	168
13	40	42	188
16	20	36	148
17	70	72	314
19	76	78	372
25	132	134	420
32	68	–	388

half that required for complex data. The number of real additions required is slightly less than half that required for complex data because $(N - 1)$ of the additions needed when N is prime add a real to an imaginary, and that addition is not actually performed. When N is 2^m, there are $(N - 2)$ of these pseudo additions. The special case for real data is discussed in [29–31].

The structure of these algorithms is in the form of $\mathbf{X}' = CDA\mathbf{X}$ or $B^T DA\mathbf{X}$ or $A^T DA\mathbf{X}$ from Eqs. (4.56) and (4.58). The $\mathbf{A}$ and $\mathbf{B}$ matrices are generally $M \times N$, with $M > N$, and have elements that are integers, generally 0 or ± 1. A pictorial description is given in Fig. 4.2.

The flow graph in Fig. 4.2(a) should be compared with the matrix description of Eqs. (4.58) and (4.75) and with the programs in [7,11,17,19]. The shape in Fig. 4.2(b) illustrates the expansion of the data by A. That is, AX has more entries than X because $M > N$. The A operator consists of additions, the D operator gives the M multiplications (some by 1), and A^T contracts the data back to N values with additions only. M is one-half the second list of multiplies in Table 4.2.

An important characteristic of the D operator in the calculation of the DFT is that its entries are either purely real or imaginary. The reduction of the $\mathbf{W}$ vector by $(s^{(N-1)/2} - 1)$ and $(s^{(N-1)/2} + 1)$ separates the real and the imaginary constants [24,26]. The number of multiplications for complex data is only twice that necessary for real data, not four times.

Although this discussion has been about the calculation of the DFT, very similar results are true for the calculation of convolution; these will be further developed in Section 4.7. The $A^T DA$ structure and the picture in Fig. 4.2 are the same for convolution. Algorithms and operation counts can be found in [7,12,17].

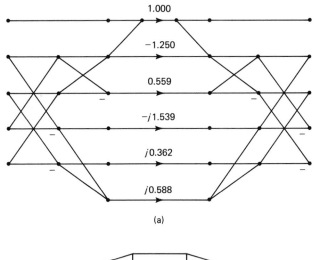

1.000

−1.250

0.559

−j1.539

j0.362

j0.588

(a)

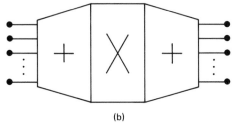

(b)

Figure 4.2 (a) Flow graph for the length-5 DFT; (b) block diagram of a Winograd short DFT.

4.4.1 The Bilinear Structure

The bilinear form introduced in Eq. (4.55) and the related linear form in Eq. (4.57) are very powerful descriptions of both the DFT and convolution.

$$\textit{Bilinear:} \qquad \mathbf{Y} = C[A\mathbf{X} \circ B\mathbf{H}] \qquad (4.76)$$

$$\textit{Linear:} \qquad \mathbf{Y} = CDA\mathbf{X} \qquad (4.77)$$

Since Eq. (4.76) is a bilinear operation defined in terms of a second bilinear operator $\circ$, this formulation can be nested. For example, if $\circ$ is itself defined in terms of a second bilinear operator @, by

$$\mathbf{X} \circ \mathbf{H} = C'[A'\mathbf{X} @ B'\mathbf{H}] \qquad (4.78)$$

then Eq. (4.76) becomes

$$\mathbf{Y} = CC'[A'A\mathbf{X} @ B'B\mathbf{H}] \qquad (4.79)$$

For convolution, if A represents the polynomial residue reduction modulo the cyclotomic polynomials, then A is square (e.g., as in Eq. 4.72), and $\circ$ represents multiplication of the residue polynomials modulo the cyclotomic polynomials. If A represents the reduction modulo the cyclotomic polynomials plus the Toom-Cook reduction, as was the case in the example of Eq. (4.75), then A is $N \times M$ and $\circ$ is term-by-term simple scalar multiplication. In this case $A\mathbf{X}$ can be thought of as a transform of $\mathbf{X}$ and C is the inverse transform. This is called a rectangular transform [19] because A is rectangular. The transform requires only additions, and convolution is done with M multiplications. The other extreme is when A represents reduction over the N complex roots of $s^N - 1$. In this case A is the DFT itself, as in the example of Eq. (4.43), and $\circ$ is point-by-point complex multiplication and C is the inverse DFT. A trivial case is where A, B, and C are identity operators and $\circ$ is the cyclic convolution.

This very general and flexible bilinear formulation coupled with the idea of nesting in Eq. (4.79) gives a description of most forms of convolution.

4.4.2 Winograd's Complexity Theorems

Because Winograd's work [7,19,24,25,32,33] has been the foundation of the modern results in efficient convolution and DFT algorithms, it is worthwhile to look at his theoretical conclusions on optimal algorithms. Most of his results are stated in terms of polynomial multiplication as in Eq. (4.30) or (4.54). The measure of computational complexity is usually the number of multiplications, and only certain multiplications are counted. This must be understood so as not to misinterpret the results.

This section simply gives a statement of the pertinent results and does not attempt to derive or prove anything. A short interpretation of each theorem is given to relate the result to the algorithms developed in this chapter. The references indicated should be consulted for background and detail.

Theorem 1.   Given two polynomials $x(s)$ and $h(s)$ of degree N and M, respectively, each with indeterminate coefficients that are elements of a field H, $N + M + 1$ multiplications are necessary to compute the coefficients of the product polynomial $x(s)h(s)$. Multiplication by elements of the field G (the field of constants), which is contained in H, are not counted, and G contains at least $N + M$ distinct elements. [24]

The upper bound in this theorem can be realized by choosing an arbitrary modulus polynomial $P(s)$ of degree $N + M + 1$ composed of $N + M + 1$ distinct linear polynomial factors with coefficients in G (which, since its degree is greater than the product $x(s)h(s)$, has no effect on the product) and by reducing $x(s)$ and $h(s)$ to $N + M + 1$ residues modulo the $N + M + 1$ factors of $P(s)$. These residues are multiplied by each other, requiring $N + M + 1$ multiplications, and the results are recombined using the Chinese remainder theorem. The operations required in the reduction and recombination are not counted, while the residue multiplications are. Since the modulus $P(s)$ is arbitrary, its factors are chosen to be simple so as to make the reduction and Chinese remainder theorem simple. Factors of zero, ± 1, and infinity are the simplest. Factors such as ± 2 complicate the actual calculations considerably, but the theorem does not take that into account. This algorithm is a form of the Toom-Cook algorithm and of Lagrange interpolation [7,17,19,24]. For our applications, H is the field of reals and G the field of rationals.

Theorem 2. If an algorithm exists that computes $x(s)h(s)$ in $N + M + 1$ multiplications, all but one of its multiplication steps must necessarily be of the form

$$m_k = [g'_k + x(g_k)][g''_k + h(g_k)] \quad \text{for} \quad k = 0, 1, \ldots, N + M$$

where g_k, g'_k, and g''_k are distinct elements of G, and g'_k and g''_k are arbitrary elements of G. [24]

Theorem 2 states that the structure of an optimal algorithm is essentially unique although the factors of $P(s)$ may be chosen arbitrarily.

Theorem 3. Let $P(s)$ be a polynomial of degree N and be of the form $P(s) = Q(s)^k$, where $Q(s)$ is an irreducible polynomial with coefficients in G, and k is a positive integer. Let $x(s)$ and $h(s)$ be two polynomials of degree at least $N - 1$ with coefficients from H; then $2N - 1$ multiplications are required to compute the product $x(s)h(s)$ modulo $P(s)$. [24]

Theorem 3 is similar to Theorem 1, with the operations of the reduction of the product modulo $P(s)$ not being counted.

Theorem 4. Any algorithm that computes the product $x(s)h(s)$ modulo $P(s)$ according to the conditions stated in Theorem 3 and requires $2N - 1$ multiplications will necessarily be of one of three structures, each of which has the form of Theorem 2 internally. [24]

As in Theorem 2, this theorem states that only a limited number of possible structures exist of optimal algorithms.

Theorem 5. If the modulus polynomial $P(s)$ has degree N and is not irreducible, it can be written in a unique factored form $P(s) = P_1^{j_1}(s) P_2^{j_2}(s) \cdots P_k^{j_k}(s)$, where each of the $P_i(s)$ is irreducible over the allowed coefficient field G. Then $2N - k$ multiplications are necessary to compute the product $x(s)h(s)$ modulo $P(s)$, where $x(s)$ and $h(s)$ have coefficients in H and are of degree at least $N - 1$. All algorithms that calculate this product in $2N - k$ multiplications must be of a form where each of the k residue polynomials of $x(s)$ and $h(s)$ is separately multiplied modulo the factors of $P(s)$ via the Chinese remainder theorem. [24]

Corollary. If the modulus polynomial is $P(s) = s^N - 1$, then $2N - \tau(N)$ multiplications are necessary to compute $x(s)h(s)$ modulo $P(s)$, where $\tau(N)$ is the number of positive divisors of N.

Theorem 5 is very general since it allows a general modulus polynomial. The proof of the upper bound involves reducing $x(s)$ and $h(s)$ modulo the k factors of $P(s)$. Each of the k irreducible residue polynomials is then multiplied using the method of Theorem 4, requiring $2N_i - 1$ multiplies, and the products are combined using the Chinese remainder theorem. The total number of multiplies from the k parts is $2N - k$. The theorem also states that the structure of these optimal algorithms is essentially unique. The special case of $P(s) = s^N - 1$ is interesting since it corresponds to cyclic convolution and, as stated in the corollary, k is easily determined. The factors of $s^N - 1$ are called *cyclotomic polynomials* and have interesting properties [7,17,19].

Theorem 6. Consider calculating the DFT of a prime-length real-valued number sequence. If G is chosen as the field of rational numbers, the number of real multiplications necessary to calculate a length-P DFT is $\mu[\text{DFT}(P)] = 2P - 3 - \tau(P - 1)$ where $\tau(P - 1)$ is the number of divisors of $P - 1$. [24,32]

Theorem 6 not only gives a lower limit on any practical prime-length DFT algorithm, but it also gives practical algorithms for $N = 3$, 5, and 7. Consider the operation counts in Table 4.2 to understand this theorem. In addition to the real multiplications counted by complexity theory, each optimal prime-length algorithm will have one multiplication by a rational constant. That constant corresponds to the residue modulo $(s - 1)$ that always exists for the modulus $P(s) = s^{N-1} - 1$. In a practical algorithm, this multiplication must be carried out, and that accounts for the difference in the prediction of Theorem 6 and the count in Table 2. In addition, another operation for certain applications must be counted as a multiplication: the calculation of the zero frequency term $X(0)$ in the first row of the example in Eq. (4.51). For applications to the WFTA discussed in Section 4.6.2, that operation must be counted as a multiply. For lengths longer than 7, optimal algorithms require too many additions, so compromise structures are used.

Theorem 7. If G is chosen as the field of rational numbers, the number of real multiplications necessary to calculate a length-N DFT where N is a prime number raised to an integer power, $N = P^m$, is given by

$$\mu[\text{DFT}(N)] = 2N - [(m^2 + m)/2]\tau(P - 1) - m - 2$$

where $\tau(P - 1)$ is the number of divisors of $(P - 1)$. [33,34]

This result seems to be practically achievable only for $N = 9$, or perhaps 25. In the case of $N = 9$, two rational multiplies must be carried out; they are counted in Table 4.2 but are not predicted by Theorem 7. Experience [35] indicates that even for $N = 25$, an algorithm based on a Cooley-Tukey FFT using a type 2 index map gives an overall more balanced result.

Theorem 8. [34] If G is chosen as the field of rational numbers, the number of real multiplications necessary to calculate a length-N DFT where $N = 2^m$ is given by

$$\mu[\text{DFT}(N)] = 2N - m^2 - m - 2$$

This result is not practically useful because the number of additions necessary to realize this minimum of multiplications becomes very large for lengths greater than 16. Nevertheless, it proves that the minimum number of multiplications required of an optimal algorithm is a linear function of N rather than of $N \log N$, which is that required of practical algorithms. The best practical power-of-two algorithm seems to be the split-radix FFT [36] discussed in Section 4.5.2.

All of these theorems use ideas based on residue reduction, multiplication of the residues, and then combination by the Chinese remainder theorem. It is remarkable that this approach finds the minimum number of required multiplications by a constructive proof that generates an algorithm that achieves this minimum; and the structure of the optimal algorithm is, within certain variations, unique. For shorter lengths, the optimal algorithms give practical programs. For longer lengths, the uncounted operations involved with the multiplication of the higher-degree residue polynomials become very large and impractical. In those cases, efficient suboptimal algorithms can be generated by using the same residue reduction as for the optimal case, but by using methods other than the Toom-Cook algorithm of Theorem 1 to multiply the residue polynomials.

Practical long DFT algorithms are produced by combining short prime-length optimal DFTs with the type 1 index map from Section 4.1 to give the prime factor algorithm and the Winograd Fourier transform algorithm discussed in Section 4.6. It is interesting to note that the index mapping technique is useful inside the short DFT algorithms to replace the Toom-Cook algorithm and outside to combine the short DFTs to calculate long DFTs.

4.5 THE COOLEY-TUKEY FAST FOURIER TRANSFORM ALGORITHM

The publication by Cooley and Tukey in 1965 [4] of an efficient algorithm for the calculation of the DFT was a major turning point in the development of digital signal processing. During the five or so years that followed, various extensions and modifications were made to the original algorithm. By the early 1970s the practical programs were basically in the form used today. The standard development presented in [1,10,37] shows how the DFT of a length-N sequence can be simply calculated from the two length-$N/2$ DFTs of the even index terms and the odd index terms. This is then applied to the two half-length DFTs to give four quarter-length DFTs and is repeated until N scalars are left, which are the DFT values. Because of alternately taking the even and odd index terms, two forms of the resulting programs are called *decimation-in-time* and *decimation-in-frequency*. For a length of 2^N, the dividing process is repeated $M = \log_2 N$ times and requires N multiplications each time. This gives the well-known formula for the computational complexity of the FFT of $N \log_2 N$ that was derived in Eq. (4.26).

Although the decimation methods are straightforward and easy to understand, they do not generalize well. For that reason we assume that the reader is familiar with that description, and here we will develop the FFT using the index map from Section 4.1.

The Cooley-Tukey FFT always uses the type 2 index map from Eq. (4.8). This is necessary for the most popular forms that have $N = R^m$ but is also used even when the factors are relatively prime and a type 1 map could be used. The time and frequency maps from Eqs. (4.4) and (4.9) are

$$n = ((K_1 n_1 + K_2 n_2))_N \qquad \text{and} \qquad k = ((K_3 k_1 + K_4 k_2))_N \qquad (4.80)$$

Type 2 conditions given in Eqs. (4.6) and (4.8) become

$$K_1 = aN_2 \qquad \text{or} \qquad K_2 = bN_1 \qquad \text{but not both}$$

and $\hspace{10cm}$ (4.81)

$$K_3 = cN_2 \qquad \text{or} \qquad K_4 = dN_1 \qquad \text{but not both}$$

where $N = N_1 N_2$. The row and column calculations in Eq. (4.11) are uncoupled by Eq. (4.12), which for this case are

$$((K_1 K_4))_N = 0 \qquad \text{or} \qquad ((K_2 K_3))_N = 0 \qquad \text{but not both} \qquad (4.82)$$

To make each short sum a DFT, the K_i must satisfy

$$((K_1 K_3))_N = N_2 \qquad \text{and} \qquad ((K_2 K_4))_N = N_1$$

To have the smallest values for K_i, the constants in Eq. (4.81) are chosen to be

$$a = d = K_2 = K_3 = 1 \qquad (4.83)$$

which makes the index maps of Eq. (4.80) become

$$n = N_2 n_1 + n_2 \qquad \text{and} \qquad k = k_1 + N_1 k_2 \qquad (4.84)$$

These index maps are all evaluated modulo N as indicated in Eq. (4.80), but in Eq. (4.84) explicit reduction is not necessary since n never exceeds N. The reduction notation will be omitted for clarity. From Eq. (4.11) and the example in Eq. (4.14), the DFT is

$$\hat{X} = \sum_{n_2=0}^{N_2-1} \sum_{n_1=0}^{N_1-1} \hat{x} \, W_{N_1}^{n_1 k_1} \, W_N^{n_2 k_1} \, W_{N_2}^{n_2 k_2} \qquad (4.85)$$

This map of Eq. (4.84) and the form of the DFT in Eq. (4.85) are the fundamentals of the Cooley-Tukey FFT.

The order of the summations using the type 2 map in Eq. (4.85) cannot be reversed as it can with the type 1 map. This is because of the W_N terms, the twiddle factors.

Turning Eq. (4.85) into an efficient program requires some care. From Section 4.1.2 we know that all the factors should be equal. If $N = R^M$, with R called the radix, N_1 is first set equal to R, and N_2 is then necessarily R^{M-1}. Consider n_1 to be the index along the rows and n_2 along the columns. The inner sum of Eq. (4.85) over n_1 represents a length-N_1 DFT for each value of n_2. These N_2 length-N_1 DFTs are the DFTs of the rows of the $\hat{x}(n_1, n_2)$ array. The resulting array of row DFTs is multiplied by an array of twiddle factors, which are the W_N terms in Eq. (4.85). The twiddle factor array for a length-8 radix-2 FFT is

$$\text{Twiddle factors:} \quad W_8^{k_1 n_2} = \begin{bmatrix} W^0 & W^0 \\ W^0 & W^1 \\ W^0 & W^2 \\ W^0 & W^3 \end{bmatrix} = \begin{bmatrix} 1 & 1 \\ 1 & W \\ 1 & -j \\ 1 & -jW \end{bmatrix} \quad (4.86)$$

The twiddle factor array will always have unity in the first row and first column.

To complete Eq. (4.85) at this point, after the row DFTs are multiplied by the twiddle factor array, the N_1 length-N_2 DFTs of the columns are calculated. However, since the column DFTs are of length R^{M-1}, they can be posed as an $R^{M-2} \times R$ array and the process repeated, again using the length-R DFTs. After M stages of length-R DFTs with twiddle factor multiplications interleaved, the DFT is complete. The flow graph of a length-2 DFT is given in Fig. 4.3(a) and is called a butterfly because of its shape. The flow graph of the complete length-8 radix-2 FFT is shown in Fig. 4.3(b). This flow graph, the twiddle factor map of Eq. (4.86), and the basic equation (4.85) should be completely understood before going further.

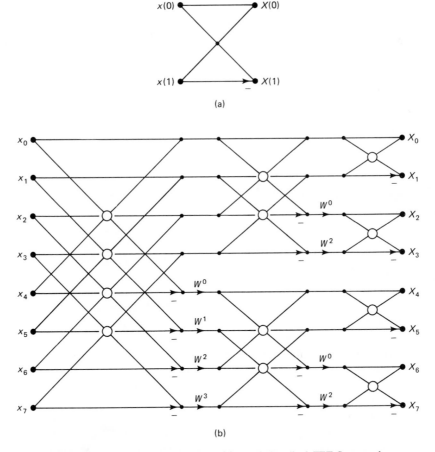

(a)

(b)

Figure 4.3   (a) A radix-2 butterfly; (b) length-8 radix-2 FFT flow graph.

A very efficient indexing scheme has evolved over the years that results in a compact and efficient computer program. Figure 4.4 gives a FORTRAN program that implements the radix-2 FFT. It should be studied [11] to see how it implements Eq. (4.85) and the flow graph representation.

This discussion, the flow graph of Fig. 4.3, and the program of Fig. 4.4 are all based on the input index map of Eqs. (4.4) and (4.80), and the calculations are performed in place. According to Section 4.1.1, this means that the output is scrambled in bit-reversed order and should be followed by an unscrambler to give the DFT in proper order. This formulation is called a decimation-in-frequency FFT [1,10,37]. A very similar algorithm based on the output index map can be derived; it is called a decimation-in-time FFT. Examples of FFT programs are found in [11].

```
N2 = N
DO 10 K = 1, M
    N1 = N2
    N2 = N2/2
    E  = 6.28318/N1
    A  = 0
    DO 20 J = 1, N2
        C = COS(A)
        S = -SIN(A)
        A = J*E
        DO 30 I = J, N, N1
            L    = I + N2
            XT   = X(I) - X(L)
            X(I) = X(I) + X(L)
            YT   = Y(I) - Y(L)
            Y(I) = Y(I) + Y(L)
            X(L) = XT*C - YT*S
            Y(L) = XT*S + YT*C
30          CONTINUE
20      CONTINUE
10  CONTINUE
```

Figure 4.4   A radix-2 Cooley-Tukey FFT program.

4.5.1 Modifications to the Basic Cooley-Tukey FFT

Soon after the paper by Cooley and Tukey [4], improvements and extensions were made in their algorithms. One very important discovery was the improvement in efficiency by using a larger radix of 4, 8, or even 16. For example, just as for the radix-2 butterfly, there are no multiplications required for a length-4 DFT, and, therefore, a radix-4 FFT would have only twiddle factor multiplications. Because there are half as many stages in a radix-4 FFT, there would be half as many multiplications as in a radix-2 FFT. In practice, because some of the multiplications are by unity, the improvement is not by a factor of two, but it is significant. A radix-4 FFT is easily developed from the basic radix-2 structure by replacing the length-2 butterfly by a length-4 butterfly and making a few other modifications. Programs can be found in [11], and operation counts will be given in Section 4.5.3.

Increasing the radix to 8 gives some improvement, but not as much as from 2 to 4. Increasing it to 16 is theoretically promising, but the small decrease in multiplications is somewhat offset by an increase in additions, and the program becomes rather long. Other radices are not attractive because they generally require a substantial number of multiplications and additions in the butterflies.

The second method of reducing arithmetic is to remove the unnecessary twiddle factor multiplications by ± 1 or by $\pm \sqrt{-1}$. This occurs when the exponent of W_N is zero or a multiple of $N/4$. A reduction of additions as well as multiplications is achieved by removing these extraneous complex multiplications since a complex multiplication requires at least two real additions. In a program, this reduction is usually achieved by having special butterflies for the cases where the twiddle factor is 1 or j. As many as four special butterflies may be necessary to remove all unnecessary arithmetic, but in many cases there will be no practical improvement above two or three butterflies.

In addition to removing multiplications by 1 or j, there can be a reduction in multiplications by using a special butterfly for twiddle factors with $W^{N/8}$, which have equal real and imaginary parts. Also, for computers or hardware with multiplication considerably slower than addition, it is desirable to use an algorithm for complex multiplication that requires three multiplications and three additions rather than the conventional four multiplications and two additions. Note that this gives no reduction in the total number of arithmetic operations but does give a trade of multiplications for additions. This is one reason not to use complex data types in programs but to explicitly program complex arithmetic.

A time-consuming and unnecessary part of the execution of an FFT program is the calculation of the sine and cosine terms that are the real and imaginary parts of the twiddle factors. There are basically three approaches to obtaining the sine and cosine values. (1) They can be calculated as needed, as is done in the sample program above. (2) One value per stage can be calculated and the others recursively calculated from those. That method is fast but suffers from accumulated roundoff errors. (3) The fastest method is to fetch precalculated values from a stored table. This has the disadvantage of requiring considerable memory space.

If all the N DFT values are not needed. special forms of the FFT can be developed using a process called pruning [38], which removes the operations concerned with the unneeded outputs.

Special algorithms are possible for cases with real data or with symmetric data [39]. The decimation-in-time algorithm can be easily modified to transform real data and save approximately half the arithmetic required for complex data [31]. There are numerous other modifications to deal with special hardware considerations such as an array processor or a special microprocessor such as the TMS 320. Examples of programs that deal with some of these items can be found in [10,11,39].

4.5.2 The Split-Radix FFT Algorithms

Recently several papers [20,29,36,40,41] have been published on algorithms to calculate a length-2^M DFT more efficiently than a Cooley-Tukey FFT of any radix. They all have the same computational complexity and are optimal for lengths up through 16 and probably give the best total add-multiply count possible for any power-of-two length. Yavne published an algorithm with the same computational complexity in 1968 [42], but it went largely unnoticed.

The basic idea behind the split-radix FFT (SRFFT) as derived by Duhamel and

Hollmann [36,41] is the application of a radix-2 index map to the even indexed terms and a radix-4 map to the odd indexed terms. The basic definition of the DFT

$$C_k = \sum_{n=0}^{N-1} x_n W^{nk} \tag{4.87}$$

with $W = e^{-j2\pi/N}$ gives

$$C_{2k} = \sum_{n=0}^{N/2-1} (x_n + x_{n+N/2})W^{2nk} \tag{4.88}$$

for the even index terms, and

$$C_{4k+1} = \sum_{n=0}^{N/4-1} [(x_n - x_{n+N/2}) - j(x_{n+N/4} - x_{n+3N/4})]W^n W^{4nk} \tag{4.89}$$

and

$$C_{4k+3} = \sum_{n=0}^{N/4-1} [(x_n - x_{n+N/2}) + j(x_{n+N/4} - x_{n+3N/4})]W^{3n} W^{4nk} \tag{4.90}$$

for the odd index terms. This results in an L-shaped "butterfly," shown in Fig. 4.5(a), which relates a length-N DFT to one length-$N/2$ DFT and two length-$N/4$ DFTs with twiddle factors. Repeating this process for the half- and quarter-length DFTs until scalars result gives the SRFFT algorithm in much the same way the decimation-in-frequency radix-2 Cooley-Tukey FFT is derived [1,10,37]. The resulting flow graph for the algorithm calculated in place looks like a radix-2 FFT except for the location of the twiddle factors. Indeed, it is the location of the twiddle factors that makes this algorithm use less arithmetic. The L-shaped SRFFT butterfly advances the calculation of the top half by one of the M stages while the lower half, like a radix-4 butterfly, calculates two stages at once. This is illustrated for $N = 8$ in Fig. 4.5(b).

Unlike the fixed-radix, mixed-radix, or variable-radix Cooley-Tukey FFT or even the prime factor algorithm or Winograd Fourier transform algorithm, the split-radix FFT does not progress completely stage by stage or, in terms of indices, does

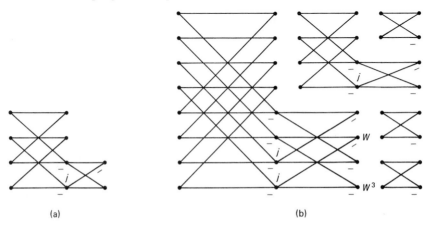

(a)   (b)

Figure 4.5   (a) Structure of the split-radix FFT butterfly and (b) transform of the length-8 split-radix FFT.

not complete each nested sum in order. This is perhaps better seen from the poly-nomial formulation of Martens [20]. Because of this, the indexing is somewhat more complicated than the conventional Cooley-Tukey program.

Figure 4.6 gives a FORTRAN program that implements the basic decimation-in-frequency split-radix FFT algorithm. The indexing scheme [41] of this program gives a structure very similar to the Cooley-Tukey programs in [11] and allows the same modifications and improvements such as decimation-in-time, multiple butterflies, table look-up of sine and cosine values, efficient complex multiply meth-ods, and real data versions [29,31].

```
          SUBROUTINE FFT (X,Y,N,M)
          N2 = 2*N
          DO 10 K = 1, M-1
              N2 = N2/2
              N4 = N2/4
              E  = 6.283185307179586/N2
              A  = 0
              DO 20 J = 1, N4
                  A3  = 3*A
                  CC1 = COS(A)
                  SS1 = SIN(A)
                  CC3 = COS(A3)
                  SS3 = SIN(A3)
                  A   = J*E
                  IS  = J
                  ID  = 2*N2
40                DO 30 I0 = IS, N-1, ID
                      I1    = I0 + N4
                      I2    = I1 + N4
                      I3    = I2 + N4
                      R1    = X(I0) - X(I2)
                      X(I0) = X(I0) + X(I2)
                      R2    = X(I1) - X(I3)
                      X(I1) = X(I1) + X(I3)
                      S1    = Y(I0) - Y(I2)
                      Y(I0) = Y(I0) + Y(I2)
                      S2    = Y(I1) - Y(I3)
                      Y(I1) = Y(I1) + Y(I3)
                      S3    = R1 - S2
                      R1    = R1 + S2
                      S2    = R2 - S1
                      R2    = R2 + S1
                      X(I2) = R1*CC1 - S2*SS1
                      Y(I2) = S2*CC1 - R1*SS1
                      X(I3) = S3*CC3 + R2*SS3
                      Y(I3) = R2*CC3 - S3*SS3
30                CONTINUE
                  IS = 2*ID - N2 + J
                  ID = 4*ID
                  IF (IS.LT.N) GOTO 40
20            CONTINUE
10        CONTINUE
          IS = 1
          ID = 4
50        DO 60 I0 = IS, N, ID
              I1    = I0 + 1
              R1    = X(I0)
              X(I0) = R1 + X(I1)
              X(I1) = R1 - X(I1)
              R1    = Y(I0)
              Y(I0) = R1 + Y(I1)
60        Y(I1) = R1 - Y(I1)
          IS = 2*ID - 1
          ID = 4*ID
          IF (IS.LT.N) GOTO 50
```

Figure 4.6   Split-radix FFT FORTRAN subroutine.

As was done for the other decimation-in-frequency algorithms, the input index map is used and the calculations are done in place, resulting in the output being in bit-reversed order. The three statements following label 30 do the special indexing required by the SRFFT. The last stage is of length 2 and, therefore, is inappropriate for the standard L-shaped butterfly, so it is calculated separately in the DO 60 loop. This program is considered a one-butterfly version. A second butterfly can be added just before statement 40 to remove the unnecessary multiplications by unity. A third butterfly can be added to reduce the number of real multiplications from four to two for the complex multiplication when W has equal real and imaginary parts. It is also possible to reduce the arithmetic for the two-butterfly case and to reduce the data transfers by directly programming a length-4 and a length-8 butterfly to replace the last three stages. This is called a two-butterfly-plus version. Operation counts for the one-, two-, two-plus-, and three-butterfly SRFFT programs are given in the next section. Some details can be found in [41].

The special case of an SRFFT for real data and symmetric data is discussed in [29]. An application of the decimation-in-time SRFFT to real data is given in [31]. Application to convolution is made in [43], to the discrete Hartley transform in [22,43], to calculating the discrete cosine transform in [40,44], and could be made to calculating number theoretic transforms.

4.5.3 Evaluation of the Cooley-Tukey FFT Algorithms

The evaluation of any FFT algorithm starts with a count of the real (or floating-point) arithmetic. Table 4.3 gives the number of real multiplications and additions required to calculate a length-N FFT of complex data. Results of programs with one, two, three, and five butterflies are given to show the improvement that can be expected from removing unnecessary multiplications and additions. Results of radices 2, 4, 8, and 16 for the Cooley-Tukey FFT as well as of the split-radix FFT are given to show the relative merits of the various structures. Comparisons of these data should be made with the table of counts for the prime factor algorithm and WFTA programs in Section 4.6.4. All programs use the four-multiply, two-add complex multiply algorithm. A similar table can be developed for the three-multiply, three-add algorithm, but the relative results are the same.

From Table 4.3 we see that a greater improvement is obtained going from radix 2 to 4 than from 4 to 8 or 16. This is partly because length-2 and -4 butterflies have no multiplications, while length-8, -16, and higher do. We also see that going from one to two butterflies gives more improvement than going from two to higher values. From the perspective of operation count and from practical experience, a three-butterfly radix-4 or a two-butterfly radix-8 FFT is a good compromise. The radix-8 and -16 programs become long, especially with multiple butterflies, and they give a limited choice of transform length unless combined with some length-2 and -4 butterflies. In Table 4.3, M_i and A_i refer to the number of real multiplications and additions used by an FFT with i butterflies. For the split-radix FFT, $M3$ and $A3$ refer to the two-butterfly-plus program and $M5$ and $A5$ refer to the three-butterfly program.

The first evaluations of FFT algorithms were in terms of the number of real

TABLE 4.3 NUMBER OF REAL MULTIPLIES AND ADDS FOR COMPLEX FFT
ALGORITHMS WITH DIFFERENT RADICES AND NUMBER OF BUTTERFLIES

N	$M1$	$M2$	$M3$	$M5$	$A1$	$A2$	$A3$	$A5$
Radix 2								
2	4	0	0	0	6	4	4	4
4	16	4	0	0	24	18	16	16
8	48	20	8	4	72	58	52	52
16	128	68	40	28	192	162	148	148
32	320	196	136	108	480	418	388	388
64	768	516	392	332	1152	1026	964	964
128	1792	1284	1032	908	2688	2434	2308	2308
256	4096	3076	2568	2316	6144	5634	5380	5380
512	9216	7172	6152	5644	13824	12802	12292	12292
1024	20480	16388	14344	13324	30720	28674	27652	27652
2048	45056	36868	32776	30732	67584	63490	61444	61444
4096	98304	81924	73736	69644	147456	139266	135172	135172
Radix 4								
4	12	0	0	0	22	16	16	16
16	96	36	28	24	176	146	144	144
64	576	324	284	264	1056	930	920	920
256	3072	2052	1884	1800	5632	5122	5080	5080
1024	15360	11268	10588	10248	28160	26114	25944	25944
4096	73728	57348	54620	53256	135168	126978	126296	126296
Radix 8								
8	32	4	4	4	66	52	52	52
64	512	260	252	248	1056	930	928	928
512	6144	4100	4028	3992	12672	11650	11632	11632
4096	65536	49156	48572	48280	135168	126978	126832	126832
Radix 16								
16	80	20	20	20	178	148	148	148
256	2560	1540	1532	1528	5696	5186	5184	5184
4096	61440	45060	44924	44856	136704	128514	128480	128480
Split radix								
2	0	0	0	0	4	4	4	4
4	8	0	0	0	20	16	16	16
8	24	8	4	4	60	52	52	52
16	72	32	28	24	164	144	144	144
32	184	104	92	84	412	372	372	372
64	456	288	268	248	996	912	912	912
128	1080	744	700	660	2332	2164	2164	2164
256	2504	1824	1740	1656	5348	5008	5008	5008
512	5688	4328	4156	3988	12060	11380	11380	11380
1024	12744	10016	9676	9336	26852	25488	25488	25488
2048	28216	22760	22076	21396	59164	56436	56436	56436
4096	61896	50976	49612	48248	129252	123792	123792	123792

multiplications required because that was the slowest operation on the computer and, therefore, controlled the execution speed. Later, with hardware arithmetic, the number of both multiplications and additions became important. Modern systems have arithmetic speeds such that indexing and data transfer times become important factors. Morris [47] has looked at some of these problems and has developed a procedure called *autogen* to write partially straight-line program code to significantly reduce overhead and speed up FFT run times. Some hardware, such as the TMS 320 signal processing chip, has the multiply and add operations combined. Some machines have vector instructions or parallel processors. Because the execution speed of an FFT depends not only on the algorithm but also on the hardware architecture and compiler, experiments must be run on the system to be used.

In many cases the unscrambler or bit-reversed counter requires 10% of the execution time; therefore, if possible, it should be eliminated. In high-speed convolution where the convolution is done by multiplication of DFTs, a decimation-in-frequency FFT can be combined with a decimation-in-time inverse FFT to require no unscrambler. It is also possible for a radix-2 FFT to do the unscrambling inside the FFT, but the structure is not very regular [10,14]. Special structures can be found in [10], and programs for data that are real or have special symmetries are in [31,39,43].

Although there can be significant differences in the efficiencies of the various Cooley-Tukey and split-radix FFTs, the number of multiplications and additions for all of them is on the order of $N \log N$. That is fundamental to the class of algorithms.

4.6 THE PRIME FACTOR AND WINOGRAD FOURIER TRANSFORM ALGORITHMS

The prime factor algorithm (PFA) and the Winograd Fourier transform algorithm (WFTA) are methods for efficiently calculating the DFT that use, and in fact depend on, the type 1 index map from Eq. (4.7). The use of this index map preceded Cooley and Tukey's paper [8,9], but its full potential was not realized until it was combined with Winograd's short DFT algorithms. The modern PFA was first presented in [28], and a program given in [13]. The WFTA was first presented in [5] and programs given in [19,39].

The number-theoretic basis for the indexing in these algorithms may at first seem more complicated than in the Cooley-Tukey FFT; however, if approached from the point of view of general index mapping in Section 4.1, it is straightforward and part of a common approach to breaking large problems into smaller ones. The development in this section will parallel that in Section 4.5.

The general index maps of Eqs. (4.4) and (4.9) must satisfy the type 1 conditions of Eqs. (4.5) and (4.7), which are

$$
\begin{aligned}
K_1 = aN_2 \quad &\text{and} \quad K_2 = bN_1 \quad \text{with} \quad (K_1, N_1) = (K_2, N_2) = 1 \\
K_3 = cN_2 \quad &\text{and} \quad K_4 = dN_1 \quad \text{with} \quad (K_3, N_1) = (K_4, N_2) = 1
\end{aligned}
\tag{4.91}
$$

where $N = N_1 N_2$. The row and column calculations in Eq. (4.11) are uncoupled by Eq. (4.12), which for this case are

$$((K_1 K_4))_N = ((K_2 K_3))_N = 0 \tag{4.92}$$

In addition, to make each short sum a DFT, the K_i must also satisfy

$$((K_1 K_3))_N = N_2 \quad \text{and} \quad ((K_2 K_4))_N = N_1 \tag{4.93}$$

To have the smallest values for K_i, the constants in Eq. (4.91) are chosen to be

$$a = b = 1, \qquad c = ((N_2^{-1}))_{N_1}, \qquad d = ((N_1^{-1}))_{N_2} \tag{4.94}$$

which gives for the index maps in Eq. (4.91)

$$n = ((N_2 n_1 + N_1 n_2))_N \qquad \text{and} \qquad k = ((K_3 k_1 + K_4 k_2))_N \tag{4.95}$$

The frequency index map is a form of the Chinese remainder theorem. Using these index maps, the DFT in Eq. (4.11) becomes

$$\hat{X} = \sum_{n_2=0}^{N_2-1} \sum_{n_1=0}^{N_1-1} \hat{x} \, W_{N_1}^{n_1 k_1} \, W_{N_2}^{n_2 k_2} \tag{4.96}$$

which is a pure two-dimensional DFT with no twiddle factors, and the summations can be done in either order. Other choices than Eq (4.94) could be used. For example, $a = b = c = d = 1$ will cause the input and output index map to be the same; therefore, there will be no scrambling of the output order. The short summations in Eq. (4.96), however, will no longer be short DFTs [13].

An important feature of the short Winograd DFTs described in Section 4.4 that is useful for both the PFA and the WFTA is the fact that the multiplier constants in Eq. (4.57) or (4.58) are either real or imaginary, never a general complex number. For that reason, multiplication by complex data requires only two real multiplications, not four. That is a very significant feature. It is also true that j multiplier can be commuted from the D operator to the last part of the A^T operator. This means the D operator has only real multipliers, and the calculations on real data remain real until the last stage. This can be seen by examining the short DFT modules in [11,26].

4.6.1 The Prime Factor Algorithm

If the DFT is calculated directly using Eq. (4.96), the algorithm is called a prime factor algorithm [8,9] (discussed in Sections 4.1 and 4.1.1). When the short DFTs are calculated by the very efficient algorithms of Winograd discussed in Section 4.4, the PFA becomes a very powerful method that is as fast as or faster than the best Cooley-Tukey FFTs [13,28].

A flow graph is not as helpful with the PFA as it was with the Cooley-Tukey FFT, but the representation in Fig. 4.7, which combines Figs. 4.1 and 4.2(b), gives a good picture of the algorithm with the example of Eq. (4.18). If N is factored into three factors, the DFT of Eq. (4.96) would have three nested summations and would be a three-dimensional DFT. This principle extends to any number of factors; however, recall that the type 1 map requires that all the factors be relatively prime. A very simple three-loop indexing scheme has been developed [13] that gives a compact, efficient PFA program for any number of factors. The basic program structure is

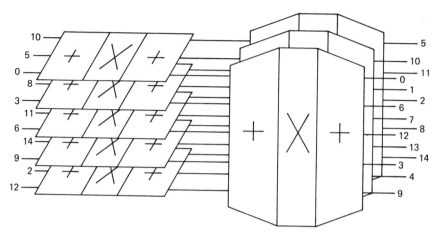

Figure 4.7 A prime factor FFT for $N = 15$.

illustrated in Fig. 4.8, with the short DFTs omitted for clarity. Complete programs are given in [11].

As in the Cooley-Tukey program, the DO 10 loop steps through the M stages (factors of N) and the DO 20 loop calculates the N_2 length-N_1 DFTs. The input index map of Eq. (4.95) is implemented in the DO 30 loop and in the statement just before label 20. In the PFA, each stage or factor requires a separately programmed module or butterfly. This lengthens the PFA program, but an efficient Cooley-Tukey program will also require three or more butterflies.

```
C ─────────────── PFA INDEXING LOOPS ───────────────
      DO 10 K = 1, M
        N1    = NI(K)
        N2    = N/N1
        I(1) = 1
        DO 20 J = 1, N2
          DO 30 L = 2, N1
            I(L) = I(L-1) + N2
            IF (I(L).GT.N) I(L) = I(L) - N
30        CONTINUE
          GOTO (20,102,103,104,105), N1
          I(1) = I(1) + N1
20      CONTINUE
10    CONTINUE
      RETURN

C ─────────────── MODULE FOR N = 2 ───────────────
102   R1       = X(I(1))
      X(I(1)) = R1 + X(I(2))
      X(I(2)) = R1 - X(I(2))
      R1       = Y(I(1))
      Y(I(1)) = R1 + Y(I(2))
      Y(I(2)) = R1 - Y(I(2))
      GOTO 20

C ─────────────── OTHER MODULES ───────────────
103   Length-3 DFT
104   Length-4 DFT
105   Length-5 DFT

      etc.
```

Figure 4.8 Part of a FORTRAN PFA program.

Because the PFA is calculated in place using the input index map, the output is scrambled. There are five approaches to dealing with this scrambled output [16]. First, there are some applications where the output does not have to be unscrambled as in the case of high-speed convolution. Second, an unscrambler can be added after the PFA to give the output in correct order just as the bit-reversed counter is used for the Cooley-Tukey FFT. A simple unscrambler is given in [11,13], but it is not in place. The third method does the unscrambling in the modules while they are being calculated. This is probably the fastest method, but the program must be written for a specific length [11,13]. A fourth, similar method achieves the unscrambling by choosing the multiplier constants in the modules properly [26]. The fifth method uses a separate indexing method for the input and output of each module [11,15].

4.6.2 The Winograd Fourier Transform Algorithm

The Winograd Fourier transform algorithm (WFTA) uses a very powerful property of the type 1 index map and the DFT to give a further reduction of the number of multiplications in the PFA. Using an operator notation where F_1 represents taking row DFTs and F_2 represents column DFTs, the two-factor PFA of Eq. (4.96) is represented by

$$X = F_2 F_1 x \tag{4.97}$$

It has been shown [25,45] that if each operator represents identical operations on each row or column, the operators commute. Since F_1 and F_2 represent length-N_1 and -N_2 DFTs, they commute, and Eq. (4.97) can also be written

$$X = F_1 F_2 x \tag{4.98}$$

If each short DFT in F is expressed by three operators as in Eq. (4.58) and Fig. 4.2, F can be factored as

$$F = A^T D A \tag{4.99}$$

where A represents the set of additions done on each row or column that performs the residue reduction as in Eq. (4.75). Because of the appearance of the flow graph of A and because it is the first operator on x, it is called a preweave operator [19]. D is the set of M multiplications, and A^T (or B^T or C^T from Eq. 4.56 or 4.57) is the reconstruction operator called the postweave. Applying Eq. (4.99) to Eq. (4.97) gives

$$X = A_2^T D_2 A_2 A_1^T D_1 A_1 x \tag{4.100}$$

This is the PFA of Eq. (4.96) and Fig. 4.7, where $A_1^T D_1 A_1$ represents the row DFTs on the array formed from x. Because these operators commute, Eq. (4.100) can also be written as

$$X = A_2^T A_1^T D_2 D_1 A_2 A_1 x \quad \text{or} \quad X = A_1^T A_2^T D_1 D_2 A_2 A_1 x \tag{4.101}$$

but the two adjacent multiplication operators can be premultiplied and the result represented by one operator $D = D_2 D_1$, which is no longer the same for each row or column. Equation (4.101) becomes

$$X = A_1^T A_2^T D A_2 A_1 x \tag{4.102}$$

This is the basic idea of the WFTA. The commuting of the multiplication operators in the center of the algorithm is called nesting, and it results in a significant decrease in the number of multiplications that must be done at the execution of the algorithm. Pictorially, the PFA of Fig. 4.7 becomes the WFTA [28] in Fig. 4.9.

The rectangular structure of the preweave addition operators causes an expansion of the data in the center of the algorithm. The 15 data points in Fig. 4.9 become 18 intermediate values. This expansion is a major problem in programming the WFTA because it prevents a straightforward in-place calculation and causes an increase in the number of required additions and in the number of multiplier constants that must be precalculated and stored.

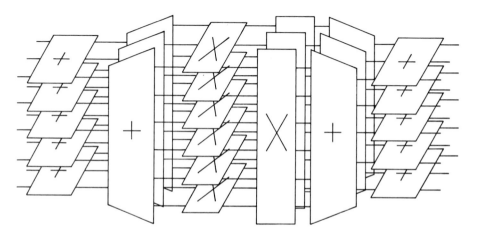

Figure 4.9 A length-15 WFTA with nested multiplications.

From Fig. 4.9 and the idea of premultiplying the individual multiplication operators, we can see why we had to consider the multiplications by unity in Table 4.2. Even if a multiplier in D_1 is unity, it may not be in $D_2 D_1$. In Fig. 4.9 with factors of three and five, there appear to be 18 multiplications required because of the expansion of the length-5 preweave operator, A_2; however, one of the multipliers in each of the length-3 and length-5 operators is unity, so one of the 18 multipliers in the product is unity. This gives 17 required multiplications—a rather impressive reduction from the $15^2 = 225$ multiplications required by direct calculation. This number of 17 complex multiplications will require only 34 real multiplications because, as mentioned earlier, the multiplier constants are purely real or imaginary while the 225 complex multiplications are general and therefore will require four times as many real multiplications.

The number of additions depends on the order of the pre- and postweave operators. For example, in the length-15 WFTA in Fig. 4.9, if the length-5 WFTA had been done first and last, there would have been six row addition preweaves in the preweave operator rather than the five shown. It is difficult to illustrate the algorithm for three or more factors of N, but the ideas apply to any number of factors. Each length has an optimal ordering of the pre- and postweave operators that will minimize the number of additions.

A program for the WFTA is not as simple as one for the FFT or PFA because of the very characteristic that reduces the number of multiplications, the nesting. A simple two-factor example program is given in [11] and a general program can be found in [19,39]. The same lengths are possible with the PFA and WFTA and the same short DFT modules can be used, but the multiplies in the modules must occur in one place for use in the WFTA.

4.6.3 Modifications of the PFA and WFTA Type Algorithms

In the previous section we saw how using the permutation property of the elementary operators in the PFA allows the nesting of the multiplications to reduce their number. We also saw that a proper ordering of the operators could minimize the number of additions. These ideas have been extended in formulating a more general algorithm-optimizing problem. If the DFT operator F in Eq. (4.99) is expressed in a still more factored form obtained from Eq. (4.75), a greater variety of ordering can be optimized. For example, if the A operators have two factors, A and A', then

$$F_1 = A_1^T A_1'^T D_1 A_1' A_1 \tag{4.103}$$

The DFT in Eq. (4.98) becomes

$$X = A_2^T A_2'^T D_2 A_2' A_2 A_1^T A_1'^T D_1 A_1' A_1 x \tag{4.104}$$

The operator notation is very helpful in understanding the central ideas, but it may hide some important facts. It has been shown [25,26] that operators in different F_i commute with each other, but the order of the operators within an F_i cannot be changed. They represent the matrix multiplications in Eq. (4.75) or (4.58) that do not commute.

This formulation allows a very large set of possible orderings; in fact, the number is so large that some automatic technique must be used to find the "best." It is possible to set up a criterion of optimality that includes not only the number of multiplications but the number of additions as well. The effects of relative multiply-add times, data transfer times, CPU register and memory sizes, and other hardware characteristics can be included in the criterion. Dynamic programming can then be applied to derive an optimal algorithm for a particular application [45]. This is a very interesting idea because there is no longer a single algorithm, but a class and an optimizing procedure. The challenge is to generate a broad enough class to result in a solution that is close to a global optimum and to have a practical scheme for finding the solution.

Results obtained applying the dynamic programming method to the design of fairly long DFT algorithms gave algorithms that had fewer multiplications and additions than either a pure PFA or WFTA [45]. It seems that some nesting is desirable, but not total nesting for four or more factors. There are also some interesting possibilities in mixing the Cooley-Tukey algorithm with this formulation. Unfortunately, the twiddle factors are not the same for all rows and columns; therefore, operations cannot commute past a twiddle factor operator. There are ways of breaking the total algorithm into horizontal paths and using different orderings along the different paths [17,26]. In a sense, this is what the split-radix FFT does with its twiddle factors when compared to a conventional Cooley-Tukey FFT.

There are other modifications of the basic structure of the type 1 index map DFT algorithm. One is to use the same index structure and conversion of the short DFTs to convolution as with the PFA but to use some other method for the high-speed convolution. Table look-up of partial products based on distributed arithmetic to eliminate all multiplications [46] looks promising for certain very specific applications, perhaps for specialized VLSI implementation. Another possibility is to calculate the short convolutions using number-theoretic transforms [7,17,19]. This would also require special hardware. Direct calculation of short convolutions is faster on certain pipelined processors such as the TMS 320 microprocessor [23].

4.6.4 Evaluation of the PFA and WFTA

As for the Cooley-Tukey FFTs, our first evaluation of these algorithms will be on the number of multiplications and additions required. The number of multiplications to compute the PFA in Eq. (4.96) is given by Eq. (4.21). Using the notation that $T(N)$ is the number of multiplications or additions necessary to calculate a length-N DFT, the total number for a four-factor PFA of length N, where $N = N_1 N_2 N_3 N_4$, is

$$T(N) = N_1 N_2 N_3 T(N_4) + N_2 N_3 N_4 T(N_1) + N_3 N_4 N_1 T(N_2) + N_4 N_1 N_2 T(N_3)$$
(4.105)

The counts of multiplies and adds in Table 4.4 are calculated from Eq. (4.105) with the counts of the factors taken from Table 4.2. The list of lengths are those possible with modules in the program of length 2, 3, 4, 5, 7, 8, 9, and 16, as is true for the PFA in [11,13] and the WFTA in [4,39]. A maximum of four relatively prime lengths can be used from this group, giving 59 different lengths over the range from 2 to 5040. The radix-2 or split-radix FFT allows 12 different lengths over the same range. If modules of length 11 and 13 from [35] are added, the maximum length becomes 720720 and the number of different lengths becomes 239. Adding modules for 17, 19, and 25 from [35] gives a maximum length of 1163962800 and a very large and dense number of possible lengths. The length of the code for the longer modules becomes excessive and should not be included unless needed.

The number of multiplications necessary for the WFTA is simply the product of those necessary for the required modules, including multiplications by unity. The total number may contain some unity multipliers, but it is difficult to remove them in a practical program. Table 4.4 contains both the total number (MULTS) and the number with the unity multiplies removed (RMULTS).

Calculating the number of additions for the WFTA is more complicated than for the PFA because of the expansion of the data moving through the algorithm. For example, the number of additions, TA, for the length-15 example in Fig. 4.9 is given by

$$TA(N) = N_2 TA(N_1) + TM_1 TA(N_2)$$
(4.106)

where $N_1 = 3$, $N_2 = 5$, and $TM_1 =$ the number of multiplies for the length-3 module and, hence, the expansion factor. As mentioned earlier, there is an optimum ordering to minimize additions. The ordering used to calculate Table 4.4 is the ordering used in [19,39], which is optimal in most cases and close to optimal in the others.

TABLE 4.4   NUMBER OF REAL MULTIPLICATIONS AND ADDITIONS FOR COMPLEX DATA

	PFA		WFTA		
Length	MULTS[1]	ADDS[2]	MULTS	RMULTS[3]	ADDS
10	20	88	24	20	88
12	16	96	24	16	96
14	32	172	36	32	172
15	50	162	36	34	162
18	40	204	44	40	208
20	40	216	48	40	216
21	76	300	54	52	300
24	44	252	48	36	252
28	64	400	72	64	400
30	100	384	72	68	384
35	150	598	108	106	666
36	80	480	88	80	488
40	100	532	96	84	532
42	152	684	108	104	684
45	190	726	132	130	804
48	124	636	108	92	660
56	156	940	144	132	940
60	200	888	144	136	888
63	284	1236	198	196	1394
70	300	1336	216	212	1472
72	196	1140	176	164	1156
80	260	1284	216	200	1352
84	304	1536	216	208	1536
90	380	1632	264	260	1788
105	590	2214	324	322	2418
112	396	2188	324	308	2332
120	460	2076	288	276	2076
126	568	2724	396	392	3040
140	600	2952	432	424	3224
144	500	2676	396	380	2880
168	692	3492	432	420	3492
180	760	3624	528	520	3936
210	1180	4848	648	644	5256
240	1100	4812	648	632	5136
252	1136	5952	792	784	6584
280	1340	6604	864	852	7148
315	2050	8322	1188	1186	10336
336	1636	7908	972	956	8508
360	1700	8148	1056	1044	8772
420	2360	10536	1296	1288	11352
504	2524	13164	1584	1572	14428
560	3100	14748	1944	1928	17168
630	4100	17904	2376	2372	21932
720	3940	18276	2376	2360	21132

(*continued*)

[1] MULTS = multiplications

[2] ADDS = additions

[3] RMULTS = unity multiplications removed

TABLE 4.4   (*cont.*)

	PFA		WFTA		
Length	MULTS[1]	ADDS[2]	MULTS	RMULTS[3]	ADDS
840	5140	23172	2592	2580	24804
1008	5804	29100	3564	3548	34416
1260	8200	38328	4752	4744	46384
1680	11540	50964	5832	5816	59064
2520	17660	82956	9504	9492	99068
5040	39100	179772	21384	21368	232668

Compared with the PFA or any of the Cooley-Tukey FFTs, the WFTA has significantly fewer multiplications. For the shorter lengths, the WFTA and the PFA have approximately the same number of additions; however, for longer lengths, the PFA has fewer and the Cooley-Tukey FFTs always have the fewest. If the total arithmetic, the number of multiplications plus the number of additions, is compared, the split-radix FFT, PFA, and WFTA all have about the same count. Special versions of the PFA and WFTA have been developed for real data [30,31].

The size of the Cooley-Tukey program is the smallest, the PFA next, and the WFTA largest. The PFA requires the smallest number of stored constants, the Cooley-Tukey or split-radix FFT next, and the WFTA requires the largest number. For a DFT of approximately 1000, the PFA stores 28 constants, the FFT 2048, and the WFTA 3564. Both the FFT and PFA can be calculated in place and the WFTA cannot. The PFA can be calculated in order without an unscrambler. The radix-2 FFT can also, but it requires additional indexing overhead [14]. The indexing and data transfer overhead is greatest for the WFTA because the separate preweave and postweave sections each require their indexing and pass through the complete data. The shorter modules in the PFA and WFTA and the butterflies in the radix-2 and radix-4 FFTs are more efficient than the longer ones because intermediate calculations can be kept in CPU registers rather than in general memory [47]. However, the shorter modules and radices require more passes through the data for a given approximate length. A proper comparison will require actual programs to be compiled and run on a particular machine. There are many open questions about the relationship of algorithms and hardware architecture.

4.7 ALGORITHMS FOR REAL DATA

Many applications involve processing real data. It is inefficient to simply use a complex FFT on real data because arithmetic would be performed on the zero imaginary parts of the input, and, because of symmetries, output values would be calculated that are redundant. There are several approaches to developing special algorithms or to modifying complex algorithms for real data.

Two methods use a complex FFT in a special way to increase efficiency [37,31]. The first method uses a length-N complex FFT to compute two length-N real FFTs by putting the two real data sequences into the real and the imaginary parts of the input to a complex FFT. Because transforms of real data have even real parts and odd imaginary parts, it is possible to separate the transforms of the two inputs with $2N - 4$ extra additions. This method requires, however, that two inputs be available at the same time.

The second method [31] uses the fact that the last stage of a decimation-in-time radix-2 FFT combines two independent transforms of length $N/2$ to compute a length-N transform. If the data are real, the two half-length transforms are calculated by the method described above, and the last stage is carried out to calculate the total length-N FFT of the real data. It should be noted that the half-length FFT does not have to be calculated by a radix-2 FFT. In fact, it should be calculated by the most efficient complex-data algorithm possible, such as the SRFFT or the PFA. The separation of the two half-length transforms and the computation of the last stage requires $N - 6$ real multiplications and $\frac{5}{2}N - 6$ real additions [31].

It is possible to derive more efficient real-data algorithms directly rather than using a complex FFT. The basic idea is from Bergland [48,49] and Sande [50], who at each stage use the symmetries of a constant-radix Cooley-Tukey FFT to minimize arithmetic and storage. In the usual derivation [1] of the radix-2 FFT, the length-N transform is written as the combination of the length-$N/2$ DFT of the even indexed data and the length-$N/2$ DFT of the odd indexed data. If the input to each half-length DFT is real, the output will have Hermitian symmetry. Hence, the output of each stage can be arranged so that the results of that stage store the complex DFT, with the real part located where half of the DFT would have gone and the imaginary part located where the conjugate would have gone. This removes most of the redundant calculations and storage but slightly complicates the addressing. The resulting butterfly structure for this algorithm [31] resembles that for the fast Hartley transform [22]. The complete algorithm has one-half the number of multiplications and $N - 2$ fewer than half the additions of the basic complex FFT. Applying this approach to the split-radix FFT gives a particularly interesting algorithm [29,31,43].

Special versions of both the PFA and WFTA can also be developed for real data. Because the operations in the stages of the PFA can be commuted, it is possible to move the combination of the transform of the real part of the input and imaginary part to the last stage. Because the imaginary part of the input is zero, half of the algorithm is simply omitted. This results in the number of multiplications required for the real transform being exactly half of that required for complex data and the number of additions being about N less than half that required for the complex case because adding a pure real number to a pure imaginary number does not require an actual addition. Unfortunately, the indexing and data transfer become somewhat more complicated [30,31]. A similar approach can be taken with the WFTA [30,31,51].

4.8 CONVOLUTION ALGORITHMS

Convolution is intimately related to the DFT. We showed in Section 4.3 that a prime-length DFT could be converted to cyclic convolution. It has been long known [1] that convolution can be calculated by multiplying the DFTs of signals.

An important question is what the fastest method for calculating digital convolution is. There are several methods that each have some advantage. The earliest method for fast convolution was the use of sectioning with overlap-add or overlap-save and the FFT [1,10,11]. In most cases the convolution is of real data, and, therefore, real-data FFTs should be used. That approach is still probably the fastest method for longer convolution on a general-purpose computer or microprocessor. The shorter convolutions should simply be calculated directly.

A relatively new approach uses index mapping directly to convert a one-dimensional convolution into a multidimensional convolution [12]. This can be done by either a type 1 or type 2 map. The short convolutions along each dimension are then done by Winograd's optimal algorithms. Unlike for the case of the DFT, there is no savings of arithmetic from the index mapping alone. All the savings come from efficient short algorithms. In the case of index mapping with convolution, the multiplications must be nested together in the center of the algorithm in the same way as for the WFTA. There is no equivalent to the PFA structure for convolution. The multidimensional convolution cannot be calculated by row and column convolutions as the DFT was by row and column DFTs.

It would first seem that applying the index mapping and optimal short algorithms directly to convolution would be more efficient than using DFTs and converting them to convolution to be calculated by the same optimal algorithms. In practical algorithms, however, the DFT method seems to be more efficient [52].

A method that is attractive for special-purpose hardware uses distributed arithmetic [53] and a table look-up of precomputed partial products to produce a system that does convolution without requiring multiplications [46].

Another method that requires special hardware uses number-theoretic transforms [7,17,19] to calculate convolution. These transforms are defined over finite fields or rings, with arithmetic performed modulo special numbers. These transforms have rather limited flexibility, but when they can be used, they are very efficient.

4.9 CONCLUSIONS

This chapter has developed a class of efficient algorithms based on index mapping and polynomial algebra. This provides a framework from which the Cooley-Tukey FFT, the split-radix FFT, the PFA, and WFTA can be derived. Even the programs implementing these algorithms can have a similar structure. Winograd's theorems were presented and shown to be very powerful both in deriving algorithms and in evaluating them. The simple radix-2 FFT provides a compact, elegant means for efficiently calculating the DFT. If some elaboration is allowed, significant improvement can be realized from the split-radix FFT, the radix-4 FFT, or the PFA. If multiplications are expensive, the WFTA requires the least of all.

Several methods for transforming real data were described that are more efficient than directly using a complex FFT. A complex FFT can be used for real data by artificially creating a complex input from two sections of real input. An alternative and slightly more efficient method is to construct a special FFT that utilizes the symmetries at each stage.

For high-speed convolution, the traditional use of the FFT or PFA with blocking is probably the fastest method, although distributed-arithmetic or number-theoretic transforms may have a future with special VLSI hardware.

The ideas presented in this chapter can also be applied to the calculation of the discrete Hartley transform [22,43,54], the discrete cosine transform [40,44,55], and number-theoretic transforms [7,17,19].

Many areas require future research. The relationship of hardware to algorithms, the proper use of multiple processors, the proper design and use of array processors and vector processors are all open to study. Many unanswered questions exist in multidimensional algorithms where a simple extension of one-dimensional methods will not suffice.

REFERENCES

1. A. V. Oppenheim and R. W. Schafer, *Digital Signal Processing*, Prentice-Hall, Englewood Cliffs, NJ, 1975.

2. M. T. Heideman, D. H. Johnson, and C. S. Burrus, "Gauss and the History of the FFT," *IEEE ASSP Magazine*, Vol. 1, pp. 14–21, Oct. 1984; also in the *Archive for History of Exact Sciences*, Vol. 34, no. 3, pp. 265–277, 1985.

3. M. T. Heideman and C. S. Burrus, "A Bibliography of Fast Transform and Convolution Algorithms: II," Dept. Electrical Engineering Technical Report No. 8402, Rice University, Houston, TX, Feb. 24, 1984.

4. J. W. Cooley and J. W. Tukey, "An Algorithm for the Machine Calculation of Complex Fourier Series," *Mathematics of Computation,* Vol. 19, pp. 297–301, April 1965; also in [9].

5. S. Winograd, "On Computing the Discrete Fourier Transform," *Proc. National Academy of Science,* USA, Vol. 73, pp. 1005–1006, April 1976.

6. C. S. Burrus, "Index Mappings for Multidimensional Formulation of the DFT and Convolution," *IEEE Trans. on ASSP*, Vol. 25, pp. 239–242, June 1977.

7. R. E. Blahut, *Fast Algorithms for Digital Signal Processing,* Addison-Wesley, Reading, MA, 1984.

8. I. J. Good, "The Interaction Algorithm and Practical Fourier Analysis," *J. Roy. Stat. Soc.,* Vol. B-20, pp. 361–372, 1958; Vol. 22, pp. 372–375, 1960.

9. L. R. Rabiner and C. M. Rader, Eds., *Digital Signal Processing*, Selected Reprints, IEEE Press, New York, 1972.

10. L. R. Rabiner and B. Gold, *Theory and Application of Digital Signal Processing,* Prentice-Hall, Englewood Cliffs, NJ, 1975.

11. C. S. Burrus and T. W. Parks, *DFT/FFT and Convolution Algorithms,* John Wiley, New York, 1985.

12. R. C. Agarwal and J. W. Cooley, "New Algorithms for Digital Convolution," *IEEE Trans. on ASSP,* Vol. 25, pp. 392–410, Oct. 1977; also in [19].

13. C. S. Burrus and P. W. Eschenbacher, "An In-Place, In-Order Prime Factor FFT Algorithm," *IEEE Trans. on ASSP*, Vol. 29, pp. 806–817, Aug. 1981; also in [47].

14. H. W. Johnson and C. S. Burrus, "An In-Place, In-Order Radix-2 FFT," *Proc. IEEE ICASSP-84*, San Diego, CA, p. 28A.2, March 1984.

15. J. H. Rothweiler, "Implementation of the In-Order Prime Factor FFT Algorithm," *IEEE Trans. on ASSP*, Vol. 30, pp. 105–107, Feb. 1982.

16. C. S. Burrus, "Unscrambling for Fast DFT Algorithms," *IEEE Trans. on ASSP*, Vol. 35, 1987.

17. H. J. Nussbaumer, *Fast Fourier Transform and Convolution Algorithms*, Springer-Verlag, Heidelberg, Germany, 1981, 1982.

18. I. Niven and H. S. Zuckerman, *An Introduction to the Theory of Numbers*, 4th Ed., John Wiley, New York, 1980.

19. J. H. McClellan and C. M. Rader, *Number Theory in Digital Signal Processing*, Prentice-Hall, Englewood Cliffs, NJ, 1979.

20. J. B. Martens, "Recursive Cyclotomic Factorization: A New Algorithm for Calculating the Discrete Fourier Transform," *IEEE Trans. on ASSP*, Vol. 32, pp. 750–762, Aug. 1984.

21. M. T. Heideman, "Applications of Multiplicative Complexity Theory to Convolution and the DFT," Ph.D. Thesis, Dept. Electrical and Computer Engineering, Rice University, Houston, TX, May 1986; also to be published as a book entitled *Multiplicative Complexity Theory, Convolution, and the DFT*, Springer-Verlag, 1987.

22. H. V. Sorensen, D. L. Jones, C. S. Burrus, and M. T. Heideman, "On Calculating the Discrete Hartley Transform," *IEEE Trans. on ASSP*, Vol. 33, pp. 1231–1238, Oct. 1985.

23. Z. Li, H. V. Sorensen, and C. S. Burrus, "FFT and Convolution Algorithms for DSP Microprocessors," *Proc. IEEE ICASSP-86*, Tokyo, pp. 289–292, 1986.

24. S. Winograd, "Arithmetic Complexity of Computation," SIAM CBMS-NSF Series, No. 33, SIAM, Philadelphia, 1980.

25. S. Winograd, "On Computing the Discrete Fourier Transform," *Math. Comp.*, Vol. 32, pp. 175–199, Jan. 1978; also in [19].

26. H. W. Johnson and C. S. Burrus, "Structures of DFT Algorithms," *IEEE Trans. on ASSP*, Vol. 33, pp. 248–254, Feb. 1985.

27. N. K. Bose, Chapter 2 in *Digital Filters, Theory, and Applications*, North-Holland, Elsevier, 1985.

28. D. P. Kolba and T. W. Parks, "A Prime Factor FFT Algorithm Using High Speed Convolution," *IEEE Trans. on ASSP*, Vol. 25, pp. 281–294, Aug. 1977; also in [19].

29. P. Duhamel, "Implementation of 'Split-Radix' FFT Algorithms for Complex, Real, and Real-Symmetric Data," *IEEE Trans. on ASSP*, Vol. 34, pp. 285–295, April 1986; a shorter version appeared in *Proc. IEEE ICASSP-85*, p. 20.6, March 1985.

30. M. T. Heideman, H. W. Johnson, and C. S. Burrus, "Prime Factor FFT Algorithms for Real-Valued Data," *Proc. IEEE ICASSP-84*, San Diego, p. 28A.7.1, March 1984.

31. H. V. Sorensen, D. L. Jones, M. T. Heideman, and C. S. Burrus, "Real-Valued Fast Fourier Transform Algorithms," *IEEE Trans. on ASSP*, Vol. 35, June 1987.

32. S. Winograd, "On the Multiplicative Complexity of the Discrete Fourier Transform," *Advances in Mathematics*, Vol. 32, pp. 83–117, May 1979.

33. S. Winograd, "Signal Processing and Complexity of Computation," *Proc. IEEE ICASSP-80*, Denver, pp. 94–101, April 1980.

34. M. T. Heideman and C. S. Burrus, "On the Number of Multiplications Necessary to Compute a Length-2^n DFT," *IEEE Transactions on ASSP*, Vol. 34, pp. 91–95, Feb. 1986.

35. H. W. Johnson and C. S. Burrus, "Large DFT Modules: $N = 11, 13, 17, 19,$ and 25," Technical Report No. 8105, Electrical Engineering Department, Rice University, Houston, TX, 1981.

36. P. Duhamel and H. Hollmann, "Split Radix FFT Algorithm," *Electronic Letters*, Vol. 20, pp. 14–16, Jan. 5, 1984.

37. E. O. Brigham, *The Fast Fourier Transform*, Prentice-Hall, Englewood Cliffs, NJ, 1974.

38. J. D. Markel, "FFT Pruning," *IEEE Trans. on Audio and Electroacoustics,* Vol. 19, pp. 305–311, Dec. 1971.

39. *Programs for Digital Signal Processing,* IEEE Press, New York, 1979.

40. M. Vetterli and H. J. Nussbaumer, "Simple FFT and DCT Algorithms with Reduced Number of Operations," *Signal Processing,* Vol. 6, pp. 267–278, Aug. 1984.

41. H. V. Sorensen, M. T. Heideman, and C. S. Burrus, "On Calculating the Split-Radix FFT," *IEEE Trans. on ASSP,* Vol. 34, pp. 152–156, Feb. 1986.

42. R. Yavne, "An Economical Method for Calculating the Discrete Fourier Transform," *Proc. Fall Joint Computer Conference*, pp. 115–125, 1968.

43. P. Duhamel and M. Vetterli, "Cyclic Convolution of Real Sequences: Hartley versus Fourier and New Schemes," *Proc. IEEE ICASSP-86,* Tokyo, pp. 229–232, 1986.

44. P. Duhamel, "New Power of 2 Discrete Cosine Transform Algorithms Based on Cyclic Convolution" (in press).

45. H. W. Johnson and C. S. Burrus, "The Design of Optimal DFT Algorithms Using Dynamic Programming," *IEEE Trans. on ASSP,* Vol. 31, pp. 378–387, April 1983.

46. S. Chu and C. S. Burrus, "A Prime Factor FFT Algorithm Using Distributed Arithmetic," *IEEE Trans. on ASSP,* Vol. 30, pp. 217–227, April 1982.

47. L. R. Morris, *Digital Signal Processing Software,* DSPSW, Inc., Toronto, Canada, 1982, 1983.

48. G. D. Bergland, "A Fast Fourier Transform Algorithm for Real-Valued Series," *Comm. ACM,* Vol. 11, pp. 703–710, Oct. 1968.

49. G. D. Bergland, "A Radix-8 Fast Fourier Transform Subroutine for Real-Valued Series," *IEEE Trans. on Audio and Electroacoustics,* Vol. 17, pp. 138–144, June 1969.

50. G. Sande, "Fast Fourier Transform—A Globally Complex Algorithm with Locally Real Implementations," *Proc. 4th Annual Princeton Conference on Information Sciences and Systems,* Princeton, pp. 136–142, March 1970.

51. T. W. Parsons, "A Winograd–Fourier Transform Algorithm for Real-Valued Data," *IEEE Trans. on ASSP,* Vol. 27, pp. 398–402, Aug. 1979.

52. I. Pitas and C. S. Burrus, "Time and Error Analysis of Digital Convolution by Rectangular Transforms," *Signal Processing,* Vol. 5, pp. 153–162, March 1983.

53. C. S. Burrus, "Digital Filter Structures Described by Distributed Arithmetic," *IEEE Trans. on CAS,* Vol. 24, pp. 674–680, Dec. 1977.

54. R. N. Bracewell, *The Hartley Transform*, Oxford Press, 1986.

55. D. F. Elliot and K. R. Rao, *Fast Transforms: Algorithms, Analysis and Application,* Academic Press, New York, 1982.

ACKNOWLEDGMENTS

The author appreciates the support for some of the results reported here by the National Science Foundation and by Texas Instruments, Inc. Thanks are also expressed to M. T. Heideman, H. W. Johnson, H. V. Sorensen, and Avideh Zakhor for their comments and contributions.

5

Fundamentals of Adaptive Signal Processing

Samuel D. Stearns
Sandia National Laboratories

5.0 INTRODUCTION

Adaptive signal processing has undergone a remarkable increase in interest and attention in recent years. This growth has been promoted by developments in microelectronics and VLSI circuit design that increased tremendously the amount of computing possible in processing digital signals. Seismic signals at 10^2 Hz and below, speech and acoustic signals at 10^2–10^5 Hz, electromagnetic signals at 10^5 Hz and beyond, as well as other similar signals appearing in time and space are now all reasonable candidates for adaptive signal processing.

Adaptive signal processing systems are mainly time-varying digital signal processing systems, that is, digital filters and similar structures with time-varying coefficients or "weights." The "adaptive" notion derives from the human desire to emulate living systems found in nature, which adapt to their environments in various remarkable ways and reflect the work of a master Designer that we have been able to copy only in very limited ways. In adaptive systems, the journey from concept to reality has taken us away from direct emulation, somewhat like the development of manned aircraft. An adaptive signal processor usually resembles its natural counterpart about as much as, or if anything less than, an airplane resembles a bird.

Adaptive signal processing has its roots in adaptive control and in the mathematics of iterative processes, where the first attempts were made to design systems that adapt to their environments. In this chapter we will not trace the early developments

Samuel D. Stearns is with Sandia National Laboratories, Division 7111, Albuquerque, NM 87185.

in adaptive control in the 1940s and 1950s, many of which are still useful. We will describe only some of the basic theory and applications of modern adaptive signal processing. Much of what we describe here is based on the more recent work of Bernard Widrow and his colleagues [1–4], which began around 1960. We use Widrow's development of the geometry of adaptation [4], and we introduce the least-mean-square (LMS) algorithm as the simplest and most widely applicable adaptive processing scheme.

We will view the adaptive signal processor as a time-varying digital structure. To simplify our discussion, we assume that the signals to be processed are digital time series with regularly spaced samples. We begin with some basic aspects and examples of such structures. In adaptation with "performance feedback" [5], the type of adaptation we discuss in this chapter, we are usually able to identify in any system the basic elements in Fig. 5.1. The adaptive system contains a digital structure, the processor, with variable, adaptive weights. The weights are adjusted repeatedly, at regular intervals, in accordance with an *adaptive algorithm*. The adaptive algorithm usually uses, either explicitly or implicitly, all of the signals shown in Fig. 5.1. The details of adaptive algorithms are given in the following sections. The main point here is that the weights in Fig. 5.1 are adjusted continually during adaptation to reduce the mean-square error $E[\epsilon_k^2]$ toward its minimum value. In this way the adaptive system seeks continually to reduce the difference between a desired response d_k and its own response y_k; thus, as we will see, the system "adapts" to its prescribed signal environment.

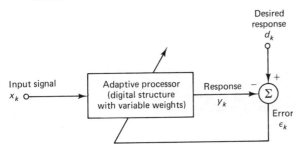

Figure 5.1   Basic elements in a single-input, performance-feedback adaptive system.

The subscript k in Fig. 5.1 denotes time and generally ranges from zero to infinity. For example, the input time series, which begins at some time which we label 0 for convenience, is $[x_0, x_1, x_2, \ldots, x_k, \ldots].^\dagger$ We have shown in Fig. 5.1 a single-input system, but multiple-input systems are also possible, as with adaptive arrays, and we will introduce a notation to cover both cases. The other three signals in Fig. 5.1 are all single time series in this chapter.

We now consider some examples of the application of Fig. 5.1, beginning with Fig. 5.2, which illustrates *adaptive prediction*. Figure 5.2(a) gives the simpler but unrealizable case, where the adaptive processor tries to minimize $E[\epsilon_k^2]$ and thereby make its output y_k as close as possible to a future value of the input, $s_{k+\Delta}$. Since the latter is not available in real time, the "realizable" form in Fig. 5.2(b), which is

---

† The subscript notation is used in this chapter in place of the "argument" notation in previous chapters to conform with the adaptive literature and to make the equations less cumbersome. Thus, x_k denotes the kth sample of x, the same as $x(k)$.

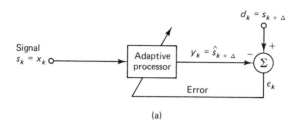

(a)

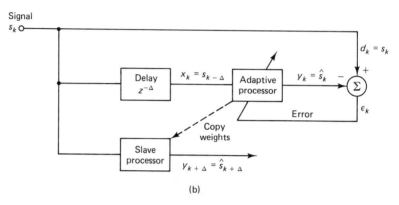

(b)

Figure 5.2 Adaptive prediction: (a) unrealizable form; (b) realizable form.

equivalent to Fig. 5.2(a), is used. The estimate, $\hat{s}_{k+\Delta}$, is produced using a slave processor. Adaptive prediction is used in speech encoding [6,7], spectral estimation [8], event detection [9–11], line enhancement [2,12], signal whitening, and other areas. Note how Fig. 5.1, the scheme common to many adaptive systems, is incorporated into the prediction schemes in Fig. 5.2.

Another application, *adaptive modeling*, is shown in Fig. 5.3. In forward modeling, illustrated in Fig. 5.3(a), the adaptive processor adjusts its weights to produce a response y_k that is as close as possible (in the least-squares sense) to the plant response d_k. If the stimulus s_k is robust in frequency content and if the internal plant noise n_k is small, then the adaptive processor will adapt to become a good model of the unknown system. Forward modeling has a wide range of applications in the biological, social, and economic sciences [5], adaptive control systems [13,14], digital filter design [15], coherence estimation [16], and geophysics [5]. With inverse modeling, shown in Fig. 5.3(b), the adaptive processor adjusts its weights to become the inverse of the plant, that is, to convert the plant output x_k back into a delayed version $s_{k-\Delta}$ of the input. The delay is usually included to account for the propagation delay through the plant and the adaptive processor, assuming that both are causal systems. As with forward modeling, if s_k is spectrally robust and if n_k is small, the adaptive processor can adapt to become an accurate inverse model of the unknown system. Inverse modeling is used in adaptive control [5,17,18], speech analysis [19], channel equalization [20,21], deconvolution [22], digital filter design [15], and other applications. Again, note the incorporation of the basic structure of Fig. 5.1 into Figs. 5.3(a) and (b).

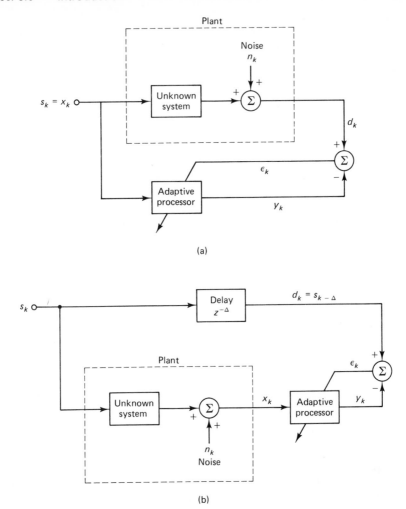

(a)

(b)

Figure 5.3   Adaptive modeling: (a) forward modeling; (b) inverse modeling.

Another application of adaptive signal processing, known as *adaptive inter-ference canceling*, is illustrated in Fig. 5.4. This simple structure appears in a wide variety of applications [2]. The desired response d_k is in this case a signal with additive noise, $s_k + n_k$. The input x_k to the adaptive processor is another noise, n_k', correlated with n_k. When s_k and n_k are uncorrelated, the adaptive processor tries to minimize $E[\epsilon_k^2]$ by making $y_k = \hat{n}_k$ an approximation to n_k, thereby making ϵ_k an approximation

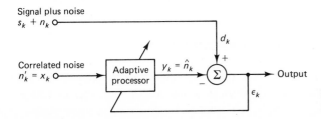

Figure 5.4   Basic structure for adaptive interference canceling.

to the signal s_k. The result is reduced noise in ϵ_k. A more quantitative analysis is given elsewhere [5]; here we wish mainly to emphasize again the incorporation of Fig. 5.1 into Fig. 5.4.

Our final application example, *adaptive array processing,* is illustrated in Fig. 5.5. The array processing system is itself a type of multiple-input interference canceler [5,23], as seen by comparing Figs. 5.4 and 5.5. As in ordinary beamforming, steering delays are used to form a beam and produce a peak array gain in a desired look direction. Thus a noisy target signal, $s_k + n_k$, is obtained through the fixed filter shown in Fig. 5.5. An estimate $\hat{n}_k$ of the noise is obtained through the multiple-input adaptive processor and is used to cancel n_k, just as in Fig. 5.4. Further details, including details on how the adaptive processor manages to exclude the signal s_k from its estimate of n_k, are given in the literature [3,5,23]; for now we see in Fig. 5.5 still another application of Fig. 5.1.

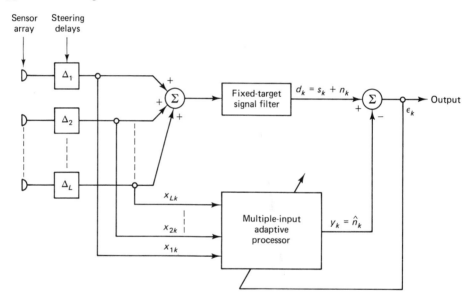

Figure 5.5   An adaptive array signal processing system.

The foregoing examples show the scope of adaptive signal processing to be considered in this chapter. We proceed to some basic concepts, adaptive algorithms, and structures that apply to all of the examples and include some specific illustrations of adaptation. Readers interested in more details can refer to a recent text [5,23,24] or a special issue of the IEEE *Proceedings* or *Transactions* devoted to adaptive signal processing, which are listed in the Reference section.

5.1 THE MEAN-SQUARE-ERROR PERFORMANCE SURFACE

Now that we have seen how Fig. 5.1 illustrates the principal signals involved in different forms of adaptation, we can discuss some basic theory. Our first objective is to see how the adaptive process, which involves minimizing the mean-square error

(MSE), $E[\epsilon_k^2]$, amounts to a search for the lowest point on a "performance surface," or "error surface," in multidimensional space.

The signal environment of an adaptive system is usually nonstationary, with the system adapting to changing signal conditions as to a slowly varying plant in Fig. 5.3 or to a drifting noise spectrum in Fig. 5.4. Changing signal properties are indeed often the principal justification for the use of adaptive processing in a given application. However, to develop basic theory, it is useful to assume temporarily that the signals are stationary and then later to study nonstationary operation [25]. We assume for now that all of the signals in Fig. 5.1 are stationary and have finite correlation functions. We also assume that the adaptive processor is a linear filter, as shown in Fig. 5.6. (Nonlinear adaptive processors [26,27] are an interesting possibility but are beyond the scope of this chapter.) The linear system has a transfer function, $W(z)$, which is adapted by adjusting weights using the adaptive algorithms described below.

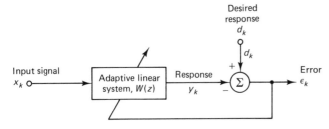

Figure 5.6   Basic adaptive system in Fig. 5.1 with a linear system as the adaptive processor.

We define the correlation functions of the signals in Fig. 5.6 as follows:[†]

$$\text{Cross-correlation:} \qquad r_{dx}(n) = E[d_k x_{k+n}] \qquad (5.1)$$

$$\text{Autocorrelation:} \qquad r_{xx}(n) = E[x_k x_{k+n}] \qquad (5.2)$$

With stationary signals we can say that the expectation ranges over the time index k. The remaining correlation functions of interest, $r_{dd}(n)$, $r_{yy}(n)$, and $r_{dy}(n)$, are defined similarly to Eqs. (5.1) and (5.2), and we note also that, by definition,

$$r_{xd}(n) = r_{dx}(-n) \qquad (5.3)$$

Referring to Fig. 5.6 and assuming fixed filter weights for the moment so that y_k is stationary, we find that the MSE is given by

$$\begin{aligned} \text{MSE} = E[\epsilon_k^2] &= E[(d_k - y_k)^2] \\ &= E[d_k^2] + E[y_k^2] - 2E[d_k y_k] \qquad (5.4) \\ &= r_{dd}(0) + r_{yy}(0) - 2r_{dy}(0) \end{aligned}$$

The z-transform of any correlation function, say $r_{uv}(n)$, is the corresponding power spectrum [28] $G_{uv}(z)$; that is,

$$G_{uv}(z) = \sum_{n=-\infty}^{\infty} r_{uv}(n)z^{-n} \qquad (5.5)$$

[†] The notation used here, although it differs slightly from that of previous chapters, is preferred for the present development.

The inverse z-transform relationship [5] also gives

$$r_{uv}(n) = \frac{1}{2\pi j} \oint G_{uv}(z)z^n \frac{dz}{z} \qquad j \triangleq \sqrt{-1} \tag{5.6}$$

Using Eq. (5.6) with $n = 0$ in Eq. (5.4), we obtain

$$\text{MSE} = r_{dd}(0) + \frac{1}{2\pi j} \oint [G_{yy}(z) - 2G_{dy}(z)] \frac{dz}{z} \tag{5.7}$$

Assuming that $z = e^{j\omega}$ is a point on the unit circle on the z-plane corresponding to a frequency of ω rad, the power spectral relationships in Fig. 5.6 are

$$G_{yy}(z) = W(z)W(z^{-1})G_{xx}(z) \tag{5.8}$$

and

$$G_{dy}(z) = W(z)G_{dx}(z) \tag{5.9}$$

Using Eqs. (5.8) and (5.9) in Eq. (5.7), we obtain the general expression for the MSE in terms of $W(z)$:

$$\boxed{\text{MSE} = r_{dd}(0) + \frac{1}{2\pi j} \oint [W(z^{-1})G_{xx}(z) - 2G_{dx}(z)]W(z) \frac{dz}{z}} \tag{5.10}$$

In this general expression we have not specified the form of the adaptive system, $W(z)$. We have only stated that $W(z)$ is linear and has adjustable weights, which for the moment are fixed. Let

$$L = \text{Number of weights in } W(z) \tag{5.11}$$

Then Eq. (5.10) describes the MSE as an L-dimensional surface in $L + 1$-dimensional space, and adaptation becomes the process of seeking the minimum point on this surface. If we allow the weights of $W(z)$ to vary, the task is to find the minimum MSE. This is essentially the unconstrained Wiener least-squares design problem [29]. In adaptive signal processing, this task becomes a continual process of updating the weights of $W(z)$, which may be constrained, in situations where the other quantities in Eq. (5.10), $r_{dd}(0)$, $G_{xx}(z)$, and $G_{dx}(z)$, may be slowly varying.[†]

We will see shortly that Eq. (5.10) is greatly simplified when $W(z)$ is a finite impulse response (FIR) filter. The integral in Eq. (5.10) can be difficult when $W(z)$ is an infinite impulse response (IIR) filter, and the task of adaptation with IIR filters is in general complicated by two factors:

1. Unconstrained poles can move outside the unit circle during adaptation, causing instability.
2. The error surface in Eq. (5.10) can have flat areas and local minima, making gradient search techniques unreliable.

[†] Strictly speaking, these quantities are defined to be constant over time, but we think of them as varying slowly.

IIR adaptive filters have, however, been analyzed to some extent and applied in situations where poles are especially useful [30–33].

Adaptive systems with FIR filters do not have the two complications mentioned. FIR filters are also simpler and more widely applicable in adaptive processing, so we will proceed now with the assumption that the single-input adaptive filter is an FIR filter. With this assumption, we will be able to include multiple-input adaptive systems in the same analysis.

5.2 THE QUADRATIC PERFORMANCE SURFACE OF THE ADAPTIVE LINEAR COMBINER

We now assume that $W(z)$ represents a causal FIR transversal filter with L weights:

$$W(z) = \sum_{i=0}^{L-1} w_i z^{-i} \tag{5.12}$$

The causal FIR filter is applicable in real-time situations and is generally suited to a wide variety of adaptive signal processing systems. Inserting Eq. (5.12) into Eq. (5.10), we obtain

$$\text{MSE} = r_{dd}(0) + \sum_{i=0}^{L-1} \sum_{m=0}^{L-1} \frac{w_i w_m}{2\pi j} \oint G_{xx}(z)\, z^{i-m} \frac{dz}{z} - 2 \sum_{i=0}^{L-1} \frac{w_i}{2\pi j} \oint G_{dx}(z)^{-i} \frac{dz}{z} \tag{5.13}$$

Using Eq. (5.6) and also Eq. (5.3) in Eq. (5.13), we obtain the MSE for the causal FIR adaptive filter in terms of correlation functions:

$$\boxed{\text{MSE} = r_{dd}(0) + \sum_{i=0}^{L-1} \sum_{m=0}^{L-1} w_i w_m r_{xx}(i - m) - 2 \sum_{i=0}^{L-1} w_i r_{xd}(i)} \tag{5.14}$$

This is the general expression for the performance surface for a causal FIR adaptive filter with given weights. We note that the MSE is a quadratic surface because the weights appear only to first and second degrees in Eq. (5.14).

We can include a class of multiple-input adaptive systems in Eq. (5.14) by using the notion of the adaptive linear combiner, illustrated in Fig. 5.7. With a single input signal, the adaptive linear combiner is an FIR filter with adjustable weights. With multiple inputs, each input signal is multiplied by its own weight. We can also imagine extending Fig. 5.7 in the multiple-input case by making each weight an adaptive FIR filter, i.e., a single-input adaptive linear combiner. This type of extension is useful, for example, in wideband adaptive array processing [5,23]. The theory covering this type of system is a fairly straightforward extension of the theory given in this chapter covering Fig. 5.7, where we use the vector $\mathbf{X}_k$ to represent either input signal at time k:

Single input: $\mathbf{X}_k = [x_k \quad x_{k-1} \quad \cdots \quad x_{k-L+1}]^T$ \tag{5.15}

Multiple inputs: $\mathbf{X}_k = [x_{0k} \quad x_{1k} \quad \cdots \quad x_{L-1,k}]^T$ \tag{5.16}

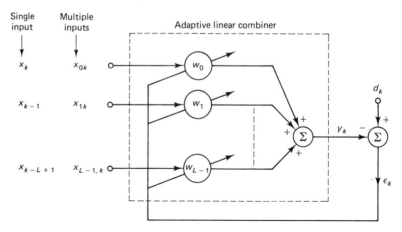

Figure 5.7 The adaptive linear combiner with single- or multiple-input signals.

To simplify Eq. (5.14) we use the symmetric input correlation matrix, or $\mathbf{R}$ matrix, given by

$$\mathbf{R} = E[\mathbf{X}_k \mathbf{X}_k^T] \tag{5.17}$$

We see from Eqs. (5.15) and (5.16) that $\mathbf{R}$ is a symmetric $L \times L$ correlation matrix and, in the single-input case, which is more widely applicable, that $\mathbf{R}$ is given by

$$\mathbf{R} = \begin{bmatrix} r_{xx}(0) & r_{xx}(1) & \cdots & r_{xx}(L-1) \\ r_{xx}(1) & r_{xx}(0) & \cdots & r_{xx}(L-2) \\ \vdots & \vdots & & \vdots \\ r_{xx}(L-1) & r_{xx}(L-2) & \cdots & r_{xx}(0) \end{bmatrix} \quad \text{(single input)} \tag{5.18}$$

The single-input $\mathbf{R}$ matrix is not only symmetric; it is also a Toeplitz matrix and relatively easy to invert if not singular [34,35]. We also need a cross-correlation vector, or $\mathbf{P}$ vector:

$$\mathbf{P} = E[d_k \mathbf{X}_k] = E[\mathbf{X}_k d_k] \tag{5.19}$$

Again, in the more common single-input case, the $\mathbf{P}$ vector is given by

$$\mathbf{P} = [r_{xd}(0) \quad r_{xd}(1) \quad \cdots \quad r_{xd}(L-1)]^T \quad \text{(single input)} \tag{5.20}$$

Finally, we need the weight vector, $\mathbf{W}$, given by

$$\mathbf{W} = [w_0 \quad w_1 \quad \cdots \quad w_{L-1}]^T \tag{5.21}$$

With these definitions we can simplify the expression in Eq. (5.14) for the quadratic performance surface and also extend the expression to cover multiple inputs. Thus, for the adaptive linear combiner in Fig. 5.7, we have

$$\boxed{\text{MSE} = r_{dd}(0) + \mathbf{W}^T \mathbf{R} \mathbf{W} - 2\mathbf{P}^T \mathbf{W}} \tag{5.22}$$

The reader should be satisfied that Eqs. (5.22) and (5.14) are equivalent in the single-input case and that in any case Eq. (5.22) describes a quadratic performance surface. Since the MSE is always positive, we also know that the performance surface, being quadratic, must be parabolic and "concave upward," that is, extending toward a positively increasing MSE. With stationary signals, the performance surface is thus a parabolic "bowl" in $L + 1$-dimensional space. In typical applications with non-stationary, slowly varying signal statistics, we think of the bowl as being ill defined, or perhaps drifting in space as the signal properties change slowly with time, and of adaptation as the process of searching for and continuously tracking the bottom of the bowl.

When searching for the bottom of the bowl we find generally that a knowledge or estimate of the gradient is useful. The gradient vector is the column vector obtained by differentiating Eq. (5.22) with respect to the weight vector $\mathbf{W}$:

$$\nabla = \frac{\partial(\text{MSE})}{\partial \mathbf{W}}$$

$$= \left[\frac{\partial(\text{MSE})}{\partial w_0} \frac{\partial(\text{MSE})}{\partial w_1} \cdots \frac{\partial(\text{MSE})}{\partial w_{L-1}} \right]^T \qquad (5.23)$$

$$= 2\mathbf{R}\mathbf{W} - 2\mathbf{P}$$

Since the bowl is quadratic, we know that the global minimum MSE is obtained where $\nabla = \mathbf{0}$. Setting $\nabla = \mathbf{0}$ in Eq. (5.23) gives the optimum weight vector, $\mathbf{W}^*$, as

$$\boxed{\mathbf{W}^* = \mathbf{R}^{-1}\mathbf{P}} \qquad (5.24)$$

This is the Wiener solution for the optimum weight vector [34] for the adaptive linear combiner. The corresponding minimum mean-square error is found by substituting Eq. (5.24) into Eq. (5.22) in the following manner:

$$(\text{MSE})_{\min} = r_{dd}(0) + (\mathbf{R}^{-1}\mathbf{P})^T\mathbf{R}\mathbf{W}^* - 2\mathbf{P}^T\mathbf{W}^*$$

$$= r_{dd}(0) - \mathbf{P}^T\mathbf{W}^* \qquad (5.25)$$

We are now ready to discuss adaptive algorithms as procedures for searching the quadratic bowl in Eq. (5.22), but first a simple two-weight example may help to clarify the nature of Eq. (5.22). We will introduce the example here and then use it throughout the remainder of this chapter in the discussion of adaptive algorithms.

5.3 AN EXAMPLE OF A QUADRATIC PERFORMANCE SURFACE

To provide an example of the quadratic performance surface of an adaptive FIR filter, we take the simple interference-canceling case illustrated in Fig. 5.8. The wideband signal and sinusoidal interference are described in the figure. We might, for instance, be trying to use this system to cancel power line interference in the signal from a biological sensor or some other type of sensor producing a low-level signal. (In a real

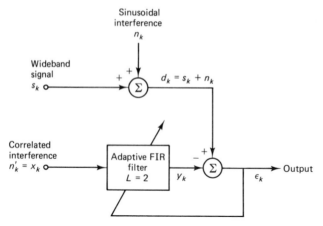

s_k = uniform white signal with $r_{ss}(0) = 0.05$

$$n_k = \sin\left(\frac{2\pi k}{16} + \frac{\pi}{10}\right)$$

$$n'_k = \sqrt{2}\sin\left(\frac{2\pi k}{16}\right)$$

Figure 5.8   A simple interference-canceling example used to illustrate the quadratic performance surface.

situation, we would not need to know the correlation properties of s_k, n_k, or n'_k exactly.) In this simple example, the signal and interference are uncorrelated. The correlation functions used to obtain the MSE are found by averaging over one period (16 samples) of the sinusoids:

$$r_{xx}(i) = \frac{1}{16}\sum_{k=0}^{15}\left(\sqrt{2}\sin\frac{2\pi k}{16}\right)\left(\sqrt{2}\sin\frac{2\pi(k+i)}{16}\right) = \cos\frac{2\pi i}{16} \quad (5.26)$$

$$r_{xd}(i) = \frac{1}{16}\sum_{k=0}^{15}\left(\sqrt{2}\sin\frac{2\pi k}{16}\right)\left[\sin\left(\frac{2\pi(k+i)}{16} + \frac{\pi}{10}\right)\right]$$

$$= \frac{1}{\sqrt{2}}\cos\left(\frac{2\pi i}{16} + \frac{\pi}{10}\right) \quad (5.27)$$

$$r_{dd}(0) = r_{ss}(0) + r_{nn}(0) = 0.55 \quad (5.28)$$

We note in these results that $r_{dd}(0)$ is just the power in s_k plus the power in n_k, i.e., $E[s_k^2] + E[n_k^2]$, and that these two quantities are 0.05 and 0.5, respectively, so the input signal-to-noise ratio is 0.1. Using Eqs. (5.26)–(5.28) in Eq. (5.14) with $L = 2$, we have the MSE in this case as follows:

$$\text{MSE} = 0.55 + w_0^2 + w_1^2 + 2w_0 w_1 \cos\frac{\pi}{8} - \sqrt{2}\,w_0\cos\frac{\pi}{10} - \sqrt{2}\,w_1\cos\frac{9\pi}{40}$$

$$(5.29)$$

A three-dimensional plot of a part of this performance surface is shown in Fig. 5.9. The surface cannot be seen very accurately, but we can note its quadratic form and also that the "bowl" in this case is very noncircular in cross section. The noncircular cross section is due to the dissimilar eigenvalues of the **R** matrix, which in this case are 1.92 and 0.08. In general, when the eigenvalues of the **R** matrix are dissimilar,

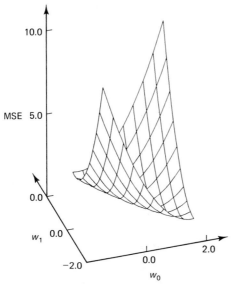

Figure 5.9 Part of the two-dimensional performance surface in Eq. (5.29) for the interference-canceling example in Fig. 5.8.

the quadratic bowl tends to be elliptical so that the gradient varies as we move around the bowl at constant MSE [5]. This property has important consequences for gradient search algorithms, as we will see.

We can view the parabolic surface in Fig. 5.9 in a different way by looking at projections of constant-MSE contours onto the $w_0 w_1$-plane, as in Fig. 5.10. Here we see that the contours are all ellipses, showing more clearly the quadratic form of Eq. (5.29). From Eq. (5.24) we can see that the optimum weight values in this example are

$$
\begin{bmatrix} w_0^* \\ w_1^* \end{bmatrix} = \begin{bmatrix} r_{xx}(0) & r_{xx}(1) \\ r_{xx}(1) & r_{xx}(0) \end{bmatrix}^{-1} \begin{bmatrix} r_{xd}(0) \\ r_{xd}(1) \end{bmatrix}
$$

$$
= \begin{bmatrix} 1 & \cos\dfrac{\pi}{8} \\ \cos\dfrac{\pi}{8} & 1 \end{bmatrix}^{-1} \begin{bmatrix} \dfrac{1}{\sqrt{2}}\cos\dfrac{\pi}{10} \\ \dfrac{1}{\sqrt{2}}\cos\dfrac{9\pi}{40} \end{bmatrix} = \begin{bmatrix} 1.20000 \\ -0.57099 \end{bmatrix} \qquad (5.30)
$$

These weight values are seen at the projection of the "bottom of the bowl" in Fig. 5.10. From Eq. (5.25), the minimum MSE when the weights are optimized as in Eq. (5.30) is

$$
(\text{MSE})_{\min} = r_{dd}(0) - \begin{bmatrix} r_{xd}(0) & r_{xd}(1) \end{bmatrix} \begin{bmatrix} w_0^* \\ w_1^* \end{bmatrix}
$$

$$
= 0.55 - \frac{1.20000}{\sqrt{2}}\cos\frac{\pi}{10} + \frac{0.57099}{\sqrt{2}}\cos\frac{9\pi}{40} = 0.05 \qquad (5.31)
$$

Thus, when the weights are optimized, y_k exactly cancels n_k as we might have anticipated from Fig. 5.8, and the output, $\epsilon_k = s_k$, is free of sinusoidal interference. We note, in this ideal example, that the signal-to-noise ratio improves from 0.1 to infinity.

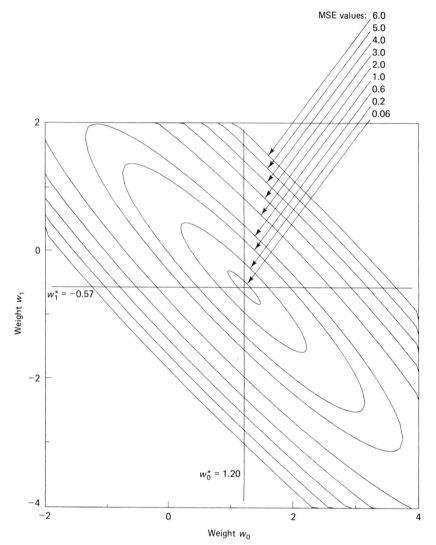

Figure 5.10   Contours of constant MSE on the performance surface in Fig. 5.9. The minimum MSE is 0.05 at $w_0^* = 1.20$ and $w_1^* = -0.57$.

5.4 SEARCH METHODS AND ADAPTIVE ALGORITHMS

So far we have viewed the adaptive process as that of searching the MSE performance surface for its global minimum or of tracking this global minimum as it changes with time. In Eq. (5.10) we saw that the performance surface could, in general, have local minima. Thus the most general adaptive algorithms for searching the performance surface under unknown signaling conditions are random search algorithms. One type of random search algorithm [36] selects a random direction in which to "look" at each iteration. An even more general random search algorithm [37] selects points at random locations on the weight plane (or hyperplane), estimates the MSE at these points, and retains points with the lowest MSE from one iteration to the next.

However, as we have stated, we are limiting our discussion in this chapter mainly to the quadratic performance surface of the adaptive linear combiner used in FIR adaptive filters. With this type of surface we can implement powerful, deterministic search methods. These methods generally are based on making local estimates of the gradient, $\hat{\nabla}$, and moving incrementally downward toward the bottom of the bowl. Thus, one adaptive cycle or adaptive iteration consists essentially of determining $\hat{\nabla}_k$ at time k and then making an appropriate adjustment in the weight vector from $\mathbf{W}_k$ to $\mathbf{W}_{k+1}$ in order to move toward $\mathbf{W}^*$ at the bottom of the bowl.

Our adaptive algorithms will be versions of two basic search techniques, steepest descent and Newton's method, which in turn result from a basic property of any parabolic surface. The property can be seen by multiplying the gradient formula in Eq. (5.23) by $\frac{1}{2}\mathbf{R}^{-1}$ to obtain

$$\tfrac{1}{2}\mathbf{R}^{-1}\nabla = \mathbf{W} - \mathbf{R}^{-1}\mathbf{P} \tag{5.32}$$

We can now combine this result with Eq. (5.24) to obtain

$$\boxed{\mathbf{W}^* = \mathbf{W} - \tfrac{1}{2}\mathbf{R}^{-1}\nabla} \tag{5.33}$$

This result is essentially Newton's root-finding method [38] applied to find the zero of a linear function (∇ in this case). It can also be derived by writing the MSE in the form of a Taylor series, as in [39]. Given any weight vector $\mathbf{W}$ along with $\mathbf{R}$ and the corresponding gradient ∇, we can move from $\mathbf{W}$ to the optimum weight vector $\mathbf{W}^*$ in a single step by adjusting $\mathbf{W}$ in accordance with Eq. (5.33).

If we could apply Eq. (5.33) in practical situations such as those illustrated in Figs. 5.2–5.5, we would always be able to adapt in one iteration to the optimal weight vector, and the adaptive process for the quadratic performance surface would be simple (but uninteresting). As in nature, however, practical adaptive signal processing systems do not have enough information to adapt perfectly in just one iteration. There are two specific problems with Eq. (5.33) in practice, under nonstationary conditions:

1. The $\mathbf{R}$ matrix is unknown and can at best only be estimated.
2. The gradient must be estimated at each adaptive iteration, using local statistics.

From Eqs. (5.17) and (5.19), we note that these two problems are directly related to the fact that the signal correlation functions are unknown and must be estimated.

To anticipate the use of noisy estimates of $\mathbf{R}^{-1}$ and ∇ in Eq. (5.33), we can modify Eq. (5.33) to obtain an algorithm that adjusts $\mathbf{W}$ in smaller increments and converges to $\mathbf{W}^*$ after many iterations. The small increments will have the effect of smoothing the noise in the estimates of $\mathbf{R}^{-1}$ and ∇, thus allowing the adaptive system to have stable behavior. The modified version of Eq. (5.33) is

$$\boxed{\mathbf{W}_{k+1} = \mathbf{W}_k - u\mathbf{R}^{-1}\nabla_k} \tag{5.34}$$

Under noise-free conditions with $\mathbf{R}$ and ∇_k known exactly, the convergence of Eq. (5.34) depends only on the scalar convergence factor, u. We can see this by substituting Eq. (5.23) and then Eq. (5.24) into Eq. (5.34):

$$
\begin{aligned}
W_{k+1} &= \mathbf{W}_k - u\mathbf{R}^{-1}(2\mathbf{R}\mathbf{W}_k - 2\mathbf{P}) \\
&= (1 - 2u)\mathbf{W}_k + 2u\mathbf{W}^*
\end{aligned}
\tag{5.35}
$$

Using Eq. (5.35) in itself, we arrive by induction at an expression for the relaxation of Eq. (5.34) from $\mathbf{W}_0$ toward the optimum weight vector, $\mathbf{W}^*$:

$$
\begin{aligned}
\mathbf{W}_{k+1} &= (1 - 2u)\mathbf{W}_k + 2u\mathbf{W}^* \\
&= (1 - 2u)^2\mathbf{W}_{k-1} + 2u\mathbf{W}^*[1 + (1 - 2u)] \\
&= (1 - 2u)^3\mathbf{W}_{k-2} + 2u\mathbf{W}^*[1 + (1 - 2u) + (1 - 2u)^2] \\
&\;\;\vdots \\
&= (1 - 2u)^{k+1}\mathbf{W}_0 + 2u\mathbf{W}^* \sum_{i=0}^{k} (1 - 2u)^i
\end{aligned}
\tag{5.36}
$$

Summing the series in Eq. (5.36) and using k in place of $k + 1$, we obtain

$$
\begin{aligned}
\mathbf{W}_k &= (1 - 2u)^k\mathbf{W}_0 + \frac{2u[1 - (1 - 2u)^k]}{1 - (1 - 2u)}\mathbf{W}^* \\
&= (1 - 2u)^k\mathbf{W}_0 + [1 - (1 - 2u)^k]\mathbf{W}^*
\end{aligned}
\tag{5.37}
$$

In this result we can see that $\mathbf{W}_k$ will converge to $\mathbf{W}^*$ at $k = \infty$ in general only if the geometric ratio $1 - 2u$ is less than 1 in magnitude. In other words, the rule is

$$
\boxed{\text{For noise-free convergence:} \quad 0 < u < 1}
\tag{5.38}
$$

In Eq. (5.33) the one-step, noise-free solution is found with $u = 1/2$. With $u > 1/2$ the convergence is oscillatory, with $\mathbf{W}_k$ jumping "back and forth" across the bowl and converging toward $\mathbf{W}^*$. In practical adaptive systems with noise, values of u well below $\frac{1}{2}$, typically on the order of 0.01, are used, giving a convergence time constant of many iterations. We define the convergence time constant to be τ iterations, where the relaxation toward $\mathbf{W}^*$ goes in proportion to $(1 - e^{-k/\tau})$. Then, using the linear approximation $(1 - 1/\tau)$ for $e^{-1/\tau}$ when τ is large and comparing with Eq. (5.37), we have

$$
\begin{aligned}
1 - e^{-k/\tau} &\approx 1 - (1 - 1/\tau)^k \\
&\approx 1 - (1 - 2u)^k
\end{aligned}
\tag{5.39}
$$

From this we have the weight-vector convergence time constant for Newton's method in terms of the adaptive gain parameter u:

> Time constant for weight-vector convergence
> under noise-free conditions:
>
> $$\tau_w \approx \frac{1}{2u}, \qquad 0 < u \ll 1/2$$

(5.40)

   In addition to the weight-vector convergence described in Eq. (5.37), the convergence of the MSE toward its minimum value is another process commonly used as a performance measure in adaptive systems. This convergence is known euphemistically as "learning," and a plot of the MSE versus the iteration number k is called a "learning curve" [4]. To derive a formula similar to Eq. (5.37) for the learning curve, it is convenient to use translated weight-vector coordinates, so we define the translated weight vector $\mathbf{V}$ as the deviation from the optimum weight vector:

$$\mathbf{V} = \mathbf{W} - \mathbf{W}^*$$

(5.41)

The minimum MSE is seen to be at $\mathbf{V} = \mathbf{0}$ in this translated system. (For example, the centered axes in Fig. 5.10 are the v_0 and v_1 axes.) Using Eq. (5.41) in Eq. (5.22) for the MSE, we obtain

$$\text{MSE} = r_{dd}(0) + (\mathbf{V} + \mathbf{W}^*)^T \mathbf{R}(\mathbf{V} + \mathbf{W}^*) - 2\mathbf{P}^T(\mathbf{V} + \mathbf{W}^*)$$
$$= r_{dd}(0) + \mathbf{W}^{*T}\mathbf{R}\mathbf{W}^* - 2\mathbf{P}^T\mathbf{W}^* + \mathbf{V}^T\mathbf{R}\mathbf{V} + \mathbf{V}^T\mathbf{R}\mathbf{W}^* + \mathbf{W}^{*T}\mathbf{R}\mathbf{V} - 2\mathbf{P}^T\mathbf{V}$$

(5.42)

From Eq. (5.24) and the first line of Eq. (5.25) we see that $(\text{MSE})_{\text{min}}$ can be substituted for the first three terms in Eq. (5.42). Also, using Eq. (5.24) plus the fact that any scalar is equal to its own transpose, we see that the last three terms in Eq. (5.42) cancel to zero. Hence, Eq. (5.42) reduces to

$$\text{MSE} = (\text{MSE})_{\text{min}} + \mathbf{V}^T\mathbf{R}\mathbf{V}$$

(5.43)

This result is often preferred over Eq. (5.22) as a general expression for the quadratic performance surface. To obtain the learning curve expression from Eq. (5.43), we use Eq. (5.41) to translate Eq. (5.37) and obtain a simpler result:

$$\mathbf{V}_k = (1 - 2u)^k \mathbf{V}_0$$

(5.44)

Substituting Eq. (5.44) into Eq. (5.43) expressed at the kth iteration, we have the learning curve formula for Newton's method:

$$(\text{MSE})_k = (\text{MSE})_{\text{min}} + (1 - 2u)^{2k} \mathbf{V}_0^T \mathbf{R} \mathbf{V}_0$$

(5.45)

This result gives us a closed expression for the learning curve with Newton's method under noise-free conditions. Comparing it with Eq. (5.37) or (5.44), we see that the geometric convergence ratio is now $(1 - 2u)^2$ instead of $(1 - 2u)$. Thus, in a development similar to Eq. (5.39), the approximate geometric ratio is $\exp(-1/\tau_{\text{MSE}}) = \exp(-2/\tau_w)$, where τ_{MSE} is the learning curve time constant. So our result similar to Eq. (5.40) for the learning curve is

$$
\boxed{
\begin{array}{c}
\text{Time constant for MSE convergence with Newton's} \\
\text{method under noise-free conditions:} \\[4pt]
\tau_{\text{MSE}} = \tau_w/2 \approx 1/4u, \qquad 0 < u \ll 1/2
\end{array}
}
\qquad (5.46)
$$

Examples of weight convergence and a learning curve are given in Figs. 5.11 and 5.12. These examples represent the noise-free application of Eq. (5.34), starting at $w_{00} = 3$ and $w_{10} = -4$ and using $u = 0.04$, well below the one-step convergence value, $u = 0.5$. The examples are for the interference-canceling example of Fig. 5.8, with performance surface illustrated in Figs. 5.9 and 5.10. In Fig. 5.11, note the

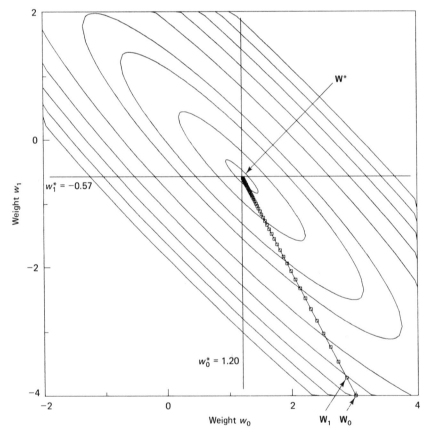

Figure 5.11 Newton weight track, showing noise-free convergence from $\mathbf{W}_0 = (3, -4)^T$ to near $\mathbf{W}^*$ for the interference-canceling system in Fig. 5.8. The adaptive gain in Eq. (5.34) was $u = 0.04$. The weight track is shown for $k = 0, 1, \ldots, 60$.

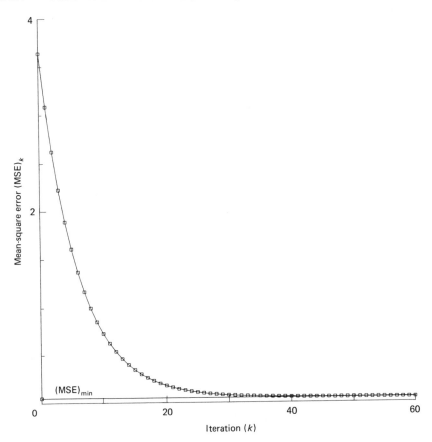

Figure 5.12 Learning curve of MSE versus iteration number, corresponding with the weight track in Fig. 5.11.

straight, noise-free track from $\mathbf{W}_0$ to $\mathbf{W}^*$ and also the translated $v_0 v_1$-coordinate system with its origin at $\mathbf{W}^*$. With $u = 0.04$, the weight convergence time constant is approximately 12 iterations in accordance with Eq. (5.40), and we note in Fig. 5.11 that there is convergence after 5 time constants, or 60 iterations. In Fig. 5.12, we see the corresponding exponential decrease of the MSE from its starting value of approximately 3.64 toward the minimum value, 0.05, in Eq. (5.31). The MSE time constant, τ_{MSE}, should be approximately 6 iterations in accordance with Eq. (5.46), and we observe in Fig. 5.12 that the approximation is valid. After 4 time constants, or 24 iterations in this example, the MSE has relaxed to approximately e^{-4}, or just under 2% of the original excess MSE, given by $(\text{MSE})_0 - (\text{MSE})_{\text{min}}$.

In this discussion of adaptive algorithms for quadratic performance surfaces, the principal formula is the Newton type of algorithm in Eq. (5.34). We must now examine how to apply this formula with imperfect knowledge of the gradient and the $\mathbf{R}$ matrix. First, we will assume that we have no knowledge of the off-diagonal elements of the $\mathbf{R}$ matrix. This assumption leads us to the "steepest-descent" class of algorithms and in turn to the widely used Widrow-Hoff LMS algorithm, which is the simplest and most useful algorithm in adaptive signal processing.

5.5 STEEPEST DESCENT AND THE LMS ALGORITHM

The least-mean-square (LMS) algorithm [4] is the simplest and most widely applicable algorithm in adaptive signal processing. It is a steepest-descent type of algorithm, that is, an algorithm whose track follows (on the average) the negative gradient of the performance surface. As mentioned above, such an algorithm does not require the complete $\mathbf{R}$ matrix. To revise Eq. (5.34) into a steepest-descent type of algorithm, we define $\hat{\nabla}_k$ to be an estimate of ∇_k and write

$$\mathbf{W}_{k+1} = \mathbf{W}_k - \mu\,\hat{\nabla}_k \tag{5.47}$$

Here we can see that the increment from $\mathbf{W}_k$ to $\mathbf{W}_{k+1}$ is in the estimated negative gradient direction, so the weight track will follow approximately a steepest-descent path on the performance surface. The parameter μ in Eq. (5.47) is the parameter used by Widrow and others [2] in the LMS algorithm. It can be related to the parameter u in Eq. (5.34) by noting that the elements of the $\mathbf{R}$ matrix have dimensions of input power, as seen in Eq. (5.17). We can define

$$\sigma^2 = \text{Average signal power to adaptive linear combiner}$$

$$= \frac{1}{L}\,E[\mathbf{X}_k^T\mathbf{X}_k] = \frac{1}{L}\,Tr[\mathbf{R}]$$

$$= E[x_k^2] \qquad \text{(single input)} \tag{5.48}$$

$$= \frac{1}{L}\sum_{i=0}^{L-1} E[x_{ik}^2] \qquad \text{(multiple inputs)}$$

Suppose we have a single white input signal with power σ^2, or multiple white inputs each with power σ^2. Then the $\mathbf{R}$ matrix, $\mathbf{R} = \sigma^2\mathbf{I}$, has all L eigenvalues equal to σ^2, and the constant-MSE contours given by Eq. (5.22) are "circular" in $(L + 1)$-dimensional space. For this case, comparing Eq. (5.47) with Eq. (5.34), we have

$$\textit{Equal eigenvalues:} \qquad \mu = \frac{u}{\sigma^2}, \qquad 0 < u < 1, \; 0 < \mu < \frac{1}{\sigma^2} \tag{5.49}$$

For the general case where the eigenvalues are not equal, it is then reasonable to use the *maximum* eigenvalue, $\lambda_{\max}$, in place of σ^2 in Eq. (5.49) to give a conservative range on μ corresponding with $0 < u < 1$ and ensure convergence of the steepest-descent algorithm, Eq. (5.47). Widrow [2] has proved this conjecture for convergence in the mean. However, the eigenvalues of $\mathbf{R}$ are not usually known in practice, so the *sum* of eigenvalues, which is the same as the trace of $\mathbf{R}$, is used instead. Thus, from Eq. (5.48), we use $L\sigma^2$ in place of $\lambda_{\max}$ and have for the general case

$$\textit{Unequal eigenvalues:} \qquad \mu = \frac{u}{Tr[\mathbf{R}]} = \frac{u}{L\sigma^2}, \qquad 0 < u < 1, \; 0 < \mu < \frac{1}{L\sigma^2}$$

$$\tag{5.50}$$

This more general bound on μ has also been derived by Widrow [3]. From Eqs. (5.50) and (5.47) the steepest-descent version of Eq. (5.34) for the general case now becomes

$$\mathbf{W}_{k+1} = \mathbf{W}_k - \frac{u}{L\sigma^2}\hat{\mathbf{V}}_k \qquad (5.51)$$

We note that the implementation of this result implies knowing the average input signal power σ^2. In practice a system for estimating σ^2 or tracking σ^2 under non-stationary conditions may be required. For example, a simple smoothing filter for estimating σ^2 in the single-input case is shown in Fig. 5.13. If we use the algorithm in the figure recursively, we obtain

$$\hat{\sigma}_k^2 = \alpha x_k^2 + (1 - \alpha)\hat{\sigma}_{k-1}^2 \qquad (5.52)$$

$$= \alpha x_k^2 + (1 - \alpha)[\alpha x_{k-1}^2 + (1 - \alpha)\hat{\sigma}_{k-2}^2]$$

$$\vdots$$

$$= \alpha \sum_{i=0}^{k} (1 - \alpha)^i x_{k-i}^2 \qquad (5.53)$$

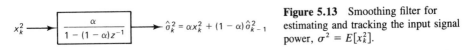

Figure 5.13 Smoothing filter for estimating and tracking the input signal power, $\sigma^2 = E[x_k^2]$.

Using a development similar to Eq. (5.39), we see from Eq. (5.53) that the operation in Fig. 5.13 amounts to "forgetting" the past values of x_k^2 exponentially, with the time constant

$$\tau_{\sigma^2} \approx \frac{1}{\alpha} \text{ iterations}, \qquad 0 < \alpha \ll 1 \qquad (5.54)$$

Unless the input signal power is known, some type of estimate like the one in Fig. 5.13 is needed to compensate for the missing $\mathbf{R}$ matrix in the steepest-descent type of algorithm in Eq. (5.51). The initial estimate, $\hat{\sigma}_0^2$, would normally be the best *a priori* estimate of σ^2.

An example of noise-free steepest descent with $\hat{\mathbf{V}}_k = \mathbf{V}_k$ is shown in Fig. 5.14 for comparison with Fig. 5.11. The only difference between the two figures is the use of Eq. (5.51) for Fig. 5.14 in place of Eq. (5.34) for Fig. 5.11. The terms in Eq. (5.51) for this case are $L = 2$ and $\sigma^2 = 1.0$ from Fig. 5.8, with $\mathbf{V}_k$ found by taking the gradient of Eq. (5.29). The value of $u = 0.2$ was used in Fig. 5.14. The principle feature of Fig. 5.14 is the direction of the weight track, which proceeds normal to the constant-MSE contours and is consequently of greater length compared with the track in Fig. 5.11. We note that when the error contours are highly elliptical, as they are here, the steepest-descent path can be much longer than the Newton path. However, if the error contours are circles, the two paths are the same. The ellipticity of the contours is determined by the eigenvalues of the $\mathbf{R}$ matrix [5], which, as we have noted, are not usually known in practice.

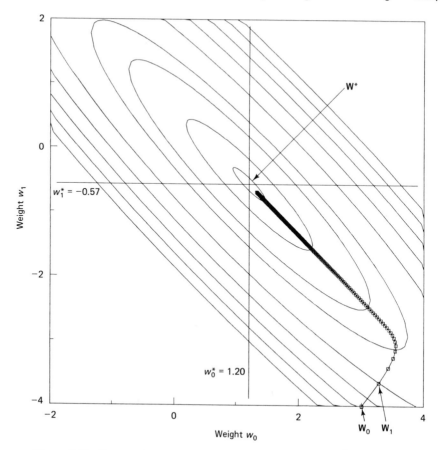

Figure 5.14 Steepest-descent weight track, showing noise-free ($\hat{\nabla}_k = \nabla_k$) convergence from $\mathbf{W}_0 = (3, -4)^T$ to near $\mathbf{W}^*$ for the interference-canceling system in Fig. 5.8. The adaptive gain parameter in Eq. (5.51) was $u = 0.2$. The weight track is shown for $k = 0, 1, \ldots, 200$.

We need one final modification of Eq. (5.47) to get from steepest descent to the LMS algorithm. Instead of the gradient at each iteration, ∇_k, which is not usually known in practice, we use a simple estimate, $\hat{\nabla}_k$. This estimate is in turn based on using the instantaneous squared error ϵ_k^2 as an estimate of the mean-square error $E[\epsilon_k^2]$. From the definition of ϵ_k as illustrated in Fig. 5.7, we can see that

$$\epsilon_k = d_k - \mathbf{W}_k^T \mathbf{X}_k \tag{5.55}$$

In Eq. (5.23) we have the gradient given as

$$\nabla_k = \frac{\partial(\text{MSE})}{\partial \mathbf{W}_k} \tag{5.56}$$

Therefore, using ϵ_k^2 as an estimate of the MSE, we have

$$\hat{\nabla}_k = \frac{\partial \epsilon_k^2}{\partial \mathbf{W}_k} = 2\epsilon_k \frac{\partial \epsilon_k}{\partial \mathbf{W}_k} = -2\epsilon_k \mathbf{X}_k \tag{5.57}$$

We can show that Eq. (5.57) is an unbiased gradient estimate by substituting Eq. (5.55) for ϵ_k and taking the expected value. The result is $E[\hat{\nabla}_k] = \nabla$ in Eq. (5.23). The result of substituting from Eq. (5.57) for $\hat{\nabla}_k$ in Eq. (5.51) is

$$\mathbf{W}_{k+1} = \mathbf{W}_k + \frac{2u}{L\sigma^2}\epsilon_k \mathbf{X}_k, \qquad 0 < u < 1 \qquad (5.58)$$

This is the LMS algorithm. We prefer the normalized constant u over μ in Eq. (5.47) to emphasize the need to estimate input power and also because the allowable range for u is easy to remember.

An example using the LMS algorithm in the system of Fig. 5.8 is shown in Fig. 5.15 for comparison with Figs. 5.11 and 5.14. The conditions are exactly the same as for Fig. 5.14 except that Eq. (5.58) is used instead of Eq. (5.51). We note in Fig. 5.15 that the weight track is now noisy due to the local gradient estimated in the LMS algorithm.

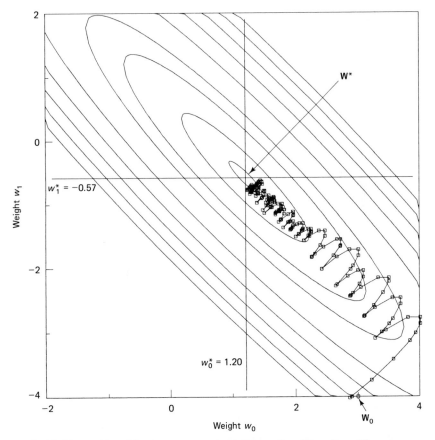

Figure 5.15 LMS weight track, showing convergence from $\mathbf{W}_0 = (3, -4)^T$ to near $\mathbf{W}^*$ for the interference-canceling system in Fig. 5.8. The adaptive gain parameter in Eq. (5.58) was $u = 0.2$. The weight track is shown for $k = 0, 1, \ldots, 200$.

A plot of the output ϵ_k versus k is shown in Fig. 5.16. This plot corresponds with the weight track in Fig. 5.15. In Fig. 5.16 we can see the sinusoidal interference gradually disappearing from the output as k increases, with $\epsilon_k \approx s_k$ when k reaches 200, which is of course the desired result in Fig. 5.8. The effect of this disappearance of the sinusoidal component of ϵ_k can also be seen in the weight track in Fig. 5.15.

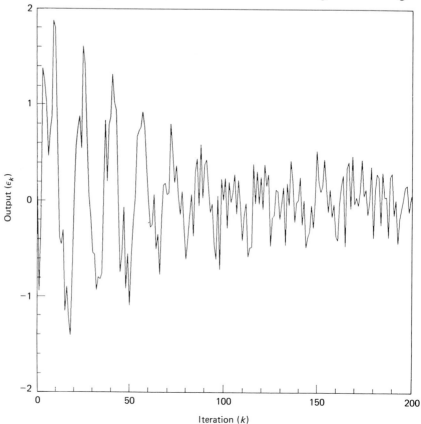

Figure 5.16   Output ϵ_k versus k for the LMS algorithm applied in Fig. 5.8, showing cancellation of the sinusoidal interference component.

The LMS algorithm is the simplest and most widely applicable adaptive algorithm. All of the terms in the weight-vector increment in Eq. (5.58) can be made available in a wide variety of applications. When the input power σ^2 to the adaptive linear combiner varies during adaptation, an algorithm such as Eq. (5.52) must be used together with Eq. (5.58) to provide a current estimate σ_k^2 to use in place of σ^2.

5.6 AN LMS-NEWTON ALGORITHM

We have seen that the LMS algorithm is a steepest-descent type of algorithm that, as in Eq. (5.47), uses an estimate of the gradient vector but no estimate of the $\mathbf{R}$ matrix. A Newton type of algorithm that "steers" the weight-vector increment and generates a straighter path to $\mathbf{W}^*$ on the performance surface (as in the ideal case of Fig. 5.11 compared with the case of Fig. 5.14) must contain an estimate of $\mathbf{R}^{-1}$ in addition to the gradient estimate. That is, from Eq. (5.34), a Newton type of algorithm using estimates of $\mathbf{R}^{-1}$ and $\mathbf{\nabla}_k$ would be

$$\mathbf{W}_{k+1} = \mathbf{W}_k - u\hat{\mathbf{R}}_k^{-1}\hat{\mathbf{\nabla}}_k \tag{5.59}$$

To see how we might compute $\hat{\mathbf{R}}_k^{-1}$ in Eq. (5.59), let us first consider a scheme [5] similar to Fig. 5.13 for estimating the entire $\mathbf{R}$ matrix instead of just the input power. The result is shown in Fig. 5.17. Just as in Eqs. (5.52)–(5.54) we have, from Fig. 5.17,

$$\hat{\mathbf{R}}_k = (1 - \alpha)\hat{\mathbf{R}}_{k-1} + \alpha\mathbf{X}_k\mathbf{X}_k^T \tag{5.60}$$

$$= \alpha \sum_{i=0}^{k} (1 - \alpha)^i\mathbf{X}_{k-i}\mathbf{X}_{k-i}^T \tag{5.61}$$

$$\tau_R \approx \frac{1}{\alpha} \text{ iterations}, \qquad 0 < \alpha \ll 1 \tag{5.62}$$

Here again, we are weighting the past input values exponentially so that they are forgotten approximately with a time constant equal to τ_R iterations. This is a natural way to estimate the $\mathbf{R}$ matrix, at least when the signal $\mathbf{X}_k$ is stationary enough to allow τ_R to be large compared with the periods of the dominant frequency components of $\mathbf{X}_k$. (For an alternative method, see [40].)

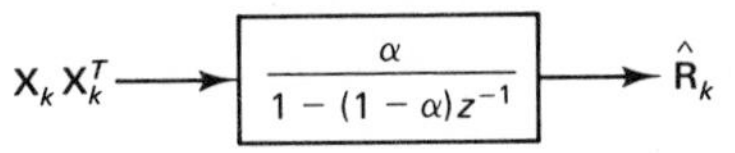

Figure 5.17   Smoothing filter for estimating the $\mathbf{R}$ matrix, similar to the filter shown in Fig. 5.13.

Assuming that Eq. (5.60) gives an acceptable estimate of $\mathbf{R}$, we now proceed to change Eq. (5.60) into an equivalent estimate of $\mathbf{R}^{-1}$ to remove the need for a matrix inversion at each iteration of Eq. (5.59). First we premultiply Eq. (5.60) by $\hat{\mathbf{R}}_k^{-1}$, then postmultiply by $\hat{\mathbf{R}}_{k-1}^{-1}$, and obtain

$$\hat{\mathbf{R}}_{k-1}^{-1} = (1 - \alpha)\hat{\mathbf{R}}_k^{-1} + \alpha\hat{\mathbf{R}}_k^{-1}\mathbf{X}_k\mathbf{X}_k^T\hat{\mathbf{R}}_{k-1}^{-1} \tag{5.63}$$

Postmultiplying this result by $\mathbf{X}_k$ gives

$$\hat{\mathbf{R}}_{k-1}^{-1}\mathbf{X}_k = (1 - \alpha)\hat{\mathbf{R}}_k^{-1}\mathbf{X}_k + \alpha\hat{\mathbf{R}}_k^{-1}\mathbf{X}_k\mathbf{X}_k^T\hat{\mathbf{R}}_{k-1}^{-1}\mathbf{X}_k$$
$$= \hat{\mathbf{R}}_k^{-1}\mathbf{X}_k(1 - \alpha + \alpha\mathbf{X}_k^T\hat{\mathbf{R}}_{k-1}^{-1}\mathbf{X}_k) \tag{5.64}$$

It will now be useful to define the auxiliary vector

$$\hat{\mathbf{S}}_k = \hat{\mathbf{R}}_{k-1}^{-1}\mathbf{X}_k \tag{5.65}$$

Then Eq. (5.64) becomes

$$\hat{\mathbf{S}}_k = \hat{\mathbf{R}}_k^{-1}\mathbf{X}_k(1 - \alpha + \alpha\mathbf{X}_k^T\hat{\mathbf{S}}_k) \tag{5.66}$$

We note that $\hat{\mathbf{R}}_k^{-1}$ is symmetric just as $\hat{\mathbf{R}}_k$ in Eq. (5.60) is symmetric and therefore the transpose of Eq. (5.65) is

$$\hat{\mathbf{S}}_k^T = \mathbf{X}_k^T\hat{\mathbf{R}}_{k-1}^{-1} \tag{5.67}$$

Dividing Eq. (5.66) by the scalar in parentheses and postmultiplying by Eq. (5.67), we obtain

$$\hat{\mathbf{R}}_k^{-1} \mathbf{X}_k \mathbf{X}_k^T \hat{\mathbf{R}}_{k-1}^{-1} = \frac{\hat{\mathbf{S}}_k \hat{\mathbf{S}}_k^T}{1 - \alpha + \alpha \mathbf{X}_k^T \hat{\mathbf{S}}_k} \tag{5.68}$$

Finally, we substitute Eq. (5.68) into the last term in Eq. (5.63) and rearrange the terms in the result to obtain

$$\hat{\mathbf{R}}_k^{-1} = \frac{1}{1 - \alpha} \left(\hat{\mathbf{R}}_{k-1}^{-1} - \alpha \frac{\hat{\mathbf{S}}_k \hat{\mathbf{S}}_k^T}{1 - \alpha + \alpha \mathbf{X}_k^T \hat{\mathbf{S}}_k} \right) \tag{5.69}$$

In Eqs. (5.65) and (5.69) we now have an iterative algorithm for $\hat{\mathbf{R}}_k^{-1}$ equivalent to Eq. (5.60), which is the desired result.

As with σ_0^2 in Eq. (5.52), the initial matrix $\hat{\mathbf{R}}_0^{-1}$ in Eq. (5.69) would normally be our best *a priori* estimate of the inverse $\mathbf{R}$ matrix. If only the signal power is known initially, then $(1/\sigma^2)\mathbf{I}$ could be used for $\hat{\mathbf{R}}_0^{-1}$.

Having the algorithm for $\hat{\mathbf{R}}_k^{-1}$ and using Eq. (5.57) as before for the gradient estimate $\hat{\nabla}_k$, we obtain for the LMS-Newton weight update formula in Eq. (5.59)

$$\boxed{\mathbf{W}_{k+1} = \mathbf{W}_k + 2u\epsilon_k \hat{\mathbf{R}}_k^{-1} \mathbf{X}_k, \qquad 0 < u < 1} \tag{5.70}$$

Just as we needed Eq. (5.52) at each iteration if σ was changing in the LMS algorithm of Eq. (5.58), we now need Eqs. (5.65) and (5.69) at each iteration when $\mathbf{R}^{-1}$ is changing in the LMS-Newton algorithm of Eq. (5.70). In contrast with Eq. (5.58), we use $\hat{\mathbf{R}}_k^{-1}$ explicitly in Eq. (5.70) instead of $\mathbf{R}^{-1}$ because $\mathbf{R}$ is usually not known accurately in adaptive situations.

A weight track produced by the LMS-Newton algorithm in Eqs. (5.65), (5.69), and (5.70) is illustrated in Fig. 5.18 for comparison with the LMS weight track in Fig. 5.15. The starting weight vector is $\mathbf{W}_0 = (3, -4)^T$ as before and the signals are as in Fig. 5.8, so this weight track is comparable with all of the previous weight tracks. For Fig. 5.18, we set $\hat{\mathbf{R}}_0^{-1} = (1/\sigma^2)\mathbf{I}$ and let the inverse-$\mathbf{R}$ algorithm in Eqs. (5.65) and (5.69) run for 500 iterations with $\alpha = 0.01$ *prior* to running the full weight-adjusting algorithm in Eqs. (5.65), (5.69), and (5.70). We did this to illustrate a "tracking" situation where $\hat{\mathbf{R}}^{-1}$ remains near $\mathbf{R}^{-1}$ during adaptation and to show a weight track that is clearly more like the Newton weight track in Fig. 5.11 and less like the steepest-descent tracks in Figs. 5.14 and 5.15. Comparing Figs. 5.15 and 5.18, we note that the latter (LMS-Newton) weights converge in fewer iterations, just as Fig. 5.11 (Newton) showed faster convergence than Fig. 5.14.

5.7 PERFORMANCE CHARACTERISTICS

We have seen examples showing how the algorithms described in the two preceding sections perform in the simple interference-canceling system of Fig. 5.8. Now we will discuss some basic performance characteristics illustrated in these examples.

First we note that considerably more computing per iteration is required for the

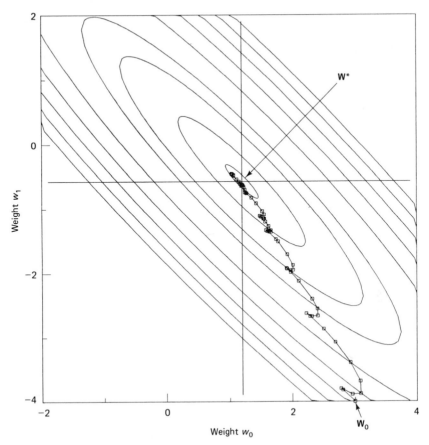

Figure 5.18 LMS-Newton weight track, showing convergence from $\mathbf{W}_0 =$ $(3, -4)^T$ to near $\mathbf{W}^*$ for the interference-canceling system in Fig. 5.8. The adaptive gain parameter in Eq. (5.70) was $u = 0.03$, α in Eq. (5.69) was 0.01, and $\mathbf{R}_0^{-1}$ in Eq. (5.69) was near $\mathbf{R}^{-1}$, as explained in the text. The weight track is shown for $k = 0, 1, \ldots , 60$.

Newton type of algorithm because the inverse-$\mathbf{R}$ estimate must be updated at each iteration, as in Eqs. (5.69) and (5.65). Also methods exist for constructing fast frequency-domain implementations of the algorithms. These implementations are called *frequency-domain adaptive filters* or *block adaptive filters* [41–46]. They are capable of producing a significant decrease in the computational burden of an adaptive processor and have been shown to have essentially the same convergence properties as the time-domain algorithm described above.

Concerning convergence, we have seen that under noise-free conditions the adaptive algorithms converge to the minimum MSE and then remain there, where the gradient of the performance surface is zero. We saw this in Eq. (5.37), for example, as long as the adaptive gain u was within its correct range from 0 to 1. We also saw in Eqs. (5.40) and (5.46) that the convergence time constants are inversely proportional to u. Convergence is faster when u is increased, at least up to $u = \frac{1}{2}$.

So why not increase u toward $\frac{1}{2}$ and have a more responsive, faster-converging adaptive system? The answer is that, in practical systems with noisy estimates of the gradient and/or the **R** matrix, increasing u means increasing the *excess MSE* after convergence. The excess MSE is the increase in the MSE over $(MSE)_{min}$ after convergence, due to these noisy estimates. From Eq. (5.43), the excess MSE may be written as

$$\text{Excess MSE} = (\text{MSE})_{\text{actual}} - (\text{MSE})_{\text{min}}$$
$$= E[\mathbf{V}_k^T \mathbf{R} \mathbf{V}_k] \quad \text{(after convergence)} \tag{5.71}$$

For example, with the noisy gradient estimate given by $\hat{\boldsymbol{\nabla}}_k$ in Eq. (5.57), the LMS algorithm adjusts the weight vector (and hence $\mathbf{V}_k$) at each iteration, causing the adaptive linear combiner to climb up the sides of its bowl-shaped performance surface in $L + 1$ dimensions. Of course, when the system discovers at the next iteration that it is at a point where the gradient is nonzero it will, on the average, adjust back toward the bottom of the bowl. With this process going on continually in a stochastic setting, the result is a nonzero, positive value for the excess MSE in Eq. (5.71).

For the LMS algorithm, assuming that **R** has equal eigenvalues and using Eq. (5.49) to relate μ and u, the excess MSE has been shown for small values of u to be approximately

$$\text{Excess MSE} \approx \mu (\text{MSE})_{\text{min}} Tr[\mathbf{R}]$$
$$\approx uL (\text{MSE})_{\text{min}}, \quad 0 < u \ll 1 \tag{5.72}$$

The derivation of this useful approximation is rather lengthy and therefore not included here. The interested reader may refer to [25] or to Chapters 5 and 6 in [5]. A related measure is the *misadjustment M*, which is a normalized version of the excess MSE:

$$\boxed{\begin{array}{c} M \triangleq \dfrac{\text{Excess MSE}}{(\text{MSE})_{\text{min}}} \\[2mm] \approx uL, \quad 0 < u \ll 1 \end{array}} \tag{5.73}$$

The approximation in Eq. (5.73) appears to be good for misadjustments up to about 0.1, even in cases where the eigenvalues of **R** are unequal, as in the example of Fig. 5.8 that we have been following in this chapter. For that example we set $\mathbf{W}_0 = \mathbf{W}^*$ and then ran the LMS algorithm of Eq. (5.58) for 15 values of u between 0.00 and 0.06, which, with $L = 2$, give theoretical misadjustments between 0.0 and 0.12 in Eq. (5.73). Each "run" actually consisted of 10 runs with $\mathbf{W}_0 = \mathbf{W}^*$ and $0 \le k \le 10^4$, using 10 different starting points on the signal waveforms. The result is plotted in Fig. 5.19, and we can see that Eq. (5.73) is a good approximation for values of M below $M = 0.1$. Above $M = 0.1$, the approximation gets worse with increasing M.

The time constants derived in Eqs. (5.40) and (5.46) were applicable to the noise-free convergence of the Newton algorithm of Eq. (5.34) and are therefore valid only for "ideal" operation of the algorithms described in this section, which use

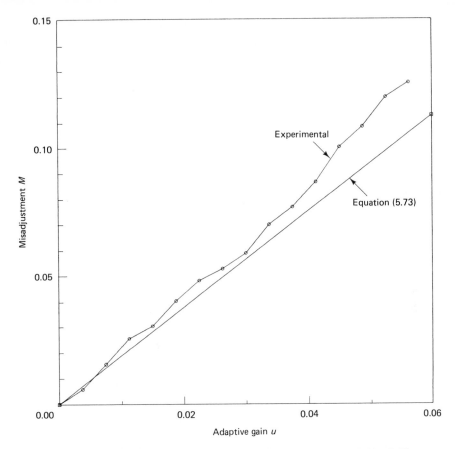

Figure 5.19   Misadjustment versus u after convergence of the adaptive filter in Fig. 5.8. Each experimental point is the average of 10^5 values of the squared error ϵ_k^2. The theoretical approximation in Eq. (5.73) is considered good up to around $M = 0.1$.

generally noisy estimates for $\mathbf{R}$ and $\mathbf{\nabla}_k$. If we assume that the noise-free formula of Eq. (5.46) is approximately correct for the learning time constant τ_{MSE}, we can combine Eqs. (5.46) and (5.73) to obtain

$$M \approx \frac{L}{4\tau_{MSE}}, \qquad 0 < u \ll 1 \tag{5.74}$$

The time constant formula, and therefore Eq. (5.74), is valid for the LMS algorithm only with circular error contours (equal eigenvalues of $\mathbf{R}$) and noise-free convergence. However, Eq. (5.74) has been found to be a good approximation with the LMS algorithm "when the eigenvalues are sufficiently similar for the learning curve to be approximately fitted by a single exponential" [25].

The main point in this discussion of convergence is summarized in Eq. (5.74). In adaptive systems, low misadjustment and fast convergence are desirable but conflicting requirements. Faster convergence means noisier performance, with greater excess MSE. In fact, noting that approximately four time constants, or $4\tau_{MSE}$, are required for convergence from an arbitrary MSE to $(MSE)_{min}$, we can restate Eq. (5.74) as follows [25]:

$$\text{Misadjustment } (M) \approx \frac{\text{Number of weights } (L)}{\text{Convergence time } (4\tau_{MSE})}, \qquad 0 < u \ll 1 \qquad (5.75)$$

This is perhaps the main performance characteristic of adaptive systems using the adaptive linear combiner structure.

Another factor affecting the stability and performance of the adaptive algorithms is the accuracy of the power and inverse-R estimates, $\hat{\sigma}_k^2$ and $\hat{R}_k^{-1}$, in nonstationary situations. If $\hat{R}_k^{-1}$ is accurate, the LMS-Newton algorithm will have superior convergence and tracking performance, as illustrated above. However, the smoothing filters in Figs. 5.13 and 5.17 for producing the estimates are simple lowpass filters, and they produce estimates of varying accuracy depending on the value of α and the frequency content of x_k. Suppose x_k is a sinusoid at frequency ω_0 radians, for example,

$$x_k = \sin k\omega_0$$
$$x_k^2 = \sin^2 k\omega_0 = \tfrac{1}{2}(1 - \cos 2k\omega_0) \qquad (5.76)$$

After the initial transient, the filter in Fig. 5.13 will pass the dc component in Eq. (5.76), which is the correct power estimate, and will reject the component at $2\omega_0$, provided that ω_0 is large enough compared with α. The power gain of the filter is

$$|H(e^{j\omega})|^2 = \left| \frac{\alpha}{1 - (1 - \alpha)e^{-j\omega}} \right|^2$$
$$= \frac{\alpha^2}{\alpha^2 + 4(1 - \alpha) \sin^2(\omega/2)} \qquad (5.77)$$

The gain is plotted in Fig. 5.20 for $\alpha = 0.1, 0.01$, and 0.001, and we can see how the rejection of the undesirable component improves as α decreases. On the other hand, the transient response of the filter decreases in duration as α increases, and α must be chosen with these factors in mind. The value 0.01 used in Fig. 5.18 is typical.

5.8 ADAPTIVE LATTICE STRUCTURE

In most of our discussion so far we have assumed that the single-input adaptive system is a transversal FIR filter. The lattice structure is an alternative to the transversal FIR filter in adaptive systems and is also potentially useful as a signal conditioner ahead of an adaptive linear combiner, as explained below. In this short section we will introduce the basic all-zero lattice structure that has been used adaptively and will briefly describe its important properties, referring to the literature for proofs and further description.

The lattice predictor structure proposed by Itakura and Saito [47] for speech analysis is shown in Fig. 5.21. We will use this structure as a basis for our discussion

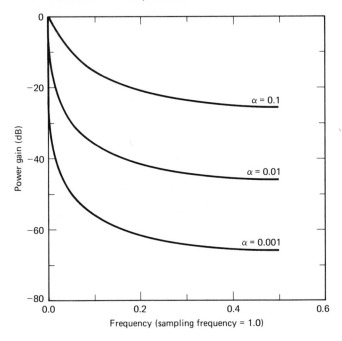

Figure 5.20 Power gain versus frequency for the smoothing filters in Figs. 5.13 and 5.17.

here. Before showing how the lattice can be made adaptive, we will discuss some of its important characteristics. Makhoul has provided an excellent basis for this discussion [48] and has also described the use of adaptive lattices in the analysis of speech [49].

A *stage* of the lattice in Fig. 5.21 is a single section with a delay, two weights, and two summing units. Each stage has two input signals and two output signals, and the inputs to the first stage are tied together as shown. In Fig. 5.21 we can see that the signals in the ith stage are related as follows:

$$\begin{bmatrix} f_{i+1}(k) \\ b_{i+1}(k) \end{bmatrix} = \begin{bmatrix} 1 & \kappa_i \\ \kappa_i & 1 \end{bmatrix} \begin{bmatrix} f_i(k) \\ b_i(k-1) \end{bmatrix}, \quad i = 0, 1, \ldots, L-1 \qquad (5.78)$$

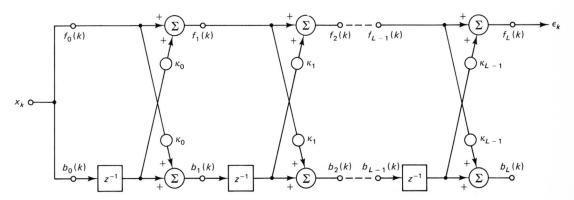

Figure 5.21 The all-zero two-multiplier lattice of Itakura and Saito. The lattice functions as a one-step predictor of x_k, and ϵ_k is the prediction error.

The weights $[\kappa_i]$ are known as reflection coefficients, or partial correlation ("parcor") coefficients. At stage 0 the inputs are $f_0(k) = b_0(k) = x_k$, and at stage $L - 1$ the outputs are $f_L(k) = \epsilon_k$ and $b_L(k)$. Taking the z-transform of Eq. (5.78) gives

$$\begin{bmatrix} F_{i+1}(z) \\ B_{i+1}(z) \end{bmatrix} = \begin{bmatrix} 1 & z^{-1}\kappa_i \\ \kappa_i & z^{-1} \end{bmatrix} \begin{bmatrix} F_i(z) \\ B_i(z) \end{bmatrix}, \qquad i = 0, 1, \ldots, L - 1 \qquad (5.79)$$

We can use Eq. (5.79) to find the overall transfer functions in Fig. 5.21. At the output of the ith stage we define

$$H_i(z) \triangleq \frac{F_i(z)}{F_0(z)} = \frac{F_i(z)}{X(z)}, \qquad G_i(z) \triangleq \frac{B_i(z)}{B_0(z)} = \frac{B_i(z)}{X(z)}, \qquad i = 0, 1, \ldots, L \qquad (5.80)$$

Using these definitions in Eq. (5.79) and using Eq. (5.79) recursively, we can write

$$\begin{bmatrix} H_i(z) \\ G_i(z) \end{bmatrix} = \begin{bmatrix} 1 & z^{-1}\kappa_{i-1} \\ \kappa_{i-1} & z^{-1} \end{bmatrix} \begin{bmatrix} 1 & z^{-1}\kappa_{i-2} \\ \kappa_{i-2} & z^{-1} \end{bmatrix} \cdots \begin{bmatrix} 1 & z^{-1}\kappa_0 \\ \kappa_0 & z^{-1} \end{bmatrix} \begin{bmatrix} 1 \\ 1 \end{bmatrix} \qquad (5.81)$$

If we set $i = L$, Eq. (5.81) gives the overall transfer functions. Thus we see that both overall transfer functions could be written as polynomials in z^{-1}. In fact, at the ith stage, we could write

$$H_i(z) = 1 + \sum_{m=1}^{i} a_{im} z^{-m}, \qquad i = 1, 2, \ldots, L \qquad (5.82)$$

The leading coefficient is seen in Eq. (5.81) to be 1, and the rest of the a's in Eq. (5.82) could be found by multiplying the terms in Eq. (5.81). In particular, if we imagine multiplying the i square matrices together in Eq. (5.81), the coefficient of z^{-i}, which is equal to a_{ii} in Eq. (5.82), is seen to be κ_{i-1}. Thus,

$$a_{ii} = \kappa_{i-1} \qquad (5.83)$$

To see how $G_i(z)$ is related to $H_i(z)$, we next revise Eq. (5.81) by replacing z with z^{-1} in the top equation to obtain

$$\begin{bmatrix} H_i(z^{-1}) \\ G_i(z) \end{bmatrix} = \begin{bmatrix} 1 & z\kappa_{i-1} \\ \kappa_{i-1} & z^{-1} \end{bmatrix} \begin{bmatrix} 1 & z\kappa_{i-2} \\ \kappa_{i-2} & z^{-1} \end{bmatrix} \cdots \begin{bmatrix} 1 & z\kappa_0 \\ \kappa_0 & z^{-1} \end{bmatrix} \begin{bmatrix} 1 \\ 1 \end{bmatrix} \qquad (5.84)$$

If we multiply the top equation in Eq. (5.84) by z^{-i}, we can write

$$\begin{bmatrix} z^{-i}H_i(z^{-1}) \\ G_i(z) \end{bmatrix} = \begin{bmatrix} z^{-1} & \kappa_{i-1} \\ \kappa_{i-1} & z^{-1} \end{bmatrix} \begin{bmatrix} z^{-1} & \kappa_{i-2} \\ \kappa_{i-2} & z^{-1} \end{bmatrix} \cdots \begin{bmatrix} z^{-1} & \kappa_0 \\ \kappa_0 & z^{-1} \end{bmatrix} \begin{bmatrix} 1 \\ 1 \end{bmatrix} \qquad (5.85)$$

If we think of carrying out the product in Eq. (5.85) from right to left, we can see that the two rows are identical, and thus the relation of $G_i(z)$ to $H_i(z)$ is

$$\boxed{G_i(z) = z^{-i}H_i(z^{-1}), \qquad i = 0, 1, \ldots, L} \qquad (5.86)$$

(For $i = 0$, this result is understood to be $G_0(z) = H_0(z) = 1$.) We can now relate the lattice in Fig. 5.21 to a specific transversal filter. We first write Eq. (5.81) in a recursive form similar to Eq. (5.79):

$$\begin{bmatrix} H_i(z) \\ G_i(z) \end{bmatrix} = \begin{bmatrix} 1 & z^{-1}\kappa_{i-1} \\ \kappa_{i-1} & z^{-1} \end{bmatrix} \begin{bmatrix} H_{i-1}(z) \\ G_{i-1}(z) \end{bmatrix}, \qquad i = 0, 1, \ldots, L-1 \qquad (5.87)$$

We next multiply the lower equation in Eq. (5.87) by κ_{i-1}, subtract the two equations, and solve for $H_{i-1}(z)$ to obtain

$$H_{i-1}(z) = \frac{H_i(z) - \kappa_{i-1} G_i(z)}{1 - \kappa_{i-1}^2}, \qquad i = L, L-1, \ldots, 1 \qquad (5.88)$$

With this result we can write an algorithm for changing a transversal filter into a lattice. We start with a filter in the form of Eq. (5.82) with $i = L$:

$$H_L(z) = 1 + \sum_{m=1}^{L} a_{Lm} z^{-m} \qquad (5.89)$$

$$= 1 - \sum_{m=1}^{L} w_m z^{-m} \qquad (5.90)$$

This we recognize as the adaptive predictor in Fig. 5.2(b) with $\Delta = 1$ and with the processor (FIR filter) weights given by $[w_m] = [-a_{Lm}]$. Now we combine Eqs. (5.82), (5.83), (5.86), and (5.88) to form the algorithm that uses Eq. (5.89) as its starting point and generates each lattice weight κ_i for $i = L - 1$ down through $i = 0$:

$$
\begin{aligned}
&\text{for } i = L, L-1, \ldots, 1: \\
&\quad \kappa_{i-1} = a_{ii} \\
&\quad G_i(z) = z^{-i} H_i(z^{-1}) \\
&\quad H_{i-1}(z) = \frac{H_i(z) - \kappa_{i-1} G_i(z)}{1 - \kappa_{i-1}^2}
\end{aligned}
\qquad (5.91)
$$

Thus we have a method for transforming a given transversal predictor, or FIR filter in the form of Eq. (5.89), into a lattice in the form of Fig. 5.21. In other words, we have shown that the lattice in Fig. 5.21 is a one-step ($\Delta = 1$) predictor and we have justified the labeling of the *forward prediction error* $f_L(k) = \epsilon_k$ as the one-step prediction error signal in Fig. 5.21.

We also note from Eqs. (5.86) and (5.89) that the transfer function from x_k to the lower output $b_L(k)$ in Fig. 5.21 is

$$G_L(z) = z^{-L}\left[1 - \sum_{m=1}^{L} w_m z^{+m} \right]$$

$$= z^{-L} - \sum_{m=0}^{L-1} w_{L-m} z^{-m} \qquad (5.92)$$

From this result, $b_L(k)$ is called the *backward prediction error* because the samples x_k through x_{k-L+1} are being used to predict the sample x_{k-L}. The forward and backward prediction errors are illustrated in Fig. 5.22, where $H_L(z)$ and $G_L(z)$ are shown in transversal form.

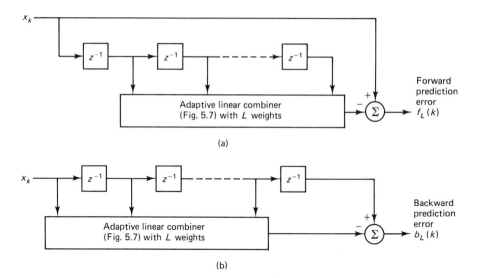

Figure 5.22 The two lattice transfer functions: (a) $H_L(z)$ in Eq. (5.89) and (b) $G_L(z)$ in Eq. (5.92), illustrated in the form of transversal filters to show the forward and backward prediction errors.

Griffiths [50] has shown how to make the lattice in Fig. 5.21 adaptive using the LMS algorithm. Griffiths allows the weights to adapt separately instead of in pairs as in Fig. 5.21, but if we keep the weights equal in pairs the results are very similar. We have seen that the output $f_{i+1}(k)$ of the ith lattice stage is the one-step prediction error at that stage. In adaptation we therefore adjust the weight value κ_i to minimize the expected value of $f_{i+1}^2(k)$. The steepest-descent formula, similar to Eq. (5.47), is

$$\kappa_i(k+1) = \kappa_i(k) - \mu_i \hat{\nabla}_k \qquad (5.93)$$

where k is the time index as usual and, similar to Eq. (5.57),

$$\hat{\nabla}_k = \text{Estimate of } \frac{\partial E[f_{i+1}^2(k)]}{\partial \kappa_i}$$

$$= \frac{\partial f_{i+1}^2(k)}{\partial \kappa_i} = 2f_{i+1}(k)\frac{\partial f_{i+1}(k)}{\partial \kappa_i} \qquad (5.94)$$

The adaptive gain for the ith stage (μ_i) is similar to the gain μ in Eq. (5.47), and the range in Eq. (5.49) applies here, with σ_i^2 the input power to the ith stage:

$$0 < \mu_i < \frac{1}{\sigma_i^2} \qquad (5.95)$$

During adaptive processing, the signal power changes throughout the lattice, and so the power (and hence μ_i) at each stage may require continual adjustment using a method such as the one in Eq. (5.52) as a part of the adaptive process.[†] Using the top

---

[†]The subscript i in Eq. (5.95) denotes the lattice stage, whereas the subscript k in Eq. (5.52) is the time index.

equation in Eq. (5.78) in Eq. (5.94) and then using Eq. (5.94) in Eq. (5.93), we have the LMS lattice algorithm:

$$\kappa_i(k + 1) = \kappa_i(k) - 2\mu_i f_{i+1}(k) b_i(k - 1), \qquad i = 0, 1, \ldots, L - 1 \tag{5.96}$$

As with the LMS algorithm discussed previously, acceptable misadjustment levels at each lattice stage are generally attained by keeping μ_i to a small fraction of its upper limit in Eq. (5.95).

An important property of the lattice shown by Jury [51] is the following:

$$\text{If } |\kappa_i| < 1 \text{ for } i = 0, 1, \ldots, L - 1, \\ \text{then } H_L(z) \text{ is minimum phase} \tag{5.97}$$

That is, the zeros of $H_L(z)$ are all inside the unit circle if the lattice weights are all less than 1 in magnitude. Therefore, the all-pole inverse, $1/H_L(z)$, is stable under these conditions. This property is very useful when applying adaptive lattices in speech communications. The latter subject has been discussed by Itakura and Saito [47] and others. In particular, Makhoul and Cosell [49] present a general class of adaptive algorithms for lattice predictors in speech analysis that goes beyond the simple LMS algorithm in Eq. (5.96). Lee, Morf, and Friedlander [52] have also presented improved adaptive algorithms for lattices.

When the adaptive lattice has converged, the optimum weight values are denoted $[\kappa_i^*]$. With stationary signals, these values turn out to be the same whether we minimize the forward or the backward prediction error. Taking the expected squared value of Eq. (5.78) and noting that $E[b_i^2(k)] = E[b_i^2(k - 1)]$ under stationary conditions, we have

$$E[f_{i+1}^2(k)] = E[f_i^2(k)] + \kappa_i^2 E[b_i^2(k)] + 2\kappa_i E[f_i(k)b_i(k - 1)] \tag{5.98}$$

$$E[b_{i+1}^2(k)] = \kappa_i^2 E[f_i^2(k)] + E[b_i^2(k)] + 2\kappa_i E[f_i(k)b_i(k - 1)] \tag{5.99}$$

At stage $i = 0$, $f_0(k) = b_0(k) = x_k$, and so Eqs. (5.98) and (5.99) are equal, and $E[f_1^2(k)] = E[b_1^2(k)]$. But then at stage 1, Eqs. (5.98) and (5.99) are equal again, and so on, and therefore we have

$$E[f_i^2(k)] = E[b_i^2(k)], \qquad i = 0, 1, \ldots, L - 1 \tag{5.100}$$

Setting the derivative of Eq. (5.98) or (5.99) with respect to κ_i equal to zero, we have the optimum weight that minimizes both prediction MSEs simultaneously:

$$\kappa_i^* = -\frac{E[f_i(k)b_i(k - 1)]}{E[f_i^2(k)]} \\ = -\frac{E[f_i(k)b_i(k - 1)]}{E[b_i^2(k)]}, \qquad i = 0, 1, \ldots, L - 1 \tag{5.101}$$

Makhoul [53] has derived the formulas in Eq. (5.101) as well as some similar formulas optimized on different criteria.

When the weights are optimized, an adaptive processor, either transversal or lattice in structure, has the property that the error and input signals are orthogonal. This property can be derived for the transversal structure from Fig. 5.7, where we have

$$\epsilon_k = d_k - \mathbf{X}_k^T \mathbf{W} \tag{5.102}$$

We multiply each scalar term in Eq. (5.102) by $\mathbf{X}_k$, take the expected value, and use Eqs. (5.17) and (5.19) to obtain

$$E[\epsilon_k \mathbf{X}_k] = \mathbf{P} - \mathbf{RW} \tag{5.103}$$

But at convergence, Eq. (5.24) gives $\mathbf{W}^* = \mathbf{R}^{-1}\mathbf{P}$, and so Eq. (5.103) becomes

$$E[\epsilon_k \mathbf{X}_k]_{\mathbf{W}=\mathbf{W}^*} = \mathbf{0} \tag{5.104}$$

Thus the error and input signals are orthogonal.

Similarly, in the lattice of Fig. 5.21 with optimal weights, Makhoul [48] has shown that the backward prediction errors (b's) are mutually orthogonal and that the set of optimal weights in Eq. (5.101), each computed using local statistics at each stage, minimizes the MSE at each stage as well as the overall MSE, i.e., the MSE at the final stage.

This mutual orthogonality of the signals in a lattice opens the possibility suggested by Griffiths [54] of using the lattice as a signal conditioner ahead of an adaptive processor, as illustrated in Fig. 5.23. Instead of feeding the signal vector $\mathbf{X}_k$ directly to the adaptive processor, we first process $\mathbf{X}_k$ with a lattice to produce a dependent signal vector $\mathbf{B}_k$, whose elements (after convergence) are mutually orthogonal. Looking again at Eq. (5.17), but now with $\mathbf{B}_k$ in place of $\mathbf{X}_k$, we see that the correlation matrix $\mathbf{R}$ is *diagonal* after the lattice has converged. That is,

$$\mathbf{R}_{(\text{Fig. 5.23})} = \text{Diag.}[r_{b_0 b_0} \quad r_{b_1 b_1} \quad \cdots \quad r_{b_{L-1} b_{L-1}}] \tag{5.105}$$

We also have, from Eq. (5.19) or (5.20),

$$\mathbf{P}_{(\text{Fig. 5.23})} = [r_{db_0} \quad r_{db_1} \quad \cdots \quad r_{db_{L-1}}]^T \tag{5.106}$$

Therefore, from Eq. (5.24), the optimum weight vector in Fig. 5.23 is

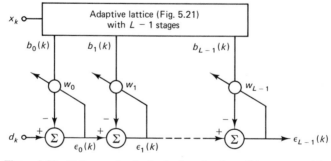

Figure 5.23 Using an adaptive lattice is a signal conditioner to provide orthogonal inputs to an adaptive linear combiner, allowing local adjustment of the combiner weights.

$$\mathbf{W}^*_{(\text{Fig. 5.23})} = \mathbf{R}^{-1}_{(\text{Fig. 5.23})} \mathbf{P}_{(\text{Fig. 5.23})}$$

$$= \left[\frac{r_{db_0}}{r_{b_0 b_0}} \quad \frac{r_{db_1}}{r_{b_1 b_1}} \quad \cdots \quad \frac{r_{db_{L-1}}}{r_{b_{L-1} b_{L-1}}} \right]^T \tag{5.107}$$

This result justifies the local weight-adjustment scheme shown in Fig. 5.23. At each stage in the adaptive linear combiner, the optimal weight is seen in Eq. (5.107) to depend only on local signals. The weight w_i^* is determined only by b_i and ϵ_{i-1}, not by the other b's. The LMS algorithm or any other desired scheme can be used to estimate continually the ratios in Eq. (5.107).

The advantage of the local weight-adjustment scheme in Fig. 5.23 is that a local adaptive gain can now be used at each stage, just as in the lattice. If $w_i(k)$ is the kth value of the ith weight in Fig. 5.21, the LMS algorithm, similar to Eq. (5.58), is

$$w_i(k + 1) = w_i(k) + \frac{2u}{\sigma_i^2} \epsilon_i(k) f_i(k), \qquad i = 0, 1, \ldots, L - 1, \tag{5.108}$$
$$0 < u < 1$$

At each stage, σ_i^2 is now $E[b_i^2]$, which must be either known or estimated as discussed previously, as in Eq. (5.52). But the steepest-descent convergence problems seen previously, in Fig. 5.14 for example, caused by differing time constants due in turn to disparate eigenvalues of the input correlation matrix, do not occur here. Thus the convergence of Eq. (5.108) is potentially faster than that of Eq. (5.58) for such cases.

A further analysis of the "orthogonalized LMS" algorithm is provided by Widrow and Walach [55]. Other methods besides the adaptive lattice, such as the discrete Fourier transform, can be used to orthogonalize the input signal and produce essentially the same effect on the performance of the adaptive linear combiner. We will conclude by showing how the application of Fig. 5.23 alters the performance surface of the interference-canceling system in Fig. 5.8.

5.9 AN EXAMPLE OF AN ADAPTIVE LATTICE

In this example we use the adaptive lattice with the interference-canceling system of Fig. 5.8 to obtain a comparison with the performances shown in Figs. 5.11, 5.14, 5.15, and 5.18. The lattice is used as in Fig. 5.23; that is, it is used to orthogonalize the signal x_k to the adaptive FIR filter in Fig. 5.8. Since the adaptive filter has only two weights, a single lattice stage suffices to provide the two orthogonal data sequences.

The system resulting from the application of Fig. 5.23 to Fig. 5.8 is shown in Fig. 5.24. The adaptive lattice has the effect of altering the $\mathbf{R}$ matrix and the $\mathbf{P}$ vector and thus the performance surface of the adaptive filter. Let us assume that the lattice is converged as in Eq. (5.101), with

$$\kappa_0 = \kappa_0^* = -\frac{E[f_0(k) b_0(k - 1)]}{E[f_0^2(k)]}$$

$$= -\frac{E[x_k x_{k-1}]}{E[x_k^2]} = -\cos \frac{\pi}{8} \tag{5.109}$$

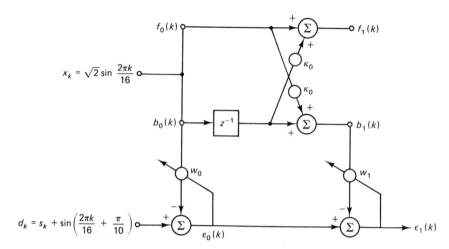

Figure 5.24   An example in which Fig. 5.23 is applied to the interference-canceling system in Fig. 5.8.

Then, with x_k as given in Fig. 5.24, the lattice signal $b_1(k)$ is

$$b_1(k) = \kappa_0 f_0(k) + b_0(k-1)$$

$$= \kappa_0 x_k + x_{k-1} \tag{5.110}$$

$$= -\sqrt{2} \sin \frac{\pi}{8} \cos \frac{2\pi k}{16}$$

We note that $b_1(k)$, a cosine function of k, is orthogonal to $b_0(k) = x_k$, a sine function of k. We also have the following correlation functions:

$$r_{b_0 b_0} = E[x_k^2] = 1 \tag{5.111}$$

$$r_{b_1 b_1} = E[b_1^2(k)] = \sin^2 \frac{\pi}{8} \tag{5.112}$$

$$r_{db_0} = \sqrt{2}\, E\left[\sin\left(\frac{2\pi k}{16} + \frac{\pi}{10}\right) \sin \frac{2\pi k}{16}\right] = \frac{1}{\sqrt{2}} \cos \frac{\pi}{10} \tag{5.113}$$

$$r_{db_1} = -\sqrt{2} \sin \frac{\pi}{8} E\left[\sin\left(\frac{2\pi k}{16} + \frac{\pi}{10}\right) \cos \frac{2\pi k}{16}\right]$$

$$= -\frac{1}{\sqrt{2}} \sin \frac{\pi}{8} \sin \frac{\pi}{10} \tag{5.114}$$

Using these results in Eqs. (5.105) and (5.106), we have

$$\mathbf{R} = \begin{bmatrix} 1 & 0 \\ 0 & \sin^2 \dfrac{\pi}{8} \end{bmatrix} \tag{5.115}$$

$$P = \frac{1}{\sqrt{2}} \begin{bmatrix} \cos \dfrac{\pi}{10} \\ -\sin \dfrac{\pi}{8} \sin \dfrac{\pi}{10} \end{bmatrix} \tag{5.116}$$

Using these results in Eq. (5.14) or (5.22), we have now an error surface for the adaptive linear combiner that differs from Eq. (5.29):

$$\text{MSE} = 0.55 + w_0^2 + w_1^2 \sin^2 \frac{\pi}{8} - \sqrt{2}\left(w_0 \cos \frac{\pi}{10} - w_1 \sin \frac{\pi}{8} \sin \frac{\pi}{10}\right) \tag{5.117}$$

Contours of this MSE, which may be compared with those in Fig. 5.10, are shown in Fig. 5.25. Figure 5.25 also shows a weight track, illustrating convergence of the adaptive filter with signals the same as in Figs. 5.11, 5.14, 5.15, and 5.18.

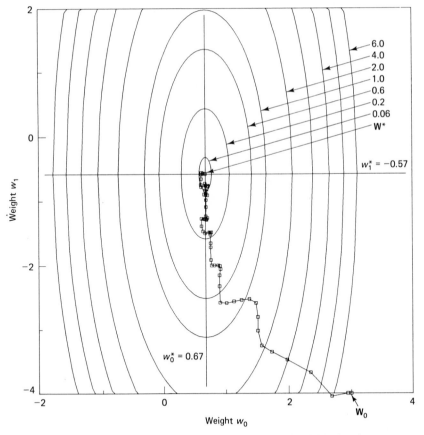

Figure 5.25   Error contours and LMS weight track for Fig. 5.24, showing convergence from $\mathbf{W}_0 = (3, -4)^T$ to near $\mathbf{W}^*$. The adaptive gain parameter was $u = 0.03$ and $k = 0, 1, \ldots, 60$ as in Fig. 5.18. MSE values from Eq. (5.117) are shown as error contours, for comparison with Fig. 5.10.

The weight track in Fig. 5.25 is not computed with a converged lattice coefficient, κ_0^*. The lattice is also allowed to converge from $\kappa_0 = 0$ at $k = 0$ to produce a more realistic example. If κ_0^* had been used instead, convergence to $\mathbf{W}^*$ would have been faster than in Fig. 5.25.

Convergence plots are shown in Fig. 5.26 for κ_0, w_0, and w_1. The adaptive gain was relatively high ($u = 0.1$) for κ_0, thus allowing the lattice to converge rapidly and produce near-orthogonal inputs to the adaptive filter. Note that while κ_0 is converging, and also after κ_0 is converged but misadjusted, the weights w_0 and w_1 are converging to their optimal values. Thus the dynamic behavior of the adaptive lattice combined with the FIR adaptive filter, although rather complex to analyze, appears to be satisfactory, and in this particular example better than that of the adaptive filter without the lattice.

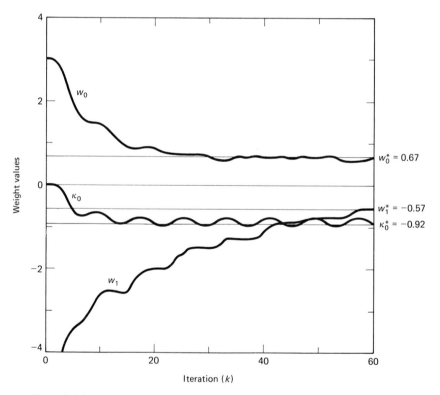

Figure 5.26   Convergence of the lattice weight κ_0 and the adaptive filter weights w_0 and w_1 in Fig. 5.24. Initial values at $k = 0$ are $\kappa_0 = 0$, $w_0 = 3$, and $w_1 = -4$. Convergence parameters (u) are 0.1 for the lattice and 0.03 for the adaptive filter.

REFERENCES

1. B. Widrow and M. E. Hoff, Jr., "Adaptive Switching Circuits," *1960 IRE WESCON Convention Record,* Part 4, pp. 96–104, 1960.

2. B. Widrow et al., "Adaptive Noise Cancelling: Principles and Applications," *Proc. IEEE,* Vol. 63, pp. 1692–1716, Dec. 1975.

3. B. Widrow et al., "Adaptive Antenna Systems," *Proc. IEEE,* Vol. 55, pp. 2143–2159, Dec. 1967.

4. B. Widrow, "Adaptive Filters," in *Aspects of Network and System Theory,* R. E. Kalman and N. DeClaris, Eds., Holt, Rinehart, and Winston, New York, 1970, p. 563.

5. B. Widrow and S. D. Stearns, *Adaptive Signal Processing,* Prentice-Hall, Englewood Cliffs, NJ, 1985.

6. B. S. Atal and S. L. Hanauer, "Speech Analysis and Synthesis by Linear Prediction of the Speech Wave," *J. Acoustical Soc. Amer.,* Vol. 50, pp. 637–755, 1971.

7. J. Makhoul, "Spectral Analysis of Speech by Linear Prediction," *IEEE Trans. Audio and Electroacoustics,* Vol. 21, pp. 140–148, June 1973.

8. L. J. Griffiths, "Rapid Measurement of Digital Instantaneous Frequency," *IEEE Trans. Acoustics, Speech, and Signal Processing,* Vol. 23, pp. 207–222, April 1975.

9. G. A. Clark and P. W. Rodgers, "Adaptive Prediction Applied to Seismic Event Detection," *Proc. IEEE,* Vol. 69, pp. 1166–1168, Sept. 1981.

10. W. P. Dove and A. V. Oppenheim, "Event Location Using Recursive Least-Squares Signal Processing," *Proc. ICASSP-80,* pp. 848–850, April 1980.

11. S. D. Stearns and L. J. Vortman, "Seismic Event Detection Using Adaptive Predictors," *Proc. ICASSP-81,* pp. 1058–1061, March 1981.

12. J. R. Zeidler, E. H. Satorius, D. M. Chabries, and H. T. Wexler, "Adaptive Enhancement of Multiple Sinusoids in Uncorrelated Noise," *IEEE Trans. Acoustics, Speech, and Signal Processing,* Vol. 26, pp. 240–254, June 1978.

13. G. F. Franklin and J. D. Powell, *Digital Control of Dynamic Systems,* Addison-Wesley, Reading, MA, 1980.

14. A. I. Landau, *Adaptive Control: The Model Reference Approach,* Dekker, 1979.

15. B. Widrow, P. F. Titchener, and R. P. Gooch, "Adaptive Design of Digital Filters," *Proc. ICASSP-81,* pp. 243–246, March 1981.

16. D. H. Youn, N. Ahmed, and G. C. Carter, "Magnitude-Squared Coherence Function Estimation: An Adaptive Approach," *IEEE Trans. Acoustics, Speech, and Signal Processing,* Vol. 31, pp. 137–142, Feb. 1983.

17. B. Widrow, J. M. McCool, and B. Medoff, "Adaptive Control by Inverse Modeling," *Conference Record, 12th Asilomar Conference on Circuits, Systems, and Computers,* pp. 90–94, Nov. 1978.

18. B. Widrow, D. Shur, and S. Shaffer, "On Adaptive Inverse Control," *Conference Record, 15th Asilomar Conference on Circuits, Systems, and Computers,* pp. 185–190, Nov. 1981.

19. J. D. Markel, "Digital Inverse Filtering: A New Tool for Formant Trajectory Estimation," *IEEE Trans. Audio and Electroacoustics,* Vol. 20, pp. 129–137, June 1972.

20. A. Gersho, "Adaptive Equalization of Highly Dispersive Channels for Data Transmission," *Bell System Tech. J.,* pp. 55–70, Jan. 1969.

21. R. W. Lucky, "Techniques for Adaptive Equalization of Digital Communication Systems," *Bell System Tech. J., pp.* 255–286, Feb. 1966.

22. L. J. Griffiths, F. R. Smolka, and L. D. Trembly, "Adaptive Deconvolution: A New Technique for Processing Time-Varying Seismic Data," *Geophysics,* June 1977.

23. R. A. Monzingo and T. W. Miller, *Introduction to Adaptive Arrays,* Wiley, New York, 1980.

24. C. F. N. Cowan and P. M. Grant, Eds., *Adaptive Filters,* Prentice-Hall, Englewood Cliffs, NJ, 1985.

25. B. Widrow, J. M. McCool, M. G. Larimore, and C. R. Johnson, Jr., "Stationary and Nonstationary Learning Characteristics of the LMS Adaptive Filter," *Proc. IEEE,* Vol. 64, pp. 1151–1162, Aug. 1976.

26. D. M. Etter and S. D. Stearns, "Adaptive Estimation of Time Delays in Sampled Data Systems," *IEEE Trans. Acoustics, Speech, and Signal Processing,* Vol. 29, pp. 582–587, June 1981.

27. S. A. White, "A Nonlinear Digital Adaptive Filter," *Conference Record, 14th Asilomar Conference on Circuits, Systems and Computers,* pp. 350–354, Nov. 1980.

28. A. V. Oppenheim and R. W. Schafer, *Digital Signal Processing,* Prentice-Hall, Englewood Cliffs, NJ, 1975, Chapter 8.

29. N. Wiener, *Extrapolation, Interpolation and Smoothing of Stationary Time Series with Engineering Applications,* Wiley, New York, 1949.

30. R. A. David, *IIR Adaptive Algorithms Based on Gradient Search Techniques,* Ph.D. Thesis, Stanford University, Stanford, CA, Aug. 1981.

31. P. L. Feintuch, "An Adaptive Recursive LMS Filter," *Proc. IEEE,* p. 1622, Nov. 1976.

32. M. G. Larimore, J. R. Treichler, and C. R. Johnson, Jr., "SHARF: An Algorithm for Adapting IIR Digital Filters," *IEEE Trans. Acoustics, Speech, and Signal Processing,* Vol. 28, pp. 428–440, Aug. 1980.

33. S. A. White, "An Adaptive Recursive Digital Filter," *Conference Record, 9th Asilomar Conference on Circuits, Systems, and Computers,* pp. 21–25, Nov. 1975.

34. N. Levinson, "The Wiener RMS Error Criterion in Filter Design and Prediction," *J. Math. and Physics,* Vol. 25, pp. 261–278, 1946.

35. S. Zohar, "Toeplitz Matrix Inversion: The Algorithm of W. F. Trench," *J. Association for Computing Machinery,* Vol. 16, pp. 592–601, Oct. 1969.

36. B. Widrow and J. M. McCool, "A Comparison of Adaptive Algorithms Based on the Methods of Steepest Descent and Random Search," *IEEE Trans. Antennas and Propagation,* Vol. 24, p. 615, Sept. 1976.

37. D. M. Etter and M. M. Masakawa, "A Comparison of Algorithms for Adaptive Estimation of the Time Delay Between Sampled Signals," *Proc. ICASSP-81,* p. 1253, March 1981.

38. G. B. Thomas, Jr., *Calculus and Analytic Geometry,* 4th ed., Addison-Wesley, Reading, MA, 1968, Section 10.3.

39. D. G. Luenberger, *Introduction to Linear and Nonlinear Programming,* Addison-Wesley, Reading, MA, 1973, Section 7.7.

40. N. Ahmed, D. R. Hummels, M. L. Uhl, and D. L. Soldan, "A Short-Term Sequential Regression Algorithm," *IEEE Trans. Acoustics, Speech, and Signal Processing,* Vol. 27, pp. 453–457, Oct. 1979.

41. N. J. Bershad and P. L. Feintuch, "Analysis of the Frequency Domain Adaptive Filter," *Proc. IEEE,* Vol. 67, pp. 1658–1659, Dec. 1979.

42. G. A. Clark, S. K. Mitra, and S. R. Parker, "Block Implementation of Adaptive Digital Filters," *IEEE Trans. Circuits and Systems,* Vol. 28, pp. 584–592, June 1981; *IEEE Trans. Acoustics, Speech, and Signal Processing,* Joint Special Issue on Adaptive Signal Processing, Vol. 29, pp. 744–752, June 1981.

43. G. A. Clark, S. R. Parker, and S. K. Mitra, "A Unified Approach to Time- and Frequency-Domain Realization of FIR Adaptive Digital Filters," *IEEE Trans. Acoustics, Speech, and Signal Processing,* Vol. 31, pp. 1073–1083, Oct. 1983.

44. M. J. Dentino, J. McCool, and B. Widrow, "Adaptive Filtering in the Frequency Domain," *Proc. IEEE,* Vol. 66, pp. 1658–1659, Dec. 1978.

45. E. R. Ferrara, "Fast Implementation of LMS Adaptive Filters," *IEEE Trans. Acoustics, Speech, and Signal Processsing,* Vol. 28, pp. 474–475, Aug. 1980.

46. F. A. Reed and P. L. Feintuch, "A Comparison of LMS Adaptive Cancellers Implemented in the Frequency Domain and the Time Domain," *IEEE Trans. Circuits and Systems,* Vol. 28, pp. 610–615, June 1981; *IEEE Trans. Acoustics, Speech, and Signal Processing,* Joint Special Issue on Adaptive Signal Processing, Vol. 29, pp. 770–775, June 1981.

47. F. Itakura and S. Saito, "Digital Filtering Techniques for Speech Analysis and Synthesis," *Proc. 7th International Congress on Acoustics,* pp. 261–264, 1971.

48. J. Makhoul, "A Class of All-Zero Lattice Digital Filters: Properties and Applications," *IEEE Trans. Acoustics, Speech, and Signal Processing,* Vol. 26, pp. 304–314, Aug. 1978.

49. J. L. Makhoul and L. K. Cosell, "Adaptive Lattice Analysis of Speech," *IEEE Trans. Acoustics, Speech, and Signal Processing,* Vol. 29, pp. 654–659, June 1981.

50. L. J. Griffiths, "A Continuously Adaptive Filter Implemented as a Lattice Structure," *Proc. ICASSP-77,* pp. 683–686, May 1977.

51. E. I. Jury, "A Note on the Reciprocal Zeros of a Real Polynomial with Respect to the Unit Circle," *IEEE Trans. Circuit Theory,* Vol. 11, p. 292, June 1964.

52. D. T. Lee, M. Morf, and B. Friedlander, "Recursive Least Squares Ladder Estimation Algorithms," *IEEE Trans. Circuits and Systems,* Vol. 28, pp. 467–481, June 1981.

53. J. Makhoul, "Stable and Efficient Lattice Methods for Linear Prediction," *IEEE Trans. Acoustics, Speech, and Signal Processing,* Vol. 25, pp. 423–428, Oct. 1977.

54. L. J. Griffiths, "An Adaptive Lattice Structure for Noise-Cancelling Applications," *Proc. ICASSP-78,* pp. 87–90, April 1978.

55. B. Widrow and E. Walach, "On the Statistical Efficiency of the LMS Algorithm with Nonstationary Inputs," *IEEE Trans. Information Theory,* Vol. 30, pp. 211–221, March 1984.

56. A. H. Gray, Jr., and J. D. Markel, "Digital Lattice and Ladder Filter Synthesis," *IEEE Trans. Audio and Electroacoustics,* Vol. 21, pp. 491–500, Dec. 1973.

57. T. Kailath, ed., "Special Issue on System Identification and Time Series Analysis," *IEEE Trans. Automatic Control,* Dec. 1974.

58. J. Makhoul and R. Viswanathan, "Adaptive Lattice Methods for Linear Prediction," *Proc. ICASSP-78,* pp. 83–86, May 1978.

59. R. S. Medaugh and L. J. Griffiths, "A Comparison of Two Fast Linear Predictors," *Proc. ICASSP-81,* pp. 293–296, April 1981.

60. S. K. Mitra and R. J. Sherwood, "Digital Ladder Networks," *IEEE Trans. Audio and Electroacoustics,* Vol. 21, pp. 30–36, Feb. 1973.

Special IEEE Issues on Adaptive Signal Processing

Proc. IEEE, Special Issue on Adaptive Systems, Vol. 64, Aug. 1976.

IEEE Trans. Acoustics, Speech, and Signal Processing, Joint Special Issue on Adaptive Signal Processing, Vol. 29, June 1981.

IEEE Trans. Antennas and Propagation, Special Issue on Active and Adaptive Antennas, Vol. 12, March 1964.

IEEE Trans. Antennas and Propagation, Special Issue on Adaptive Antennas, Vol. 24, Sept. 1976.

IEEE Trans. Circuits and Systems, Joint Special Issue on Adaptive Signal Processing, Vol. 28, June 1981.

IEEE Trans. Information Theory, Special Issue on Adaptive Filtering, Vol. 30, March 1984.

IEEE Trans. Circuits and Systems, Special Issue on Adaptive Systems and Applications, Vol. 34, in press.

ACKNOWLEDGMENTS

The author wishes to thank the following people who criticized and helped develop this chapter: Ruth A. David, Glenn R. Elliott, Delores M. Etter, Donald Hush, Kevin Keisner, John Shynk, Otis M. Solomon, and Bernard Widrow.

6

Short-Time Fourier Transform

S. Hamid Nawab
Boston University

Thomas F. Quatieri
Lincoln Laboratory

6.0 INTRODUCTION

In this chapter, we examine the short-time Fourier transform (STFT), which has played a significant role in digital signal processing, including speech, music, and sonar applications. It has been found that in such applications it is advantageous to combine traditional time-domain and frequency-domain concepts into a single framework. For example, the physical phenomenon of Doppler shift in signals from moving sources is generally characterized as a change in center frequency over time. If this center frequency is defined in terms of the Fourier transform, we encounter the problem that there is only one Fourier transform for the *entire* signal. It is thus impossible to characterize a change in the center frequency over time. In contrast, the short-time Fourier transform consists of a separate Fourier transform for each instant in time. In particular, we associate with each instant the Fourier transform of the signal in the neighborhood of that instant. To illustrate this, we consider the Fourier transform plots in temporal order in Fig. 6.1. These plots were obtained from successive 2-s intervals of an acoustic recording of a moving helicopter. Each plot contains a gap whenever the Fourier transform magnitude exceeds a threshold. By following such gaps from plot to plot, we can track the frequencies of highest energy. In particular,

S. Hamid Nawab is with the Electrical, Computer, and Systems Engineering Department, Boston University, Boston, MA 02215. Thomas F. Quatieri is with Lincoln Laboratory, Massachusetts Institute of Technology, Lexington, MA 02173. This work was sponsored by the Department of the Air Force. The views expressed are those of the authors and do not reflect the official policy or position of the U.S. government.

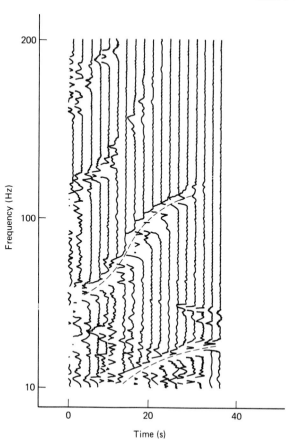

Figure 6.1 Consecutive Fourier transform magnitudes of an acoustic recording of a moving helicopter. Each plot contains a gap whenever the Fourier transform magnitude exceeds a threshold. By following such gaps from plot to plot, we can track the frequencies of highest energy. The frequency tracks of the two highest-energy harmonics are indicated by dashed lines.

we can clearly see the Doppler shift in the various harmonics of the underlying periodic signal. As we will see later in this chapter, the STFT is also useful for characterizing other time-dependent frequency changes in applications such as sonar, music, and speech processing. In fact, it has even been suggested that the human ear performs this type of time-frequency analysis on speech.

The rigorous development of the STFT originally took place in the context of analog signals through the works of Fano [1] and Schroeder and Atal [2]. This work was motivated by previous experimental work for measuring time-dependent spectra with analog devices such as the sound spectrograph [3]. The time-dependent spectrum at the output of such a device is generally displayed in a form known as a spectrogram, an example of which is illustrated in Fig. 6.2 for a segment of speech. In this two-dimensional display, the horizontal axis is time and the vertical axis denotes frequency. The gray level indicates the spectral magnitude, with the darkest regions corresponding to the highest energy. The real-time constraint for such analysis means that the transform at any time should depend only on the past values of the signal. Fano thus defined and developed formal properties of a transform that at any instant weighted the past values of the signal with a decaying exponential and took the squared magnitude of the Fourier transform of the result. Schroeder and Atal extended the concept by using arbitrary weighting functions instead of the exponential. The

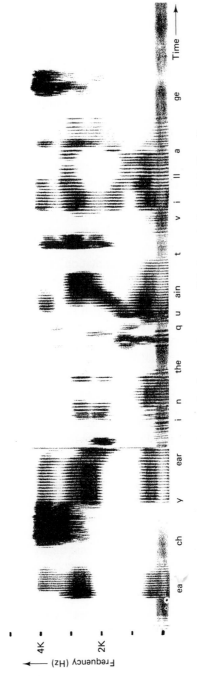

Figure 6.2 The spectrogram for the speech utterance "Each year in the quaint village."

resulting time-frequency function, as we will see, is the magnitude squared of the STFT.

For application in digital signal processing it is necessary to extend the STFT framework to discrete-time signals. An early example of such an extension was the digital spectrogram [4]. This was a digital system for generating speech spectrograms of the type produced by analog devices such as the sound spectrograph. Just as the analog spectrogram can be related to the analog STFT, the digital spectrogram can be related to the concept of a digital or discrete-time STFT. Since in this chapter we are primarily interested in the discrete-time case, we will assume that the STFT corresponds to a discrete-time signal unless stated otherwise. The underlying idea for the discrete-time STFT is once again to take a separate (discrete-time) Fourier transform in the neighborhood of each time sample. These Fourier transforms can be displayed for analysis as in the case of the digital spectrogram. Alternatively, these Fourier transforms can be individually processed and then recombined to form a new processed signal. This enables the signal processing to be adapted to the individual spectral characteristics of each short-time region. Examples of such adaptive processing for the purpose of speech coding and time-scale modification, beamforming in sonar, and image enhancement will be discussed at the end of the chapter.

For practical implementation, each Fourier transform in the STFT has to be replaced by the discrete Fourier transform (DFT). The resulting STFT is discrete in *both* time and frequency and thus is suitable for digital implementation. We call this the *discrete STFT* to distinguish it from the *discrete-time STFT*, which is continuous in frequency. These two transforms, their properties, interrelationships, and applications are the primary focus of this chapter. The discrete-time STFT is particularly useful as a conceptual and analytical tool, while the discrete STFT helps us understand the specific computational details of the algorithms based on the STFT.

This chapter can roughly be divided into two parts, the first four sections dealing with fundamental concepts and issues and the remaining sections dealing with the extension and application of the basic ideas. We begin in Section 6.1 with a formal introduction to the discrete-time and the discrete STFT of a sequence. In particular, we emphasize the similarities and differences in the various properties of the two transforms. In Section 6.2 we illustrate the issues involved in choosing an appropriate framework for computing the STFT in any given situation. We then consider in Section 6.3 the problem of obtaining a sequence back from its STFT. While this is straightforward for the discrete-time STFT, a number of important STFT concepts are introduced for explaining the more complicated area of sequence recovery or synthesis from the discrete STFT. The basic theory part of the chapter is essentially concluded in Section 6.4, which deals with concepts that have been developed for treating the magnitude of the STFT as a transform in its own right. In Section 6.5, we consider the very important practical problem of estimating a signal from a processed STFT that does not satisfy the definitional constraints of the STFT. This area has led to many practical applications of the STFT. However, the STFT is not the only transform to have been considered for time-dependent frequency analysis. In Section 6.6, we consider the relationship of the STFT to some of the other transforms that have been proposed for time-frequency analysis. Finally, in Section 6.7, we illustrate the role the STFT has played in application areas such as speech processing, sensor array processing, and image processing.

6.1 SHORT-TIME FOURIER TRANSFORM OF A SEQUENCE

In this section, we define the STFT representation for a sequence and show how this representation is related to the time- and frequency-domain properties of the original sequence. A major theme used throughout this section is that the representation of the STFT of a sequence is analogous to the Fourier transform representation of a sequence. This analogy is extensively used for deriving STFT properties, including the existence of inverse relations for obtaining a sequence back from its STFT.

6.1.1 Fourier Transform View

The STFT is presented in this section as an extension to the basic Fourier transform definitions for a sequence. In particular, we introduce the *discrete-time STFT* and the *discrete STFT* as counterparts to the discrete-time Fourier transform and the discrete Fourier transform, respectively.

The discrete-time STFT is related to the discrete-time Fourier transform, which is given by

$$X(\omega) = \sum_{n=-\infty}^{\infty} x(n)e^{-j\omega n} \tag{6.1}$$

where ω is a continuous variable denoting frequency. The discrete-time STFT of $x(n)$ is a set of such discrete-time Fourier transforms corresponding to different time sections of $x(n)$. The time section for time n_0 is obtained by multiplying $x(n)$ with a shifted sequence $w(n_0 - n)$. The expression for the discrete-time STFT at time n_0 is therefore given by

$$X(n_0, \omega) = \sum_{n=-\infty}^{\infty} x(n)w(n_0 - n)e^{-j\omega n} \tag{6.2}$$

where $w(n)$ is referred to as the *analysis window* or sometimes as the *analysis filter* for reasons that will become clear later in this chapter. The sequence $f_{n_0}(n) = x(n)w(n_0 - n)$ is generally called a *short-time section* of $x(n)$ at time n_0. This sequence is obtained by time-reversing the analysis window $w(n)$, shifting the result by n_0 points, and multiplying it with $x(n)$. This series of operations is illustrated in Fig. 6.3. Once we have the short-time section for time n_0, we can take its Fourier transform to obtain the frequency function $X(n_0, \omega)$ with n_0 fixed. To obtain $X(n_0 + 1, \omega)$, we slide the time-reversed analysis window one point from its previous position, multiply it with $x(n)$, and take the Fourier transform of the resulting short-time section. Continuing this way, we generate a set of discrete-time Fourier transforms that together constitute the discrete-time STFT. We obtain the mathematical representation for the STFT by replacing the fixed n_0 of Eq. (6.2) by the variable n. To avoid confusion, we rename the variable of summation in Eq. (6.2) as m. We thus obtain the STFT definition:

$$X(n, \omega) = \sum_{m=-\infty}^{\infty} x(m)w(n - m)e^{-j\omega m} \tag{6.3}$$

The analysis window is generally considered to be part of the specification of the

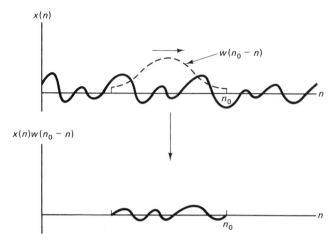

Figure 6.3 The series of operations required to compute a short-time section.

STFT. Since a short-time section of $x(n)$ is the product $x(n)w(n_0 - n)$, it is clear that changing the analysis window will generally change all the short-time sections and therefore the STFT. Typically, the analysis window is selected to have a much shorter duration than the signal $x(n)$ for which the STFT is computed. For example, Fig. 6.4 gives an illustration of an analysis window that is commonly used in speech applications. It is a 256-point window known as a *Hamming window*. In contrast, if $x(n)$ is obtained from a speech sentence lasting 3 s and sampled at 10 kHz, $x(n)$ is a 30,000-point sequence. The shorter duration of the analysis window is what constitutes the *short-time* nature of the STFT.

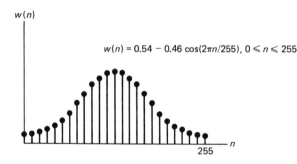

Figure 6.4 A 256-point Hamming window.

For digital processing, we use the discrete STFT, which is related to the discrete-time STFT in the same manner as the DFT is related to the discrete-time Fourier transform. Recall that the DFT $X(k)$ of a finite-duration sequence $x(n)$ is obtained by sampling the discrete-time Fourier transform over one period. That is,

$$X(k) = X(\omega)\big|_{\omega=2\pi k/N} R_N(k) \tag{6.4}$$

where N is the frequency sampling factor and $R_N(k)$ is an N-point rectangular sequence given by

$$R_N(k) = u(k) - u(k - N) \tag{6.5}$$

In analogy, the discrete STFT is obtained from the discrete-time STFT through the following relation:

$$X(n, k) = X(n, \omega)\big|_{\omega=2\pi k/N} R_N(k) \tag{6.6}$$

where we have sampled the discrete-time STFT with a frequency sampling interval of $2\pi/N$ to obtain the discrete STFT. Substituting Eq. (6.3) into Eq. (6.6), we obtain the following relation between the discrete STFT and its corresponding sequence $x(n)$:

$$X(n, k) = \sum_{m=-\infty}^{\infty} x(m)w(n - m)e^{-j2\pi km/N} R_N(k) \tag{6.7}$$

In many applications, the time variation (the n dimension) of $X(n, k)$ is decimated by a temporal decimation factor L to yield the function $X(nL, k)$.

Just as the discrete-time STFT can be viewed as a set of Fourier transforms of the short-time sections $f_n(m)$, the discrete STFT in Eq. (6.7) is easily seen to be a set of DFTs of the short-time sections $f_n(m)$. When the time dimension of the discrete STFT is decimated, the corresponding short-time sections $f_{nL}(m)$ are a subset of $f_n(m)$ obtained by incrementing n by multiples of L. This notion is illustrated in Fig. 6.5.

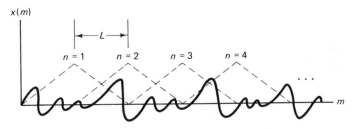

Figure 6.5   The analysis window positions used in computing $X(nL, k)$.

6.1.2 Filtering View

The STFT can also be viewed as the output of a filtering operation where the analysis window $w(n)$ plays the role of the filter impulse response; hence the alternative name *analysis filter* for $w(n)$. For the filtering view of the STFT, we fix the value of ω at ω_0 and rewrite Eq. (6.3) as

$$X(n, \omega_0) = \sum_{m=-\infty}^{\infty} [x(m)e^{-j\omega_0 m}]w(n - m) \tag{6.8}$$

We then recognize from the form of Eq. (6.8) that it represents the convolution of the sequence $x(n)e^{-j\omega_0 n}$ with the sequence $w(n)$. We rewrite Eq. (6.8) as

$$X(n, \omega_0) = [x(n)e^{-j\omega_0 n}] * w(n) \tag{6.9}$$

where $*$ denotes convolution. Furthermore, the product $x(n)e^{-j\omega_0 n}$ can be interpreted as the modulation of $x(n)$ up to frequency ω_0. Thus, $X(n, \omega_0)$ for each ω_0 is a sequence in n that is the output of the process illustrated in Fig. 6.6. The signal $x(n)$ is modulated

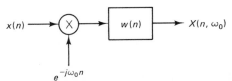

Figure 6.6   Filtering view of STFT analysis at frequency ω_0. Complex exponential modulation is followed by a lowpass filter.

with $e^{-j\omega_0 n}$ and the result passed through a filter whose impulse response is the analysis window, $w(n)$.

A slight variation on the filtering and modulation view of the STFT is obtained by manipulating Eq. (6.9) into the following form:

$$X(n, \omega_0) = e^{-j\omega_0 n}[x(n) * w(n)e^{+j\omega_0 n}] \tag{6.10}$$

In this case, the sequence $x(n)$ is first passed through the same filter as in the previous case except for a linear phase factor. The filter output is then modulated by $e^{-j\omega_0 n}$. This view of the time variation of the STFT for a fixed frequency is illustrated in the block diagram of Fig. 6.7.

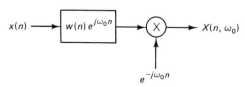

Figure 6.7 Alternative filtering view of STFT analysis at frequency ω_0. A bandpass filter is followed by complex exponential modulation.

The discrete STFT of Eq. (6.7) can also be interpreted from the filtering viewpoint. In particular, having a finite number of frequencies allows us to view the discrete STFT as the output of the filter bank shown in Fig. 6.8. Note that each filter is acting as a bandpass filter centered around its selected frequency. Thus the discrete STFT can be viewed as a collection of sequences, each corresponding to the frequency components of $x(n)$ falling within a particular frequency band.

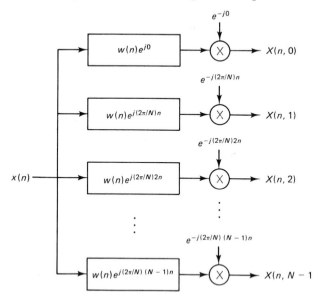

Figure 6.8 The discrete STFT as the output of a filter bank consisting of bandpass filters.

6.1.3 Properties

In this section we develop a number of properties of the discrete and the discrete-time STFT. These properties provide insight into the characteristics of these transforms for various classes of signals and certain commonly used signal manipulations (e.g., time shifting). The sampling relationship between the discrete-time STFT, $X(n, \omega)$, and the discrete STFT, $X(n, k)$, makes it convenient to first investigate the properties of $X(n, \omega)$ alone and then to see how they are affected by the sampling. We will see that most of the properties of the discrete-time STFT have similar counterparts for the discrete STFT. A number of properties of the discrete-time STFT are easily obtained by using the Fourier transform viewpoint, while a different set of properties is most convenient to derive from the filtering viewpoint.

We begin with a set of properties based on the interpretation of the STFT as a collection of Fourier transforms corresponding to the short-time sections of the sequence. In particular, we view $X(n, \omega)$ as a Fourier transform for each fixed n. Therefore, the frequency function $X(n, \omega)$ for each n has all the general properties of a Fourier transform. Table 6.1 lists a number of these properties.

TABLE 6.1   PROPERTIES OF THE DISCRETE-TIME STFT BASED ON THE FOURIER TRANSFORM INTERPRETATION

Property 1:	$X(n, \omega) = X(n, \omega + 2\pi)$
Property 2:	$x(n)$ real $\longleftrightarrow X(n, \omega) = X^*(n, -\omega)$
Property 3:	$x(n)$ real $\longleftrightarrow |X(n, \omega)| = |X(n, -\omega)|$
Property 4:	$x(n)$ real $\longleftrightarrow \arg[X(n, \omega)] = -\arg[X(n, -\omega)]$
Property 5:	$x(n - n_0) \longleftrightarrow e^{-j\omega n_0} X(n - n_0, \omega)$

The first property follows from the periodic nature of the discrete-time Fourier transform, whereas the next three properties follow from the conjugate-symmetric property of the Fourier transform of real sequences. The fifth property is analogous to the Fourier transform property that a time shift in a sequence leads to a linear phase factor in the frequency domain. A shift by n_0 in the original time sequence also means that we obtain the same short-time sections as before except that each short-time section has also been shifted in time by n_0. For this reason there is also a time shift indicated in the STFT of property 5.

The properties of the discrete-time STFT given in Table 6.1 can be extended to the discrete STFT using the relationship in Eq. (6.6) where $2\pi/N$ is the sampling interval in frequency. The resulting properties of the discrete STFT are shown in Table 6.2.

The first property emphasizes the aperiodic nature of the discrete STFT in the frequency dimension. The next three symmetry properties are natural counterparts to the properties of the discrete-time STFT. The shifting property is also a straightforward extension of the corresponding property of the discrete-time STFT. It should

TABLE 6.2   PROPERTIES OF THE DISCRETE STFT

Property 1:	$X(n, k)$ is zero outside $0 \leq k < N$
Property 2:	$x(n)$ real $\longleftrightarrow X(n, k) = X^*(n, N - k)$ for $0 < k < N$
Property 3:	$x(n)$ real $\longleftrightarrow |X(n, k)| = |X(n, N - k)|$ for $0 < k < N$
Property 4:	$x(n)$ real $\longleftrightarrow \arg[X(n, k)] = -\arg[X(n, N - k)]$ for $0 < k < N$
Property 5:	$x(n - n_0) \longleftrightarrow e^{-j(2\pi n_0 k/N)} X(n - n_0, k)$

be noted that if the discrete STFT is decimated in time by a factor L, the first four properties still hold. However, for the fifth property, when the shift is not an integer multiple of L, there is no general relationship between the discrete STFTs of $x(n)$ and $x(n - n_0)$. This happens because the short-time sections corresponding to $X(nL, k)$ cannot be expressed as shifted versions of the short-time sections corresponding to the discrete STFT of $x(n - n_0)$.

As with the Fourier transform interpretation, the filtering view also allows us to easily deduce a number of STFT properties. In particular, we view $X(n, \omega)$ as a filter output for each fixed frequency. Therefore, the time variation of $X(n, \omega)$ for each ω has all the general properties of a filtered sequence. Table 6.3 lists a number of these properties for the discrete-time STFT.

TABLE 6.3   PROPERTIES OF DISCRETE-TIME STFT BASED ON THE FILTERING INTERPRETATION

Property 1:	$X(n, 0) = x(n) * w(n)$
Property 2:	$x(n)$ length N, $w(n)$ length M, $X(n, \omega)$ length $N + M - 1$ along n
Property 3:	Bandwidth of sequence $X(n, \omega_0) \le$ Bandwidth of $w(n)$
Property 4:	The sequence $X(n, \omega_0)$ has spectrum centered at the origin
Property 5:	$x(n)$ causal, $w(n)$ causal, $X(n, \omega)$ causal in time

---

The first property is obtained by substituting $\omega_0 = 0$ in Eq. (6.9). In the second property, we make use of a standard result for the length of a sequence obtained through the convolution of any two sequences of lengths N and M. For the third property, we note that $X(n, \omega)$ as a function of n is the output of a filter whose bandwidth is the bandwidth of the analysis window. The fourth property follows from the modulation step used in obtaining $X(n, \omega_0)$. The fifth property is a standard result on the convolution of causal sequences. The STFT properties from the filtering viewpoint remain the same for the discrete STFT since, for a fixed frequency, the time variation of the discrete STFT is the same as the time variation of the discrete-time STFT at that frequency.

We have seen in this section that a number of STFT properties can be derived from either the Fourier transform viewpoint or the filtering viewpoint. There are of course many other properties of the STFT, but in this section we have concentrated on the ones typically encountered in various application areas.

6.1.4 Invertibility

In this section we consider the problem of obtaining a sequence back from its discrete or discrete-time STFT. Whereas the discrete-time STFT is always invertible in this sense, the discrete STFT requires certain constraints on its sampling rate for invertibility.

The invertibility of the discrete-time STFT is best seen in analogy with the discrete-time inverse Fourier transform. In the Fourier transform interpretation, the discrete-time STFT is viewed for each value of n as a function of frequency obtained by taking the Fourier transform of the short-time section $f_n(m) = x(m)w(n - m)$. It follows that if for each n we take the inverse Fourier transform of the corresponding function of frequency, then we obtain the sequence $f_n(m)$. If we evaluate this short-time section at $m = n$ we obtain the value $x(n)w(0)$. Assuming that $w(0)$ is nonzero, we can divide by $w(0)$ to recover $x(n)$. The process of taking the inverse Fourier transform of $X(n, \omega)$ for a specific n and then dividing by $w(0)$ is represented by the following relation:

$$x(n) = \left[\frac{1}{2\pi w(0)}\right] \int_{2\pi} X(n, \omega) e^{j\omega n} \, d\omega \tag{6.11}$$

This equation represents a *synthesis equation* for the discrete-time STFT. In fact, there are numerous synthesis equations that map $X(n, \omega)$ uniquely back to $x(n)$. We will discuss these in Section 6.4.

In contrast to the discrete-time STFT, the discrete STFT $X(n, k)$ is not always invertible. For example, consider the case when $w(n)$ is bandlimited with bandwidth B. Figure 6.9 shows the filter regions used to obtain $X(n, k)$ for the case when the sampling interval $2\pi/N$ is greater than B. Note that in this case there are frequency components of $x(n)$ that do not pass through any of the filter regions of the discrete STFT. Those frequency components can have arbitrary values and yet the discrete STFT would be the same. Thus in such cases the discrete STFT is not a unique representation of $x(n)$ and therefore cannot be invertible. The invertibility problem is also of interest when the discrete STFT has been decimated in time. For example, consider the case when the analysis window $w(n)$ is nonzero over its finite length N_w. Figure 6.10 shows the case when temporal decimation factor L is greater than N_w. Note that in this case there are samples of $x(n)$ that are not included in any short-time section of the discrete STFT. These samples can have arbitrary values and yet the time-decimated discrete STFT would be the same. Thus in such cases the discrete STFT is not a unique representation of $x(n)$ and therefore cannot be invertible.

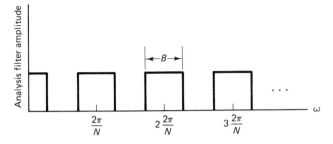

Figure 6.9 Undersampled STFT when the frequency sampling interval $2\pi/N$ is greater than the analysis filter bandwidth B.

By selecting appropriate constraints on the frequency-sampling and time-decimation rates, the discrete STFT becomes invertible. For example, let's consider the case of a finite-length analysis window. We have already seen that in such cases the discrete STFT is not invertible if the temporal decimation factor L is greater than the analysis window length N_w. We will now see that if the temporal decimation factor is less than or equal to the analysis window length, then the discrete STFT is invertible provided we impose constraints on the frequency sampling interval. Suppose that the temporal decimation factor is equal to the analysis window length. The discrete STFT

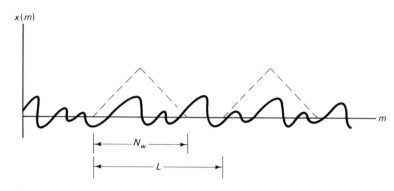

Figure 6.10   Underdecimated STFT when the decimation factor L is larger than the length N_w of the analysis window.

in this case consists of the DFTs of adjacent but nonoverlapping short-time sections, as illustrated in Fig. 6.11. It is clear that the signal $x(n)$ can be obtained only if each of these short-time sections is known for all time. Thus to reconstruct $x(n)$ from its discrete STFT, we must require that each N_w-point short-time section be completely recoverable from its DFT. However, it is well known that the DFT of an N_w-point sequence is invertible provided its frequency sampling interval is less than $2\pi/N_w$. It follows from this discussion that the discrete STFT is invertible for situations where the analysis window is nonzero over its finite-length N_w, the temporal decimation factor L is less than N_w, and the frequency sampling interval $2\pi/N$ is less than $2\pi/N_w$. Even tighter bounds than these can be derived, as we will see in Section 6.4. Furthermore, we will also see that a variety of other assumptions about the analysis window lead to different tradeoffs between time and frequency sampling.

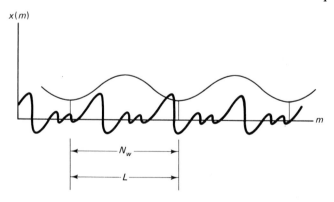

Figure 6.11   The decimation factor L equals the analysis window length N_w. The window positions are adjacent without any overlap or missed regions.

6.2 SHORT-TIME FOURIER ANALYSIS

As we discussed in Section 6.0, the time-varying spectral characteristics of a sequence can be analyzed from its STFT. To carry out such short-time Fourier analysis we must select the analysis window as well as the algorithm for computing the discrete STFT. In Section 6.2.1, we discuss the factors influencing analysis window selection. Some of these factors depend on the algorithm used to compute the discrete STFT. There are two basic approaches to computing the discrete STFT that correspond to the Fourier transform and filtering interpretations of the STFT discussed in Section 6.1. In Sections 6.2.2 and 6.2.3, we examine each of these approaches and the computational issues associated with them.

6.2.1 Selection of Analysis Window

The properties of the STFT are sensitive to the selection of the analysis window. To analyze the output of STFT analysis we should have an understanding of the effects of the analysis window on the characteristics of the STFT. In this section we discuss the nature of the dependence of the STFT properties on the analysis window.

A basic issue in analysis window selection is the compromise required between a long window for frequency resolution and a short window for not allowing the temporal properties of the signal to vary appreciably. To see this, we first recall that the STFT $X(n, \omega)$ is the Fourier transform of the short-time section $f_n(m) = x(m)w(n - m)$. From Fourier transform theory, we know that the Fourier transform of the product of two sequences is given by the convolution of their respective Fourier transforms. With $X(\omega)$ as the Fourier transform of $x(n)$, and $W(-\omega)e^{-j\omega n}$ as the Fourier transform of $w(n - m)$ with respect to the variable m, we can write the STFT as

$$X(n, \omega) = \left(\frac{1}{2\pi}\right)\int_{-\pi}^{\pi} W(\theta)e^{j\theta n}X(\omega + \theta)\, d\theta \tag{6.12}$$

Thus, any frequency variation of the STFT for any fixed time may be interpreted as a smoothed version of the Fourier transform of the underlying signal, as illustrated in Fig. 6.12. Thus, for faithful reproduction of the properties of $X(\omega)$ in $X(n, \omega)$, the function $W(\omega)$ should appear as an impulse with respect to $X(\omega)$. The closer $W(\omega)$ is to an impulse (i.e., narrow bandwidth), $X(n, \omega)$ is said to have better frequency resolution. However, for a given window, frequency resolution varies inversely with the length of the window. Thus, better frequency resolution requires longer analysis windows, whereas the desire for short-time analysis requires shorter analysis windows. An example of this tradeoff is shown in Fig. 6.13. Here we have the Fourier transform of a section of a chirp signal whose frequency is a linear function of time. The aim is to measure the instantaneous frequency at time t_0. This is done using analysis windows of various lengths. Very short windows give low resolution because

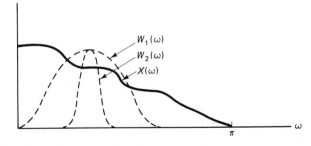

Figure 6.12   The window Fourier transform as a smoothing function for the Fourier transform of the underlying signal. A narrowband and a wideband window are illustrated.

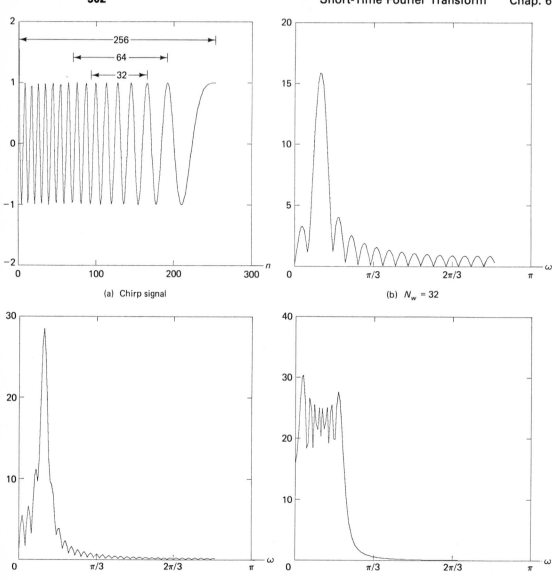

Figure 6.13   Effect of the length of the analysis window on the Fourier transform of a linearly frequency-modulated sinusoid of length 256 samples whose frequency increases from $\pi/4$ to $\pi/8$. The Fourier transform is taken for sections centered around the 128th sample (frequency $3\pi/16$). Transforms are shown for window lengths 32, 64, and 256.

of window spectrum smoothing, whereas very long windows yield a wideband spectrum because of the time-varying frequency of the chirp.

The selection of analysis window is also important when the STFT is to be used for synthesizing the original signal. As we will see in Section 6.3, the STFT synthesis methods impose various constraints on the analysis window to guarantee reconstruction of the original signal.

6.2.2 Fast Fourier Transform Implementation

We have seen that the discrete STFT can be viewed as the DFT of each short-time section corresponding to a different shift in the finite-length analysis window $w(n)$. Thus one approach to discrete STFT analysis is to implement a DFT process that computes equally spaced frequency samples of $X(n, \omega)$ for each possible value of n. The DFT computation is generally on the order of the square of the DFT length. However, if the number of frequency samples N is chosen to be a highly composite number (usually a power of two) then, as we will see in this section, the fast Fourier transform (FFT) algorithm can be used to efficiently carry out the DFT computations in the discrete STFT.

To see how the FFT can be used for short-time Fourier analysis, let us assume that the analysis window $w(n)$ has length N_w. The function we are interested in computing for $k = 0, 1, \ldots, N - 1$, is given by

$$X(n, k) = \sum_{m=-\infty}^{\infty} w(n - m)x(m)e^{-j2\pi mk/N} = \sum_{m=-\infty}^{\infty} f_n(m)e^{-j2\pi mk/N} \qquad (6.13)$$

where $f_n(m)$ are the short-time sections of $x(n)$. First let us consider the case where the desired number of frequency samples N is greater than or equal to N_w. In this case, each short-time section $f_n(m)$, with n fixed, is a finite-length sequence whose length is less than or equal to N. It follows that $X(n, k)$ for a fixed n can be computed as an N-point FFT applied to $f_n(m)$ padded with the appropriate number of zeros.

Now we consider the case where the number of frequency samples N is less than the analysis window length N_w. Since $X(n, \omega)$ is the Fourier transform of $f_n(m)$, uniform frequency sampling of $X(n, \omega)$ with sampling interval $2\pi/N$ corresponds in the time domain to convolution of $f_n(m)$ with a periodic impulse train with period N. The result of this convolution is given by

$$\tilde{f}_n(m) = \sum_{k=-\infty}^{\infty} f_n(m + kN) \qquad (6.14)$$

The sequence $\tilde{f}_n(m)$, known as the time-aliased version of $f_n(m)$, is periodic with period N, and the N-point FFT of one period of $\tilde{f}_n(m)$ yields the desired N frequency samples of the discrete STFT. If $X(nL, k)$ is the desired function, we compute the $\tilde{f}_n(m)$ only for every Lth short-time section. That is,

$$\tilde{f}_{nL}(m) = \sum_{k=-\infty}^{\infty} f_{nL}(m + kN) \qquad (6.15)$$

In summary, the FFT can be used to efficiently compute $X(n, k)$ by computing the time-aliased version of each short-time section and then applying the N-point FFT to each of those sections. Note that if N is greater than or equal to the analysis window length, the computation of the time-aliased version is eliminated.

6.2.3 The Filter Bank Approach

Short-time Fourier analysis can also be carried out through computations suggested by the filtering interpretation of the discrete STFT. In this section we develop the formal details of the filter bank analysis. In particular, we present several alternatives for such implementation. The filter bank analysis is advantageous in applications requiring infinite or long-duration windows or where the STFT is to be computed at a small number of frequencies. These advantages will be discussed in more detail at the end of the section.

To develop the filter bank implementation techniques, we begin with the filtering view of the discrete STFT:

$$X(n, k) = e^{-j\omega_k n}[x(n) * w(n)e^{j\omega_k n}] \qquad (6.16)$$

where $\omega_k = 2\pi k/N$ for $k = 0, 1, \ldots, N - 1$. The idea is to pass the signal $x(n)$ through a bank of bandpass filters, shown in Fig. 6.8, where the output of each filter is the time variation of the STFT at the frequency ω_k. If the output of each filter is decimated in time by a factor L, we obtain the discrete STFT, $X(nL, k)$. Each branch of the filter bank corresponds to a bandpass filter centered around the frequency $\omega_k = 2\pi k/N$.

An alternative implementation of the filter bank analysis replaces the bandpass filters by lowpass filters as in Fig. 6.14. The equation for each branch of the lowpass filter bank is given by

$$X(n, k) = x(n)e^{-j\omega_k n} * w(n) \qquad (6.17)$$

where $\omega_k = 2\pi k/N$ for $k = 0, 1, \ldots, N - 1$. As in the bandpass implementation, we can obtain $X(nL, k)$ by decimating each filter output by the decimation factor L.

Both the bandpass and the lowpass techniques of Figs. 6.8 and 6.14 involve complex exponential modulations. To carry out this complex modulation in terms of real operations, a more detailed specification of the filter bank implementations is needed. For the lowpass implementation, we note that

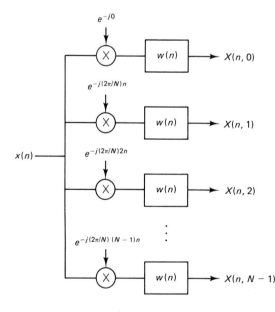

Figure 6.14   The discrete STFT as the output of a filter bank consisting of lowpass filters.

$$X(n, k) = x(n)e^{-j\omega_k n} * w(n)$$

$$= [x(n)\cos(\omega_k n) * w(n)] + j[x(n)\sin(\omega_k n) * w(n)] \qquad (6.18)$$

$$= X_r(n, k) + jX_i(n, k)$$

where $X_r(n, k)$ and $X_i(n, k)$ can be implemented as shown in Fig. 6.15. Similarly, we can derive a real implementation of the bandpass analysis. The computation for this implementation is represented in the block diagram of Fig. 6.16.

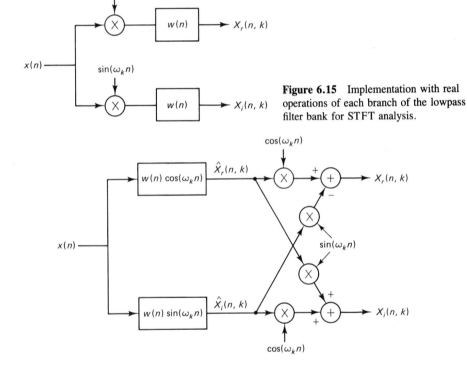

Figure 6.15   Implementation with real operations of each branch of the lowpass filter bank for STFT analysis.

Figure 6.16   Implementation with real operations of each branch of the bandpass filter bank for STFT analysis.

If the bandpass filter bank has the modulation components removed, then the overall structure becomes much simpler and it computes samples of $X(n, \omega)e^{j\omega n}$ rather than the samples of $X(n, \omega)$. Such an output often suffices for STFT analysis. In particular, the linear phase factor does not affect the kinds of spectral characteristics (e.g., bandwidth, LPC parameters) that are generally of interest in short-time Fourier analysis of signals like speech. Another reason, which we will explain further in Section 6.3, is that under certain conditions the outputs of the filter bank in Fig. 6.8 can be simply summed to retrieve the original sequence $x(n)$. In particular, for this purpose it is required that the frequency responses of the filters should add up to give unity across the entire bandwidth of $x(n)$.

The filter bank implementations are particularly useful when the analysis window is infinitely long (e.g., an exponential) or well approximated by an infinitely long window. In such cases the filters can be implemented by recursive (infinite impulse response) techniques. It is noted that the FFT approach to STFT analysis can handle infinite-duration windows only by approximating them with finite-length windows. As an example, consider a window $w(n) = (0.9)^n u(n)$. A recursive implementation of this analysis filter is given by

$$y(n) = 0.9y(n - 1) + x(n) \qquad (6.19)$$

where $x(n)$ and $y(n)$ are the input and output of the filter with impulse response $w(n)$. This requires on the order of two operations per sample of the discrete STFT. On the other hand, using an N-point FFT implementation, where N is the frequency sampling factor, requires Log N operations per sample of the discrete STFT. In our example, the exponential window may be approximated to within 16-bit precision by a 128-point truncation. In this case, the FFT analysis would require Log(128) = 7 operations per sample of the discrete STFT. This compares with 2 operations for the filter bank analysis.

The filter bank approach may also be advantageous when the STFT is to be computed at a relatively small number of frequencies. In some applications, one is interested in the STFT over only a small band of frequencies. For example, with a 256-point analysis window one may desire four consecutive frequencies with resolution $2\pi/256$. In this case the FFT approach would require $N \log N = 256$ log(256) = 2048 operations per unit time. On the other hand, the filter bank approach would use four finite impulse response (FIR) filters of length 256 points each. This means a total of 1024 operations per unit time. Furthermore, if the filters can be suitably approximated by infinite impulse reponse (IIR) filters, the computation may require fewer operations.

6.3 SHORT-TIME FOURIER SYNTHESIS

As discussed in Section 6.0, many digital processing applications of the STFT employ methods for synthesizing a sequence from its discrete STFT. We have already seen in Section 6.1 that such synthesis is not always possible since the discrete STFT is not invertible under all conditions. On the other hand, we have seen that the *discrete-time* STFT is always invertible. A common approach to developing synthesis methods for the discrete STFT has been to start from one of the many *synthesis equations* that express a sequence in terms of its discrete-time STFT. A discretized version of such an equation is then considered as the basis for a candidate synthesis method. Conditions are derived under which such a method can synthesize a sequence from its discrete STFT. The various synthesis methods differ not only with respect to the conditions under which they are valid but also in terms of their computational properties.

In this section we present a number of synthesis methods as well as their underlying theory. In particular, we begin with two classical methods that have been widely used for short-time synthesis. They are the filter bank summation (FBS) method and the overlap-add (OLA) method. There are other useful synthesis methods that are best understood by introducing the concept of a *synthesis filter* for the STFT. This concept and the associated methods are also discussed in this section.

6.3.1 Filter Bank Summation (FBS) Method

In this section we present a traditional short-time synthesis method that is commonly referred to as the filter bank summation (FBS) method. This method is best described in terms of the filtering interpretation of the discrete STFT. In this interpretation, the discrete STFT is considered to be the set of outputs of a bank of filters. In the FBS method, the output of each filter is modulated with a complex exponential and these modulated filter outputs are summed at each instant of time to obtain the corresponding time sample of the original sequence, as shown in Fig. 6.17. For dealing with temporal decimation, the traditional strategy is to perform a temporal interpolation filtering on the discrete STFT to restore the temporal decimation factor to unity. The FBS method is then performed on the interpolated output.

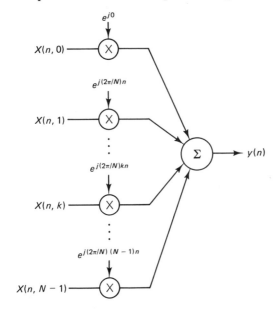

Figure 6.17 Filter bank summation procedure for signal synthesis from the discrete STFT.

The FBS method is motivated by the following relation between a sequence and its discrete-time STFT:

$$x(n) = \left[\frac{1}{2\pi w(0)}\right] \int_{-\pi}^{\pi} X(n, \omega)e^{j\omega n} \, d\omega \qquad (6.20)$$

where without loss of generality we assume that $w(0)$ is nonzero. We derived this equation in Section 6.2 in the context of the invertibility of the discrete-time STFT. The FBS method carries out a discretized version of the operations suggested on the right side of Eq. (6.20). That is, given a discrete STFT $X(n, k)$, the FBS method synthesizes a sequence $y(n)$ satisfying the following equation:

$$y(n) = \left[\frac{1}{Nw(0)}\right] \sum_{k=0}^{N-1} X(n, k)e^{j2\pi nk/N} \qquad (6.21)$$

We are of course interested in the FBS method for those situations where the sequence $y(n)$ in Eq. (6.21) is the same as the sequence $x(n)$ corresponding to the discrete STFT $X(n, k)$. Substituting $X(n, k)$ in Eq. (6.21) for the FBS method, we obtain

$$y(n) = \left[\frac{1}{Nw(0)}\right] \sum_{k=0}^{N-1} \sum_{m=-\infty}^{\infty} x(m)w(n-m)e^{-j2\pi km/N}e^{j2\pi nk/N} \qquad (6.22)$$

Using the linear filtering interpretation of the STFT, this equation reduces to

$$y(n) = \left[\frac{1}{Nw(0)}\right]x(n) * \sum_{k=0}^{N-1} w(n)e^{j2\pi nk/N} \qquad (6.23)$$

Taking $w(n)$ out of the summation and noting that the finite sum over the complex exponentials reduces to an impulse train with period N, we obtain

$$y(n) = \left[\frac{1}{Nw(0)}\right]x(n) * w(n)N \sum_{r=-\infty}^{\infty} \delta(n-rN) \qquad (6.24)$$

In Eq. (6.24) we note that $y(n)$ is obtained by convolving $x(n)$ with a sequence that is the product of the analysis window with a periodic impulse train. It follows that if we desire $y(n) = x(n)$, then the product of $w(n)$ and the periodic impulse train must reduce to $Nw(0)\delta(n)$. That is,

$$w(n)N \sum_{r=-\infty}^{\infty} \delta(n-rN) = Nw(0)\delta(n) \qquad (6.25)$$

This will clearly be satisfied for any causal analysis window whose length N_w is less than the number of analysis filters N. That is, *any* finite-length analysis window can be used in the FBS method provided the length of the window is less than the frequency sampling factor N. We can even have $N < N_w$ provided $w(n)$ is chosen such that every Nth sample is zero. That is,

$$w(rN) = 0 \qquad \text{for} \qquad r = -1, 1, -2, 2, -3, 3, \ldots \qquad (6.26)$$

as illustrated in Fig. 6.18.

Equation (6.25) is known as the FBS constraint because this is the requirement on the analysis window that ensures exact signal synthesis with the FBS method. This constraint is more commonly expressed in the frequency domain. Taking the Fourier transform of both sides of Eq. (6.25), we obtain

$$\sum_{k=0}^{N-1} W(\omega - 2\pi k/N) = Nw(0) \qquad (6.27)$$

This constraint essentially states that the frequency responses of the analysis filters should sum to a constant across the entire bandwidth, as shown in Fig. 6.19. We have already seen that any finite-length analysis window whose length is less than the frequency sampling factor N satisfies this constraint. We conclude that a filter bank with N filters, based on an analysis filter of length less than N, is *always* an allpass system.

As noted before, the FBS method is just one of the many different methods available for synthesizing a sequence from its discrete STFT. Each such method imposes its own set of constraints on the analysis window as for the FBS method in

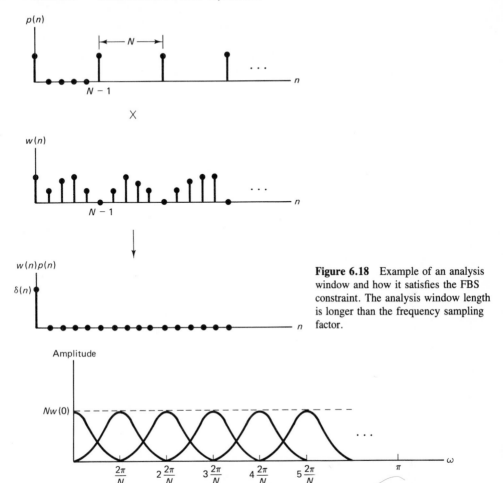

Figure 6.18 Example of an analysis window and how it satisfies the FBS constraint. The analysis window length is longer than the frequency sampling factor.

Figure 6.19 The FBS constraint visualized in the frequency domain.

Eq. (6.27). Additionally, each method has its own computational characteristics. The FBS method for decimation factor $L = 1$ and an analysis filter of length N_w requires on the order of N_w operations (complex additions and multiplications) per sample of $x(n)$. On the other hand, if the decimation factor $L \gg 1$, the FBS method must be extended to include interpolation of each analysis filter output. This leads to a computational requirement of order N_w^2 operations per sample of $x(n)$.

6.3.2 Overlap-Add (OLA) Method

We now consider another classical method, the overlap-add (OLA) method for short-time synthesis. Just as the FBS method was motivated from the filtering view of the STFT, the OLA method is motivated from the Fourier transform view of the STFT.

The simplest method obtainable from the Fourier transform view is in fact not the OLA method. It is instead a method known as the inverse discrete Fourier transform (IDFT) method. In this method, for each fixed time, we take the inverse DFT of the corresponding frequency function and divide the result by the analysis window. This method is generally not favored in practical applications because the slightest perturbation in the STFT can result in a synthesized signal very different from the original. For example, consider the case where the STFT is multiplied by a linear phase factor of the form $e^{j\omega n_0}$ with n_0 unknown. Then the IDFT for each fixed time results in a shifted version of the corresponding short-time section. Since the shift n_0 is unknown, dividing by the analysis window without taking the shift into account introduces a distortion in the resulting synthesized signal. In contrast, the OLA method, which we describe next, results in a shifted version of the original signal without distortion.

The OLA method is also best described in terms of the Fourier transform view. In the OLA method, we take the inverse DFT for each fixed time in the discrete STFT. However, instead of dividing out the analysis window from each of the resulting short-time sections, we perform an overlap-and-add operation between the short-time sections. This method works provided the analysis window is designed such that the overlap-and-add operation effectively eliminates the analysis window from the synthesized sequence.

The OLA method is motivated by the following relation between a sequence and its discrete-time STFT:

$$x(n) = \left[\frac{1}{2\pi W(0)}\right] \int_{-\pi}^{\pi} \sum_{r=-\infty}^{\infty} X(r, \omega)e^{j\omega n}\, d\omega \tag{6.28a}$$

where

$$W(0) = \sum_{n=-\infty}^{\infty} w(n) \tag{6.28b}$$

To derive this synthesis equation, we note that if we take the Fourier transform of both sides of Eq. (6.9) with respect to the variable n and evaluate the result at frequency zero, we obtain

$$\tilde{X}(\phi, \omega)\big|_{\phi=0} = X(\omega)W(0) \tag{6.29}$$

where we denote the Fourier transform in n of $X(n, \omega)$ by $\tilde{X}(\phi, \omega)$. But because a Fourier transform evaluated at zero frequency is equal to the sum of all the samples of the time-domain sequence, we have

$$\tilde{X}(\phi, \omega)\big|_{\phi=0} = \sum_{r=-\infty}^{\infty} X(r, \omega) \tag{6.30}$$

Now, dividing Eq. (6.29) by $W(0)$ and substituting for $\tilde{X}(\phi, \omega)\big|_{\phi=0}$ the expression in Eq. (6.30), we obtain

$$X(\omega) = \left[\frac{1}{W(0)}\right] \sum_{r=-\infty}^{\infty} X(r, \omega) \tag{6.31}$$

Taking the inverse Fourier transform of Eq. (6.31), which maps ω to n, we obtain the desired relation in Eq. (6.28).

The OLA method carries out a discretized version of the operations suggested on the right of Eq. (6.28a). That is, given a discrete STFT $X(n, k)$, the OLA method synthesizes a sequence $y(n)$ satisfying the following equation:

$$y(n) = \left[\frac{1}{W(0)}\right] \sum_{p=-\infty}^{\infty} \left[\frac{1}{N} \sum_{k=0}^{N-1} X(p, k)e^{j2\pi kn/N}\right] \tag{6.32}$$

The term inside the rectangular brackets on the right is an inverse DFT that for each p gives

$$y_p(n) = x(n)w(p - n) \tag{6.33}$$

The expression for $y(n)$ therefore becomes

$$y(n) = \left[\frac{1}{W(0)}\right] \sum_{p=-\infty}^{\infty} x(n)w(p - n) \tag{6.34}$$

which then reduces to

$$y(n) = x(n)\left[\frac{1}{W(0)}\right] \sum_{p=-\infty}^{\infty} w(p - n) \tag{6.35}$$

In Eq. (6.35) we note that $y(n)$ will be equal to $x(n)$ provided

$$\sum_{p=-\infty}^{\infty} w(p - n) = W(0) \tag{6.36}$$

Furthermore, if the discrete STFT has been decimated in time by a factor L, it can be similarly shown that if the analysis window satisfies

$$\sum_{p=-\infty}^{\infty} w(pL - n) = \frac{W(0)}{L} \tag{6.37}$$

then $x(n)$ can be synthesized using the following relation:

$$x(n) = \left[\frac{L}{W(0)}\right] \sum_{p=-\infty}^{\infty} \left[\frac{1}{N} \sum_{k=0}^{N-1} X(pL, k)e^{j2\pi kn/N}\right] \tag{6.38}$$

Equation (6.37) is the general constraint imposed by the OLA method on the analysis window. It requires the sum of all the analysis windows (obtained by sliding $w(n)$ with L-point increments at a time) to add up to a constant, as shown in Fig. 6.20. It is

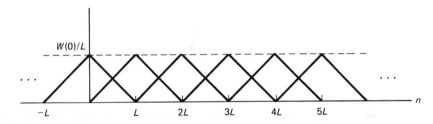

Figure 6.20   The OLA constraint visualized in the time domain.

interesting to note the duality between this constraint and the FBS constraint in Eq. (6.27) where the shifted versions of the Fourier transform of the analysis window were required to add up to a constant. For the FBS method we also saw that all finite-length analysis windows whose length is less than the frequency sampling factor satisfy the FBS constraint. Analogously, we can show that the OLA constraint in Eq. (6.37) is satisfied by all finite-bandwidth analysis windows whose maximum frequency is less than $2\pi/L$, where L is the temporal decimation factor.

To see how finite-bandwidth analysis windows satisfy the OLA constraint, suppose that the analysis window has maximum frequency ω_c, and consequently bandwidth $2\omega_c$, as illustrated in Fig. 6.21. If we let $\hat{w}(p)$ denote the sequence $w(pL - n)$, then the OLA constraint in Eq. (6.37) can be rewritten as

$$\hat{W}(0) = W(0)/L \tag{6.39}$$

where $\hat{W}(\omega)$ denotes the Fourier transform of $\hat{w}(p)$. Noting that $\hat{w}(p)$ is a sampled version of $w(p - n)$, we can easily show that

$$\hat{W}(\omega) = \frac{1}{L} \sum_{k=-\infty}^{\infty} e^{-j(\omega - k2\pi/L)n} W(\omega - k2\pi/L) \tag{6.40}$$

If there is no overlap between $W(\omega)$ and $W(\omega - k2\pi/L)$ at $\omega = 0$, then Eq. (6.40) gives the OLA constraint expressed in Eq. (6.39). To have no overlap at $\omega = 0$ between $W(\omega)$ and $W(\omega - k2\pi/L)$ it is easy to see that we must have $\omega_c < 2\pi/L$, where ω_c is the maximum frequency in $W(\omega)$. We conclude that any finite-bandwidth window whose maximum frequency is less than $2\pi/L$ will satisfy the OLA constraint in Eq. (6.37).

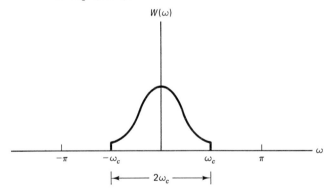

Figure 6.21 Example of a Fourier transform of an analysis window with bandwidth $2\omega_c$.

The OLA method has computational properties that differ from those of the FBS method. Recall that with $L = 1$, the FBS method with an analysis filter of length N_w requires on the order of N_w operations per sample of $x(n)$. In contrast, the OLA method for $L = 1$ requires on the order of $N_w \, \text{Log} \, N_w$ operations per sample of $x(n)$. For $L \gg 1$, the FBS method requires on the order of N_w^2 operations per sample of $x(n)$, while the OLA method requires on the order of $\text{Log} \, N_w$ operations per sample of $x(n)$. That is, for a large decimation rate, the OLA method is significantly more efficient than the FBS method. The difference is that FBS has to carry out an interpolation to reduce the decimation factor L to unity while the OLA method uses the decimated STFT directly.

6.3.3 Generalized Filter Bank Summation

In this section we present the generalized FBS method, which allows a wider range of analysis windows than the ordinary FBS method discussed in Section 6.3.1. It also includes the OLA method as a special case. The generalized FBS method consists of a two-dimensional smoothing of the discrete STFT followed by the application of the ordinary FBS method.

The generalized FBS method is motivated by the following relation between a sequence and its discrete-time STFT:

$$x(n) = \left[\frac{1}{2\pi w(0)}\right] \int_{-\pi}^{\pi} Y(n, \omega)e^{j\omega n} \, d\omega \qquad (6.41a)$$

where

$$Y(n, \omega) = \frac{1}{2\pi} \int_{-\pi}^{\pi} \sum_{r=-\infty}^{\infty} F(n - r, \omega - \phi)X(r, \phi) \, d\phi \qquad (6.41b)$$

and $F(n, \omega)$ is an arbitrary smoothing function. It should be noted that Eq. (6.41a) has the same form as the motivating equation (6.20) for the FBS method except that $X(n, \omega)$ has been smoothed to form $Y(n, \omega)$. If we combine Eqs. (6.41a) and (6.41b) and denote by $f(n, m)$ the inverse Fourier transform (mapping ω to m) of $F(n, \omega)$, then we obtain

$$x(n) = \left(\frac{1}{2\pi}\right) \int_{-\pi}^{\pi} \left[\sum_{r=-\infty}^{\infty} f(n, n - r)X(r, \omega)\right] e^{j\omega n} \, d\omega \qquad (6.42)$$

where the function $f(n, m)$ is generally referred to as the *synthesis filter*. Note that the motivating equation for ordinary FBS can be obtained by setting the synthesis filter to be a nonsmoothing filter, i.e., $f(n, m) = \delta(m)$ in Eq. (6.41). It can also be easily shown that $f(n, m) = 1/W(0)$ yields the synthesis equation in Eq. (6.28) for the OLA method. In fact, it can be shown [5] that any $f(n, m)$ that satisfies

$$\sum_{m=-\infty}^{\infty} f(n, -m)w(m) = 1 \qquad (6.43)$$

will make Eq. (6.41) a valid synthesis equation.

The generalized FBS method carries out a discretized version of the operations suggested by Eq. (6.42). That is, given a discrete STFT that has been decimated in time by a factor L, the generalized FBS method synthesizes a sequence $y(n)$ satisfying the following equation:

$$y(n) = \frac{L}{N} \sum_{i=-\infty}^{\infty} \sum_{k=0}^{N-1} f(n, n - iL)X(iL, k)e^{-j2\pi nk/N} \qquad (6.44)$$

Although Eq. (6.44) contains the time-varying synthesis filter $f(n, n - iL)$, in this section we consider only the time-invariant case $f(n, n - iL) = f(n - iL)$ because of its practical importance. For example, we have already seen that the synthesis equations for both the FBS and OLA methods can be described through time-invariant synthesis filters. For a time-invariant synthesis filter, Eq. (6.44) reduces to

$$y(n) = \frac{L}{N} \sum_{i=-\infty}^{\infty} \sum_{k=0}^{N-1} f(n - iL)X(iL, k)e^{-j2\pi nk/N} \qquad (6.45)$$

This equation holds only when certain constraints are satisfied by the analysis and synthesis filters as well as the temporal decimation and frequency sampling factors. To find these constraints, we first rewrite Eq. (6.45) as

$$y(n) = L \sum_{i=-\infty}^{\infty} f(n - iL)\frac{1}{N} \sum_{k=0}^{N-1} X(iL, k)e^{-j2\pi nk/N} \qquad (6.46)$$

and note that the second summation represents an inverse DFT that evaluates as follows:

$$\frac{1}{N} \sum_{k=0}^{N-1} X(iL, k)e^{-j2\pi nk/N} = \sum_{p=-\infty}^{\infty} x(n - pN)w(iL - n + pN) \qquad (6.47)$$

Equation (6.46) therefore becomes

$$y(n) = L \sum_{p=-\infty}^{\infty} \left[\sum_{i=-\infty}^{\infty} f(n - iL)w(iL - n + pN) \right] x(n - pN) \qquad (6.48)$$

For the right side of Eq. (6.48) to reduce to $x(n)$, we clearly require the term in rectangular brackets to reduce to $\delta(p)$. This condition can then be stated as

$$L \sum_{i=-\infty}^{\infty} f(n - iL)w(iL - n + pN) = \delta(p) \qquad \text{for all } n \qquad (6.49)$$

which is a discretized version of the constraint of Eq. (6.43) for a time-invariant synthesis filter. This constraint essentially states that the product of $f(n)$ and $w(-n)$ should be such that samples that are an integer multiple of L points apart from each other add up to unity. Furthermore, N should be such that whenever $w(-n)$ is shifted by an integer multiple of N, then its product with $f(n)$ should be such that samples that are an integer multiple of L points apart from each other add up to zero. An easy way to satisfy the latter constraint is to set N to be larger than or equal to the analysis window length and then to restrict the synthesis filter length to be the analysis window length.

The constraint in Eq. (6.49) for generalized FBS reduces to the OLA and FBS constraints when the appropriate synthesis window is used for each method. For example, let us consider how the constraint in Eq. (6.49) reduces to the constraint in Eq. (6.27) that we derived for the ordinary FBS method. The synthesis filter for the ordinary FBS method is $f(n, m) = \delta(m) = f(m)$ regardless of the analysis filter. Substituting $f(n) = \delta(n)$ in Eq. (6.48), with $L = 1$, we obtain the synthesis equation

$$y(n) = \sum_{p=-\infty}^{\infty} \left[\sum_{i=-\infty}^{\infty} \delta(n - i)w(i - n + pN) \right] x(n - pN) \qquad (6.50)$$

and the corresponding constraint of Eq. (6.49) becomes

$$\sum_{i=-\infty}^{\infty} \delta(n - i)w(i - n + pN) = \delta(p) \qquad (6.51)$$

Note that the impulse function on the left side of Eq. (6.51) is nonzero only if $n = i$. It is easy to see that in this case, the left-hand side reduces to $w(pN)$. Therefore, Eq. (6.51) becomes

$$w(pN) = \delta(p) \tag{6.52}$$

This can be rewritten as

$$w(m) \sum_{p=-\infty}^{\infty} \delta(m - pN) = w(0)\delta(m) \tag{6.53}$$

which in the frequency domain is given by

$$\sum_{k=0}^{N-1} W(\omega - 2\pi k/N) = Nw(0) \tag{6.54}$$

This is the same as the condition in Eq. (6.27) for the FBS method. It should also be noted that if $L \neq 1$, and if we let $h(n)$ be the interpolating filter preceding the FBS method, then the synthesis equation (6.45) with the synthesis filter $f(n) = h(n)$ turns out to be equivalent to the entire process of interpolation and filter bank summation.

Let us now reconsider in detail the general FBS method as an implementation of Eq. (6.45). Assuming a finite-length interpolation filter and interchanging the summations, Eq. (6.45) can be rewritten as

$$y(n) = \frac{L}{N} \sum_{k=0}^{N-1} \left[\sum_{i=N_\ell}^{N_u} f(n - iL)X(iL, k) \right] e^{-j2\pi nk/N} \tag{6.55}$$

where N_ℓ and N_u are determined by the region of support of the interpolation filter. The term within the rectangular brackets

$$\text{rec}(n, k) = \sum_{i=N_\ell}^{N_u} f(n - iL)X(iL, k) \tag{6.56}$$

represents the temporal convolution of $f(n)$ and $X(nL, k)$. The computation in Eq. (6.56) has the same general form as the interpolation required before application of the FBS method to the discrete STFT with temporal decimation factor $L > 1$. The remainder of Eq. (6.55) (i.e., outside of $\text{rec}(n, k)$) is identical to the FBS method; it requires modulation followed by summation.

The amount of computation required by the generalized FBS method is linearly proportional to the frequency sampling rate, and it is inversely proportional to the temporal decimation factor. Let N be the frequency sampling factor for the STFT, L the temporal decimation factor for the STFT, and K the length of the interpolation filter. For each sample of $x(n)$ we require N interpolations, followed by a modulation and summation of the resulting N samples. Each interpolation requires on the order of K/L operations. Thus, the N interpolations require a total of NK/L operations. The modulation and summation involves on the order of N operations. Thus the total number of operations required per sample of $x(n)$ is on the order of $(NK/L) + N = N(K + L)/L$ operations. In the next section we will consider another synthesis method that also utilizes the synthesis filter but offers a different computational tradeoff.

6.3.4 Helical Interpolation Method

We will now investigate the helical interpolation synthesis method for implementing
Eq. (6.45), which is a discretized version of the STFT synthesis equation (6.42) with
a time-invariant synthesis filter. In the previous section we saw how Eq. (6.45) could
be implemented as a generalized FBS method. In particular, Eq. (6.45) was interpreted as an interpolation with a synthesis filter followed by a modulation and summation of the interpolated output. In contrast, for the helical interpolation method we
interpret Eq. (6.45) as a series of inverse discrete Fourier transforms whose outputs
are combined through a sophisticated interpolation procedure generally known as
helical interpolation. More specifically, we rewrite Eq. (6.45) as follows:

$$y(n) = L \sum_{i=N_\ell}^{N_u} f(n - iL) \left[\frac{1}{N} \sum_{k=0}^{N-1} X(iL, k) e^{-j2\pi nk/N} \right] \qquad (6.57)$$

where we have assumed that the synthesis filter is finite length. The term inside the
rectangular brackets in Eq. (6.57) is an inverse DFT that can be efficiently computed
using the FFT. Denoting the result of this inverse DFT by $f_{iL}(n)$, we can rewrite Eq.
(6.57) as

$$y(n) = L \sum_{i=N_\ell}^{N_u} f(n - iL) \tilde{f}_{iL}(n) \qquad (6.58)$$

where $\tilde{f}_{iL}(n)$ represents a periodic extension of the short-time section $f_{iL}(n)$. This
equation represents the computation of $y(n)$ for some particular n as the interpolation
of the nth sample of each of the $(N_u - N_\ell)/L$ functions $\tilde{f}_{iL}(n)$ centered about n. Since
the functions $\tilde{f}_{iL}(n)$ are periodic with period N, it follows that $\tilde{f}_{iL}(n) =
\tilde{f}_{iL}(n \text{ modulo } N)$. Thus Eq. (6.58) can be rewritten as

$$y(n) = L \sum_{i=N_\ell}^{N_u} f(n - iL) \tilde{f}_{iL}(n \text{ modulo } N) \qquad (6.59)$$

The fact that n modulo N when plotted as a function of n resembles a helical function
leads to the name *helical interpolation* for the operation suggested by Eq. (6.59).

    The helical interpolation method for short-time synthesis is computationally
more efficient than the generalized FBS method for realistic situations. If the temporal
decimation factor is L, the frequency sampling factor is N, and the synthesis filter is
K points long, then we have already seen that the generalized FBS method requires
on the order of $N(K + L)/L$ operations per sample of $x(n)$. On the other hand, with
the helical interpolation method we require an FFT with $N \log N$ operations for every
L samples of $x(n)$ and K/L operations per sample of $x(n)$ for the helical interpolation.
Thus, the helical interpolation method requires a total of $(K + N \log N)/L$ operations per sample of $x(n)$. If we hold K and L constant, we see that generalized FBS
computation is on the order of N operations per sample while helical interpolation is
on the order of $N \log N$ operations per sample. However, for realistic values of K,
L, and N, helical interpolation is always preferred. To illustrate this, Fig. 6.22 presents
a plot of the number of operations per sample of $x(n)$ as a function of the frequency
sampling factor N. For this figure we have selected typical values for L and K as 32

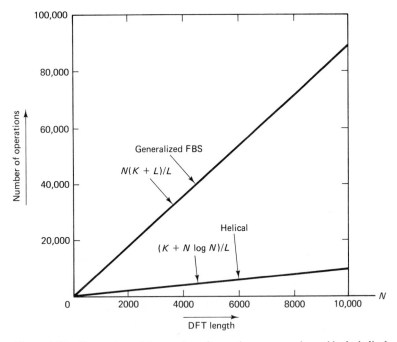

Figure 6.22   Comparison of the number of operations per sample used in the helical interpolation method and the generalized FBS method. It is assumed that the interpolation filter is 256 points long and the decimation factor $L = 32$.

and 256. From the figure we see that helical interpolation is always better than the generalized FBS method for realistic values of N.

6.3.5 Weighted Overlap-Add

We will now investigate a third synthesis method for implementing Eq. (6.45). This method is known as the weighted overlap-add (WOLA) synthesis method and it uses the same number of arithmetic operations as the helical interpolation method. However, the order in which the operations are performed is different for the two methods and thus leads to different space-time tradeoffs.

As with the helical interpolation method, the WOLA method is best viewed in terms of Eq. (6.55) when it is rewritten in the form of Eq. (6.57). The WOLA method, like the helical interpolation method, uses the FFT to compute the inverse DFT represented in Eq. (6.57) by the term in rectangular brackets. If the inverse DFT of $X(iL, k)$ is denoted by $\tilde{f}_{iL}(n)$, then Eq. (6.57) can be written as in Eq. (6.59). The difference between the helical interpolation method and the WOLA method lies in the algorithm used to compute the right side of Eq. (6.59). In the helical interpolation method, the algorithm was based on an interpretation of that equation as a filtering operation with respect to the index i. That is, for each n, $f(n - iL)$ as a sequence in i is multiplied with $\tilde{f}_{iL}(n)$ and the result is summed over all i to obtain $x(n)$ for the particular n. If the interpolation filter is K points long, this means that to compute the result for a particular n, we need access to approximately K/L short-time sections. This implies a storage requirement of $N_w K/L$ samples, where N_w is the length of each short-time section. In contrast, the WOLA method is based on interpreting Eq. (6.59) as an overlap-add procedure with respect to the index n. In particular, for each i, $f(n - iL)$ as a sequence in n is used to weight the sequence $\tilde{f}_{iL}(n)$. The weighted sequences that overlap each other are then added to obtain $x(n)$ for all n. Since the weighted sequences are at most K points long, the overlap-add procedure requires at most $K - 1$ points from the past to be used for the computation of any particular sample. This implies a storage requirement of K samples in the WOLA method in contrast to $N_w K/L$ samples needed to be stored for the helical interpolation method. On the other hand, since both the methods are implementing the same equation (6.59), they both perform the same number of operations.

It is interesting to note the parallels between the WOLA method and the OLA method discussed previously. The two methods are essentially the same except that the WOLA method includes the extra step of multiplying each short-time section with the synthesis window before performing the overlap-add procedure. The extra step is required because unlike the OLA method, the WOLA method does not place restrictions on the analysis window. In fact, the multiplication in the WOLA method of each short-time section with the synthesis window can be viewed as a transformation on the analysis window in order to make it satisfy the OLA requirement of Eq. (6.37).

6.4 SHORT-TIME FOURIER TRANSFORM MAGNITUDE (STFTM) OF A SEQUENCE

In speech applications, the spectrogram that can be related to the magnitude of the STFT has played a major role. For example, visual cues in the spectrogram have been related to parameters important for speech perception. In fact, it has been suggested [6] that the human ear extracts perceptual information strictly from a spectrogram-like representation of speech. In particular, this representation is a nonnegative time-frequency function. On the other hand, the STFT is generally a complex-valued function and for applications such as time-scale modification of speech, estimation of the phase of this function is computationally difficult [7]. In contrast, a number of techniques have been developed where the processed signal is estimated from only the STFT magnitude (STFTM), thus circumventing the phase estimation problem.

In this section we introduce the magnitude of the STFT as an alternative time-frequency signal representation. We will see in this section that many signals can be represented by the real-valued and nonnegative STFTM. Furthermore, we can develop analysis and synthesis techniques for the STFTM just as we did for the STFT. We will see that while STFTM analysis is similar to STFT analysis, short-time synthesis is very different for the two transforms.

As with the discrete-time STFT, the discrete-time STFTM is symmetric and periodic with period 2π. However, unlike the STFT, the STFTM is a real representation of the signal $x(n)$. The STFTM can be related to another function, the short-time autocorrelation, $r(n, m)$, through the following Fourier relationships:

$$r(n, m) = \frac{1}{2\pi} \int_{-\pi}^{\pi} |X(n, \omega)|^2 e^{j\omega m} \, d\omega \tag{6.60a}$$

$$|X(n, \omega)|^2 = \sum_{m=-\infty}^{\infty} r(n, m)e^{-j\omega m} \tag{6.60b}$$

where $r(n, m)$ is given by the convolution of a short-time section with its time-reversed version:

$$r(n, m) = [x(m)w(m - n)] * [x(-m)w(-m - n)]$$

$$= \sum_{p=-\infty}^{\infty} x(p)w(p - n)x(p - m)w(p - m - n) \tag{6.60c}$$

with $*$ denoting convolution. Generally, the short-time section $x(m)w(m - n)$ cannot be obtained from its short-time autocorrelation function [8]. This relationship between the short-time section, its Fourier transform magnitude, and its autocorrelation is illustrated in Fig. 6.23. However, as we will see shortly, the autocorrelations of short-time sections that have partial overlap in time can be used jointly to solve for the underlying short-time sections. This will enable us under certain conditions to use the STFTM as a unique representation of the underlying signal.

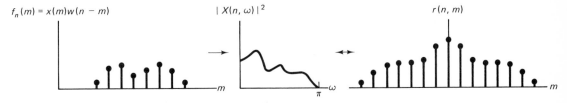

Figure 6.23   Illustration of the noninvertibility of the short-time Fourier transform magnitude for a fixed n. The Fourier transform magnitude is one-to-one with the autocorrelation.

6.4.1 Signal Representation

We will now consider the problem of determining when the discrete-time STFTM can be used to represent a sequence uniquely. That the STFTM is not a unique representation in all cases is easily seen from the simple observation that $x(n)$ and its negative, $-x(n)$, have the same STFTM. We will in fact demonstrate that there are other kinds of situations where the STFTM is not a unique signal representation. We will then proceed to show that by imposing certain mild restrictions on the analysis window and the signal, unique signal representation is indeed possible with the discrete-time STFTM. In particular, if the analysis window is nonzero over its length N_w, then one-sided sequences can be represented uniquely to within a sign factor except for those sequences that have $N_w - 2$ or more consecutive zero samples between two nonzero samples. Furthermore, if we know the sign of just one sample of the sequence to be represented, the STFTM representation becomes completely unique.

To develop insight into the kinds of situations where a sequence cannot be represented uniquely by its discrete-time STFTM, let us consider the case of a sequence $x(n)$ with a gap of zero samples between two nonzero portions. That is, suppose $x(n)$ is the sum of two signals $x_1(n)$ and $x_2(n)$ occupying different regions of the n-axis as depicted in Fig. 6.24(a). Suppose further that the gap of zeros between $x_1(n)$ and $x_2(n)$ is large enough so that there is no analysis window position for which the corresponding short-time section includes nonzero contribution from $x_1(n)$ as well as $x_2(n)$. Clearly, in such a situation, the STFTM of $x(n)$ is the sum of the STFT magnitudes of $x_1(n)$ and $x_2(n)$. However, we have previously observed that a signal and its negative have the same STFTM. It follows that $x(n)$ has the same STFTM as the signals obtained from the differences $x_1(n) - x_2(n)$ and $x_2(n) - x_1(n)$ shown in Figs. 6.24(b) and (c). We conclude that if there is a large enough gap of zero samples, there will be sign ambiguities on either side of the gap. Consequently, any uniqueness conditions must include a restriction on the length of the zero gaps between nonzero portions of the signal.

Let us now see how a one-sided sequence $x(n)$ can be recovered from its discrete-time STFTM when the analysis window is nonzero over its finite duration and $x(n)$ satisfies the appropriate zero-gap restriction. The key to recovering $x(n)$ is the observation that $|X(n, \omega)|$ has additional information about the short-time sections of $x(n)$ besides their spectral magnitudes. This information is contained in the overlap of the analysis window positions. If the short-time section at time n is known, then the signal corresponding to the spectral magnitude of the adjacent section at time $n + 1$ must be *consistent* in the region of overlap with the known short-time section. In particular, if the analysis window were nonzero and of length N_w, then, as illustrated

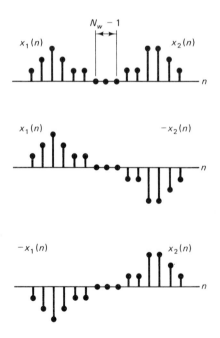

Figure 6.24 Three sequences with the same STFTM.

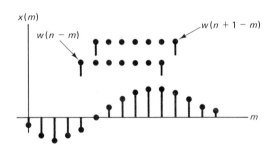

Figure 6.25 Illustration of the consistency required among adjacent short-time sections. Note the samples that are common to the adjacent sections.

in Fig. 6.25, after dividing out the analysis window, the first $N_w - 1$ samples of the segment at time $n + 1$ must equal the last $N_w - 1$ samples of the segment at time n. Therefore, if we could *extrapolate* the last sample of a segment from its first $N_w - 1$ values, we could repeat this process to obtain the entire signal $x(n)$.

To develop the procedure for extrapolating the next sample of a sequence using its STFTM, assume that the sequence $x(n)$ has been obtained up to some time $n - 1$. Thus, as illustrated in Fig. 6.26, the first $N_w - 1$ samples under the analysis window positioned at time n are known. The goal is to compute the sample $x(n)$ from these initial samples and the STFT magnitude $|X(n + 1, \omega)|$ or, equivalently, $r(n, m)$. Note that the last value of the short-time autocorrelation function $r(n, N_w - 1)$ is given by the product of the first and last value of the segment

$$r(n, N_w - 1) = [w(0)x(n)][w(N_w - 1)x(n - (N_w - 1))] \tag{6.61}$$

as illustrated in Fig. 6.27. Therefore $x(n)$ is given by

$$x(n) = \frac{r(n, N_w - 1)}{w(0)w(N_w - 1)x(n - (N_w - 1))} \tag{6.62}$$

If the first value of the short-time section, i.e., $x(n - (N_w - 1))$, happens to equal zero, we must then find the first nonzero value within the section and again use the

$x(n)$

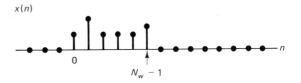

$x(n - N_w + 1)$

$$\sum_{m = -\infty}^{\infty} x(n)x(n - N_w + 1) = x(0)x(N_w - 1)$$

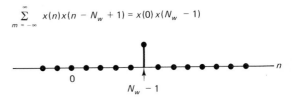

$x(m)$

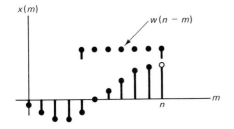

Figure 6.26 Illustration of the samples under $w(n - m)$.

Figure 6.27 Computation of the last nonzero autocorrelation sample (assuming a rectangular analysis window).

product relation given by Eq. (6.62). We can always find such a sample since we have assumed at most $N_w - 2$ consecutive zero samples between any two nonzero samples of $x(n)$.

6.4.2 STFTM Analysis

Like the STFT, the STFTM can be used for analyzing the time-varying spectral characteristics of a sequence. To carry out such STFTM analysis on a digital computer, we need to introduce the *discrete* STFTM. By sampling the frequency dimension of the STFTM, $|X(n, \omega)|$, we obtain the discrete STFTM, which is defined as $|X(n, k)|$, the magnitude of the discrete STFT. In the last section, we saw that under certain conditions, the discrete-time STFTM is a unique signal representation. That theory can be easily extended to the discrete STFTM. In particular, the uniqueness conditions of the previous section relied on using the short-time autocorrelation functions of adjacent short-time sections that are overlapping in time. These autocorrelation functions can be obtained even if the STFTM is sampled in frequency. That is, if the analysis window is N_w points long, then each short-time autocorrelation function is at most $2N_w - 1$ points long and thus can be obtained (without aliasing) from $2N_w - 1$ frequency samples of the STFTM. Therefore, the uniqueness conditions of the discrete-time STFTM extend without change to the discrete STFTM with adequate frequency sampling. To consider the effects of temporal decimation with factor L, we note that adjacent short-time sections now have an overlap of $N_w - L$ instead of $N_w - 1$. The successive extrapolation procedure discussed in the previous section can be extended to this case by requiring the extrapolation of the L last samples of a short-time section, using the first $N_w - L$ samples and the short-time autocorrelation function of that section. This can be accomplished provided the overlap between adjacent short-time sections is greater than $N_w/2$ and there are no zero gaps of length greater than $N_w - 2L$. In addition, to initialize the extrapolation procedure, L initial samples of the underlying sequence have to be known. There are also a variety of other uniqueness conditions that express the tradeoff between time decimation and frequency sampling [9,10].

   Having established the kinds of conditions required for the discrete STFTM to be a valid signal representation, we now discuss the various implementation approaches for the discrete STFTM. Since the STFTM and the STFT are closely related, their analysis implementations are similar. In particular, we can use the FFT and filter bank approaches of STFT analysis for STFTM analysis.

   The FFT implementation of STFT analysis can be extended to STFTM analysis in a straightforward manner. Recall that the FFT implementation of the discrete STFT involved the computation of a time-aliased version of each short-time section followed by an FFT. For STFTM analysis, we can follow the FFT by the magnitude operation applied to each output sample from the FFT.

   The filter bank implementation of STFTM analysis also parallels the STFT implementations. From Section 6.2.3, we know that there are two types of filter bank implementations for the STFT—the lowpass filter and bandpass filter implementations. Each of these implementations has a corresponding implementation for the discrete STFTM. The STFTM lowpass implementation shown in Fig. 6.28 merely

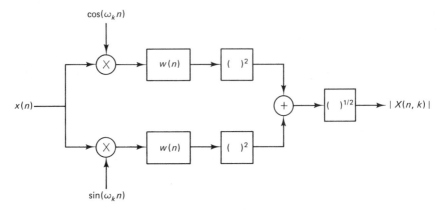

Figure 6.28 Implementation with real operations of each branch of the lowpass filter bank for STFTM analysis.

involves cascading the magnitude operation after the computation of the STFT. However, the bandpass implementation shown in Fig. 6.29 is considerably simplified for the STFTM. This happens because the complex modulation following the bandpass filter is eliminated by the magnitude operation.

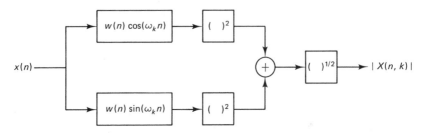

Figure 6.29 Implementation with real operations of each branch of the bandpass filter bank for STFTM analysis.

6.4.3 STFTM Synthesis

A number of methods [9] are available for synthesizing a sequence from its discrete STFTM. These methods are not related in any simple way to the STFT synthesis methods because of the nonlinear mapping from the STFT to the STFTM. In this section, we will discuss only one of these methods as a way of illustrating the basic issues involved.

The synthesis method we will now discuss is based on the *sequential extrapolation approach*, illustrated in Fig. 6.30. For synthesizing a sequence $x(n)$ from $|X(nL, k)|$, we assume that the analysis window $w(n)$ is a known sequence with no zero samples over its finite length. Furthermore, these nonzero samples are in the region $0 \leq n < N_w$. The signal $x(n)$ has no more than $N_w - 2L$ consecutive zeros separating any two nonzero samples. It is also assumed that the first nonzero sample of $x(n)$ falls at $n = 0$. Finally, we assume that the L samples of $x(n)$ for $0 \leq n < L$ are known. The L known samples of $x(n)$ completely determine the short-time section

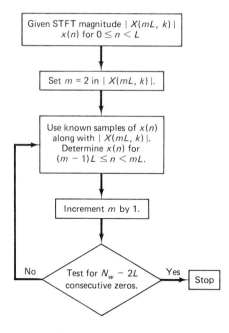

Figure 6.30   The sequential extrapolation approach for STFTM synthesis.

corresponding to $|X(nL, k)|$ for $n = 1$. The short-time section corresponding to $|X(nL, k)|$ for $n = 2$ can then be extrapolated from its DFT magnitude and its known samples in the region of overlap with the previously determined short-time section. This process continues as the complete extrapolation of each new short-time section makes possible the extrapolation of the next overlapping short-time section. The synthesis stops when a short-time section is encountered for which the known samples are not sufficient to complete the extrapolation. For finite-length signals the synthesis stops after all the nonzero short-time sections have been extrapolated.

6.5 SIGNAL ESTIMATION FROM MODIFIED STFT OR STFTM

In many applications it is desired to synthesize a signal from a time-frequency function formed by modifying an STFT or STFTM of a signal we wish to process. Such modifications may arise due to quantization errors in, for example, speech coding or *purposeful* time-varying filtering for signal processing applications such as speech enhancement. An arbitrary function of time and frequency, however, does not necessarily represent the STFT or STFTM of a signal. This is because the definitions of these transforms impose a structure on their time and frequency variations. In particular, because of the overlap between short-time sections, adjacent short-time segments cannot have arbitrary variations. A necessary but not sufficient condition on these variations is that the short-time section corresponding to each time instant must lie within the duration of the corresponding analysis window. For example, the short-time section corresponding to $X(0, \omega)$ is given by $f_0(n) = x(n)w(-n)$ and therefore it must lie within the duration of $w(-n)$. Even if this time-placement constraint is satisfied, a further condition that the STFT or STFTM must satisfy is that adjacent short-time sections should be consistent in their region of overlap. When the STFT or STFTM of a signal is modified, the resulting time-frequency function does not generally satisfy such constraints. In this section we consider various ways of estimating signals from such arbitrary time-frequency functions.

The synthesis methods we discussed in Sections 6.3 and 6.4 were derived with the assumption that the time-frequency functions to which they are applied satisfy the constraints in the definitions of the STFT or STFTM. Given a function that does not satisfy those constraints, the synthesis methods have no theoretical validity for their application. However, under certain conditions, those methods can be shown to yield *reasonable* results in the presence of modifications. For example, in Section 6.5.1 we will illustrate conditions under which the FBS and OLA methods yield intuitively satisfying results when the STFT has been modified with a multiplicative factor. In Section 6.5.2 we discuss a theoretically based approach to signal synthesis from modified STFT. A similar approach is then discussed in Section 6.5.3 for signal estimation from the modified STFTM.

6.5.1 Heuristic Application of STFT Synthesis Methods

Historically, signal estimation from modified STFT has been performed by applying the FBS and OLA synthesis methods of Section 6.3 on time-frequency functions that are not valid STFT functions. For example, if the valid STFT $X(n, \omega)$ is multiplied by a linear phase factor $e^{j\omega n_0}$ the resulting time-frequency function $Y(n, \omega)$ is not a valid STFT. The reason for this is that the time-placement constraint imposed by the STFT definition is violated by the linear phase modification. In particular, if $w(-n)$ falls between $n = N_1$ and $n = N_2$, then the time-placement constraint requires that the inverse Fourier transform of $Y(0, \omega)$ should fall between $n = N_1$ and N_2. However, because of the linear phase factor the inverse Fourier transform of $Y(0, \omega)$ is shifted by n_0. Even though $Y(n, \omega)$ is not a valid STFT, it is desirable that applying a synthesis method to $Y(n, \omega)$ should yield a reasonable result.

Such heuristic application of the synthesis methods has been practically utilized in many signal processing applications. Since the synthesis methods have no theoretical basis for their application in such situations, it is common to analyze the effects that the methods have on the synthesized signal [11].

In this section, we will contrast the FBS and OLA synthesis methods when they are applied to the STFT that has been modified through multiplication with another time-frequency function. For both methods the resulting synthesized signal can be shown to be a time-varying convolution between $x(n)$ and a function $\hat{p}(n, m)$ as illustrated in Fig. 6.31(a). Let us assume that the STFT $X(n, k)$ has been modified to give the function $Y(n, k)$:

$$Y(n, k) = X(n, k)P(n, k) \tag{6.63}$$

where $P(n, k)$ is the modifying function. Also let $2\pi/N$ be the frequency sampling factor for $X(n, k)$ and let $p(n, m)$ for each n be the N-point inverse DFT of $P(n, k)$. For the FBS method it can be easily shown that $\hat{p}(n, m)$ can be obtained by multiplying $p(n, m)$ for each n by the window function $w(m)$, as shown in Fig. 6.31(b). On the other hand, for the OLA method it can be similarly shown that $\hat{p}(n, m)$ can be obtained

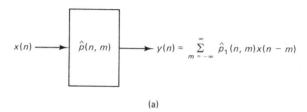

(a)

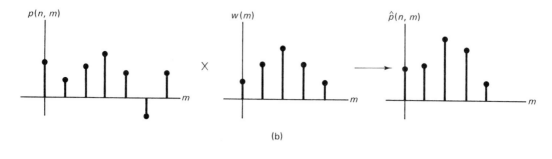

(b)

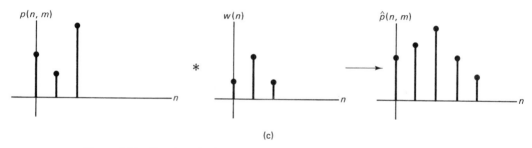

(c)

Figure 6.31 Signal synthesis from a multiplicatively modified STFT using FBS and OLA methods. (a) $\hat{p}(m, n)$ relates the original $x(n)$ to the synthesized signal $y(n)$; (b) and (c) illustration of how $\hat{p}(m, n)$ is obtained for FBS and OLA, respectively.

by convolving $p(n, m)$ for each m by the window function $w(n)$, as illustrated in Fig. 6.31(c). It is interesting to note that if $p(n, m)$ is independent of n, i.e., $p(n, m) = p(m)$, then the FBS method results in a synthesized signal that is the convolution of $x(n)$ with $p(n)w(n)$. On the other hand, the time-invariant case for the OLA method results in a synthesized signal that is the convolution of $x(n)$ with $p(n)$.

In this section we have seen how the effects of applying the FBS and OLA methods to the modified STFT may be analyzed for the case of multiplicative modifications. A similar analysis may also be carried out for situations where a time-frequency function has been added to a valid STFT [11].

6.5.2 Least-Squares Signal Estimation from Modified STFT

Rather than applying the FBS and OLA methods in a brute force manner, we will now consider a different approach that is specifically designed for signal estimation from the modified STFT. In this approach we estimate a signal whose STFT is closest in a least-squares sense to the modified STFT. More specifically, we wish to minimize the mean-square error between the discrete-time STFT $X_e(n, \omega)$ of the signal estimate and the modified discrete-time STFT, which we denote by $Y(n, \omega)$. This optimization results in the following solution for the estimated signal $x_e(n)$:

$$x_e(n) = \frac{\displaystyle\sum_{m=-\infty}^{\infty} w(m - n)f_m(n)}{\displaystyle\sum_{m=-\infty}^{\infty} w^2(m - n)} \tag{6.64}$$

where $f_m(n)$ is the inverse Fourier transform of the frequency variation at time m of the modified STFT $Y(m, \omega)$. Since in practice we have only the discrete function $Y(n, k)$, the short-time sections $f_n(m)$ can be obtained provided the frequency sampling factor N is large enough to avoid aliasing in the short-time sections. The specific distance measure used in the minimization is the squared error between $X_e(n, \omega)$ and $Y(n, \omega)$ integrated over all ω and summed over all n:

$$D[X_e(n,\omega),Y(n,\omega)] = \sum_{m=-\infty}^{\infty} \frac{1}{2\pi} \int_{-\pi}^{\pi} |X_e(m, \omega) - Y(m, \omega)|^2 \, d\omega \tag{6.65}$$

Note that although the distance measure is defined over continuous frequency, the solution for $x_e(n)$ that minimizes the distance measure does not involve continuous frequency. However, it is required that the frequency sampling be large enough so that unaliased versions of the short-time sections are obtained.

The solution in Eq. (6.64) extends in a simple manner to the case involving temporal decimation. Specifically, if L is the temporal decimation factor, then the solution in Eq. (6.64) becomes

$$x_e(n) = \frac{\displaystyle\sum_{m=-\infty}^{\infty} w(mL - n)f_{mL}(n)}{\displaystyle\sum_{m=-\infty}^{\infty} w^2(mL - n)} \tag{6.66}$$

In general, the sum in the denominator of the right side of Eq. (6.66) is a function of n. However, there exist analysis windows $w(n)$ such that the sum in the denominator is independent of n. It should be noted that the sum in the denominator has the same form as the sum in the constraint equation (6.37) for the OLA method except that the analysis window is replaced by its square. That is, any window whose square satisfies the OLA constraint will make the denominator sum in Eq. (6.66) independent of n. If this happens, then Eq. (6.66) can be simply interpreted as an overlap-add operation among the short-time sections corresponding to $Y(n, \omega)$. That is, if the square of the analysis window satisfies the OLA constraint of Eq. (6.37), then the solution for $x_e(n)$ that minimizes Eq. (6.65) is obtained by essentially applying the WOLA synthesis method to the modified STFT. In particular, if the analysis window is rectangular, the WOLA method reduces to the OLA method. We can thus conclude that applying the OLA method with brute force to a modified STFT with a rectangular analysis window will indeed give the solution that minimizes Eq. (6.65).

6.5.3 Least-Squares Signal Estimation from Modified STFTM

The least-squares approach can also be used for signal estimation from the modified STFTM. The resulting method estimates a sequence $x_e(n)$ from a desired time-frequency function $|X_d(n, \omega)|$, which is a modified version of an original STFTM, $|X(n, \omega)|$. The method iteratively reduces the following distance measure between the STFTM $|X_e(n, \omega)|$ of the signal estimate and the modified STFTM $|X_d(n, \omega)|$:

$$D[|X_e(n, \omega)|, |X_d(n, \omega)|] = \sum_{m=-\infty}^{\infty} \frac{1}{2\pi} \int_{-\pi}^{\pi} [|X_e(m, \omega)| - |X_d(m, \omega)|]^2 \, d\omega \qquad (6.67)$$

The solution is found iteratively because as yet no closed-form solution has been discovered for $x_e(n)$ using the distance criterion in Eq. (6.67). The iteration takes place as follows. An arbitrary sequence (usually white noise) is selected as the first estimate $x_e^1(n)$ of $x_e(n)$. We then compute the STFT of $x_e^1(n)$ and modify it by replacing its magnitude by the desired magnitude $|X_d(n, \omega)|$. From the resulting modified STFT, we can obtain a signal estimate using the method based on Eq. (6.64) in the previous section. This process continues iteratively, as shown in Fig. 6.32. In particular, the $(i + 1)$st estimate $x_e^{i+1}(n)$ is first obtained by computing the STFT $X_e^i(n, \omega)$ of $x_e^i(n)$

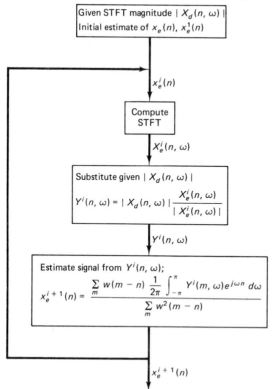

Figure 6.32 Least-squares signal estimation from modified STFTM.

and replacing its magnitude by $|X_d(n, \omega)|$ to obtain $Y^i(n, \omega)$. The signal with the STFT closest to $Y^i(n, \omega)$ is found by using Eq. (6.64). All steps in the iteration can be summarized in the following update equation:

$$x_e^{i+1}(n) = \frac{\displaystyle\sum_{m=-\infty}^{\infty} w(m-n) \frac{1}{2\pi} \int_{-\pi}^{\pi} Y^i(m, \omega) e^{j\omega n} \, d\omega}{\displaystyle\sum_{m=-\infty}^{\infty} w^2(m-n)} \qquad (6.68a)$$

where

$$Y^i(m, \omega) = |X_d(m, \omega)| \frac{X_e^i(m, \omega)}{|X_e^i(m, \omega)|} \qquad (6.68b)$$

It has been shown [12] that this iterative procedure reduces the distance measure of Eq. (6.67) on every iteration. Furthermore, the process converges to one of the critical points, not necessarily the global minimum of that distance measure. Although we restricted the preceding discussion to the discrete-time STFT, these results are easily extendable to the case where the STFT has been decimated in time. Furthermore, even with discrete frequency the method appears to iteratively reduce the distance measure in Eq. (6.67) provided the frequency sampling factor is sufficiently large to avoid aliasing when determining the short-time sections corresponding to $Y^i(n, \omega)$.

6.6 TIME-FREQUENCY DISTRIBUTIONS

Over the years, various frameworks have been proposed [13,14] for capturing the largely intuitive notions of "instantaneous" or "time-varying" spectra. These frameworks generally have as their central component a signal transform (or "distribution") that maps the original signal into a two-dimensional space with one dimension associated with time and the other with frequency. Clearly, the STFT and STFTM share these characteristics. Furthermore, as we illustrate in this section, the STFT and STFTM are closely related to well-known time-frequency distributions such as the complex energy density (CED) [13], the Wigner distribution [14] and the ambiguity function [15].

The discrete-time CED of a sequence $x(n)$ is defined by

$$E_x(n, \omega) = x(n)X^*(\omega)e^{-j\omega n} \qquad (6.69)$$

where * denotes complex conjugation. The CED has a number of properties that are often associated with "instantaneous" spectra. For example, integration along the frequency dimension of the CED gives the instantaneous power $|x(n)|^2$. Conversely, summation along the time dimension of the CED gives the spectral power $|X(\omega)|^2$ at the corresponding frequency in the underlying signal. Furthermore, the CED is invertible to within a sign factor. The basic idea is to let ω be zero in Eq. (6.69). We thus obtain $x(n)X^*(0)$. The magnitude of $X^*(0)$ can be obtained by summing $E_x(n, 0)$ for all n. In summary, $x(n)$ to within a sign factor is represented by

$$y(n) = \frac{E_x(n, 0)}{\sum\limits_{n=-\infty}^{\infty} E_x(n, 0)} \qquad (6.70)$$

The CED is related to the STFTM through a convolution process. In particular, it can be shown that the square of the STFTM is equal to the two-dimensional convolution of the CED of the original signal with the CED of the analysis window used in the STFT. For example, if $E_x(n, \omega)$ is the CED of $x(n)$, $E_w(n, \omega)$ is the CED of $w(n)$, and $|X(n, \omega)|$ is the STFTM of $x(n)$, then

$$|X(n, \omega)|^2 = \sum\limits_{k=-\infty}^{\infty} \int_0^{2\pi} E_x(k, v) E_w(n - k, \omega + v) \, dv \qquad (6.71)$$

To derive this relationship, we observe that the CED of the convolution of two sequences is equal to the convolution (in time) of the CED of each sequence. That is, if $y(n) = x_1(n) * x_2(n)$, then

$$E_y(n, \omega) = \sum\limits_{k=-\infty}^{\infty} E_{x_1}(k, \omega) E_{x_2}(n - k, \omega) \qquad (6.72)$$

Integrating both sides of Eq. (6.72) with respect to frequency, we obtain

$$\int_0^{2\pi} E_y(n, \omega) \, d\omega = \int_0^{2\pi} \sum\limits_{k=-\infty}^{\infty} E_{x_1}(k, \omega) E_{x_2}(n - k, \omega) \, d\omega \qquad (6.73)$$

If we let $x_1(n) = x(n)e^{jvn}$ and $x_2(n) = w(n)$, then the left-hand side of Eq. (6.73) becomes the square of the STFTM of $x(n)$, and Eq. (6.73) can be rewritten as

$$|X(n, v)|^2 = \sum\limits_{k=-\infty}^{\infty} \int_0^{2\pi} E_x(k, \omega) E_w(n - k, \omega + v) \, d\omega \qquad (6.74)$$

where we have used the fact that the CED of $x(n)e^{j\omega n}$ is $E_x(n, \omega - v)$. Interchanging v and ω, we see that Eq. (6.74) becomes identical to Eq. (6.71). We have thus established the convolutional relationship between the CED and the STFTM. Similar convolutional relationships exist between the STFTM and other time-frequency functions such as the Wigner distribution [14].

It is also interesting to note that the ambiguity function (AF), well known in the radar field, is also closely related to time-frequency functions such as the STFT, the STFTM, and the CED. The ambiguity function (AF) of a sequence $x(n)$ is expressed as

$$A_x(n, \omega) = \sum\limits_{m=-\infty}^{\infty} x(m)x(m - n)e^{j\omega m} \qquad (6.75)$$

This function can be viewed as the STFT of $x(n)$ obtained with respect to the analysis window $x(-n)$. Furthermore, the two-dimensional Fourier transform of the AF is the CED of $x(n)$. Since the CED is invertible to within a sign factor, it follows that the same is true for the ambiguity function. We can also show that the two-dimensional Fourier transform of the STFTM of a sequence is equal to the product of the ambiguity functions of the sequence and analysis window, respectively. This follows easily by taking the two-dimensional Fourier transform of Eq. (6.71) and mapping the CED convolution into a product of ambiguity functions. In fact, this relationship can be used to show that the discrete-time STFTM is invertible to within a sign factor provided the length of the analysis window is longer than that of the signal being represented [16]. In particular, the recovery procedure consists of taking the two-dimensional Fourier transform of the square of the STFTM and dividing the result by the AF of the analysis window. We thus obtain the AF of the original sequence. We can obtain the original sequence (to within a sign ambiguity) from this AF by taking its Fourier transform and using the result in Eq. (6.70). The restriction to long analysis windows results from the division by the analysis window AF in this procedure. For situations involving shorter analysis windows, a different approach such as the one described in Section 6.4 has to be adopted.

6.7 APPLICATIONS

The STFT concepts introduced in this chapter are used in a number of signal processing applications. Most prominent among these applications is speech processing; however, the STFT is also useful in such diverse areas as acoustic beamforming and image restoration. In this section, we present some examples that illustrate the role of the STFT in such applications.

6.7.1 Speech Processing

The STFT has played a major role in speech processing applications such as time-scale modification [7] and bandwidth reduction [17]. In time-scale modification, we are interested in changing the apparent rate of articulation of the original speech while maintaining its perceptual quality. Controls for time-scale modification on a tape recorder, for example, would allow users to rapidly scan large quantities of material or slowly play back difficult to understand speech such as a foreign language. For the blind, this is a particularly encouraging prospect since even normal recorded speech offers a reading rate two to three times that for Braille. In bandwidth reduction, we are also interested in preserving the perceptual quality of the speech in a parsimonious digital representation for limited-bandwidth transmission. To understand why the STFT is suitable for such applications, it is useful to examine the nature of the speech waveform in terms of a simplified model for speech production.

Speech can be modeled as the output of a linear time-varying filter that approximates the transmission characteristics of the vocal tract, as illustrated in Fig. 6.33. The input to the filter is constantly changing. Voiced sounds (e.g., the vowel *e*) are produced by exciting the vocal tract with approximately periodic pulses of airflow caused by vibration of the vocal cords at some time-varying fundamental frequency or "pitch." Unvoiced or what is often referred to as fricative sounds (e.g., *s*) are produced by exciting the vocal tract with a noiselike excitation generated by forcing air through the constricted vocal cavity. Thus the STFT of a voiced sound takes on a harmonic appearance, while the STFT of an unvoiced sound is noiselike, as depicted

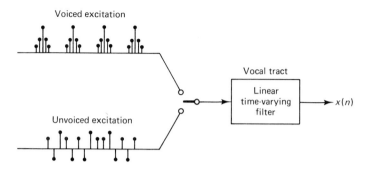

Figure 6.33 Speech production model.

in Fig. 6.34. More generally, the speech production mechanism can generate various combinations of these sounds as, for example, voiced fricatives. Typically, each of these sounds lasts for about 20 ms, during which spectral content of the speech remains stationary except for gradual changes leading into the next sound. Stationarity of spectral content is important for bandwidth reduction, while the rate of change of spectral content is important for time-scale modification. In particular, bandwidth reduction of speech is often achieved by efficiently representing the spectral content of each stationary sound. On the other hand, time-scale modification is obtained by increasing or decreasing the rate at which the spectral content changes from one sound to another.

Various speech researchers have established a relationship between the STFT and time-scale modification of speech. In particular, it is generally agreed that time-scale modification of speech leads to a linear time scaling of the STFTM. The effect of time-scale modification on the STFT phase has also been modeled, but this turns out to be a rather complex and nonlinear effect. However, this effect has been explicitly taken into account in the work of Portnoff [7]. In his technique, the linearly time-scaled STFTM and the nonlinearly processed STFT phase are combined to obtain a modified STFT. This modified STFT is used in the helical interpolation method of Section 6.3.4 to synthesize the desired speech with time-scale modification. An alternative approach to time-scale modification ignores the STFT phase entirely. Instead, the STFTM is linearly time-scaled and used as input to the least-squares technique of Section 6.5.3 for signal estimation from the modified STFTM. The resulting signal has an STFTM that is closest in the least-squares sense to the linearly time-scaled STFTM.

Another common speech application of the STFT is in the area of bandwidth reduction. The digital representation of the speech waveform is often impractical for speech transmission over a limited-bandwidth channel. For example, speech sampled at 8000 samples/s with a 12-bit accuracy contains 96,000 bits/s, and hence a bandwidth requirement too large for most practical channels. The purpose of the vocoder (*voice coder*) is to reduce the bandwidth requirement by reducing the required number of bits transmitted. One approach to bandwidth reduction is to quantize either the discrete STFT or parameters of an STFT model, according to their relative perceptual importance. These quantized values are transmitted and handed to a synthesizer to

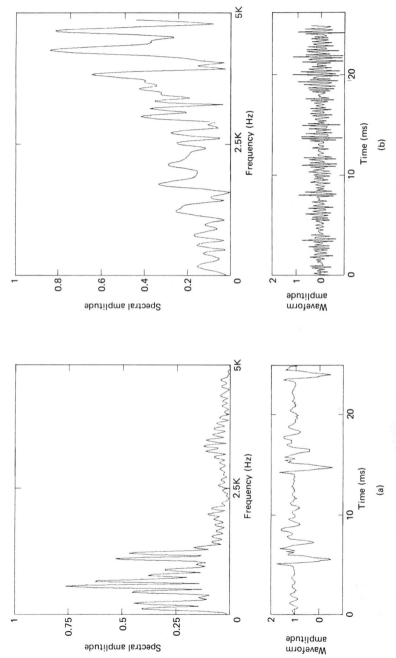

Figure 6.34 Fourier transform magnitudes of (a) typical voiced and (b) typical unvoiced sounds.

obtain a signal estimate. Two illustrative examples of bandwidth reduction techniques are based respectively on the filter bank and the Fourier transform interpretations of the STFT.

A bandwidth reduction method [18] based on the filter bank interpretation of the STFT divides the speech band (0–3.2 kHz) into 4–8 "subbands." A filter bank with a filter at the center frequency of each of these subbands is used to compute a discrete STFT. The output of each of these filters is then quantized (bits per sample) differently for each subband. In particular, subbands of less perceptual importance are quantized more coarsely. Generally, the temporal decimation rate and frequency sampling rate are chosen such that the number of samples in the discrete STFT is equal to the number of samples in the original sequence. Thus, the key to bandwidth reduction lies in the coarse quantization of various subbands. The resulting modified STFT is then transmitted over a limited-bandwidth channel. At the receiving end we can use any of the synthesis techniques in Section 6.5 for signal estimation from a modified STFT.

Another approach to bandwidth reduction is based on the Fourier transform view of the STFT and is known as *adaptive transform coding* [19]. In this method a DFT computation results in typically 64–512 uniform frequency samples, thus requiring a much larger time decimation interval L than does the subband technique. The large number of frequency samples in this method make evident important perceptual information about the underlying speech. This information is then used to adaptively quantize the discrete STFT. The signal estimate at the receiver end is obtained by the weighted overlap-add synthesis method of Section 6.3.5.

6.7.2 Sensor Array Processing

The short-time Fourier transform has also been utilized in applications such as sonar and geophysical exploration [20]. In these applications, arrays of spatially distributed sensors such as microphones are often used to determine various characteristics of propagating waves. In particular, the problem of isolating wave components from a particular direction has received considerable attention. This *beamforming* problem can be addressed in a number of ways, including a frequency-domain method in which the short-time Fourier transform plays a central role.

To isolate the wave components from a particular direction, the basic idea is to delay each sensor signal in such a way that the wave components of interest are time-aligned over all sensors. Thus if we sum the sensor signals, the desired wave components will add coherently while energy from other directions will add incoherently. The signal resulting from this delay-and-sum operation is thus considered an estimate of the wave components from the desired direction. If the signal from the ith sensor is digitized to form the sequence $x_i(n)$, the delay-and-sum operation is expressed as

$$x(n) = \frac{1}{N} \sum_{i=0}^{N-1} x_i(n - n_i) \tag{6.76}$$

where n_i is the delay applied to the ith sensor signal and there are a total of N sensors. For the frequency-domain method of beamforming, we express Eq. (6.76) in the frequency domain. That is, taking the Fourier transform of both sides of Eq. (6.76), we obtain

$$X(\omega) = \frac{1}{N} \sum_{i=0}^{N-1} X_i(\omega) e^{-j\omega n_i} \tag{6.77}$$

In practice, the frequency content of the sensor signals slowly changes as a function of time. For example, the propagating waves may be originating from sources whose directions are slowly changing with respect to the array. To capture this time dependence of the frequencies coming from a particular direction, the frequency-domain beamforming method employs the STFT for each sensor signal instead of the Fourier transform in Eq. (6.77). We thus obtain

$$X_n(\omega) = \frac{1}{N} \sum_{i=0}^{N-1} X_i(n, \omega) e^{-j\omega n_i} \tag{6.78}$$

where $X_i(n, \omega)$ is the STFT of $x_i(n)$ and $X_n(\omega)$ is the final result of the frequency-domain beamforming method. The time-frequency function $X_n(\omega)$ represents the time-varying frequency content of the wave components coming from a particular direction. For digital implementations, the frequency variable is discretized, resulting in the use of the discrete STFT.

6.7.3 Image Processing

Image processing techniques in the frequency domain such as Wiener filtering for noise reduction generally are based on the assumption that the frequency content of an image is the same over all subsections of the image. However, this assumption is typically not valid for many practical images since they usually consist of many different regions with disparate characteristics. To overcome this problem, one approach is to divide the image into smaller sections and process each of them separately; the size of the sections is chosen to be small enough so that most sections consist of a uniform pattern. A convenient mechanism for implementing this idea is a two-dimensional extension of the short-time Fourier transform. This extension of the STFT to a two-dimensional signal $x(n_1, n_2)$ is given by

$$X(n_1, n_2, \omega_1, \omega_2) = \sum_{m_1} \sum_{m_2} x(m_1, m_2) w(n_1 - m_1, n_2 - m_2) e^{-j(\omega_1 m_1 + \omega_2 m_2)} \tag{6.79}$$

where $w(n_1, n_2)$ is the two-dimensional analysis window. As in the case of the one-dimensional STFT, the two-dimensional STFT can also be interpreted as a set of Fourier transforms obtained by sliding the analysis window (in two dimensions now) over the original signal and taking the Fourier transform of the product at each new position. Thus the analysis window helps divide the original image into separate but overlapping sections, each the same size as the analysis window. The convenience of the STFT derives from the fact that after we have processed the separate sections of the image, it is necessary to combine the sections to form the entire processed image. The overlap-add synthesis technique for the STFT is ideal for accomplishing this purpose. In particular, for the OLA method to succeed we require that the overlapping analysis window positions should sum to a constant over the entire image. A particularly convenient way of generating two-dimensional analysis windows with such a property makes use of one-dimensional windows that satisfy the OLA constraint. In particular, if $w_1(n)$ and $w_2(n)$ are any two analysis windows satisfying the OLA constraint, then it can be shown [21] that the window $w(n_1, n_2) = w_1(n_1)w_2(n_2)$ satisfies the two-dimensional OLA constraint.

REFERENCES

1. R. M. Fano, "Short-Time Autocorrelation Functions and Power Spectra," *J. Acoustical Soc. Amer.*, Vol. 22, pp. 546–550, Sept. 1950.

2. M. R. Schroeder and B. S. Atal, "Generalized Short-Time Power Spectra and Autocorrelation Functions," *J. Acoustical Soc. Amer.*, Vol. 34, pp. 1679–1683, Nov. 1962.

3. W. Koening, H. K. Dunn, L. Y. Lacey, "The Sound Spectrograph," *J. Acoustical Soc. Amer.*, Vol. 18, pp. 19–49, 1946.

4. A. V. Oppenheim, "Speech Spectrograms Using the Fast Fourier Transform," *IEEE Spectrum,* Vol. 7, pp. 57–62, August 1970.

5. M. R. Portnoff, "Representation of Digital Signals and Systems Based on the Short-Time Fourier Transform," *IEEE Trans. Acoustics, Speech, and Signal Processing*, Vol. 28, pp. 55–69, Feb. 1980.

6. J. C. Anderson, "Speech Analysis/Synthesis Based on Perception," Ph.D. Thesis, MIT, Cambridge, MA, Sept. 1984.

7. M. R. Portnoff, "Time-Scale Modification of Speech Based on Short-Time Fourier Analysis, *IEEE Trans. Acoustics, Speech, and Signal Processing*, Vol. 30, pp. 374–390, June 1981.

8. A. V. Oppenheim and R. W. Schafer, *Digital Signal Processing*, Prentice-Hall, Englewood Cliffs, NJ, 1983.

9. S. H. Nawab, T. F. Quatieri, and J. S. Lim, "Signal Reconstruction from Short-Time Fourier Transform Magnitude," *IEEE Trans. Acoustics, Speech, and Signal Processing,* Vol. 31, pp. 986–998, Aug. 1983.

10. D. Israelevitz, "Some Results on the Time-Frequency Sampling of the Short-Time Fourier Transform Magnitude," *IEEE Trans. Acoustics, Speech, and Signal Processing*, Vol. 33, pp. 1611–1613, Dec. 1985.

11. J. B. Allen and L. R. Rabiner, "A Unified Theory of Short-Time Spectrum Analysis and Synthesis," *Proc. IEEE*, Vol. 65, pp. 1558–1564, Nov. 1977.

12. D. Griffin and J. S. Lim, "Signal Estimation from Modified Short-Time Fourier Transform," *IEEE Trans. Acoustics, Speech, and Signal Processing*, Vol. 32, pp. 236–243, April 1984.

13. A. W. Rihaczek, "Signal Energy Distribution in the Time and Frequency Domain," *IEEE Trans. Information Theory*, Vol. 10, May 1968.

14. E. Wigner, *Physical Review*, No. 40, p. 749, 1932.

15. A. V. Oppenheim, *Applications of Digital Signal Processing*, Prentice-Hall, Englewood Cliffs, NJ, 1978.

16. R. A. Altes, "Detection, Estimation, and Classification with Spectrograms," *J. Acoustical Soc. Amer.*, Vol. 67, pp. 1232–1246, April 1980.

17. L. R. Rabiner and R. W. Schafer, *Digital Processing of Speech Signals,* Prentice-Hall, Englewood Cliffs, NJ, 1978.

18. R. E. Crochiere, "On the Design of Sub-Band Coders for Low-Bit Rate Speech Communication," *Bell Syst. Tech. J.,* Vol. 65, pp. 747–770, May–June 1977.

19. J. S. Tribolet and R. E. Crochiere, "Frequency Domain Coding of Speech," *IEEE Trans. Acoustics, Speech, and Signal Processing,* Vol. 27, pp. 512–530, Oct. 1979.

20. D. E. Dudgeon and R. W. Mersereau, *Multidimensional Signal Processing,* Prentice-Hall, Englewood Cliffs, NJ, 1984.

21. J. S. Lim, "Image Restoration by Short-Space Spectral Subtraction," *IEEE Trans. Acoustics, Speech, and Signal Processing,* Vol. 29, pp. 191–197, Aug. 1980.

7

Two-Dimensional Signal Processing

Jae S. Lim
Massachusetts Institute of Technology

7.0 INTRODUCTION

At a conceptual level, there is a great deal of similarity between two-dimensional (2-D) signal processing and one-dimensional (1-D) signal processing. In 1-D signal processing, the concepts discussed are filtering, Fourier transforms, discrete Fourier transforms, fast Fourier transforms, etc. In 2-D signal processing, we again are concerned with concepts such as filtering, Fourier transforms, discrete Fourier transforms, and fast Fourier transforms. As a consequence, the general concepts that we develop in 2-D signal processing can be viewed, in many cases, as straightforward extensions of the results in 1-D signal processing.

At a more detailed level, however, considerable differences exist between 1-D and 2-D signal processing. One major difference is the amount of data involved in typical applications. In speech processing, an important 1-D signal processing application, speech is typically sampled at a 10-kHz rate and we have 10,000 data points to process in a second. However, in video processing, where processing an image frame is an important 2-D signal processing application, we may have 30 frames/s, with each frame consisting of 500×500 pixels (picture elements). In this case, we would have 7.5 million data points to process per second, which is orders of magnitude greater than the case of speech processing. Due to this difference in data rate requirements, the computational efficiency of a signal processing algorithm plays a much more important role in 2-D signal processing, and advances in hardware tech-

Jae S. Lim is with the Research Laboratory of Electronics and the Department of Electrical Engineering and Computer Science, Massachusetts Institute of Technology, Cambridge, MA 02139. This work has been supported in part by the National Science Foundation under Grant ECS-8407285 and in part by the Advanced Research Projects Agency monitored by ONR under contract N00014-81-K-0742.

nology will have a much greater impact on 2-D signal processing applications in the future.

Another major difference comes from the fact that there is less complete mathematics for 2-D signal processing than for 1-D signal processing. For example, many 1-D systems are described by differential equations, while many 2-D systems can be described by partial differential equations. We know a great deal more about differential equations than about partial differential equations. Another example is the absence of the fundamental theorem of algebra for 2-D polynomials. For 1-D polynomials, the fundamental theorem of algebra states that any 1-D polynomial can be factored as a product of first-order polynomials. A 2-D polynomial, however, generally cannot be factored as a product of lower-order polynomials. This difference has a major impact on many results in signal processing. For example, an important structure for realizing a 1-D digital filter is the cascade structure. In the cascade structure, the z-transform of the digital filter unit sample response is factored as a product of lower-order polynomials and the realizations of these lower-order factors are cascaded. The z-transform of a 2-D digital filter unit sample response cannot, in general, be factored as a product of lower-order polynomials and the cascade structure therefore is not a general structure for a 2-D digital filter realization. Another consequence of the nonfactorability of a 2-D polynomial is the difficulty associated with issues related to the system stability. In a 1-D system, the pole locations can be easily determined, and an unstable system can be stabilized without affecting the magnitude response by simple manipulation of pole locations. In a 2-D system, because poles are surfaces rather than points and because of the absence of the fundamental theorem of algebra, it is extremely difficult to determine the pole locations. As a result, checking the stability of a 2-D system and stabilizing an unstable 2-D system without affecting the magnitude response is extremely difficult.

Another difference between 1-D and 2-D signal processing is the notion of causality. In a typical 1-D application such as speech processing, a system is generally required to be causal to avoid delay since there is a well-defined notion of past, present, and future. In a typical 2-D application such as image processing, it may not be necessary to impose the causality constraint. An image frame, for example, can be processed from top to bottom, from left to right, in diagonal directions, etc. As a result, there is generally more flexibility in designing a 2-D system than in designing a 1-D system.

As we have seen, there is considerable similarity and at the same time considerable difference between 1-D and 2-D signal processing. In this chapter, we will study the results in 1-D signal processing that can be extended to 2-D signal processing. Our discussion will rely heavily on the reader's knowledge of 1-D signal processing theory (a brief summary of which is in the Appendix to this book). We will also study, with much greater emphasis, the results in 2-D signal processing that are significantly different from those in 1-D signal processing. We will study what the differences are, where they come from, and what impacts they have on 2-D signal processing applications. Since we will study the similarities and differences between 1-D and 2-D signal processing and since 1-D signal processing is a special case of 2-D signal processing, this chapter will help us understand not only 2-D signal processing theories but also 1-D signal processing theories at a much deeper level.

7.1 SIGNALS AND SYSTEMS

7.1.1 Signals

The signals that we deal with in developing 2-D signal processing theories are discrete space signals, which are discrete in space[†] and continuous in amplitude. Discrete space signals are also referred to as *sequences*.

A two-dimensional (2-D) discrete space signal (sequence) will be denoted by functions whose two arguments are integers. For example, $x(n_1, n_2)$ represents a sequence that is defined for all integer values of n_1 and n_2. Note that $x(n_1, n_2)$ for a noninteger n_1 or n_2 is not zero but is undefined. The notation $x(n_1, n_2)$ refers to the discrete space function x or to the value of the function x at a specific (n_1, n_2). The distinction between these two will be obvious from the context.

One reasonable way to sketch a sequence $x(n_1, n_2)$ is to use a three-dimensional (3-D) perspective plot, where the height at (n_1, n_2) represents the amplitude at (n_1, n_2). Sketching a 3-D perspective plot, however, is often very tedious. An alternative way to sketch a 2-D sequence, which we'll use in this chapter, is with a 2-D plot where open circles represent the amplitude of 0 and the solid circles represent nonzero amplitudes, with the value in parentheses representing the amplitude. An example of a 2-D sequence sketched in this way is shown in Fig. 7.1. In this figure, $x(3, 1)$ is 0 and $x(1, 1)$ is 2. Many sequences that we use will have amplitudes 0 or 1 for large regions of (n_1, n_2). In such instances, for convenience, the open circles and parentheses will be eliminated. If there is neither an open circle nor a filled circle at a particular (n_1, n_2), the sequence has zero amplitude at that point. If there is a filled circle with

[†]Even though we refer to "space," the independent variable can represent other quantities, such as time.

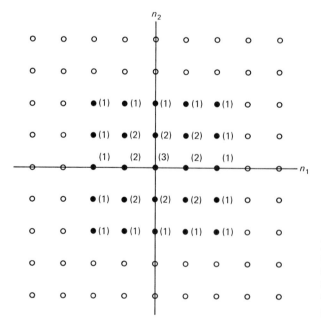

Figure 7.1   An example of a 2-D sequence. Open circles represent the amplitude of 0, and solid circles represent nonzero amplitudes, with the values in parentheses representing the amplitude.

no amplitude specification at a particular (n_1, n_2) then the sequence has amplitude 1 at that point. Fig. 7.2 shows the result when this additional simplification is made to the sequence in Fig. 7.1.

Some sequences and classes of sequences that play a particularly important role in 2-D digital signal processing are discussed next.

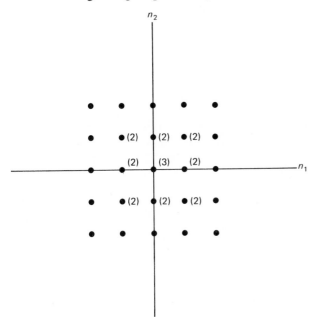

Figure 7.2 The sequence in Fig. 7.1 sketched with some simplifications. The open circles have been eliminated, and solid circles with amplitude of 1 have no amplitude specifications.

Impulses.   The impulse or unit sample sequence, denoted by $\delta(n_1, n_2)$, is defined as

$$\delta(n_1, n_2) = \begin{cases} 1, & n_1 = n_2 = 0 \\ 0, & \text{otherwise} \end{cases} \tag{7.1}$$

The sequence $\delta(n_1, n_2)$, sketched in Fig. 7.3, plays a role similar to the impulse $\delta(n)$ in 1-D signal processing.

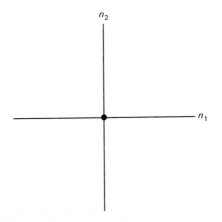

Figure 7.3 Unit sample sequence $\delta(n_1, n_2)$.

Any sequence $x(n_1, n_2)$ can be represented as a linear combination of delayed impulses as follows:

$$
\begin{aligned}
x(n_1, n_2) = \ \cdots\ & + x(-1, -1) \cdot \delta(n_1 + 1, n_2 + 1) + x(0, -1) \cdot \delta(n_1, n_2 + 1) \\
& + x(1, -1) \cdot \delta(n_1 - 1, n_2 + 1) + \cdots + x(-1, 0) \cdot \delta(n_1 + 1, n_2) \\
& + x(0, 0) \cdot \delta(n_1, n_2) + x(1, 0) \cdot \delta(n_1 - 1, n_2) + \cdots \\
& + x(-1, 1) \cdot \delta(n_1 + 1, n_2 - 1) + x(0, 1) \cdot \delta(n_1, n_2 - 1) \\
& + x(1, 1) \cdot \delta(n_1 - 1, n_2 - 1) + \cdots \\
= \ & \sum_{k_1=-\infty}^{\infty} \sum_{k_2=-\infty}^{\infty} x(k_1, k_2) \cdot \delta(n_1 - k_1, n_2 - k_2)
\end{aligned}
\tag{7.2}
$$

The representation of $x(n_1, n_2)$ by Eq. (7.2) is very useful in system analysis.

One class of impulses that do not have any counterparts in 1-D processing are line impulses. An example of a line impulse is the 2-D sequence $\delta_T(n_1)$, which is sketched in Fig. 7.4 and is defined as

$$
x(n_1, n_2) = \delta_T(n_1) = \begin{cases} 1, & n_1 = 0 \\ 0, & \text{otherwise} \end{cases}
\tag{7.3}
$$

Other examples of line impulses include $\delta_T(n_2)$ and $\delta(n_1 - n_2)$, which are defined similarly to $\delta_T(n_1)$. The subscript T in $\delta_T(n_1)$ indicates that $\delta_T(n_1)$ is a 2-D sequence. When the 2-D sequence is a function of only one variable, it may be confused with a 1-D sequence. For example, $\delta_T(n_1)$ without the subscript T may be confused with the 1-D unit sample sequence $\delta(n_1)$. To avoid this confusion, whenever a 2-D sequence is a function of one variable, the subscript T will be used to note that it is a 2-D sequence. The sequence $x_T(n_1)$, for example, is a 2-D sequence, while $x(n_1)$ is a 1-D sequence.

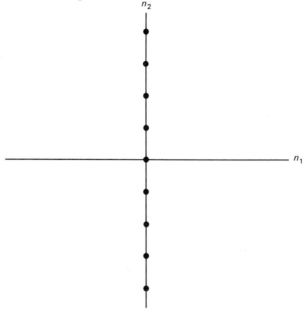

Figure 7.4   Line impulse $\delta_T(n_1)$.

Step Sequences.   The unit step sequence, denoted by $u(n_1, n_2)$, is defined as

$$u(n_1, n_2) = \begin{cases} 1, & n_1, n_2 \geq 0 \\ 0, & \text{otherwise} \end{cases} \tag{7.4}$$

The sequence $u(n_1, n_2)$, sketched in Fig. 7.5, is related to $\delta(n_1, n_2)$ as

$$u(n_1, n_2) = \sum_{k_1=-\infty}^{n_1} \sum_{k_2=-\infty}^{n_2} \delta(k_1, k_2) \tag{7.5}$$

or

$$\delta(n_1, n_2) = u(n_1, n_2) - u(n_1 - 1, n_2) - u(n_1, n_2 - 1) + u(n_1 - 1, n_2 - 1) \tag{7.6}$$

Some step sequences do not have any counterparts in 1-D processing, such as the 2-D sequence $u_T(n_1)$, which is defined as

$$x(n_1, n_2) = u_T(n_1) = \begin{cases} 1, & n_1 \geq 0 \\ 0, & \text{otherwise} \end{cases} \tag{7.7}$$

Other examples include $u_T(n_2)$ and $u(n_1 - n_2)$, which are defined similarly to $u_T(n_1)$.

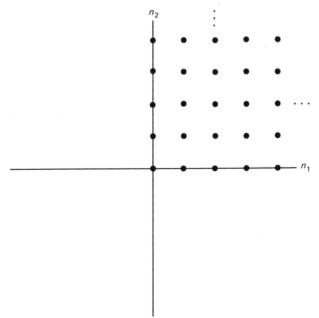

Figure 7.5   Unit step sequence $u(n_1, n_2)$.

Exponential Sequences.   Exponential sequences of the type $A \cdot \alpha^{n_1} \cdot \beta^{n_2}$ are an important class of sequences for system analysis. As we will see later, these sequences are eigenfunctions of linear shift-invariant systems.

Separable Sequences.   A 2-D sequence $x(n_1, n_2)$ is called a separable sequence if it can be expressed in the form

$$x(n_1, n_2) = f(n_1) \cdot g(n_2) \tag{7.8}$$

where $f(n_1)$ is a function of only n_1, and $g(n_2)$ is a function of only n_2. Even though it is possible to view $f(n_1)$ and $g(n_2)$ as 2-D sequences, it is more convenient to consider them as 1-D sequences. For that reason, we use the notations $f(n_1)$ and $g(n_2)$ rather than $f_T(n_1)$ and $g_T(n_2)$.

The unit sample sequence $\delta(n_1, n_2)$ is a separable sequence since $\delta(n_1, n_2)$ can be expressed as

$$\delta(n_1, n_2) = \delta(n_1) \cdot \delta(n_2) \tag{7.9}$$

where $\delta(n_1)$ and $\delta(n_2)$ are 1-D unit sample sequences. Other examples of separable sequences include $u(n_1, n_2)$ and $a^{n_1} \cdot b^{n_2} + b^{n_1+n_2}$, which can be written as $(a^{n_1} + b^{n_1}) \cdot b^{n_2}$.

Separable sequences form a very special class of 2-D sequences; a typical 2-D sequence is not, in general, a separable sequence. To illustrate this, we consider a sequence $x(n_1, n_2)$ that is zero outside $0 \leq n_1 \leq N_1 - 1$ and $0 \leq n_2 \leq N_2 - 1$. A general sequence $x(n_1, n_2)$ of this type has $N_1 \cdot N_2$ degrees of freedom. If $x(n_1, n_2)$ is a separable sequence, $x(n_1, n_2)$ is completely specified by some $f(n_1)$ that is zero outside $0 \leq n_1 \leq N_1 - 1$ and some $g(n_2)$ that is zero outside $0 \leq n_2 \leq N_2 - 1$; consequently, it has only $N_1 + N_2$ degrees of freedom.

Even though separable sequences form a very special class of 2-D sequences, they play an important role in 2-D signal processing. In those cases when the results that apply to 1-D sequences do not extend to general 2-D sequences in a straightforward manner, they often extend to separable 2-D sequences. In addition, the separability of a sequence can be exploited in reducing computations in various contexts such as digital filtering and discrete Fourier transform computation. This will be discussed in later sections.

Periodic Sequences. A sequence $x(n_1, n_2)$ is called periodic with a period $N_1 \times N_2$ if $x(n_1, n_2)$ satisfies the following condition:

$$x(n_1, n_2) = x(n_1 + N_1, n_2) = x(n_1, n_2 + N_2) \qquad \text{for all } (n_1, n_2) \tag{7.10}$$

where N_1 and N_2 are integers. For example, $\cos[\pi n_1 + (\pi/2)n_2]$ is a periodic sequence with period 2×4 since $\cos[\pi n_1 + (\pi/2)n_2] = \cos[\pi(n_1 + 2) + (\pi/2)n_2] = \cos[\pi n_1 + (\pi/2)(n_2 + 4)]$ for all (n_1, n_2). A periodic sequence is often denoted by adding a ~ (tilde) on the sequence, e.g., $\tilde{x}(n_1, n_2)$, to distinguish it from an aperiodic sequence.

7.1.2 Systems

If there is a unique output for any given input, the input-output mapping is called a system. A system $\mathbf{T}$ that maps an input $x(n_1, n_2)$ to an output $y(n_1, n_2)$ is represented by

$$y(n_1, n_2) = \mathbf{T}[x(n_1, n_2)] \tag{7.11}$$

This definition of a system is very broad. Without some restrictions, the characterization of a system requires a complete input-output relationship—knowing the output of a system to a certain set of inputs does not allow us to determine the output of the system to other sets of inputs. Two types of restrictions that greatly simplify the characterization and analysis of a system are linearity and shift invariance. Fortunately, many systems in practice can often be approximated by a linear and shift-invariant system.

The linearity of a system $\mathbf{T}$ is defined as

$$\text{Linearity} \quad \longleftrightarrow \quad \mathbf{T}[a \cdot x_1(n_1, n_2) + b \cdot x_2(n_1, n_2)]$$
$$= a \cdot y_1(n_1, n_2) + b \cdot y_2(n_1, n_2) \tag{7.12}$$

where $\mathbf{T}[x_1(n_1, n_2)] = y_1(n_1, n_2)$, $\mathbf{T}[x_2(n_1, n_2)] = y_2(n_1, n_2)$, a and b are any scalar constants, and $A \leftrightarrow B$ means that A implies B and B implies A. The condition in Eq. (7.12) is called the principle of *superposition*.

The shift invariance of a system is defined as

$$\text{Shift invariance} \quad \longleftrightarrow \quad \mathbf{T}[x(n_1 - m_1, n_2 - m_2)] = y(n_1 - m_1, n_2 - m_2) \tag{7.13}$$

where $y(n_1, n_2) = \mathbf{T}[x(n_1, n_2)]$ and m_1 and m_2 are any integers.

For a linear and shift-invariant system, we can derive the following input-output relation using Eqs. (7.2), (7.12), and (7.13):

$$y(n_1, n_2) = \mathbf{T}[x(n_1, n_2)] = \sum_{k_1=-\infty}^{\infty} \sum_{k_2=-\infty}^{\infty} x(k_1, k_2) \cdot h(n_1 - k_1, n_2 - k_2) \tag{7.14}$$

where $h(n_1, n_2) = \mathbf{T}[\delta(n_1, n_2)]$, the response of the system when the input is $\delta(n_1, n_2)$. Equation (7.14) states that the unit sample response of a linear shift-invariant system is completely characterized by the unit sample response $h(n_1, n_2)$. Specifically, for a linear shift-invariant system, knowledge of $h(n_1, n_2)$ alone allows us to determine the output of the system to any input from Eq. (7.14). Equation (7.14) is referred to as *convolution* and is denoted by the convolution operator $*$. For a linear shift-invariant system,

$$y(n_1, n_2) = x(n_1, n_2) * h(n_1, n_2) \tag{7.15}$$
$$= \sum_{k_1=-\infty}^{\infty} \sum_{k_2=-\infty}^{\infty} x(k_1, k_2) \cdot h(n_1 - k_1, n_2 - k_2)$$

Note that the unit sample response $h(n_1, n_2)$, which plays such an important role for a linear shift-invariant system, loses its significance for a nonlinear or shift-variant system. All the results in this section are straightforward extensions of 1-D results.

7.1.3 Convolution

The convolution operator in Eq. (7.15) has a number of properties that are straightforward extensions of 1-D results. Some of the more important ones are listed below.

Commutativity:

$$x(n_1, n_2) * y(n_1, n_2) = y(n_1, n_2) * x(n_1, n_2) \tag{7.16}$$

Associativity:

$$[x(n_1, n_2) * y(n_1, n_2)] * z(n_1, n_2) = x(n_1, n_2) * [y(n_1, n_2) * z(n_1, n_2)] \qquad (7.17)$$

Distributivity:

$$x(n_1, n_2) * [y(n_1, n_2) + z(n_1, n_2)]$$
$$= [x(n_1, n_2) * y(n_1, n_2)] + [x(n_1, n_2) * z(n_1, n_2)] \qquad (7.18)$$

Convolution with Delayed Unit Sample Sequence:

$$x(n_1, n_2) * \delta(n_1 - m_1, n_2 - m_2) = x(n_1 - m_1, n_2 - m_2) \qquad (7.19)$$

The convolution of two sequences $x(n_1, n_2)$ and $h(n_1, n_2)$ can be obtained by explicitly evaluating Eq. (7.15). It is often simpler and more instructive, however, to evaluate Eq. (7.15) graphically. Specifically, the convolution sum in Eq. (7.15) can be interpreted as multiplying two sequences $x(k_1, k_2)$ and $h(n_1 - k_1, n_2 - k_2)$ that are functions of the variables (k_1, k_2) and summing the product over all integer values of (k_1, k_2). The result, which is a function of (n_1, n_2), is the result of convolving $x(n_1, n_2)$ and $h(n_1, n_2)$. As an example, consider the two sequences $x(n_1, n_2)$ and $h(n_1, n_2)$, shown in Figs. 7.6(a) and (b). From $x(n_1, n_2)$ and $h(n_1, n_2)$, we can obtain $x(k_1, k_2)$ and $h(n_1 - k_1, n_2 - k_2)$ as functions of k_1 and k_2, as shown in Figs. 7.6(c) and (d). Note that we can obtain $h(n_1 - k_1, n_2 - k_2)$ as a function of k_1 and k_2 from $h(n_1, n_2)$ by first changing the variables n_1 and n_2 to k_1 and k_2, flipping the sequence with respect to the origin, and then shifting the result in the positive k_1 and k_2 directions by n_1 and n_2 points, respectively. Once we have obtained $x(k_1, k_2)$ and $h(k_1 - n_1, k_2 - n_2)$, we can multiply and sum them for each different set of n_1 and n_2. The result is shown in Fig. 7.6(e).

A linear shift-invariant system is called separable, if its unit sample response $h(n_1, n_2)$ is a separable sequence. For a separable system, it is possible to reduce the number of arithmetic operations in computing the convolution sum. For this reason, separable systems are sometimes used in processing images. To illustrate this, we consider an input sequence $x(n_1, n_2)$ of size $N \times N$ points and a unit sample response $h(n_1, n_2)$ of size $M \times M$ points, as follows:

$$x(n_1, n_2) = 0 \text{ outside } 0 \leq n_1 \leq N - 1, 0 \leq n_2 \leq N - 1,$$
$$h(n_1, n_2) = 0 \text{ outside } 0 \leq n_1 \leq M - 1, 0 \leq n_2 \leq M - 1 \qquad (7.20)$$

where $N \gg M$ in typical cases. The size of nonzero values of $x(n_1, n_2)$ and $h(n_1, n_2)$ is shown in Figs. 7.7(a) and (b). Denoting the output of the system with $y(n_1, n_2)$, $y(n_1, n_2)$ can be expressed as

$$y(n_1, n_2) = x(n_1, n_2) * h(n_1, n_2)$$
$$= \sum_{k_1=-\infty}^{\infty} \sum_{k_2=-\infty}^{\infty} x(k_1, k_2) \cdot h(n_1 - k_1, n_2 - k_2) \qquad (7.21)$$

The size of nonzero values of $y(n_1, n_2)$ is shown in Fig. 7.7(c).

If Eq. (7.21) is used directly to compute $y(n_1, n_2)$, approximately $(N + M - 1)^2 \cdot M^2$ arithmetic operations (defined as one multiplication and one

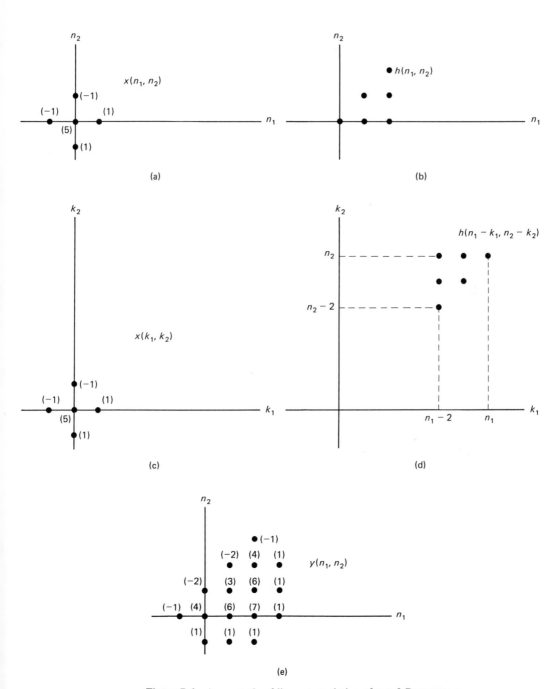

Figure 7.6   An example of linear convolution of two 2-D sequences.

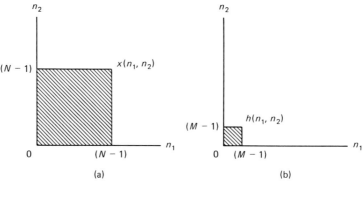

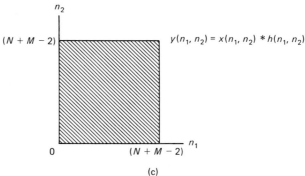

Figure 7.7   Regions for which the signal values have nonzero amplitude.

addition) are required since the number of nonzero output points is $(N + M - 1)^2$ and computing each output point requires approximately M^2 arithmetic operations. If $h(n_1, n_2)$ is a separable sequence, it can be expressed as

$$h(n_1, n_2) = h_1(n_1) \cdot h_2(n_2), \tag{7.22}$$

where

$$h_1(n_1) = 0 \text{ outside } 0 \leq n_1 \leq N - 1$$

$$h_2(n_2) = 0 \text{ outside } 0 \leq n_2 \leq M - 1$$

From Eqs. (7.21) and (7.22),

$$
\begin{aligned}
y(n_1, n_2) &= \sum_{k_1=-\infty}^{\infty} \sum_{k_2=-\infty}^{\infty} x(k_1, k_2) \cdot h_1(n_1 - k_1) \cdot h_2(n_2 - k_2) \\
&= \sum_{k_1=-\infty}^{\infty} h_1(n_1 - k_1) \cdot \sum_{k_2=-\infty}^{\infty} x(k_1, k_2) \cdot h_2(n_2 - k_2)
\end{aligned}
\tag{7.23}
$$

For a fixed k_1, $\sum_{k_2=-\infty}^{\infty} x(k_1, k_2) \cdot h_2(n_2 - k_2)$ in Eq. (7.23) corresponds to a 1-D convolution of $x(k_1, n_2)$ and $h_2(n_2)$. For example, using the notation

$$f(k_1, n_2) = \sum_{k_2=-\infty}^{\infty} x(k_1, k_2) \cdot h_2(n_2 - k_2), \tag{7.24}$$

$f(0, n_2)$ is the 1-D convolution of $x(0, n_2)$ with $h_2(n_2)$. Since there are N different values of k_1 for which $x(k_1, k_2)$ is nonzero, computing $f(k_1, n_2)$ requires N 1-D convolutions and therefore requires approximately $N \cdot (N + M - 1) \cdot M$ arithmetic operations. Once $f(k_1, n_2)$ is computed, $y(n_1, n_2)$ can be computed from Eqs. (7.23) and (7.24) as

$$y(n_1, n_2) = \sum_{k_1=-\infty}^{\infty} h_1(n_1 - k_1) \cdot f(k_1, n_2) \tag{7.25}$$

For a fixed n_2, Eq. (7.25) is a 1-D convolution of $h_1(n_1)$ and $f(n_1, n_2)$. For example, $y(n_1, 1)$ is a 1-D convolution of $f(n_1, 1)$ and $h_1(n_1)$. Since there are $(N + M - 1)$ different values of n_2 for which $f(k_1, n_2)$ is nonzero, computing $y(n_1, n_2)$ requires $(N + M - 1)$ 1-D convolutions and therefore requires approximately $(N + M - 1)^2 \cdot M$ arithmetic operations. Assuming $M \ll N$, the total number of arithmetic operations required is approximately $(N + M - 1)^2 \cdot 2M$, which compares favorably with $(N + M - 1)^2 \cdot M^2$. When $M = 10$, exploiting the separability of $h(n_1, n_2)$ reduces the number of arithmetic operations by approximately a factor of 5.

7.1.4 Stable Systems and Special Support Systems

For practical considerations, it is often appropriate to impose additional constraints on the class of systems we consider. Systems with those constraints are stable systems and special support systems.

A system is considered *stable* in the bounded input–bounded output (BIBO) sense if and only if a bounded input always leads to a bounded output. Stability is often a desirable constraint to impose since an unstable system can generate an unbounded output, which can cause system overload or other difficulties. From this definition and Eq. (7.15), it can be shown that a necessary and sufficient condition for a linear shift-invariant system to be stable is that its unit sample response $h(n_1, n_2)$ be absolutely summable. For a linear shift-invariant system,

$$\text{Stability} \quad \longleftrightarrow \quad \sum_{n_1=-\infty}^{\infty} \sum_{n_2=-\infty}^{\infty} |h(n_1, n_2)| < \infty \tag{7.26}$$

Even though Eq. (7.26) is a straightforward extension of 1-D results, issues related to stability, such as testing the stability of a system, are quite different between 1-D and 2-D results, as we will explore further in Section 7.5. Because of Eq. (7.26), an absolutely summable sequence is defined to be a stable sequence.

Special support systems can be viewed as extensions of 1-D causal systems. Specifically, a 1-D system is *causal* if and only if the current output $y(n)$ does not depend on any future values of input, e.g., $x(n + 1), x(n + 2), x(n + 3), \ldots$. With this definition, it can be shown that a necessary and sufficient condition for a 1-D linear shift-invariant system to be causal is that its unit sample response $h(n)$ be zero for $n < 0$. Causality is often a desirable constraint to impose in designing 1-D systems. In typical 1-D signal processing applications such as speech processing, there is a well-defined time reference, and a noncausal system requires delay, which is undesirable in many real-time applications. In typical 2-D signal processing applications such as image processing, the causality may not be necessary. At a given time, a whole image may be available for processing, and it may be processed from left to right, from top to bottom, or in any direction. Even though the notion of causality may not be as important a constraint in 2-D signal processing, it is useful to extend the notion that a 1-D causal linear shift-invariant system has its unit sample response $h(n)$ whose nonzero values lie in a particular region. A 2-D linear shift-invariant system whose unit sample response $h(n_1, n_2)$ has all its nonzero values in a particular region is called a *special support system.*

A 2-D linear shift-invariant system with its unit sample response $h(n_1, n_2)$ is called a *quadrant support system* when $h(n_1, n_2)$ is a quadrant support sequence. A sequence is called a *quadrant support sequence,* or quadrant sequence, when all its nonzero values lie in one quadrant. An example of a first-quadrant sequence is the unit step sequence $u(n_1, n_2)$.

A 2-D linear shift-invariant system with its unit sample response $h(n_1, n_2)$ is called a *wedge support system* when $h(n_1, n_2)$ is a wedge support sequence. Consider two lines emanating from the origin. If all the nonzero values in a sequence lie in the region bounded by the two lines with an angle less than 180° between the two lines, the sequence is called a *wedge support sequence,* or wedge sequence. An example of a wedge sequence is shown in Fig. 7.8.

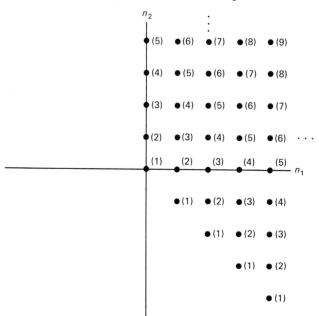

Figure 7.8 An example of a wedge sequence.

Quadrant sequences and wedge sequences are closely related. A quadrant sequence is always a wedge sequence. In addition, it can be shown that any wedge sequence can always be transformed into a first-quadrant sequence by a linear change of variables. To illustrate this, we consider the wedge sequence $x(n_1, n_2)$ shown in Fig. 7.8. Suppose we obtain a new sequence $y(n_1, n_2)$ from $x(n_1, n_2)$ by the following linear change of variables:

$$y(n_1, n_2) = x(m_1, m_2)\big|_{m_1 = n_1, m_2 = n_2 - n_1} \tag{7.27}$$

The sequence $y(n_1, n_2)$ we obtained is shown in Fig. 7.9, and it is clearly a first-quadrant sequence. For this example, the stability of $x(n_1, n_2)$ is equivalent to the stability of $y(n_1, n_2)$, since $\sum_{n_1} \sum_{n_2} |x(n_1, n_2)| = \sum_{n_1} \sum_{n_2} |y(n_1, n_2)|$. It is possible to show that a proper choice of linear change of variables maps any wedge sequence to a first-quadrant sequence without affecting the stability.

The notions that a wedge sequence can always be transformed to a first-quadrant sequence by a simple linear mapping of variables and that the stability of these two sequences is equivalent with a proper choice of linear mapping are very useful in discussing the stability of a 2-D system. As we discuss later, our primary concern in checking the stability of a 2-D system will be limited to a class of systems known as "recursively computable" systems. For a recursively computable system, the stability depends on the stability of a wedge sequence $h(n_1, n_2)$. Our approach to checking the stability of a wedge sequence $h(n_1, n_2)$ will be to transform $h(n_1, n_2)$ to a first-quadrant sequence $h'(n_1, n_2)$ by a linear transformation of variables and then to check the stability of $h'(n_1, n_2)$. It is much easier to develop stability theorems that apply to first-quadrant sequences than to wedge sequences. This is discussed further in Section 7.5.

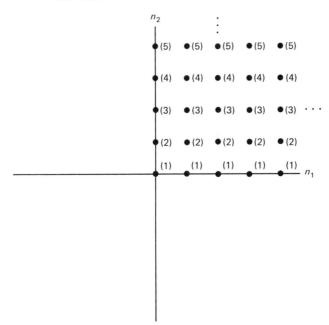

Figure 7.9 First-quadrant sequence obtained by linear mapping of variables of the wedge sequence in Fig. 7.8.

7.2 FOURIER TRANSFORM

7.2.1 Fourier Transform Pair

It is remarkable that any stable sequence $x(n_1, n_2)$ can be obtained by appropriately combining complex exponentials of the form $X(\omega_1, \omega_2) \cdot e^{j\omega_1 n_1} \cdot e^{j\omega_2 n_2}$. The function $X(\omega_1, \omega_2)$ that represents the amplitude associated with the complex exponential

$e^{j\omega_1 n_1} \cdot e^{j\omega_2 n_2}$ can be obtained from $x(n_1, n_2)$. The relationships between $x(n_1, n_2)$ and $X(\omega_1, \omega_2)$ are given by

---

Discrete Space Fourier Transform Pair

$$X(\omega_1, \omega_2) = \sum_{n_1=-\infty}^{\infty} \sum_{n_2=-\infty}^{\infty} x(n_1, n_2) \cdot e^{-j\omega_1 n_1} \cdot e^{-j\omega_2 n_2} \qquad (7.28)$$

$$x(n_1, n_2) = \frac{1}{(2\pi)^2} \int_{\omega_1=-\pi}^{\pi} \int_{\omega_2=-\pi}^{\pi} X(\omega_1, \omega_2) \cdot e^{j\omega_1 n_1} \cdot e^{j\omega_2 n_2} \qquad (7.29)$$

Equation (7.28) shows how the amplitude $X(\omega_1, \omega_2)$ associated with the exponential $e^{j\omega_1 n_1} \cdot e^{j\omega_2 n_2}$ can be determined from $x(n_1, n_2)$. The function $X(\omega_1, \omega_2)$ is called the *discrete space Fourier transform,* or Fourier transform, of $x(n_1, n_2)$. The sequence $x(n_1, n_2)$ is called the *inverse discrete space Fourier transform,* or inverse Fourier transform, of $X(\omega_1, \omega_2)$. The consistency of Eqs. (7.28) and (7.29) can be easily shown by combining them.

From Eq. (7.28), we can see that $X(\omega_1, \omega_2)$ is in general complex, even though $x(n_1, n_2)$ may be real, and that $X(\omega_1, \omega_2)$ is a function of continuous variables ω_1 and ω_2, even though $x(n_1, n_2)$ is a function of discrete variables n_1 and n_2. In addition, $X(\omega_1, \omega_2)$ is always periodic with period 2π with respect to each of the two variables ω_1 and ω_2; i.e., $X(\omega_1, \omega_2) = X(\omega_1 + 2\pi, \omega_2) = X(\omega_1, \omega_2 + 2\pi)$ for all ω_1 and ω_2. We can also show that $X(\omega_1, \omega_2)$ uniformly converges for stable sequences.

The 2-D complex exponential $e^{j\omega_1 n_1} \cdot e^{j\omega_2 n_2}$ is an eigenfunction of a 2-D linear shift-invariant system. Specifically, when $e^{j\omega_1 n_1} \cdot e^{j\omega_2 n_2}$ is used as an input $x(n_1, n_2)$, the output of the system $y(n_1, n_2)$ is given by

$$y(n_1, n_2) = H(\omega_1, \omega_2) \cdot e^{j\omega_1 n_1} \cdot e^{j\omega_2 n_2}$$

where

$$H(\omega_1, \omega_2) = \sum_{k_1=-\infty}^{\infty} \sum_{k_2=-\infty}^{\infty} h(k_1, k_2) \cdot e^{-j\omega_1 k_1} \cdot e^{-j\omega_2 k_2} \qquad (7.30)$$

From Eq. (7.30), the output is a scaled version of the input, and the scalar $H(\omega_1, \omega_2)$ is called the *frequency response* of the linear shift-invariant system.

7.2.2 Properties

We can derive a number of useful properties from the Fourier transform pair in Eqs. (7.28) and (7.29). Some of the more important properties, often useful in practice, are listed in Table 7.1. Most of these properties are essentially straightforward extensions of 1-D Fourier transform properties. The only exception is property 4, which applies to separable sequences. If a 2-D sequence $x(n_1, n_2)$ can be written as $x_1(n_1) \cdot x_2(n_2)$, then its Fourier transform $X(\omega_1, \omega_2)$ is given by $X_1(\omega_1) \cdot X_2(\omega_2)$, where $X_1(\omega_1)$ and $X_2(\omega_2)$ represent the 1-D Fourier transforms of $x_1(n_1)$ and $x_2(n_2)$ respectively. This property follows directly from the Fourier transform pair of equations (7.28) and (7.29).

TABLE 7.1   PROPERTIES OF THE FOURIER TRANSFORM

$$x(n_1, n_2) \longleftrightarrow X(\omega_1, \omega_2) \qquad y(n_1, n_2) \longleftrightarrow Y(\omega_1, \omega_2)$$

Property 1: Linearity

$$a \cdot x(n_1, n_2) + b \cdot y(n_1, n_2) \longleftrightarrow a \cdot X(\omega_1, \omega_2) + b \cdot Y(\omega_1, \omega_2)$$

Property 2: Convolution

$$x(n_1, n_2) * y(n_1, n_2) \longleftrightarrow X(\omega_1, \omega_2) \cdot Y(\omega_1, \omega_2)$$

Property 3: Modulation

$$x(n_1, n_2)y(n_1, n_2) \longleftrightarrow X(\omega_1, \omega_2) \circledast Y(\omega_1, \omega_2)$$

$$= \frac{1}{4\pi^2} \cdot \int_{\theta_1 = -\pi}^{\pi} \int_{\theta_2 = -\pi}^{\pi} X(\theta_1, \theta_2) \cdot Y(\omega_1 - \theta_1, \omega_2 - \theta_2) \cdot d\theta_1 \cdot d\theta_2$$

Property 4: Separable sequence

$$x_1(n_1)x_2(n_2) \longleftrightarrow X_1(\omega_1)X_2(\omega_2)$$

Property 5: Shift of a sequence or Fourier transform

(a) $x(n_1 - m_1, n_2 - m_2) \longleftrightarrow X(\omega_1, \omega_2) \cdot e^{-j\omega_1 m_1} \cdot e^{-j\omega_2 m_2}$

(b) $e^{j\nu_1 n_1} \cdot e^{j\nu_2 n_2} \cdot x(n_1, n_2) \longleftrightarrow X(\omega_1 - \nu_1, \omega_2 - \nu_2)$

Property 6: Differentiation

(a) $-jn_1 \cdot x(n_1, n_2) \longleftrightarrow \dfrac{\partial X(\omega_1, \omega_2)}{\partial \omega_1}$

(b) $-jn_2 \cdot x(n_1, n_2) \longleftrightarrow \dfrac{\partial X(\omega_1, \omega_2)}{\partial \omega_2}$

Property 7: Parseval's theorem

$$\sum_{n_1 = -\infty}^{\infty} \sum_{n_2 = -\infty}^{\infty} |x(n_1, n_2)|^2 = \frac{1}{(2\pi)^2} \cdot \int_{\omega_1 = -\pi}^{\pi} \int_{\omega_2 = -\pi}^{\pi} |X(\omega_1, \omega_2)|^2 \cdot d\omega_1 \cdot d\omega_2$$

Property 8: Symmetry properties

(a) $x(n_1, n_2)$: real $\longleftrightarrow X(\omega_1, \omega_2) = X^*(-\omega_1, -\omega_2)$
     $X_R(\omega_1, \omega_2), |X(\omega_1, \omega_2)|$: even (symmetric with respect to the origin)
     $X_I(\omega_1, \omega_2), \theta_x(\omega_1, \omega_2)$: odd (antisymmetric with respect to the origin)

(b) $x(n_1, n_2)$: real and even $\longleftrightarrow X(\omega_1, \omega_2)$: real and even

(c) $x(n_1, n_2)$: real and odd $\longleftrightarrow X(\omega_1, \omega_2)$: pure imaginary and odd

7.2.3 Examples

Example 1

We wish to determine $H(\omega_1, \omega_2)$ for the sequence $h(n_1, n_2)$ shown in Fig. 7.10. From Eq. (7.28),

$$H(\omega_1, \omega_2) = \sum_{n_1 = -\infty}^{\infty} \sum_{n_2 = -\infty}^{\infty} h(n_1, n_2) \cdot e^{-j\omega_1 n_1} \cdot e^{-j\omega_2 n_2}$$

$$= \frac{1}{3} + \frac{1}{6} \cdot e^{-j\omega_1} + \frac{1}{6} \cdot e^{-j\omega_2} + \frac{1}{6} \cdot e^{j\omega_1} + \frac{1}{6} \cdot e^{j\omega_2}$$

$$= \frac{1}{3} + \frac{1}{3} \cos \omega_1 + \frac{1}{3} \cos \omega_2$$

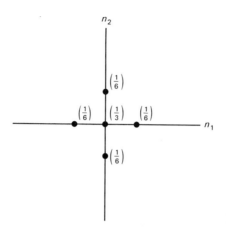

Figure 7.10   A 2-D sequence $h(n_1, n_2)$.

The function $H(\omega_1, \omega_2)$ for this example is real, and $|H(\omega_1, \omega_2)|$ is shown in Fig. 7.11. If $H(\omega_1, \omega_2)$ were the frequency response of a linear shift-invariant system, the system would correspond to a lowpass filter. The function $H(\omega_1, \omega_2)$ has smaller values in frequency regions away from the origin. A lowpass filter when applied to an image blurs the image. Figure 7.12(a) shows an image of 512×512 pixels. Figure 7.12(b) shows the image obtained by processing the image in Fig. 7.12(a) with a lowpass filter whose unit sample response $h(n_1, n_2)$ is shown in Fig. 7.10.

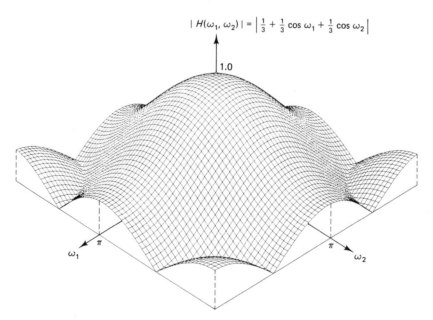

Figure 7.11   The magnitude of the Fourier transform of $h(n_1, n_2)$ shown in Fig. 7.10.

(a)

Figure 7.12   (a) An original image of
256 × 256 pixels; (b) the image in part
(a) filtered by the lowpass filter shown
in Fig. 7.11.

(b)

Example 2

We wish to determine $h(n_1, n_2)$ for the Fourier transform $H(\omega_1, \omega_2)$ shown in Fig. 7.13.
The function $H(\omega_1, \omega_2)$ is given by

$$H(\omega_1, \omega_2) = \begin{cases} 1, & |\omega_1| \le a \text{ and } |\omega_2| \le b \quad \text{(shaded region)} \\ 0, & a < |\omega_1| \le \pi \text{ or } b < |\omega_2| \le \pi \quad \text{(unshaded region)} \end{cases}$$

Since $H(\omega_1, \omega_2)$ is always periodic with period 2π along each of the variables ω_1 and
ω_2, $H(\omega_1, \omega_2)$ is shown only for $|\omega_1| \le \pi$ and $|\omega_2| \le \pi$. The function $H(\omega_1, \omega_2)$ can
be expressed as $H_1(\omega_1) \cdot H_2(\omega_2)$, where one possible choice of $H_1(\omega_1)$ and $H_2(\omega_2)$ is also
shown in Fig. 7.13. Computing the 1-D inverse Fourier transforms of $H_1(\omega_1)$ and $H_2(\omega_2)$

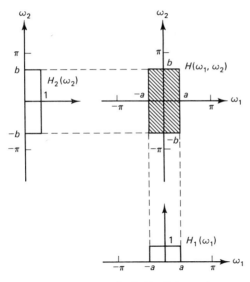

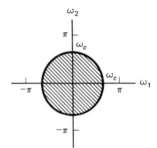

Figure 7.13   Frequency response of a separable ideal lowpass filter.

and using property 4 in Table 7.1, we have

$$h(n_1, n_2) = h_1(n_1) \cdot h_2(n_2) = \frac{\sin an_1}{\pi n_1} \cdot \frac{\sin bn_2}{\pi n_2}$$

Example 3

We wish to determine $h(n_1, n_2)$ for the Fourier transform $H(\omega_1, \omega_2)$ shown in Fig. 7.14. The function $H(\omega_1, \omega_2)$ is given by

$$H(\omega_1, \omega_2) = \begin{cases} 1, & \sqrt{\omega_1^2 + \omega_2^2} \le \omega_c \quad \text{(shaded region)} \\ 0, & \sqrt{\omega_1^2 + \omega_2^2} > \omega_c \text{ and } |\omega_1|, |\omega_2| < \pi \quad \text{(unshaded region)} \end{cases}$$

If $H(\omega_1, \omega_2)$ above is the frequency response of a 2-D linear shift-invariant system, the system is called a *circularly symmetric ideal lowpass filter*. The inverse Fourier transform of $H(\omega_1, \omega_2)$ in this example requires a fair amount of algebra, and the result is

$$h(n_1, n_2) = \frac{\omega_c}{2\pi \sqrt{n_1^2 + n_2^2}} \cdot J_1\left(\omega_c \cdot \sqrt{n_1^2 + n_2^2}\right) \qquad (7.31)$$

where $J_1(x)$ represents the Bessel function of the first kind and the first order. This example shows that 2-D Fourier transform or inverse Fourier transform operations can

Figure 7.14   Frequency response of a circularly symmetric ideal lowpass filter.

become quite complex algebraically relative to 1-D Fourier transform or inverse Fourier transform operations, even though the 2-D Fourier transform pair and many 2-D Fourier transform properties are straightforward extensions of 1-D results. From Eq. (7.31), we see that the unit sample response of a 2-D circularly symmetric ideal lowpass filter is also circularly symmetric, e.g., it is a function of $n_1^2 + n_2^2$. This is a special case of the result that circular symmetry of $H(\omega_1, \omega_2)$ implies circular symmetry of $h(n_1, n_2)$. We note, however, that circular symmetry of $h(n_1, n_2)$ does not imply circular symmetry of $H(\omega_1, \omega_2)$. The function $J_1(x)/x$ is sketched in Fig. 7.15. The sequence $h(n_1, n_2)$ in Eq. (7.31) is sketched in Fig. 7.16 for the case $\omega_c = 0.4\pi$.

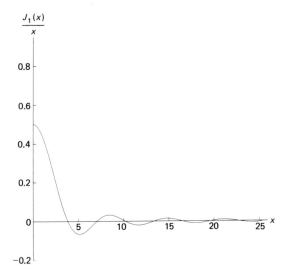

Figure 7.15 Sketch of $J_1(x)/x$, where $J_1(x)$ is the Bessel function of the first kind and the first order.

7.3 z-TRANSFORM

7.3.1 z-Transform

The Fourier transform discussed in Section 7.2 uniformly converges for stable sequences, and many interesting classes of unstable sequences, such as the unit step sequence $u(n_1, n_2)$, cannot be represented by their Fourier transforms. In this section, we discuss the z-transform representation of a sequence, which converges for a much wider class of signals.

The z-transform of a sequence $x(n_1, n_2)$ is denoted by $X(z_1, z_2)$ and is defined by

$$X(z_1, z_2) = \sum_{n_1=-\infty}^{\infty} \sum_{n_2=-\infty}^{\infty} x(n_1, n_2) \cdot z_1^{-n_1} \cdot z_2^{-n_2} \qquad (7.32)$$

where z_1 and z_2 are complex variables. Since each of the variables z_1 and z_2 represents 2-D space, the space represented by (z_1, z_2) is four-dimensional (4-D). As a result, it is extremely difficult to visualize points or segments of $X(z_1, z_2)$.

$h(n_1, n_2)$

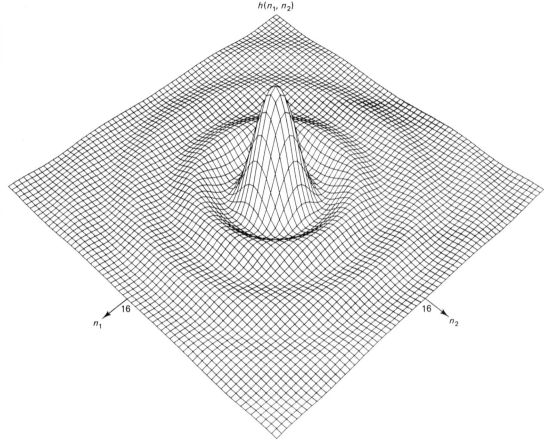

Figure 7.16 The unit sample response of a circularly symmetric ideal lowpass filter with $\omega_c = 0.4\pi$ in Eq. (7.31). The value at the origin, $h(0,0)$, is 0.126.

From Eqs. (7.28) and (7.32), we see that $X(z_1, z_2)$ is related to $X(\omega_1, \omega_2)$ by

$$X(z_1, z_2)\big|_{z_1 = e^{j\omega_1}, z_2 = e^{j\omega_2}} = \sum_{n_1=-\infty}^{\infty} \sum_{n_2=-\infty}^{\infty} x(n_1, n_2) \cdot e^{-j\omega_1 n_1} \cdot e^{-j\omega_2 n_2} = X(\omega_1, \omega_2)$$

$$(7.33)$$

Equation (7.33) states that $X(\omega_1, \omega_2)$ is $X(z_1, z_2)$ evaluated at $z_1 = e^{j\omega_1}$ and $z_2 = e^{j\omega_2}$. This is one reason why the z-transform is considered a generalization of the Fourier transform. The 2-D space represented by $(z_1 = e^{j\omega_1}, z_2 = e^{j\omega_2})$ is called the *unit surface*.

Suppose $X(z_1, z_2)$ in Eq. (7.32) is evaluated along $(z_1 = r_1 \cdot e^{j\omega_1}, z_2 = r_2 \cdot e^{j\omega_2})$, where r_1 and ω_1 are the radius and argument in the z_1-plane and r_2 and ω_2 are the radius and argument in the z_2-plane. The function $X(z_1, z_2)$ can be expressed as

$$X(z_1, z_2)\big|_{z_1 = r_1 \cdot e^{j\omega_1}, z_2 = r_2 \cdot e^{j\omega_2}} = \sum_{n_1=-\infty}^{\infty} \sum_{n_2=-\infty}^{\infty} x(n_1, n_2) \cdot r_1^{-n_1} \cdot r_2^{-n_2} \cdot e^{-j\omega_1 n_1} \cdot e^{-j\omega_2 n_2}$$

$$= F\big[x(n_1, n_2) \cdot r_1^{-n_1} \cdot r_2^{-n_2}\big] \qquad (7.34)$$

where $F[x(n_1, n_2) \cdot r_1^{-n_1} \cdot r_2^{-n_2}]$ represents the Fourier transform of $x(n_1, n_2) \cdot r_1^{-n_1} \cdot r_2^{-n_2}$. Since $X(\omega_1, \omega_2)$ uniformly converges for an absolutely summable sequence, from Eq. (7.34) $X(z_1, z_2)$ uniformly converges when $r_1^{-n_1} \cdot r_2^{-n_2} \cdot x(n_1, n_2)$ is absolutely summable:

$$\sum_{n_1=-\infty}^{\infty} \sum_{n_2=-\infty}^{\infty} |r_1^{-n_1} \cdot r_2^{-n_2} \cdot x(n_1, n_2)| < \infty, \qquad \text{where } r_1 = |z_1|, r_2 = |z_2| \qquad (7.35)$$

From Eq. (7.35), the convergence of $X(z_1, z_2)$ will generally depend on the value of $r_1 = |z_1|$ and $r_2 = |z_2|$. For example, for the unit step sequence $u(n_1, n_2)$, $r_1^{-n_1} \cdot r_2^{-n_2} \cdot u(n_1, n_2)$ is absolutely summable only for $(|z_1| > 1, |z_2| > 1)$, and its z-transform converges only for $(|z_1| > 1, |z_2| > 1)$. The region in the (z_1, z_2)-plane where $X(z_1, z_2)$ uniformly converges is called the *region of convergence (ROC)*.

For 1-D signals, the ROC is typically bounded by two concentric circles whose origin is at $|z| = 0$, as shown in Fig. 7.17(a). For 2-D signals, (z_1, z_2) represents a 4-D space and therefore the ROC cannot be sketched in a way analogous to that in Fig. 7.17(a). Fortunately, however, the ROC depends only on $|z|$ for 1-D signals and on $|z_1|$ and $|z_2|$ for 2-D signals. Therefore, an alternative way to sketch the ROC for 1-D signals is to use the $|z|$-axis. The ROC in Fig. 7.17(a) sketched using the $|z|$-axis is shown in Fig. 7.17(b). For 2-D signals, we can use the $(|z_1|, |z_2|)$-axes to sketch the ROC. An example of the ROC sketched in the $(|z_1|, |z_2|)$-plane is shown in Fig. 7.17(c). In this sketch, each point in the $(|z_1|, |z_2|)$-plane corresponds to a 2-D subspace in the 4-D (z_1, z_2) space. The ROC plays an important role in the z-transform representation of a sequence, as we will see shortly.

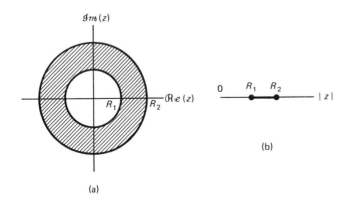

(a)

(b)

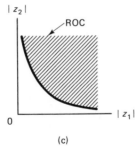

(c)

Figure 7.17   The representation of ROCs for 1-D and 2-D z-transforms. (a) A typical ROC for a 1-D z-transform; (b) the ROC in part (a) sketched using the $|z|$-axis; (c) a typical ROC for a 2-D z-transform using the $|z_1|$- and $|z_2|$-axes.

7.3.2 z-Transform Examples

Example 1

We wish to determine the z-transform and its ROC for the following sequence:

$$x(n_1, n_2) = a^{n_1} \cdot b^{n_2} \cdot u(n_1, n_2)$$

The sequence is sketched in Fig. 7.18(a). Using the z-transform definition in Eq. (7.32), we have

$$X(z_1, z_2) = \sum_{n_1=-\infty}^{\infty} \sum_{n_2=-\infty}^{\infty} a^{n_1} \cdot b^{n_2} \cdot u(n_1, n_2) \cdot z_1^{-n_1} \cdot z_2^{-n_2}$$

$$= \sum_{n_1=0}^{\infty} (a \cdot z_1^{-1})^{n_1} \cdot \sum_{n_2=0}^{\infty} (b \cdot z_2^{-1})^{n_2}$$

$$= \frac{1}{1 - a \cdot z_1^{-1}} \cdot \frac{1}{1 - b \cdot z_2^{-1}}, \qquad |z_1| > |a| \text{ and } |z_2| > |b|$$

The ROC is sketched in Fig. 7.18(b).

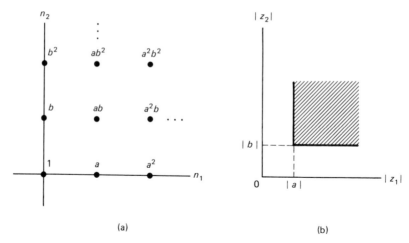

(a)   (b)

Figure 7.18   (a) A sequence $x(n_1, n_2) = a^{n_1} \cdot b^{n_2} \cdot u(n_1, n_2)$ and (b) the ROC of its z-transform.

For 1-D signals, poles of $X(z)$ are points in the z-plane. For 2-D signals, poles of $X(z_1, z_2)$ are 2-D surfaces in the 4-D (z_1, z_2) space. In Example 1, for instance, the poles of $X(z_1, z_2)$ can be represented as follows: $(z_1 = a, \text{ any } z_2)$, $(\text{any } z_1, z_2 = b)$. Each of the two pole representations corresponds to a 2-D surface in the 4-D (z_1, z_2) space.

For the 1-D case, the ROC is bounded by poles. For the 2-D case, the ROC is bounded by pole surfaces. To illustrate this, we consider the pole surfaces in Example 1. Taking the magnitudes of z_1 and z_2 that correspond to the pole surfaces, we have $|z_1| = |a|$ and $|z_2| = |b|$. These are the solid lines that bound the ROC, as shown in Fig. 7.18(b). It should be noted that each of the two solid lines in Fig. 7.18(b) represents a 3-D space since each point in the $(|z_1|, |z_2|)$-plane corresponds to a 2-D space. The pole surfaces, therefore, lie in the 2-D subspace within the 3-D spaces corresponding to the two solid lines in Fig. 7.18(b). To see this point more clearly, consider the 3-D space corresponding to $|z_1| = |a|$. This space can be represented as $(z_1 = |a| \cdot e^{j\omega_1}$, any $z_2)$, which is shown in Fig. 7.19. The pole surface corresponding to $(z_1 = a$, any $z_2)$ is sketched in the shaded region in Fig. 7.19.

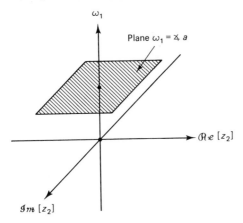

Figure 7.19   Pole surface within the 3-D space of $(|z_1| = |a|, z_2)$ for the sequence in Fig. 7.18.

Example 2

We wish to determine the z-transform and its ROC for the following sequence:

$$x(n_1, n_2) = -a^{n_1} \cdot b^{n_2} \cdot u(-n_1 - 1, n_2)$$

The sequence is sketched in Fig. 7.20(a). Using the z-transform definition in Eq. (7.32) and after a little algebra, we have

$$X(z_1, z_2) = \frac{1}{1 - a \cdot z_1^{-1}} \cdot \frac{1}{1 - b \cdot z_2^{-1}}$$

ROC: $|z_1| < |a|, |z_2| > |b|$

Pole Surfaces: $(z_1 = a$, any $z_2), ($any $z_1, z_2 = b)$

The ROC is sketched in Fig. 7.20(b).

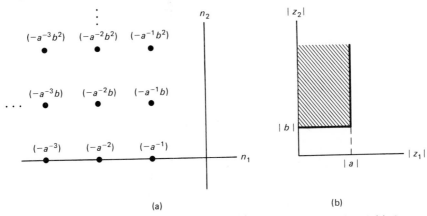

(a)   (b)

Figure 7.20   (a) A sequence $x(n_1, n_2) = -a^{n_1} \cdot b^{n_2} \cdot u(-n_1 - 1, n_2)$ and (b) the ROC of its z-transform.

Examples 1 and 2 show the importance of the ROC in the z-transform representation of a sequence. Specifically, even though the two sequences in the examples are very different, their z-transforms are exactly the same. Given only the z-transform, therefore, it is not possible to uniquely determine the sequence. The unique determination of the sequence requires not only the z-transform but its ROC.

Example 3

We wish to determine the z-transform and its ROC for the following sequence:

$$x(n_1, n_2) = a^{n_1} \cdot \delta(n_1 - n_2) \cdot u(n_1, n_2)$$

The sequence is sketched in Fig. 7.21(a). Using the z-transform definition in Eq. (7.32) and after a little algebra, we obtain

$$X(z_1, z_2) = \frac{1}{1 - a \cdot z_1^{-1} \cdot z_2^{-1}}$$

$$\text{ROC:} \quad |z_1| \cdot |z_2| > |a|$$

$$\text{Pole Surfaces:} \quad (\text{any } z_1 \neq 0, \ z_2 = a/z_1)$$

The ROC is sketched in Fig. 7.21(b). The pole surface is again a 2-D plane in the 4-D (z_1, z_2) space.

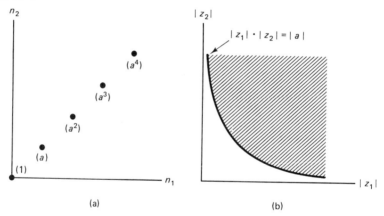

(a)   (b)

Figure 7.21 (a) The sequence $x(n_1, n_2) = a^{n_1} \cdot \delta(n_1 - n_2) \cdot u(n_1, n_2)$ and (b) the ROC of its z-transform.

7.3.3 Properties of the ROC

Many useful properties of the ROC can be obtained from the z-transform definition of Eq. (7.32). Some important properties are listed in Table 7.2.

Property 2 provides a necessary and sufficient condition for a sequence to be a first-quadrant sequence. The condition is that for any (z_1', z_2') in the ROC, all (z_1, z_2), including $|z_1| = \infty$ and $|z_2| = \infty$, in the shaded region in Fig 7.22 are also in the region of convergence.

The sketch in Fig. 7.22 is called a *constraint map* since it shows the constraints that the ROC of any first-quadrant sequence has to satisfy. Two examples of ROCs

TABLE 7.2   PROPERTIES OF THE ROC

Property 1:   A ROC is bounded by pole surfaces and is a connected region with no pole surfaces inside the ROC.

Property 2:   First-quadrant support sequence $\longleftrightarrow$ For any $(|z_1'|, |z_2'|)$ in the ROC, all $(|z_1| \geq |z_1'|, |z_2| \geq |z_2'|)$ are in the ROC.

Property 3:   Finite extent sequence $\longleftrightarrow$ ROC is everywhere except possibly $(|z_1| = 0$ or $\infty, |z_2| = 0$ or $\infty)$.

Property 4:   Stable sequence $\longleftrightarrow$ ROC includes the unit surface, $(|z_1| = 1, |z_2| = 1)$.

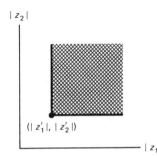

Figure 7.22   Constraint map for the ROC of first-quadrant sequences. For any (z_1', z_2') in the ROC, all (z_1, z_2) in the shaded region is also in the ROC.

that satisfy the constraint map in Fig. 7.22 are those shown in Figs. 7.18 and 7.21. They are the ROCs of $x(n_1, n_2) = a^{n_1} \cdot b^{n_2} \cdot u(n_1, n_2)$ and $x(n_1, n_2) = a^{n_1} \cdot \delta(n_1 - n_2) \cdot u(n_1, n_2)$, both of which are first-quadrant sequences. Constraint maps can also be obtained for other quadrant sequences.

7.3.4 Properties of the *z*-Transform

Many properties of the z-transform can be obtained from the z-transform definition of Eq. (7.32). Some important properties are listed in Table 7.3. All the properties, except properties 3 and 7, can be viewed as straightforward extensions of the 1-D case. Property 3 applies to separable sequences, and property 7 can be used in determining the z-transform of a first-quadrant sequence obtained by linearly mapping the variables of a wedge sequence.

7.3.5 Inverse *z*-Transform

The z-transform definition in Eq. (7.32) can be used to determine the z-transform and ROC of a 2-D sequence. As in the 1-D case, using Eq. (7.32) and Cauchy's integral theorem, we can determine the inverse z-transform relation that expresses $x(n_1, n_2)$ as a function of $X(z_1, z_2)$ and its ROC. The inverse z-transform relation is given by

$$x(n_1, n_2) = \frac{1}{(2\pi j)^2} \cdot \oint_{C_1} \oint_{C_2} X(z_1, z_2) \cdot z_1^{n_1 - 1} \cdot z_2^{n_2 - 1} \cdot dz_1 \cdot dz_2 \qquad (7.36)$$

TABLE 7.3   PROPERTIES OF THE z-TRANSFORM

$$x(n_1, n_2) \longleftrightarrow X(z_1, z_2), \text{ROC: } R_x \qquad y(n_1, n_2) \longleftrightarrow Y(z_1, z_2), \text{ROC: } R_y$$

Property 1: Linearity

$$a \cdot x(n_1, n_2) + b \cdot y(n_1, n_2) \longleftrightarrow a \cdot X(z_1, z_2) + b \cdot Y(z_1, z_2)$$
ROC: at least $R_x \cap R_y$

Property 2: Convolution

$$x(n_1, n_2) * y(n_1, n_2) \longleftrightarrow X(z_1, z_2) \cdot Y(z_1, z_2)$$
ROC: at least $R_x \cap R_y$

Property 3: Separable sequence

$$x_1(n_1) \cdot x_2(n_2) \longleftrightarrow X_1(z_1) \cdot X_2(z_2)$$
ROC: $|z_1| \in$ ROC of $X_1(z_1)$ and $|z_2| \in$ ROC of $X_2(z_2)$

Property 4: Shift of a sequence

$$x(n_1 - m_1, n_2 - m_2) \longleftrightarrow X(z_1, z_2) \cdot z_1^{-m_1} \cdot z_2^{-m_2}$$
ROC: R_x with possible exception of $|z_1| = 0, \infty, |z_2| = 0, \infty$

Property 5: Differentiation

(a) $-n_1 \cdot x(n_1, n_2) \longleftrightarrow \dfrac{\partial X(z_1, z_2)}{\partial z_1} z_1,$ ROC: R_x

(b) $-n_2 \cdot x(n_1, n_2) \longleftrightarrow \dfrac{\partial X(z_1, z_2)}{\partial z_2} z_2,$ ROC: R_x

Property 6: Symmetry properties
(a) $x^*(n_1, n_2) \longleftrightarrow X^*(z_1^*, z_2^*),$ ROC: R_x
(b) $x(-n_1, -n_2) \longleftrightarrow X(z_1^{-1}, z_2^{-1}),$ ROC: $|z_1^{-1}|, |z_2^{-1}|$ in R_x

Property 7: Linear mapping of variables

$$x(n_1, n_2) = y(m_1, m_2)|_{(m_1 = I \cdot n_1 + J \cdot n_2, \, m_2 = K \cdot n_1 + L \cdot n_2)} \longleftrightarrow Y(z_1, z_2) = X(z_1^I \cdot z_2^K, z_1^J \cdot z_2^L)$$
ROC: $(|z_1^I \cdot z_2^K|, |z_1^J \cdot z_2^L|)$ in R_x
Note: For those points of $y(m_1, m_2)$ that do not correspond to any $x(n_1, n_2)$, $y(m_1, m_2)$ is taken to be zero.

where C_1 and C_2 are both in the ROC of $X(z_1, z_2)$, C_1 is a closed contour that encircles in the counterclockwise direction the origin in the z_1-plane for any fixed z_2 on C_2, and C_2 is a closed contour that encircles in the counterclockwise direction the origin in the z_2-plane for any fixed z_1 on C_1.

The conditions that the contours C_1 and C_2 in Eq. (7.36) must satisfy appear to be quite complex, but there is a simple way to determine the contours C_1 and C_2 given the ROC. Suppose $(|z_1'|, |z_2'|)$ lies in the ROC. One set of contours C_1 and C_2 that satisfies the conditions in Eq. (7.36) is

$$C_1: \quad z_1 = |z_1'| \cdot e^{j\omega_1}, \qquad \omega_1: \ 0 \text{ to } 2\pi$$
$$C_2: \quad z_2 = |z_2'| \cdot e^{j\omega_2}, \qquad \omega_2: \ 0 \text{ to } 2\pi$$
(7.37)

If the sequence is stable, so that the ROC includes the unit surface, one possible set of contours is

$$C_1: \quad z_1 = e^{j\omega_1}, \qquad \omega_1: \quad 0 \text{ to } 2\pi$$
$$C_2: \quad z_2 = e^{j\omega_2}, \qquad \omega_2: \quad 0 \text{ to } 2\pi$$

(7.38)

If this set of contours is chosen in Eq. (7.36), then Eq. (7.36) reduces to the inverse Fourier transform relation in Eq. (7.29). This again shows that the Fourier transform representation is a special case of the z-transform representation.

From Eq. (7.36), the result of the contour integration differs depending on the contours. Therefore, a sequence $x(n_1, n_2)$ is not uniquely specified by $X(z_1, z_2)$ alone. Since the contours C_1 and C_2 can be determined from the ROC of $X(z_1, z_2)$, a sequence $x(n_1, n_2)$ is uniquely specified by both $X(z_1, z_2)$ and its ROC:

$$x(n_1, n_2) \quad \longleftrightarrow \quad X(z_1, z_2), \quad \text{ROC} \tag{7.39}$$

In theory, Eq. (7.36) can be used directly in determining $x(n_1, n_2)$ from $X(z_1, z_2)$ and its ROC. In practice, however, it is extremely tedious to perform the contour integration even for simple problems.

For 1-D signals, the approaches that have been used in performing the inverse z-transform operation without explicit evaluation of a contour integral are series expansion, partial fraction expansion, and inverse z-transform using z-transform properties. Among these approaches, the partial fraction expansion method is the only one that can always be used in performing the z-transform operation for any rational z-transform. In this method, the z-transform $X(z)$ is first expressed as a sum of simpler z-transforms by factoring the denominator polynomial as a product of lower-order polynomials. The inverse z-transform is then performed for each of the simpler z-transforms, and the results are added to obtain $x(n)$.

For 2-D signals, unfortunately, the partial fraction expansion method is not a general procedure that can be used to perform the inverse z-transform operation for a rational z-transform. In the 1-D case, the factorization of any 1-D polynomial as a product of lower-order polynomials is guaranteed by the fundamental theorem of algebra. A 2-D polynomial, however, cannot in general be factored as a product of lower-order polynomials. Therefore, a procedure analogous to the 1-D case cannot generally be used. Partly due to the difficulty involved in the partial fraction expansion method, no known practical method exists that can be used to perform the inverse z-transform of a general 2-D rational z-transform.

7.4 DIFFERENCE EQUATION

7.4.1 Linear Constant-Coefficient Difference Equation

Difference equations play a more important role for discrete space systems than differential equations do for analog systems. In addition to representing a wide class of discrete space systems, difference equations can be used to recursively generate their solutions. This use can be exploited in realizing digital filters with infinite-extent unit sample responses.

In this section we consider the class of linear constant coefficient difference equation (LCCDE) of the following form:

$$\sum_{(k_1,k_2)\in R_A}\sum a(k_1,k_2)\cdot y(n_1-k_1,n_2-k_2) = \sum_{(k_1,k_2)\in R_B}\sum b(k_1,k_2)\cdot x(n_1-k_1,n_2-k_2)$$

$$(7.40)$$

where $a(k_1,k_2)$ and $b(k_1,k_2)$ are known sequences, R_A represents the region in (k_1,k_2) such that $a(k_1,k_2)$ is nonzero, and R_B is similarly defined.

The LCCDE alone does not specify a system since there are many solutions of $y(n_1,n_2)$ in Eq. (7.40) for a given $x(n_1,n_2)$. For example, if $y_1(n_1,n_2)$ is a solution to $y(n_1,n_2) = \frac{1}{2}y(n_1-1,n_2+1) + \frac{1}{2}y(n_1+1,n_2-1) + x(n_1,n_2)$, then so is $y_1(n_1,n_2) + f(n_1+n_2)$ for any function f. To uniquely specify a solution, we need a set of boundary conditions. Since the boundary conditions have to determine specific functional forms as well as the constants associated with the functions, they are typically an infinite number of points in the output $y(n_1,n_2)$. This aspect differs fundamentally from the 1-D case. In the 1-D case, the LCCDE specifies a solution within arbitrary constants. For example, an Nth-order 1-D LCCDE specifies a solution within N constants, and therefore N initial conditions (N points in the output $y(n)$) are generally sufficient to uniquely specify a solution. In the 2-D case, we need a set of boundary conditions that are typically an infinite number of points in the output $y(n_1,n_2)$.

7.4.2 Difference Equation with Boundary Conditions

The problem of solving an LCCDE with boundary conditions can be stated as follows: Given $x(n_1,n_2)$, and $y(n_1,n_2)$ for $(n_1,n_2)\in R_{BC}$, find the solution to

$$\sum_{(k_1,k_2)\in R_A}\sum a(k_1,k_2)\cdot y(n_1-k_1,n_2-k_2)$$

$$(7.41)$$

$$= \sum_{(k_1,k_2)\in R_B}\sum b(k_1,k_2)\cdot x(n_1-k_1,n_2-k_2)$$

One approach that can be used in solving a 1-D LCCDE with initial conditions is to obtain a homogeneous solution and a particular solution, determine the total solution as a sum of these, and then impose initial conditions. For the 2-D case, unfortunately, this approach cannot be used. First, there is no general procedure to obtain the homogeneous solution. The homogeneous solution consists of unknown functions, and the specific functional form of $k\cdot a^n$, used in the 1-D case, cannot be used for the 2-D case. Second, the particular solution cannot generally be obtained by inspection or by the z-transform method since there is no practical procedure for performing the inverse z-transform operation for the 2-D case. Furthermore, determining the unknown functions in the homogeneous solution by imposing the boundary conditions (an infinite number of known values of $y(n_1,n_2)$) is not a simple linear problem.

Another approach in solving Eq. (7.41) is to compute $y(n_1,n_2)$ recursively, which is how it is typically done on a computer. To illustrate this approach, consider the following 2-D LCCDE with boundary conditions (BC):

$$y(n_1, n_2) = y(n_1 - 1, n_2) + y(n_1, n_2 - 1) + y(n_1 - 1, n_2 - 1) + x(n_1, n_2)$$

$$x(n_1, n_2) = \delta(n_1, n_2) \tag{7.42}$$

BC: $y(n_1, n_2) = 1,$ for $n_1 < 0$ or $n_2 < 0$

The output $y(n_1, n_2)$ can be obtained recursively as follows:

$$y(0, 0) = y(-1, 0) + y(0, -1) + y(-1, -1) + x(0, 0) = 4$$

$$y(1, 0) = y(0, 0) + y(1, -1) + y(0, -1) + x(1, 0) = 6 \tag{7.43}$$

$$y(0, 1) = y(-1, 1) + y(0, 0) + y(-1, 0) + x(0, 1) = 6$$

$$y(2, 0) = \cdots$$
$$\vdots$$

Even though the approach to computing the output recursively appears to be quite simple, the proper choice of the boundary conditions so that the LCCDE with boundary conditions has a unique solution is not straightforward. In the 1-D case, N initial conditions are typically both necessary and sufficient for an Nth-order difference equation. In the 2-D case, the choice of the boundary conditions so that the LCCDE will have a unique solution is not too obvious.

Suppose we have chosen the boundary conditions such that the LCCDE has a unique solution and therefore we can consider the LCCDE with boundary conditions as a system. In both the 1-D and 2-D cases, the system is in general neither linear nor shift-invariant. The difference equation is of interest to us primarily because it is the only practical way to realize an infinite impulse response (IIR) digital filter. Since a digital filter is a linear shift-invariant system, we need to force the difference equation to become a linear shift-invariant system. We can do this by choosing a proper set of boundary conditions. In the 1-D case, one way to force a difference equation to be a linear shift-invariant system is to impose an initial condition known as an initial rest condition. An initial rest condition is defined to be an initial condition obtained by requiring the output $y(n)$ to be zero for $n < n_0$ whenever the input $x(n)$ is zero for $n < n_0$, and it also forces the resulting system to be causal. Even though the extension is not quite straightforward, it is possible to extend the general idea behind the initial rest condition to the 2-D case. This is discussed in the following section.

7.4.3 Difference Equation as a Linear Shift-Invariant System

One way to force an LCCDE with boundary conditions to be a linear shift-invariant system so that it can be used in realizing an IIR filter is to choose the boundary conditions using the following three steps.

1. Interpret the LCCDE as a specific computational procedure.
2. Determine R_{BC}, the region (n_1, n_2) for which the boundary conditions are applied, as follows:
   (a) Determine R_h, the region of support of the unit sample response $h(n_1, n_2)$.

(b) Determine R_y, the region of support for the output $y(n_1, n_2)$, from R_h and R_x, where R_x is similarly defined.

(c) R_{BC}: all $(n_1, n_2) \notin R_y$.

3. Boundary conditions: Set $y(n_1, n_2) = 0$ for $(n_1, n_2) \in R_{BC}$.

Step 1. In this step, we interpret a difference equation as a specific computational procedure. The best way to explain this step is by looking at a specific example. Consider the following LCCDE:

$$y(n_1, n_2) + 2 \cdot y(n_1 - 1, n_2) + 3y(n_1, n_2 - 1) + 4(n_1 - 1, n_2 - 1) = x(n_1, n_2)$$

(7.44)

Since there are four terms of the form $y(n_1 - k_1, n_2 - k_2)$, we obtain four equations by leaving only one of the four terms on the left-hand side of the equation, as follows:

$$y(n_1, n_2) = -2y(n_1 - 1, n_2) - 3y(n_1, n_2 - 1) - 4 \cdot y(n_1 - 1, n_2 - 1) + x(n_1, n_2)$$

(7.45a)

$$y(n_1 - 1, n_2) = -\tfrac{1}{2}y(n_1, n_2) - \tfrac{3}{2}y(n_1, n_2 - 1) - 2y(n_1 - 1, n_2 - 1) + \tfrac{1}{2}x(n_1, n_2)$$

(7.45b)

$$y(n_1, n_2 - 1) = -\tfrac{1}{3}y(n_1, n_2) - \tfrac{2}{3}y(n_1 - 1, n_2) - \tfrac{4}{3}y(n_1 - 1, n_2 - 1) + \tfrac{1}{3}x(n_1, n_2)$$

(7.45c)

$$y(n_1 - 1, n_2 - 1) = -\tfrac{1}{4}y(n_1, n_2) - \tfrac{1}{2}y(n_1 - 1, n_2) - \tfrac{3}{4}y(n_1, n_2 - 1) + \tfrac{1}{4}(n_1, n_2)$$

(7.45d)

By a simple change of variables, Eqs. (7.45) can be rewritten so that the left-hand side of each equation has the form $y(n_1, n_2)$:

$$y(n_1, n_2) = -2y(n_1 - 1, n_2) - 3(n_1, n_2 - 1) - 4y(n_1 - 1, n_2 - 1) + x(n_1, n_2)$$

(7.46a)

$$y(n_1, n_2) = -\tfrac{1}{2}y(n_1 + 1, n_2) - \tfrac{3}{2}y(n_1 + 1, n_2 - 1) - 2y(n_1, n_2 - 1)$$
$$+ \tfrac{1}{2}x(n_1 + 1, n_2)$$

(7.46b)

$$y(n_1, n_2) = -\tfrac{1}{3}y(n_1, n_2 + 1) - \tfrac{2}{3}y(n_1 - 1, n_2 + 1) - \tfrac{4}{3}y(n_1 - 1, n_2)$$
$$+ \tfrac{1}{3}(n_1, n_2 + 1)$$

(7.46c)

$$y(n_1, n_2) = -\tfrac{1}{4}y(n_1 + 1, n_2 + 1) - \tfrac{1}{2}y(n_1, n_2 + 1) - \tfrac{3}{4}y(n_1 + 1, n_2)$$
$$+ \tfrac{1}{4}x(n_1 + 1, n_2 + 1)$$

(7.46d)

Even though all four equations in Eq. (7.46) are the same LCCDEs, they correspond to four different specific computational procedures by proper interpretation. The interpretation we use is that the left-hand side $y(n_1, n_2)$ is always computed from the right-hand side expression for all (n_1, n_2). When this interpretation is strictly followed, then each of the four equations in Eq. (7.46) corresponds to a different computational procedure. This will become clearer when we discuss step 2.

It is often convenient to represent a specific computational procedure pictorially. Specifically, to pictorially represent the specific computational procedure corresponding to Eq. (7.46a), we consider computing $y(0,0)$. Since $y(n_1, n_2)$ on the left-hand side is always computed from the right-hand side, $y(0,0)$ in this case is computed as

$$y(0,0) \quad \longleftarrow \quad -2y(-1,0) - 3y(0,-1) - 4y(-1,-1) + x(0,0) \qquad (7.47)$$

We have used $\leftarrow$ to emphasize that $y(0,0)$ is always computed from the right-hand side. Equation (7.47) is represented in Fig. 7.23. The value $y(0,0)$ that is computed is denoted by $\times$ in Fig. 7.23(a). The values $y(-1,0)$, $y(0,-1)$, and $y(-1,-1)$ that are used in obtaining $y(0,0)$ are marked as solid dots ($\bullet$) in the figure, with the proper coefficient attached to the corresponding point. The value $x(0,0)$ used in obtaining $y(0,0)$ is marked as a solid dot in Fig. 7.23(b). To compute $y(0,0)$, therefore, we look at $y(n_1, n_2)$ and $x(n_1, n_2)$ at the points marked by solid dots, multiply each value by the corresponding scaling factor indicated, and sum all the terms. Figure 7.23(a) is called an *output mask* and Fig. 7.23(b) is called an *input mask* since they are "masks" that are applied to the output and input to compute $y(0,0)$. We note that the output mask always has $\times$ at the origin, but the input mask does not have $\times$ anywhere.

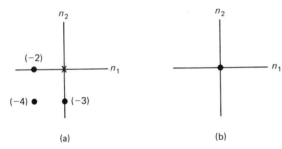

Figure 7.23   (a) Output mask and (b) input mask corresponding to Eq. (7.47). Note that the output mask always has an $\times$ at the origin, but the input mask does not have an $\times$ anywhere.

Even though the output and input masks are sketched for the case when $y(0,0)$ is computed, they are also very useful in visualizing what happens when other points of $y(n_1, n_2)$ are computed. Suppose we wish to compute $y(5,3)$ using the same computational procedure as in Fig. 7.23. The points that are used in determining $y(5,3)$ are indicated in Fig. 7.24, with the proper scaling factors attached. Figure 7.24 is simply a shifted version of Fig. 7.23.

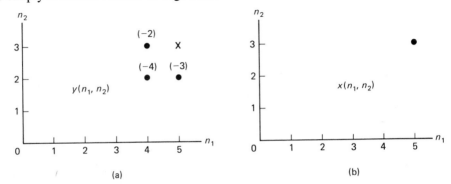

Figure 7.24   (a) Output mask and (b) input mask in Fig. 7.23 shifted to illustrate the computation of $y(5,3)$.

From the above discussion, an LCCDE can lead to many different specific computational procedures. Which procedure is chosen from these many possibilities is determined from the specific application context. We discuss this point in the specific example following steps 2 and 3.

Step 2. In this step, we determine R_y, the region of support of the output $y(n_1, n_2)$. To determine R_y, we first determine R_h, the region of support of the unit sample response. To illustrate how R_h is determined, consider the following computational procedure:

$$y(n_1, n_2) \longleftarrow -2y(n_1 - 1, n_2) - 3y(n_1, n_2 - 1)$$
$$- 4y(n_1 - 1, n_2 - 1) + x(n_1, n_2) \tag{7.48}$$

The output and input masks for this computational procedure were shown in Fig. 7.23. The region of support R_h is the region of (n_1, n_2) for which $y(n_1, n_2)$ is influenced by the pulse $\delta(n_1, n_2)$ when we set $x(n_1, n_2) = \delta(n_1, n_2)$. Consider $y(0, 0)$. From Eq. (7.48) or Fig. 7.23, we see that $y(0, 0)$ is influenced by $y(-1, 0)$, $y(0, -1)$, $y(-1, -1)$, and $\delta(0, 0)$. Clearly $y(0, 0)$ is influenced by the pulse $\delta(n_1, n_2)$. Now consider $y(1, 0)$, $y(0, 1)$, and $y(1, 1)$. From Eq. (7.48) or Fig. 7.23, we have

$$y(1, 0) \longleftarrow -2 \cdot y(0, 0) - 3 \cdot y(1, -1) - 4 \cdot y(0, -1) + \delta(1, 0)$$
$$y(0, 1) \longleftarrow -2 \cdot y(-1, 1) - 3 \cdot y(0, 0) - 4 \cdot y(-1, 0) + \delta(0, 1) \tag{7.49}$$
$$y(1, 1) \longleftarrow -2 \cdot y(0, 1) - 3 \cdot y(1, 0) - 4 \cdot y(0, 0) + \delta(1, 1)$$

Since $\delta(n_1, n_2)$ has already influenced $y(0, 0)$, and $y(0, 0)$ in turn influences $y(1, 0), y(0, 1)$, and $y(1, 1)$, the above three output values in Eq. (7.49) are influenced by the pulse $\delta(n_1, n_2)$. Now consider $y(-1, 0)$. From Eq. (7.48) or Fig. 7.23, we see that $y(-1, 0)$ is obtained from

$$y(-1, 0) \longleftarrow -2 \cdot y(-2, 0) - 3 \cdot y(-1, -1) - 4 \cdot y(-2, -1) + \delta(-1, 0) \tag{7.50}$$

This is shown in Fig. 7.25. The terms that influence $y(-1, 0)$ in Eq. (7.50) are obtained from

$$y(-2, 0) \longleftarrow -2 \cdot y(-3, 0) - 3 \cdot y(-2, -1) - 4 \cdot y(-3, -1) + \delta(-2, 0)$$
$$y(-1, -1) \longleftarrow -2 \cdot y(-2, -1) - 3 \cdot y(-1, -2) - 4 \cdot y(-2, -2) + \delta(-1, -1)$$
$$y(-2, -1) \longleftarrow -2 \cdot y(-3, -1) - 3 \cdot y(-2, -2) - 4 \cdot y(-3, -2) + \delta(-2, -1) \tag{7.51}$$

These are also shown in Fig. 7.25. From Fig. 7.25, the points that influence $y(-1, 0)$ are shown as $y(n_1, n_2)$ in the shaded region in Fig. 7.26. Since the pulse $x(n_1, n_2) = \delta(n_1, n_2)$ has its first impact on $y(0, 0)$, and $y(0, 0)$ does not in any way affect $y(-1, 0)$, $y(-1, 0)$ is not influenced by the pulse $\delta(n_1, n_2)$. If we consider all other points of $y(n_1, n_2)$ analogously, we can easily argue that the region of (n_1, n_2) for which $y(n_1, n_2)$ is influenced by $x(n_1, n_2) = \delta(n_1, n_2)$ is only the first-quadrant region. This is the region R_h. In essence, the pulse $\delta(n_1, n_2)$ has a direct effect only on $y(0, 0)$.

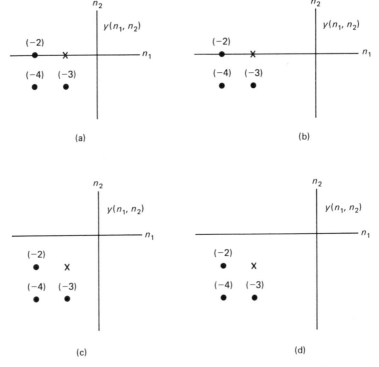

Figure 7.25 Computation of $y(-1,0)$ and its neighborhood values: (a) $y(-1,0)$; (b) $y(-2,0)$; (c) $y(-1,-1)$; (d) $y(-2,-1)$.

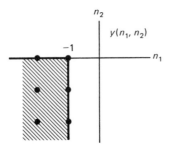

Figure 7.26 The region that influences $y(-1,0)$.

Because of the specific computational procedure in Eq. (7.48) and Fig. 7.23, $y(0,0)$ influences only the first-quadrant region.

Once R_h is determined, R_y can be obtained from R_h and R_x. Suppose $x(n_1, n_2)$ is the sequence shown in Fig. 7.27(a). By convolving $x(n_1, n_2)$ and $h(n_1, n_2)$, R_y can be easily determined. For R_h considered above, R_y is given by the shaded region in Fig. 7.27(b). R_{BC}, the region (n_1, n_2) for which the boundary conditions are applied, is determined to be all (n_1, n_2) outside R_y. In the current example, R_{BC} is given by the shaded region in Fig. 7.27(c).

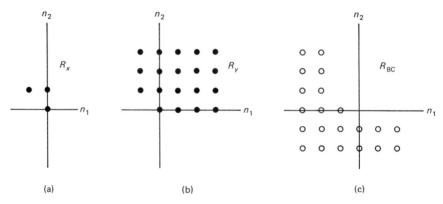

Figure 7.27   Region of support for (a) $x(n_1, n_2)$, (b) $y(n_1, n_2)$ and (c) boundary conditions for the computational procedure in Eq. (7.48).

Step 3.   In this step, we choose the boundary conditions such that $y(n_1, n_2) = 0$ for all $(n_1, n_2) \in R_{BC}$. In the example considered in Step 2, we choose the boundary conditions as follows:

$$y(n_1, n_2) = 0 \quad \text{for } n_1 < -1 \quad \text{or} \quad n_2 < 0 \quad \text{or} \quad (n_1, n_2) = (-1, 0)$$

To illustrate the three steps further, we consider one specific example in the next section.

7.4.4 Example

Suppose we have designed an IIR digital filter whose system function $H(z_1, z_2)$ is given by

$$H(z_1, z_2) = \frac{1}{1 + \frac{1}{2}z_1^{-1} \cdot z_2^{-1}} \tag{7.52}$$

The IIR filter was designed by making the unit sample response of the designed system as close as possible in some sense to an ideal unit sample response that is a first-quadrant sequence. Therefore, we know that the filter is a first-quadrant support system. We wish to determine the output of the filter when the input $x(n_1, n_2)$ is given by

$$x(n_1, n_2) = 1, \quad -1 \le n_1 \le 1 \text{ and } -1 \le n_2 \le 1$$
$$0, \quad \text{otherwise} \tag{7.53}$$

as shown in Fig. 7.28.

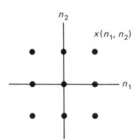

Figure 7.28   An input sequence $x(n_1, n_2)$ to the system of Eq. (7.52).

Since the only practical way to implement a general IIR filter is by using a difference equation, we first convert Eq. (7.52) to a difference equation as follows:

$$H(z_1, z_2) = \frac{Y(z_1, z_2)}{X(z_1, z_2)} = \frac{1}{1 + \frac{1}{2}z_1^{-1} \cdot z_2^{-1}}$$

$$Y(z_1, z_2) + \frac{1}{2} \cdot Y(z_1, z_2) \cdot z_1^{-1} \cdot z_2^{-1} = X(z_1, z_2) \qquad (7.54)$$

$$y(n_1, n_2) + \frac{1}{2} \cdot y(n_1-1, n_2-1) = x(n_1, n_2)$$

Since the IIR filter is a linear shift-invariant system, we choose the proper set of boundary conditions so that the LCCDE becomes a linear shift-invariant system. Two specific computational procedures correspond to the LCCDE in Eq. (7.54):

$$y(n_1, n_2) \quad \longleftarrow \quad -\frac{1}{2}y(n_1 - 1, n_2 - 1) + x(n_1, n_2) \qquad (7.55)$$

$$y(n_1, n_2) \quad \longleftarrow \quad -2 \cdot y(n_1 + 1, n_2 + 1) + 2 \cdot x(n_1 + 1, n_2 + 1) \qquad (7.56)$$

The output and input masks corresponding to each of the two computational procedures in Eqs. (7.55) and (7.56) are shown in Fig. 7.29. The region of support of $h(n_1, n_2)$ for each of the two computational procedures is shown in Fig. 7.30. Since we know that the filter is a first-quadrant support system, we choose the computational procedure given by Eq. (7.55). To determine the boundary conditions for the computational procedure chosen, we determine R_y, the region of support for the output sequence $y(n_1, n_2)$, from R_x in Fig. 7.28 and R_h in Fig. 7.30(a). The region R_y is shown in Fig. 7.31(a). The boundary conditions that we use are shown in Fig. 7.31(b). With the boundary conditions shown in Fig. 7.31(b), the output $y(n_1, n_2)$ can be recursively computed from Eqs. (7.53) and (7.55). The result is shown in Fig. 7.32. If we double the input, the output will also double. If we shift the input, then the output will be shifted by a corresponding amount. This is consistent with the fact that the computational procedure is a linear shift-invariant system.

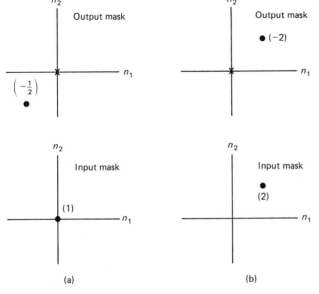

Figure 7.29 Output and input masks for the computational procedure given by (a) Eq. (7.55) and (b) Eq. (7.56).

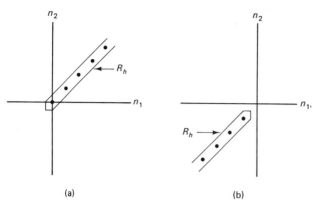

(a)   (b)

Figure 7.30   The regions of support for $h(n_1, n_2)$ corresponding to the two computational procedures given by (a) Eq. (7.55) and (b) Eq. (7.56).

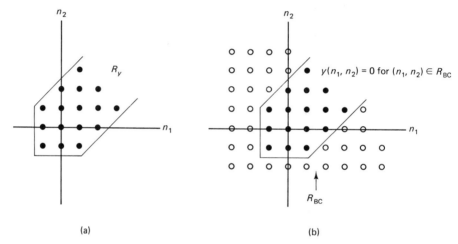

(a)   (b)

Figure 7.31   The regions of support for (a) $y(n_1, n_2)$ and (b) boundary conditions, for the computational procedure given by Eq. (7.55) and the input $x(n_1, n_2)$ in Fig. 7.28.

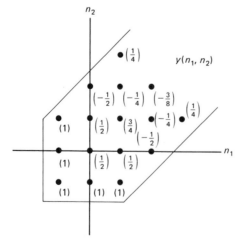

Figure 7.32   The output $y(n_1, n_2)$ for the computational procedure given by Eq. (7.55) and input $x(n_1, n_2)$ in Fig. 7.28.

7.4.5 Recursive Computability

The LCCDE with boundary conditions plays a particularly important role in digital signal processing since it is the only practical way to realize an IIR filter. As we discussed in the previous sections, the LCCDE can be used as a recursive procedure in computing the output. We define a system to be recursively computable when there exists a path we can follow in computing every output point recursively, one at a time. The example in Section 7.4.4 corresponds to a recursively computable system.

From the definition of a recursively computable system, we can easily show that not all computational procedures resulting from LCCDEs are recursively computable. For example, consider a computational procedure whose output mask is shown in Fig. 7.33. From the output mask, it is clear that computing $y(0,0)$ requires $y(1,0)$ and computing $y(1,0)$ requires $y(0,0)$. Therefore, we cannot compute $y(0,0)$ and $y(1,0)$ one at a time recursively, and the computational procedure in Fig. 7.33 is not a recursively computable system. For a finite-extent input $x(n_1, n_2)$, we can show that a system is recursively computable if the output mask has a wedge support. Examples of wedge support output masks are shown in Fig. 7.34. The types of output masks shown in Figs. 7.34(b) and (c) are called "nonsymmetric half-plane" output masks. Examples of output masks that do not have wedge support are shown in Fig. 7.35.

For a recursively computable system, we can follow many different paths in computing all the output values we need. Even though $y(n_1, n_2)$ can be computed in many different orders, the result does not depend on the specific order that is used.

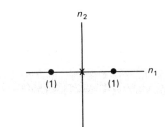

Figure 7.33 An example of an output mask of a system that is not recursively computable.

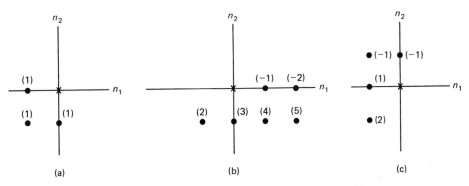

Figure 7.34 Examples of output masks of recursively computable systems.

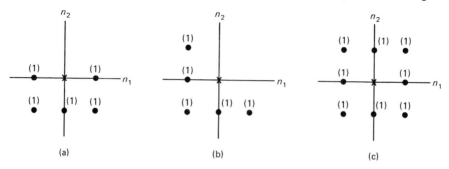

Figure 7.35 Examples of output masks of systems that are not recursively computable.

7.5 STABILITY

7.5.1 Stability Problem

When a discrete space system such as a digital filter is designed, an important consideration is the stability of the system. In this section, we consider the problem of testing the stability of a discrete space system.

As discussed in Section 7.1.4, a system is considered stable in the bounded input–bounded output (BIBO) sense if and only if a bounded input always leads to a bounded output. For a linear shift-invariant system, a necessary and sufficient condition for the system to be stable is that its unit sample response $h(n_1, n_2)$ be absolutely summable:

$$\sum_{n_1=-\infty}^{\infty} \sum_{n_2=-\infty}^{\infty} |h(n_1, n_2)| < \infty$$

Even though this condition is a straightforward extension of 1-D results, issues related to stability are quite different for 1-D and 2-D results.

Often we are given the system function $H(z_1, z_2)$ and information about its ROC, and we wish to determine the system stability. If the ROC is explicitly given, determining the system stability is straightforward because a system is stable if and only if the ROC includes the unit surface ($|z_1| = |z_2| = 1$). Unfortunately, however, the ROC is seldom explicitly given; typically, only implicit information about the ROC is available. When the system function corresponds to a digital filter, for example, the region of support of its unit sample response $h(n_1, n_2)$ is usually known from the filter design step. Since our main interest is testing the stability of digital filters, the stability problem we consider in this section is to determine the system stability given $H(z_1, z_2)$ and the region of support of $h(n_1, n_2)$.

When the system function $H(z_1, z_2)$ corresponds to a digital filter, restrictions are imposed on $H(z_1, z_2)$ and the region of support of $h(n_1, n_2)$. One restriction is that $H(z_1, z_2)$ is a rational z-transform, which can be expressed as

$$H(z_1, z_2) = \frac{B(z_1, z_2)}{A(z_1, z_2)} \tag{7.57}$$

where $A(z_1, z_2)$ and $B(z_1, z_2)$ are finite-order polynomials in z_1 and z_2. Another restriction is that the system is recursively computable. With these two restrictions, the system can be realized by an LCCDE with boundary conditions, and the output can be computed recursively one at a time.

In the 1-D case, when the system function $H(z)$ is expressed as $B(z)/A(z)$ where there are no common factors between $A(z)$ and $B(z)$, $B(z)$ does not affect the system stability. In the 2-D case, however, the presence of $B(z_1, z_2)$ in Eq. (7.57) can stabilize an otherwise unstable system, even when there are no common factors between $A(z_1, z_2)$ and $B(z_1, z_2)$. This case occurs very rarely, there is no known procedure that can be used when such a case does occur, and an unstable system stabilized by $B(z_1, z_2)$ is unstable for all practical purposes. We will therefore assume that the numerator polynomial $B(z_1, z_2)$ does not affect the system stability. To make this explicit, we'll assume that $B(z_1, z_2) = 1$.

When the input $x(n_1, n_2)$ is a finite-extent sequence, which is the case of our primary interest, the recursive computability requires that the output mask has wedge shape support. This in turn requires the unit sample response $h(n_1, n_2)$ to have wedge shape support when $B(z_1, z_2) = 1$. As discussed in Section 7.1.4, it is always possible to find a linear mapping of variables that transforms a wedge sequence to a first-quadrant sequence without affecting the stability of the sequence. Therefore, stability results that apply to first-quadrant sequences can be used for all recursively computable systems. In our approach, we will first transform a recursively computable system to a first-quadrant support system by a linear mapping of variables. This transformation changes the system function $H(z_1, z_2)$ to a new system function $H'(z_1, z_2)$. We will then apply to $H'(z_1, z_2)$ stability results that apply to first-quadrant support systems. As we have discussed, the reason for first transforming a wedge support system to a first-quadrant support system is that developing stability results that apply to first-quadrant support systems is much easier notationally and conceptually than developing stability results that apply to wedge support systems.

Given the preceding discussion, the stability problem we will consider can be stated as follows.

Problem: Given $H(z_1, z_2) = 1/[A(z_1, z_2)]$ and assuming that $h(n_1, n_2)$ is a first-quadrant sequence, determine the system stability.

In the following sections, we discuss several stability theorems and use them to solve the problem of determining the system stability.

7.5.2 Stability Theorems

In the 1-D case, the problem of testing the stability of a causal system whose system function is given by $H(z) = 1/[A(z)]$ is quite straightforward. Since a 1-D polynomial $A(z)$ can always be factored as a product of first-order polynomials, we can always determine the poles of $H(z)$. The stability of the causal system is equivalent to the condition that all the poles are inside the unit circle.

A similar approach cannot be used in testing the stability of a 2-D first-quadrant support system. The approach just described requires the specific locations of all poles. Since a 2-D polynomial $A(z_1, z_2)$ in general cannot be factored as a product of lower-order polynomials, it is extremely difficult to determine all the pole surfaces of

$H(z_1, z_2) = 1/[A(z_1, z_2)]$, and the approach based on explicit determination of all pole surfaces has not led to successful practical procedures of testing the system stability. In this section, we will discuss theorems that can be used in developing practical procedures for testing the stability of a 2-D first-quadrant support system without explicit determination of all pole surfaces.

Shanks' Theorem.   Shanks' theorem does not directly lead to practical stability testing procedures, but it is one of the earliest such theorems, is conceptually very simple, and has led to other stability theorems, which we will discuss later. In addition, this theorem shows that a result that is not useful for 1-D signals can be very useful when it is extended to the 2-D case.

In the 1-D case, the stability of a causal system with system function $H(z) = 1/[A(z)]$ is equivalent to requiring all the poles of $H(z)$ to be within the unit circle:

$$\text{Stability} \quad \longleftrightarrow \quad \text{All poles (solutions to } A(z) = 0) \text{ are inside } |z| = 1 \qquad (7.58)$$

An equivalent statement to Eq. (7.58) is given by

$$\text{Stability} \quad \longleftrightarrow \quad A(z) \neq 0 \qquad \text{for } |z| \geq 1 \qquad (7.59)$$

If all the poles are inside $|z| = 1$, then $A(z)$ cannot be zero for any $|z| \geq 1$. If $A(z) \neq 0$ for any $|z| \geq 1$, then all the poles must be inside $|z| = 1$. Therefore, Eqs. (7.58) and (7.59) are equivalent conditions.

Even though Eqs. (7.58) and (7.59) are equivalent statements, their implications for testing system stability are quite different. The condition in Eq. (7.58) suggests a procedure where we explicitly determine all pole locations and see if they all are inside $|z| = 1$. The condition in Eq. (7.59), however, suggests a procedure where we evaluate $A(z)$ for each z such that $|z| \geq 1$ and see if $A(z)$ is zero for any $|z| \geq 1$. This requires a search in the 2-D plane. In the 1-D case, the procedure suggested by Eq. (7.58) is extremely simple. Therefore, the procedure suggested by Eq. (7.59), which requires a 2-D search, is not useful. In the 2-D case, however, the procedure suggested by Eq. (7.58) is extremely difficult. The extension of Eq. (7.59) to the 2-D case is Shanks' theorem.

Shanks' theorem can be stated as follows:

$$\text{Stability} \quad \longleftrightarrow \quad A(z_1, z_2) \neq 0 \qquad \text{for any } |z_1|, |z_2| \geq 1 \qquad (7.60)$$

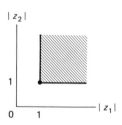

The result in Eq. (7.60) is simple to demonstrate. To show that stability implies the condition in Eq. (7.60), we first note that stability implies that the unit surface $(|z_1| = 1, |z_2| = 1)$ is in the inside of the ROC of $H(z_1, z_2)$. Because $h(n_1, n_2)$ is a first-quadrant sequence, the ROC of $H(z_1, z_2)$ has to satisfy the conditions given by the constraint map in Fig. 7.22. Therefore all (z_1, z_2) such that $|z_1| \geq 1, |z_2| \geq 1$ has to be in the ROC. Since the ROC cannot have any pole surfaces $(A(z_1, z_2) = 0)$, $A(z_1, z_2) \neq 0$ for any $|z_1|, |z_2| \geq 1$. To show that the condition in Eq. (7.60) implies the system stability, we note that this condition implies that there are no pole surfaces for any $|z_1|, |z_2| \geq 1$. Because $h(n_1, n_2)$ is a first-quadrant sequence, the ROC of $H(z_1, z_2)$ has to satisfy the constraint map in Fig. 7.22. This requirement combined with the property that the ROC is bounded by pole surfaces implies that the ROC includes the unit surface, which in turn implies the system stability.

The condition in Eq. (7.60) suggests a procedure where we evaluate $A(z_1, z_2)$ in the 4-D space $(|z_1| \geq 1, |z_2| \geq 1)$ shown in the shaded region in Eq. (7.60). The search in 4-D space is, of course, a tremendous amount of work, and this procedure itself cannot be used in practice. Other theorems, however, state that the space where we search is considerably smaller than the 4-D space in Shanks' theorem. Since the proofs of these theorems are quite involved and the proofs themselves do not provide much insight into the theorems, we state the theorems with our interpretations but without proof.

Huang's Theorem.   Huang's theorem can be stated as follows:

$$\text{Stability} \longleftrightarrow \begin{array}{lll} \text{(a)} & A(z_1, z_2) \neq 0 & \text{for } |z_1| = 1, |z_2| \geq 1 \\ \text{(b)} & A(z_1, z_2) \neq 0 & \text{for } |z_1| \geq 1, z_2 = 1 \end{array} \qquad (7.61)$$

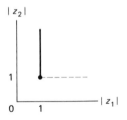

To satisfy condition (a), we need to ensure that $A(z_1, z_2)$ is not zero for any (z_1, z_2) such that $|z_1| = 1$ and $|z_2| \geq 1$. This requires a 3-D search, where the space to be searched is shown by a solid dark line in the figure in Eq. (7.61). To satisfy condition (b), we need to ensure that $A(z_1, z_2)$ is not zero for any (z_1, z_2) such that $|z_1| \geq 1$ and $z_2 = 1$. This requires a 2-D search, where the space to be searched is shown by the dotted line in the figure. The dotted line is used to emphasize that this is a 2-D search problem, where the search is required in the 2-D subspace of the 3-D space corresponding to $(|z_1| \geq 1, |z_2| = 1)$.

The 3-D search problem corresponding to condition (a) can be substituted for by many 1-D stability tests. To satisfy condition (a), we need to make sure that $A(z_1, z_2) \neq 0$ in the 3-D space corresponding to $(|z_1| = 1, |z_2| \geq 1)$. One approach to satisfy this condition consists of two steps:

1. Solve all (z_1, z_2) such that $A(|z_1| = 1, z_2) = 0$. This is equivalent to solving all (ω_1, z_2) such that $A(e^{j\omega_1}, z_2) = 0$.
2. Check if all $|z_2|$ obtained in step 1 are less than 1.

In step 1, we determine all ($|z_1| = 1, z_2$) such that $A(|z_1| = 1, z_2) = 0$. The solutions obtained in this step will contain all solutions to $A(z_1, z_2) = 0$ in the 3-D space of ($|z_1| = 1, z_2$). If none of these solutions has $|z_2|$ greater than or equal to 1, then $A(z_1, z_2) \neq 0$ for any ($|z_1| = 1, |z_2| \geq 1$), which satisfies condition (a). Step 2 is clearly a trivial operation. In step 1, we have to solve the following equation:

$$A(e^{j\omega_1}, z_2) = 0$$

Suppose we consider a fixed value of ω_1, say ω_1'. Then $A(e^{j\omega_1'}, z_2)$ is a 1-D polynomial in the variable z_2, and solving for all z_2 such that $A(e^{j\omega_1'}, z_2) = 0$ is equivalent to a 1-D stability test. If we vary ω_1 continuously from 0 to 2π and we perform the 1-D stability test for each ω_1, we will find all possible values of ($e^{j\omega_1}, z_2$) such that $A(e^{j\omega_1}, z_2) = 0$. In practice, we cannot change ω_1 continuously, and we have to consider discrete values of ω_1. By performing many 1-D stability tests, we can obtain a table such as Table 7.4. By choosing Δ sufficiently small, it is possible to essentially determine all possible values of (ω_1, z_2) such that $A(e^{j\omega_1}, z_2) = 0$. By checking if all the values of $|z_2|$ in Table 7.4 are smaller than 1, we can satisfy condition (a) without a 3-D search.

TABLE 7.4   SOLUTION TO $A(e^{j\omega_1}, z_2) = 0$

ω_1	z_2
0	$a_0, b_0, c_0, d_0, \ldots$
Δ	$a_1, b_1, c_1, d_1, \ldots$
2Δ	$a_2, b_2, c_2, d_2, \ldots$
$\vdots$	$\vdots$
.	
.	
2π	$a_0, b_0, c_0, d_0, \ldots$

The 2-D search problem corresponding to condition (b) can be substituted for by one 1-D stability test. To satisfy condition (b), we need to make sure that $A(z_1, z_2) \neq 0$ in the 2-D space corresponding to ($|z_1| \geq 1, z_2 = 1$). One approach to satisfy this condition consists of the following two steps.

1. Solve all (z_1, z_2) such that $A(z_1, z_2 = 1) = 0$.
2. Check if all $|z_1|$ obtained in step 1 are less than 1.

Step 1 determines all ($z_1, z_2 = 1$) such that $A(z_1, 1) = 0$. If all $|z_1|$ are less than 1, then $A(z_1, 1)$ cannot be zero for $|z_1| \geq 1$, which satisfies condition (b). Step 2 is a trivial operation. Step 1 is equivalent to a 1-D stability test since $A(z_1, 1)$ is a 1-D polynomial in the variable z_1.

From the preceding discussion, it is clear that a 2-D stability test can be performed by many 1-D stability tests and one 1-D stability test. This fact can help us develop a procedure to be used in practice to test the stability of a 2-D system.

Among the many variations to Huang's theorem is the following:

$$\text{Stability} \longleftrightarrow \quad \text{(a)} \quad A(z_1, z_2) \neq 0 \qquad \text{for } |z_1| \geq 1, |z_2| = 1$$
$$\text{(b)} \quad A(z_1, z_2) \neq 0 \qquad \text{for } z_1 = 1, |z_2| \geq 1$$

(7.62)

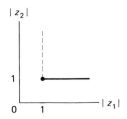

This variation is the same as Huang's theorem except that the roles of z_1 and z_2 have been interchanged.

DeCarlo-Strintzis Theorem.   The DeCarlo-Strintzis theorem can be stated as follows:

$$\text{Stability} \longleftrightarrow \quad \text{(a)} \quad A(z_1, z_2) \neq 0, \qquad |z_1| = |z_2| = 1$$
$$\text{(b)} \quad A(z_1, 1) \neq 0, \qquad |z_1| \geq 1$$
$$\text{(c)} \quad A(1, z_2) \neq 0, \qquad |z_2| \geq 1$$

(7.63)

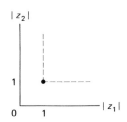

Each of the three conditions in this theorem corresponds to a 2-D search problem. Condition (a) requires $A(z_1, z_2)$ to be nonzero on the 2-D unit surface. Conditions (b) and (c) require $A(z_1, z_2)$ to be nonzero in the 2-D spaces shown by the dotted lines in the figure in Eq. (7.63). From the search point of view, therefore, the conditions imposed by the DeCarlo-Strintzis theorem are considerably simpler than those in Huang's theorem, which requires a 3-D search.

In practice, however, this theorem is not much simpler than Huang's theorem. Specifically, condition (b) in the DeCarlo-Strintzis theorem is the same as condition (b) in Huang's theorem, which can be checked by a 1-D stability test. Condition (c) in the DeCarlo-Strintzis theorem is the same as condition (b) with the roles of z_1 and

z_2 interchanged. Condition (c) can therefore be checked by one 1-D stability test. In the case of condition (a), however, the 2-D search problem cannot be simplified by one 1-D stability test, and a full 2-D search on the unit surface is generally necessary. Computations involved in this 2-D search are often comparable to those of many 1-D stability tests that can be used in testing condition (a) of Huang's theorem.

7.5.3 Methods for Stability Test

From the two theorems discussed in Section 7.5.2, we can develop many different methods to check the stability of a 2-D system. One such method is shown in Fig. 7.36. Test 1 in the figure checks condition (b) of Huang's theorem and the DeCarlo-Strintzis theorem. Test 2 checks condition (c) of the DeCarlo-Strintzis theorem. Test 3 checks condition (a) of the DeCarlo-Strintzis theorem. If a system passes all three tests, the system is stable. Otherwise, it is unstable.

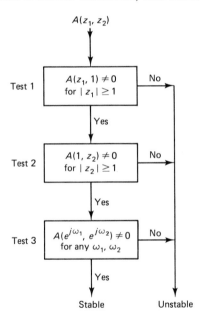

Figure 7.36 One approach to test the stability of $H(z_1, z_2) = 1/[A(z_1, z_2)]$.

To illustrate how the procedure in Fig. 7.36 can be used in testing the stability of a 2-D system, we consider two examples:

Example 1

$$H(z_1, z_2) = \frac{1}{A(z_1, z_2)} = \frac{1}{1 - 0.6z_1^{-1} - 0.6z_2^{-1}}$$

where $h(n_1, n_2)$ is a first-quadrant sequence.

Test 1

$$A(z_1, 1) = 1 - 0.6z_1^{-1} - 0.6 = 0$$
$$z_1 = \tfrac{3}{2} \geq 1$$

The system fails test 1 and therefore is unstable.

Example 2

$$H(z_1, z_2) = \frac{1}{A(z_1, z_2)} = \frac{1}{1 - \frac{1}{2}z_1^{-1} - \frac{1}{4}z_1^{-1} \cdot z_2^{-1}},$$

where $h(n_1, n_2)$ is a first-quadrant sequence.

Test 1

$$A(z_1, 1) = 1 - \frac{1}{2}z_1^{-1} - \frac{1}{4}z_1^{-1}$$

$$z_1 = \frac{3}{4} < 1$$

Test 1 passed.

Test 2

$$A(1, z_2) = 1 - \frac{1}{2} - \frac{1}{4} \cdot z_2^{-1}$$

$$z_2 = \frac{1}{2} < 1$$

Test 2 passed.

Test 3

$$A(\omega_1, \omega_2) = A(z_1, z_2)\big|_{z_1 = e^{j\omega_1}, z_2 = e^{j\omega_2}}$$

$$= 1 - \frac{1}{2}e^{-j\omega_1} - \frac{1}{4}e^{-j\omega_1} \cdot e^{-j\omega_2} \neq 0 \qquad \text{for any } (\omega_1, \omega_2)$$

Test 3 passed.

The system passes all three tests and is therefore stable. In this example, test 3 could be performed by inspection. In typical cases, however, test 3 requires a considerable amount of computation.

From the preceding discussion, it is clear that testing the stability of a 2-D system is considerably more complex than testing the stability of a 1-D system. A 2-D stability test problem typically corresponds to many 1-D stability tests. In addition, the stability of a 2-D system cannot in general be absolutely guaranteed by a finite number of 1-D stability tests since $A(e^{j\omega_1}, z_2) = 0$ has to be solved for every possible ω_1. The complexity of testing the stability of a 2-D system and the lack of simple procedures to design a stable filter and to stabilize an unstable filter explain, in part, why 2-D finite impulse response (FIR) digital filters, which are always stable, are much preferred in practice over 2-D infinite impulse response (IIR) digital filters.

7.6 DISCRETE FOURIER TRANSFORM AND FAST FOURIER TRANSFORM

7.6.1 Discrete Fourier Transform (DFT)

In many signal processing applications, such as image processing, we deal with sequences of finite extent. For such sequences, the Fourier transform and z-transform uniformly converge and are well defined. The Fourier transform and z-transform representations $X(\omega_1, \omega_2)$ and $X(z_1, z_2)$ are functions of continuous variables (ω_1, ω_2) and (z_1, z_2). For finite-extent sequences, which can be represented by a finite number of values, then the Fourier transform and z-transform are not computationally convenient frequency-domain representations. The discrete Fourier transform (DFT) is a frequency-domain representation of finite-extent sequences, where a finite-extent sequence is represented in the frequency domain by a finite number of values.

The 2-D DFT representation can be derived by a straightforward extension of the 1-D result. Specifically, the 2-D DFT pair is given by

Discrete Fourier Transform Pair

$$X(k_1, k_2) = \begin{cases} \sum_{n_1=0}^{N_1-1} \sum_{n_2=0}^{N_2-1} x(n_1, n_2) \cdot e^{-j(2\pi/N_1)k_1 \cdot n_1} \cdot e^{-j(2\pi/N_2)k_2 \cdot n_2}, \\ \qquad\qquad \text{for} \quad 0 \le k_1 \le N_1 - 1, \\ \qquad\qquad\qquad\qquad 0 \le k_2 \le N_2 - 1 \\ 0, \qquad\qquad\qquad\qquad \text{otherwise} \end{cases} \qquad (7.64)$$

$$x(n_1, n_2) = \begin{cases} \dfrac{1}{N_1 \cdot N_2} \cdot \sum_{k_1=0}^{N_1-1} \sum_{k_2=0}^{N_2-1} X(k_1, k_2) \cdot e^{j(2\pi/N_1)k_1 \cdot n_1} \cdot e^{j(2\pi/N_2)k_2 \cdot n_2}, \\ \qquad\qquad \text{for} \quad 0 \le n_1 \le N_1 - 1, \\ \qquad\qquad\qquad\qquad 0 \le n_2 \le N_2 - 1 \\ 0 \qquad\qquad\qquad\qquad \text{otherwise} \end{cases} \qquad (7.65)$$

From Eqs. (7.64) and (7.65), an $N_1 \times N_2$-point sequence $x(n_1, n_2)$ is represented by an $N_1 \times N_2$-point sequence $X(k_1, k_2)$ in the frequency domain. The sequence $X(k_1, k_2)$ is called the DFT of $x(n_1, n_2)$, and $x(n_1, n_2)$ is called the inverse DFT of $X(k_1, k_2)$. The DFT pair given in the box is defined only for a finite-extent first-quadrant sequence. This is not a serious restriction in practice since a finite-extent sequence can always be shifted, and the shift can easily be accounted for in typical applications.

For a finite-extent first-quadrant sequence $x(n_1, n_2)$ that is zero outside $0 \le n_1 \le N_1 - 1$ and $0 \le n_2 \le N_2 - 1$, the DFT $X(k_1, k_2)$ is related in a straightforward way to the discrete-space Fourier transform $X(\omega_1, \omega_2)$. From Eqs. (7.28) and (7.64), it is clear that

$$X(k_1, k_2) = X(\omega_1, \omega_2)\big|_{\omega_1=(2\pi/N_1)k_1, \, \omega_2=(2\pi/N_2)k_2}$$
$$\text{for } 0 \le n_1 \le N_1 - 1, \, 0 \le n_2 \le N_2 - 1 \qquad (7.66)$$

Equation (7.66) states that the DFT coefficients of $x(n_1, n_2)$ are samples of $X(\omega_1, \omega_2)$ at equally spaced points on the Cartesian grid, beginning at $\omega_1 = \omega_2 = 0$. Since $X(k_1, k_2)$ completely specifies $x(n_1, n_2)$, which in turn completely specifies $X(\omega_1, \omega_2)$, $X(\omega_1, \omega_2)$ has considerable redundant information; e.g., $N_1 \times N_2$ samples of $X(\omega_1, \omega_2)$ completely specify $X(\omega_1, \omega_2)$.

We can derive a number of useful properties from the DFT pair of equations (7.64) and (7.65). Some of the important properties, often useful in practice, are listed in Table 7.5. Most of these properties are straightforward extensions of 1-D results.

Analogous to the 1-D case, property 2 and property 4 present alternative ways to perform linear convolution of two finite-extent sequences. To linearly convolve $f(n_1, n_2)$ and $g(n_1, n_2)$, we could assume the proper periodicity $N_1 \times N_2$, determine

TABLE 7.5   PROPERTIES OF THE DISCRETE FOURIER TRANSFORM

$$x(n_1, n_2), y(n_1, n_2) = 0 \quad \text{outside } 0 \le n_1 \le N_1 - 1, 0 \le n_2 \le N_2 - 1$$

$$x(n_1, n_2) \longleftrightarrow X(k_1, k_2) \qquad y(n_1, n_2) \longleftrightarrow Y(k_1, k_2)$$

$$R_{N_1 \times N_2}(n_1, n_2) = \begin{cases} 1, & 0 \le n_1 \le N_1 - 1, 0 \le n_2 \le N_2 - 1 \\ 0, & \text{otherwise} \end{cases}$$

$N_1 \times N_2$-point DFT and inverse DFT are assumed.

Property 1: Linearity

$$a \cdot x(n_1, n_2) + b \cdot y(n_1, n_2) \longleftrightarrow a \cdot X(k_1, k_2) + b \cdot Y(k_1, k_2)$$

Property 2: Circular convolution

$$x(n_1, n_2) \circledast y(n_1, n_2) \longleftrightarrow X(k_1, k_2) \cdot Y(k_1, k_2)$$

Property 3: Separable sequence

$$x(n_1, n_2) = x_1(n_1) \cdot x_2(n_2) \longleftrightarrow X(k_1, k_2) = X_1(k_1) \cdot X_2(k_2),$$
$X_1(k_1)$: N_1-point 1-D DFT, $X_2(k_2)$: N_2-point 1-D DFT

Property 4: Relation between circular and linear convolution

$$f(n_1, n_2) = 0 \quad \text{outside } 0 \le n_1 \le N_1' - 1, 0 \le n_2 \le N_2' - 1$$
$$g(n_1, n_2) = 0 \quad \text{outside } 0 \le n_1 \le N_1'' - 1, 0 \le n_2 \le N_2'' - 1$$
$$f(n_1, n_2) * g(n_1, n_2) = f(n_1, n_2) \circledast g(n_1, n_2)$$
with periodicity $N_1 \ge N_1' + N_1'' - 1$, $N_2 \ge N_2' + N_2'' - 1$

$F(k_1, k_2)$ and $G(k_1, k_2)$, multiply $F(k_1, k_2)$ and $G(k_1, k_2)$, and then perform the inverse DFT operation of $F(k_1, k_2) \cdot G(k_1, k_2)$. Even though this approach appears to be quite cumbersome, it sometimes reduces the amount of computations involved in performing linear convolution in practical applications. The 1-D methods for performing convolution such as the overlap-add method and the overlap-save method, which are based on this notion of performing convolution by computing DFTs, extend to the problem of performing 2-D convolution in a straightforward manner.

7.6.2 Fast Fourier Transform (FFT)

Row-column decomposition.   The DFT discussed in previous sections is used in a variety of signal processing applications, so it is of considerable interest to efficiently compute the DFT and inverse DFT. One efficient way to compute the 2-D DFT is known as the fast Fourier transform by row-column decomposition. This simple method uses 1-D FFT algorithms and offers considerable computational savings compared with direct computation. It is also the most popular 2-D FFT algorithm.

To appreciate the computational efficiency of the row-column decomposition method, we first consider computing the DFT directly (the inverse DFT computation is essentially the same). Consider an $N_1 \times N_2$-point complex sequence $x(n_1, n_2)$ that is zero outside $0 \le n_1 \le N_1 - 1, 0 \le n_2 \le N_2 - 1$. The DFT of $x(n_1, n_2)$, $X(k_1, k_2)$, is related to $x(n_1, n_2)$ by

$$X(k_1, k_2) = \sum_{n_1=0}^{N_1-1} \sum_{n_2=0}^{N_2-1} x(n_1, n_2) \cdot e^{-j(2\pi/N_1)k_1 \cdot n_1} \cdot e^{-j(2\pi/N_2)k_2 \cdot n_2},$$
$$0 \le k_1 \le N_1 - 1, \, 0 \le k_2 \le N_2 - 1$$
$$(7.67)$$

From Eq. (7.67), directly computing $X(k_1, k_2)$ for each (k_1, k_2) requires $N_1 \cdot N_2 - 1$ complex additions and $N_1 \cdot N_2$ complex multiplications. Since there are $N_1 \cdot N_2$ different values of (k_1, k_2), the total number of arithmetic operations required in computing $X(k_1, k_2)$ from $x(n_1, n_2)$ is $N_1^2 \cdot N_2^2$ complex multiplications and $N_1 \cdot N_2 \cdot (N_1 \cdot N_2 - 1)$ complex additions.

To develop the row-column decomposition method, we rewrite Eq. (7.67) as follows:

$$X(k_1, k_2) = \sum_{n_2=0}^{N_2-1} \underbrace{\sum_{n_1=0}^{N_1-1} x(n_1, n_2) \cdot e^{-j(2\pi/N_1)k_1 \cdot n_1}}_{f(k_1, n_2)} \cdot e^{-j(2\pi/N_2)k_2 \cdot n_2}, \qquad \begin{matrix} 0 \le k_1 \le N_1 - 1, \\ 0 \le k_2 \le N_2 - 1 \end{matrix}$$
$$(7.68)$$

Consider a fixed n_2, say $n_2 = 0$. Then $x(n_1, n_2)|_{n_2=0}$ represents a row of $x(n_1, n_2)$, and $f(k_1, n_2)|_{n_2=0}$ is nothing but the 1-D N_1-point DFT of $x(n_1, n_2)|_{n_2=0}$ with respect to the variable n_1. Therefore, we can compute $f(k_1, 0)$ from $x(n_1, n_2)$ by computing one 1-D N_1-point DFT. Since there are N_2 different values for n_2 in $f(k_1, n_2)$ that are of interest to us, we can compute $f(k_1, n_2)$ from $x(n_1, n_2)$ by computing N_2 1-D N_1-point DFTs. This process is illustrated in Fig. 7.37.

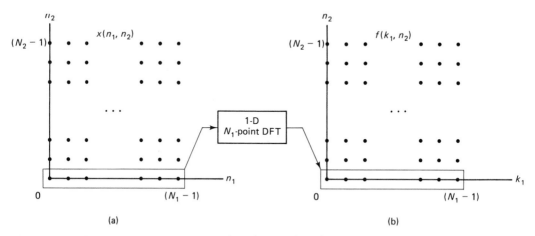

Figure 7.37 Computation of $f(k_1, n_2)$ from $x(n_1, n_2)$ by performing N_2 1-D N_1-point DFTs.

Once we compute $f(k_1, n_2)$ from Eq. (7.68), we can compute $X(k_1, k_2)$ from $f(k_1, n_2)$ as follows:

$$X(k_1, k_2) = \sum_{n_2=0}^{N_2=1} f(k_1, n_2) \cdot e^{-j(2\pi/N_2)k_2 \cdot n_2}, \qquad 0 \le k_1 \le N_1 - 1, \, 0 \le k_2 \le N_2 - 1$$
$$(7.69)$$

To compute $X(k_1, k_2)$ from $f(k_1, n_2)$, consider a fixed k_1, say $k_1 = 0$. Then $f(k_1, n_2)|_{k_1=0}$ represents one column of $f(k_1, n_2)$, and $X(k_1, k_2)|_{k_1=0}$ in Eq. (7.69) is nothing but the 1-D N_2-point DFT of $f(k_1, n_2)|_{k_1=0}$ with respect to the variable n_2. Therefore, we can compute $X(0, k_2)$ from $f(k_1, n_2)$ by computing one 1-D N_2-point DFT. Since there are N_1 different values for k_1 in $X(k_1, k_2)$ that are of interest to us, we can compute $X(k_1, k_2)$ from $x(n_1, n_2)$ by computing N_1 1-D N_2-point DFTs. This process is illustrated in Fig. 7.38.

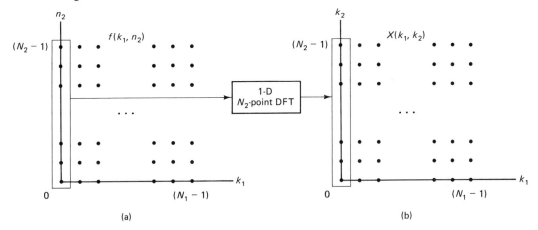

Figure 7.38   Computation of $X(k_1, k_2)$ from $f(k_1, n_2)$ by performing N_1 1-D N_2-point DFTs.

From the preceding discussion, we can compute $X(k_1, k_2)$ from $x(n_1, n_2)$ with a total of N_2 1-D N_1-point DFTs for the row operations and N_1 1-D N_2-point DFTs for the column operations. Suppose we compute the 1-D DFTs directly. Since direct computation of one 1-D N-point DFT requires N^2 multiplications and about N^2 additions, the total number of arithmetic operations involved in computing $X(k_1, k_2)$ is $N_1 \cdot N_2(N_1 + N_2)$ multiplications and $N_1 \cdot N_2(N_1 + N_2)$ additions. This is a significant computational saving relative to the $N_1^2 \cdot N_2^2$ multiplications and $N_1^2 \cdot N_2^2$ additions required for direct computation of $X(k_1, k_2)$.

To further reduce the number of arithmetic operations, we can of course use 1-D FFT algorithms to compute the 1-D DFTs in the preceding discussion. When $N = 2^M$, an N-point 1-D FFT algorithm requires $(N/2) \cdot \log_2^N$ multiplications and $N \cdot \log_2^N$ additions. To compute N_2 1-D N_1-point DFTs and N_1 1-D N_2-point DFTs using 1-D FFT algorithms when $N_1 = 2^{M_1}$ and $N_2 = 2^{M_2}$, we need a total of $[(N_1 \cdot N_2)/2] \log_2 N_1 \cdot N_2$ multiplications and $N_1 \cdot N_2 \log_2 N_1 \cdot N_2$ additions. This is a significant computational saving relative to direct computation of the 1-D DFTs. If we represent the total number of points $N_1 \cdot N_2$ as N, then the number of computations involved in the preceding case can be expressed as $(N/2) \cdot \log_2^N$ multiplications and $N \cdot \log_2^N$ additions. These are exactly the same expressions as those for the 1-D N-point DFT computations, using an FFT algorithm such as a decimation-in-time algorithm.

To appreciate the computational saving involved, Table 7.6 shows the relative amount of computations for the three methods. When $N_1 = N_2 = 512$, row-column decomposition alone reduces the number of multiplications and additions by a factor

TABLE 7.6 COMPARISON OF THREE METHODS TO COMPUTE A 2-D
$N_1 \times N_2$-POINT DFT IN THE REQUIRED NUMBER OF MULTIPLICATIONS
AND ADDITIONS

	Number of multiplications ($N_1 = N_2 = 512$)	Number of additions ($N_1 = N_2 = 512$)
Direct computation	$N_1^2 \cdot N_2^2$ (100%)	$N_1^2 \cdot N_2^2$ (100%)
Row-column decomposition with direct 1-D DFT composition	$N_1 \cdot N_2 \cdot (N_1 + N_2)$ (0.4%)	$N_1 \cdot N_2 \cdot (N_1 + N_2)$ (0.4%)
Row-column decomposition with 1-D FFT algorithm	$\dfrac{N_1 \cdot N_2}{2} \cdot \log_2 N_1 \cdot N_2$ (0.0035%)	$N_1 \cdot N_2 \cdot \log_2(N_1 \cdot N_2)$ (0.007%)

of 250 relative to direct 2-D DFT computation. The reduction by an additional factor of 110 for multiplications and of 55 for additions is obtained by using 1-D FFT algorithms. The total reduction in computations by row-column decomposition and 1-D FFT algorithms combined is by a factor of 30,000 for multiplications and of 15,000 for additions for $N_1 = N_2 = 512$ compared with the direct 2-D DFT computation.

In the derivation of the row-column decomposition approach, we expressed $X(k_1, k_2)$ in the form of Eq. (7.68). This led to the procedure where we performed row operations before column operations. An alternative way to write Eq. (7.68) is as follows:

$$X(k_1, k_2) = \sum_{n_1=0}^{N_1-1} \underbrace{\sum_{n_2=0}^{N_2-1} x(n_1, n_2) \cdot e^{-j(2\pi/N_2)k_2 \cdot n_2}}_{g(n_1, k_2)} \cdot e^{-j(2\pi/N_1)k_1 \cdot n_1}, \qquad \begin{aligned} 0 \le k_1 \le N_1 - 1, \\ 0 \le k_2 \le N_2 - 1 \end{aligned} \qquad (7.70)$$

If Eq. (7.70) is used in computing $X(k_1, k_2)$, then the column operations are performed before row operations. The computations involved in this case remain the same as in the previous case, where the row operations are performed first.

The computation of a reasonable-size 2-D DFT requires a fair amount of memory. When we compute the 2-D DFT of size 512×512, we need about a quarter of a million memory locations. As the cost of memory becomes cheaper, this amount of memory probably will not be a major issue in practical applications. If the memory size is an important consideration, however, the data may have to be stored on slow memory such as disk memory, and the number of I/O operations must be reduced. If rows of the data $x(n_1, n_2)$ are stored as blocks of data on the disk, one efficient approach is to perform the row operations first, transpose the result so that columns become rows, and then perform row operations on the transposed result. The result can then be transposed to have $X(k_1, k_2)$ stored in the proper order. Procedures for data transposition that are efficient in the required number of I/O operations can be found in [1].

The row-column decomposition approach discussed in this section is very efficient computationally, is conceptually simple to understand, and can be implemented using existing 1-D FFT algorithms. For these reasons, it is probably the most popular method used in the 2-D DFT computation.

Vector radix FFT.   In the row-column decomposition method, the 2-D DFT computation is transformed to many 1-D DFT computations, and 1-D FFT algorithms are used efficiently to compute the 1-D DFTs. An alternative approach is to extend the idea behind 1-D FFT algorithm development directly to the 2-D case. This extension leads to vector radix FFT algorithms.

Although there are many variations, all 1-D FFT algorithms are based on one simple principle: an N-point DFT can be computed by two $(N/2)$-point DFTs or three $(N/3)$-point DFTs, etc. This simple principle can be extended to the 2-D case in a straightforward manner. Specifically, an $N_1 \times N_2$-point DFT can be computed by four $(N_1/2) \times (N_2/2)$-point DFTs, or six $(N_1/2) \times (N_2/3)$-point DFTs, or nine $(N_1/3) \times (N_2/3)$-point DFTs, etc. Using this extension, various 1-D FFT algorithms such as decimation-in-time and decimation-in-frequency algorithms can be directly extended to the 2-D case, and all the properties also extend directly from the 1-D case to the 2-D case. In the 2-D decimation-in-space algorithm, for example, in-place computation is possible, and bit reversal of input is necessary to have correct output and in-place computation. Compared with the row-column decomposition method, vector radix FFT algorithms are roughly the same in various aspects and do not offer any significant advantages. The amount of computations and memory locations required, for example, are roughly the same in both cases.

7.7 FINITE IMPULSE RESPONSE DIGITAL FILTERS

Three steps are generally followed in using digital filters. In the first step, we specify the characteristics required of the filter. The filter specification depends, of course, on the application for which the filter is used. For example, in a case where we wish to restore a signal that has been degraded by background noise, the filter characteristics we require depend on the spectral characteristics of the signal and the background noise. The second step is the filter design step, where we determine $h(n_1, n_2)$, the unit sample response of the filter, or its system function $H(z_1, z_2)$ that meets the design specification. The third step is the filter implementation step, in which we realize a discrete space system with the given $h(n_1, n_2)$ or $H(z_1, z_2)$.

The three steps are closely related. For example, it does not make much sense to specify a filter that cannot be designed. Neither does it make much sense to design a filter that cannot be implemented. Despite the close relationship among the three steps, we discuss them separately and point out the interrelationships as appropriate.

We restrict ourselves to a certain class of digital filters for practical reasons. One restriction is that $h(n_1, n_2)$ be real. In practice, we often deal with real data. So to ensure that the processed signal is real when the input signal is real, we require $h(n_1, n_2)$ to be real. Another restriction is the stability of $h(n_1, n_2)$, i.e., $\sum_{n=-\infty}^{\infty} |h(n_1, n_2)| < \infty$. In practice, an unbounded output can cause many difficulties such as system overload. For these practical reasons, we restrict our discussion to the class of digital filters whose unit sample response $h(n_1, n_2)$ is real and stable.

Digital filters can often be classified into two groups. In the first group $h(n_1, n_2)$ is a finite-extent sequence, and the filters in this group are called finite impulse response (FIR) filters. In the second group, $h(n_1, n_2)$ is of infinite extent, and the filters in this group are called infinite impulse response (IIR) filters. In this section we concentrate on FIR filters and in Section 7.8 on IIR filters. As in the 1-D case, the design and implementation of FIR filters differ considerably from those of IIR filters.

7.7.1 Zero-Phase Filters

A digital filter $h(n_1, n_2)$ is said to have zero phase when its frequency response $H(\omega_1, \omega_2)$ is a real function, so that

$$H(\omega_1, \omega_2) = H^*(\omega_1, \omega_2) \tag{7.71}$$

Strictly speaking, the filter whose frequency response is real may not be a zero-phase filter, since $H(\omega_1, \omega_2)$ can be negative. In practice, the frequency regions for which $H(\omega_1, \omega_2)$ is negative typically correspond to the stopband regions, and a phase of $180°$ in the stopband regions has little significance.

From the symmetry properties of the Fourier transform, Eq. (7.71) is equivalent in the spatial domain to the following expression:

$$h(n_1, n_2) = h^*(-n_1, -n_2) \tag{7.72}$$

Since we consider only real $h(n_1, n_2)$, Eq. (7.72) reduces to

$$h(n_1, n_2) = h(-n_1, -n_2) \tag{7.73}$$

Equation (7.73) states that the unit sample response of a zero-phase filter is symmetric with respect to the origin.

One characteristic of a zero-phase filter is its tendency to preserve the shape of the signal component in the passband region of the filter. This characteristic is quite useful in applications such as image processing, where the shape of the signal is very important. In addition, from Eq. (7.73) it is very easy to require zero phase for FIR filters, and design and implementation are often simplified if we require zero phase. For these reasons, we restrict our discussion of FIR filters to zero-phase filters.

7.7.2 Filter Specification

Like 1-D digital filters, 2-D digital filters are generally specified in the frequency domain. Since $H(\omega_1, \omega_2) = H(\omega_1 + 2\pi, \omega_2) = H(\omega_1, \omega_2 + 2\pi)$ for all (ω_1, ω_2), $H(\omega_1, \omega_2)$ for $-\pi \leq \omega_1, \omega_2 \leq \pi$ completely specifies $H(\omega_1, \omega_2)$. In addition, since $h(n_1, n_2)$ is assumed real, $H(\omega_1, \omega_2) = H^*(-\omega_1, -\omega_2)$. Specifying $H(\omega_1, \omega_2)$ for $-\pi \leq \omega_1 \leq \pi$, $0 \leq \omega_2 \leq \pi$ therefore completely specifies $H(\omega_1, \omega_2)$ for all (ω_1, ω_2).

Since $H(\omega_1, \omega_2)$ is in general a complex function of (ω_1, ω_2), we need to specify both the magnitude and the phase of $H(\omega_1, \omega_2)$. For FIR filters, we require zero phase and therefore need to specify only the magnitude response. Like the 1-D filter specification, one scheme that is sometimes used for the magnitude specification is a "tolerance" scheme. An example of a lowpass filter specified using a tolerance scheme is shown in Fig. 7.39. The filter has a passband region where we require $1 - \delta_p \le |H(\omega_1, \omega_2)| \le 1 + \delta_p$ and a stopband region where we require $|H(\omega_1, \omega_2)| \le \delta_s$. The variables δ_p and δ_s are "passband tolerance" and "stopband tolerance" respectively. Other filters can also be specified analogously.

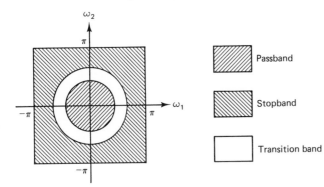

Passband

Stopband

Transition band

Figure 7.39 An example of a 2-D lowpass filter specification using a tolerance scheme. In the passband region, $1 - \delta_p \le H(\omega_1, \omega_2) \le 1 + \delta_p$. In the stopband region, $H(\omega_1, \omega_2) \le \delta_s$.

7.7.3 FIR Filter Design

The problem of designing a filter is basically that of determining $h(n_1, n_2)$ or $H(z_1, z_2)$ that meets the design specification. The four standard approaches to designing FIR filters are the window method, frequency sampling method, the optimal filter design, and the transformation method. The window method and frequency sampling method are straightforward extensions of 1-D results. The optimal design problem differs significantly between the 1-D and the 2-D case. In the 1-D case, practical algorithms exist for the design of optimal filters. In the 2-D case, practical algorithms for designing optimal filters have not yet been developed. In the transformation method, a 2-D filter is designed from a 1-D filter. There is no counterpart of this method in the design of 1-D FIR filters. We discuss each of the four methods, with much greater emphasis on the methods with significant differences between the 1-D and the 2-D case.

Window method.   The window method is a straightforward extension of the 1-D results. In the window method, the desired frequency response $H_d(\omega_1, \omega_2)$ is assumed to be known. By performing an inverse Fourier transform on $H_d(\omega_1, \omega_2)$, we can determine the desired unit sample response of the filter, $h_d(n_1, n_2)$. In general $h_d(n_1, n_2)$ is an infinite-extent sequence. In the window method, we obtain an FIR filter by applying a window $w(n_1, n_2)$ to $h_d(n_1, n_2)$. If $h_d(n_1, n_2)$ and $w(n_1, n_2)$ are symmetric with respect to the origin, then $h_d(n_1, n_2) \cdot w(n_1, n_2)$ is also symmetric with respect to the origin, and therefore the resulting filter is zero phase.

A 2-D window used in the filter design is typically obtained from a 1-D window by using one of the following two methods:

$$w(n_1, n_2) = w_a(t_1) \cdot w_b(t_2)\big|_{t_1 = n_1, t_2 = n_2} \qquad \text{or} \qquad (7.74)$$

$$w(n_1, n_2) = w_a(t)\big|_{t = \sqrt{n_1^2 + n_2^2}} \qquad (7.75)$$

where $w_a(t)$ and $w_b(t)$ are 1-D analog windows. Equation (7.74) leads to a separable rectangularly shaped window, and Eq. (7.75) leads to a circularly symmetric window. The shape and effective size of the 1-D windows used in Eqs. (7.74) and (7.75) are determined by recognizing that the sidelobe behavior and therefore the passband and stopband tolerances are affected primarily by the window shape only and that the

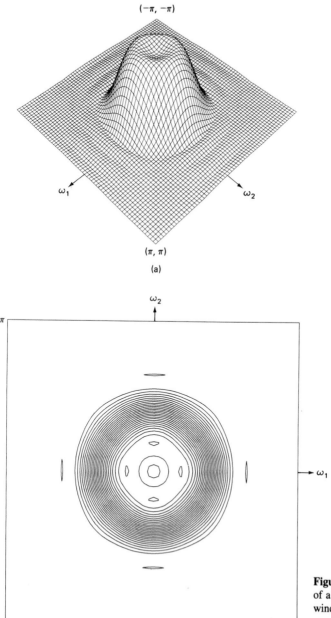

(a)

(b)

Figure 7.40   The frequency response of a lowpass filter designed by the window method. The window is a separable Kaiser window, of 9×9 points with $\omega_c = 0.4\pi$ in Eq. (7.78). (a) Perspective plot; (b) contour plot.

mainlobe behavior and therefore the transition width are affected by both the window shape and the effective window size. In a typical design, therefore, the window shape is chosen first based on the passband and stopband tolerance requirements and then the window size is determined based on the transition width requirements. Two examples of digital filters designed by the window method are shown in Figs. 7.40 and 7.41.

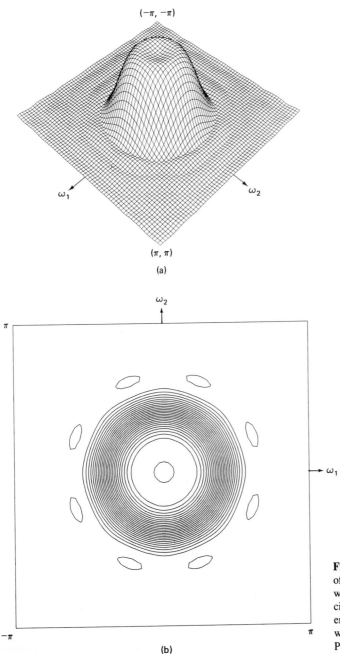

Figure 7.41 The frequency response of a lowpass filter designed by the window method. The window is a circularly symmetric Kaiser window, enclosed in a square of 9×9 points, with $\omega_c = 0.4\pi$ in Eq. (7.78). (a) Perspective plot; (b) contour plot.

In Fig. 7.40, Eq. (7.74) was used, and in Fig. 7.41, Eq. (7.75) was used. In both cases, the 1-D analog Kaiser window and $h_d(n_1, n_2)$ used are

$$w_a(t) = w_b(t) = \begin{cases} \dfrac{I_0\left(0.4 \sqrt{1 - \dfrac{t^2}{N^2}}\right)}{I_0(0.4)}, & t \le N \\ 0, & \text{otherwise} \end{cases} \tag{7.76}$$

$$h_d(n_1, n_2) = \frac{\omega_c}{2\pi \sqrt{n_1^2 + n_2^2}} \cdot J_1(\omega_c \cdot \sqrt{n_1^2 + n_2^2}) \tag{7.77}$$

where $I_0(x)$ is the modified Bessel function of the first kind of zeroth order. The sequence $h_d(n_1, n_2)$ used corresponds to the circularly symmetric ideal lowpass filter with cutoff frequency $\omega_c = 0.4\pi$. In each type, the filter frequency response is displayed by a perspective plot and contour plot.

The window method is not optimal, in the sense that there exists in general a filter that meets the given design specification and whose size is smaller than the filter designed by the window method. For an arbitrary $H_d(\omega_1, \omega_2)$, determining $h_d(n_1, n_2)$ from $H_d(\omega_1, \omega_2)$ may require a large inverse DFT computation. In addition, due to a lack of control over the frequency-domain specification parameters, it is sometimes necessary to design several filters to meet the given design specification. Despite these disadvantages, the window method is often used in practice because of its conceptual and computational simplicity.

Frequency sampling method. The frequency sampling method is also a straightforward extension of the 1-D results. In the frequency sampling method, the desired frequency response $H_d(\omega_1, \omega_2)$ is sampled at equally spaced points on the Cartesian grid and the inverse DFT of the result is computed. Specifically, let $H(k_1, k_2)$ be obtained by

$$H(k_1, k_2) = H_d(\omega_1, \omega_2)\big|_{\omega_1 = (2\pi/N_1)k_1, \, \omega_2 = (2\pi/N_2)k_2} \tag{7.78}$$

The unit sample response of the filter, $h(n_1, n_2)$, is obtained from

$$h(n_1, n_2) = \text{IDFT}[H(k_1, k_2)] \tag{7.79}$$

where IDFT is the inverse DFT. If $H_d(\omega_1, \omega_2)$ is zero phase and N_1 and N_2 are odd integers, then the resulting $h(n_1, n_2)$ is also zero phase, i.e., $h(n_1, n_2)$ is symmetric with respect to the origin. When $H_d(\omega_1, \omega_2)$ is sampled exactly, it has been observed that the stopband and passband behavior are rather poor. They can be improved considerably if some transition samples are taken in the frequency region where $H_d(\omega_1, \omega_2)$ has a sharp transition.

As with the window method, the filter designed by the frequency sampling method is not optimal, in the sense that there exists in general a filter that meets the same design specification and whose size is smaller than the filter designed by the frequency sampling method. In addition, due to a lack of control over the frequency-domain specification parameters, we may have to design several filters to meet the given design specification. Despite these disadvantages, the frequency sampling method is sometimes used in practice because of its conceptual and computational simplicity. Determining specific values and the region of the transition samples is more cumbersome than with the window method, but an inverse transform of $H_d(\omega_1, \omega_2)$ is not needed in the frequency sampling method. Performance, measured in terms of the filter size needed to meet a given design specification, appears to be similar for both the window method and the frequency sampling method.

Optimal filter design.   Unlike the 1-D case, no practical procedures have been developed to reliably design a 2-D optimal FIR filter. A detailed discussion of the 2-D optimal FIR filter design problem requires considerable effort, so we only contrast some major differences between the 1-D and 2-D cases to suggest the complexity of the 2-D case relative to the 1-D case.

We first review briefly the 1-D optimal filter design problem. For simplicity, we concentrate on the design of a zero-phase lowpass filter with design specification parameters δ_p (passband tolerance), δ_s (stopband tolerance), ω_p (passband frequency), and ω_s (stopband frequency). The problem of designing an optimal filter can be stated as follows.

Problem 1:   Given ω_p, ω_s, $k = \delta_p/\delta_s$, and N (filter length), determine $h(n)$ such that the design specification is satisfied with the smallest δ_s.

This problem can be shown to be a special case of a weighted Chebyshev approximation problem, which is a functional approximation problem. The weighted Chebyshev approximation problem has been studied extensively in mathematics. One theorem, the alternation theorem, states that the problem has a unique solution. The theorem provides a necessary and sufficient condition for the unique solution. The Remez multiple exchange algorithm exploits this necessary and sufficient condition to solve the weighted Chebyshev approximation problem. The Remez exchange algorithm was first used by Parks and McClellan [2] to solve the optimal filter design problem stated as Problem 1. The optimal filter design algorithm based on the Remez exchange algorithm is an iterative procedure in which the filter is improved in each iteration. Each iteration consists of two steps. One step is the determination of candidate filter coefficients $h(n)$ from candidate "alternation frequencies," which involves solving a set of linear equations. The other step is the determination of candidate alternation frequencies from the candidate filter coefficients. This step involves evaluating $H(\omega)$ on a dense grid of ω and looking for local extrema of $H(\omega)$. Once the local extrema are found, candidate alternation frequencies can be determined straightforwardly from the local extrema and bandedge frequencies $(0, \pi, \omega_p, \text{ and } \omega_s)$. The iterative algorithm is guaranteed to converge to the desired solution. Experience has shown that the algorithm converges very fast, and it is widely used in practice to design optimal filters.

A 2-D zero-phase optimal lowpass filter corresponding to Problem 1 can be stated as follows.

Problem 2:   Given R_p (passband region), R_s (stopband region), $k = \delta_p/\delta_s$, R_h (support region of $h(n_1, n_2)$), determine $h(n_1, n_2)$ such that the design specification is satisfied with the smallest δ_s.

To solve this problem, an approach similar to the 1-D case has been considered. A theorem, analogous to the alternation theorem, applies to Problem 2. Unlike the 1-D case, the theorem states that the problem does not have a unique solution. This is not much of an issue since we can obtain any one of the many possible solutions. The theorem also provides a necessary and sufficient condition for the solutions to the problem. An iterative algorithm similar to the Remez multiple exchange algorithm exploits this necessary and sufficient condition. As in the 1-D case, the iterative algorithm attempts to improve the filter in each iteration. Each iteration consists of two steps. One step is the determination of candidate filter coefficients $h(n_1, n_2)$ from candidate "critical-point frequencies," which are analogous to alternation frequencies. This step involves solving a set of linear equations. The other step is the determination of candidate critical-point frequencies from the candidate filter coefficients. This step involves evaluating $H(\omega_1, \omega_2)$ on a dense grid of (ω_1, ω_2) and looking for local extrema of $H(\omega_1, \omega_2)$. Compared with the 1-D case, evaluation of $H(\omega_1, \omega_2)$ on a dense grid requires computations that are typically several orders of magnitude greater than required by evaluation of $H(\omega)$. In addition, finding local extrema of $H(\omega_1, \omega_2)$ requires searching the function along many directions at many frequencies, while finding the local extrema of $H(\omega)$ requires searching the function only in one direction. Once the local extrema are found, candidate critical-point frequencies are determined from the local extrema. In the 1-D case, this is straightforward since all local extrema, with a possible exception of $\omega = 0$, π, are alternation frequencies. In the 2-D case, however, choosing a set of frequencies that form a critical set from the local extrema is quite involved and complex. Partly because of the difficulties already cited, the iterative algorithm developed so far is very expensive computationally, has not been demonstrated to converge to a correct solution, and is seldom used in practice. Developing a computationally efficient algorithm for the 2-D optimal FIR filter remains as an area for further research.

Transformation method. In the transformation method, a 2-D zero-phase FIR filter is designed from 1-D zero-phase FIR filter. To illustrate the basic idea, consider the following transformation:

$$H(\omega_1, \omega_2) = H(\omega)\big|_{\omega=G(\omega_1, \omega_2)} \tag{7.80}$$

where $H(\omega)$ is a 1-D digital filter frequency response and $H(\omega_1, \omega_2)$ is the frequency response of the resulting 2-D digital filter. Suppose $H(\omega)$ is a bandpass filter as shown in Fig. 7.42. Consider one particular frequency, $\omega = \omega_0'$. Suppose the function $\omega_0' = G(\omega_1, \omega_2)$ represents a contour in the (ω_1, ω_2)-plane shown in Fig. 7.42. Then, according to Eq. (7.80), $H(\omega_1, \omega_2)$ evaluated on the contour equals $H(\omega_0')$. If we now consider other frequencies $\omega_1', \omega_2', \ldots, \omega_N'$ and if their corresponding contours are as shown in Fig. 7.42, then the resulting 2-D filter will be a bandpass filter.

Several important issues need to be considered in this method. One issue is whether or not the resulting 2-D filter is a zero-phase FIR filter. The second issue is whether or not a transformation function $G(\omega_1, \omega_2)$ exists such that there will be a nice mapping between ω and $\omega = G(\omega_1, \omega_2)$ such as shown in Fig. 7.42. Both of these issues are resolved by using a 1-D zero-phase filter and the appropriate transformation function.

Consider a 1-D zero-phase FIR filter $h(n)$ with length $2N + 1$. The frequency response $H(\omega)$ can be expressed as

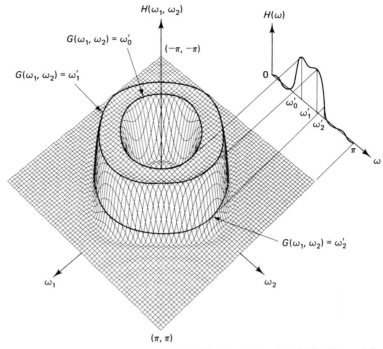

Figure 7.42   Illustration of the principle behind the design of a 2-D filter from a 1-D filter by frequency transformation.

$$H(\omega) = \sum_{n=-N}^{N} h(n) \cdot e^{-j\omega n} = h(0) + \sum 2 \cdot h(n) \cdot \cos \omega n$$

$$= \sum_{n=0}^{N} a(n) \cdot \cos \omega n = \sum_{n=0}^{N} b(n) \cdot (\cos \omega)^n \qquad (7.81)$$

In Eq. (7.81), the sequence $b(n)$ is not the same as $h(n)$, but it can be obtained simply from $h(n)$. The 2-D frequency response $H(\omega_1, \omega_2)$ is then obtained by

$$H(\omega_1, \omega_2) = H(\omega)\big|_{\cos \omega = T(\omega_1, \omega_2)} = \sum_{n=0}^{N} b(n) \cdot [T(\omega_1, \omega_2)]^n \qquad (7.82)$$

where $T(\omega_1, \omega_2)$ is the Fourier transform of a finite-extent sequence symmetric with respect to the origin, so that $T(\omega_1, \omega_2)$ can be expressed as

$$T(\omega_1, \omega_2) = \sum\sum_{(n_1, n_2) \in R_T} t(n_1, n_2) \cdot e^{-j\omega_1 n_1} \cdot e^{-j\omega_2 n_2}$$

$$= \sum\sum_{(n_1, n_2) \in R_C} c(n_1, n_2) \cdot \cos \omega_1 n_1 \cdot \cos \omega_2 n_2 \qquad (7.83)$$

where R_T and R_C represent the region of support of $t(n_1, n_2)$ and $c(n_1, n_2)$, respectively. The sequence $c(n_1, n_2)$ is simply related to $t(n_1, n_2)$ and can be easily obtained from $t(n_1, n_2)$. An example of $T(\omega_1, \omega_2)$ often used in practice is

$$T(\omega_1, \omega_2) = \tfrac{1}{2} \cos \omega_1 + \tfrac{1}{2} \cos \omega_2 + \tfrac{1}{2} \cos \omega_1 \cdot \cos \omega_2 - \tfrac{1}{2} \qquad (7.84)$$

The sequences $t(n_1, n_2)$ and $c(n_1, n_2)$ that correspond to $T(\omega_1, \omega_2)$ in Eq. (7.84) are shown in Fig. 7.43.

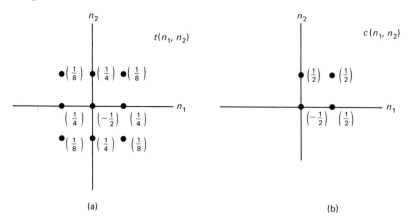

(a)   (b)

Figure 7.43   (a) A transformation sequence $t(n_1, n_2)$ and (b) the corresponding sequence $c(n_1, n_2)$ in Eq. (7.83) often used in the transformation method.

From Eqs. (7.82) and (7.83), $H(\omega_1, \omega_2)$ is always real, and therefore the resulting 2-D filter is a zero-phase filter. In addition, it is an FIR filter. For example, when $t(n_1, n_2)$ has a region of support of size $(2M_1 + 1) \times (2M_2 + 1)$ and the length of $h(n)$ is $2N + 1$, the resulting 2-D filter $H(\omega_1, \omega_2)$ is a finite-extent sequence of size $(2NM_1 + 1) \times (2NM_2 + 1)$. If $N = 10$ and $M_1 = M_2 = 1$, the region of support of $t(n_1, n_2)$ is 3×3 and the region of support of $h(n_1, n_2)$ is 21×21. In this example, the 2-D filter obtained has a large region of support for a short 1-D filter and $t(n_1, n_2)$ with a small region of support. This is typically the case.

By choosing $T(\omega_1, \omega_2)$ in Eq. (7.83) properly, we can obtain many different sets of contours that can be used for the 2-D filter design. For the transformation function $T(\omega_1, \omega_2)$ in Eq. (7.84), the set of contours obtained by $\cos \omega = T(\omega_1, \omega_2)$ for $\omega = 0$, $\tfrac{1}{10} \pi, \tfrac{2}{10} \pi, \ldots, \pi$ are shown in Fig. 7.44. This can be used to design many different 2-D FIR filters. From a lowpass 1-D filter of size 21 points, whose $H(\omega)$ is shown in Fig. 7.45(a), we can obtain a 2-D filter whose frequency response is shown in Fig. 7.45(b). If we begin with a 1-D highpass or bandpass filter, the resulting 2-D filter based on Eq. (7.84) would be a highpass or bandpass filter. Additional transformations and examples of filters designed by the transformation method can be found in [3].

Even though the transformation method is somewhat more complex conceptually than the window method or the frequency sampling method, its performance appears to be better than either of the two methods. In a certain restricted set of cases, the filter designed by the transformation method has been shown to be optimal. Since practical procedures do not exist for the design of optimal filters, the transformation method is one to consider in sophisticated applications that require high performance.

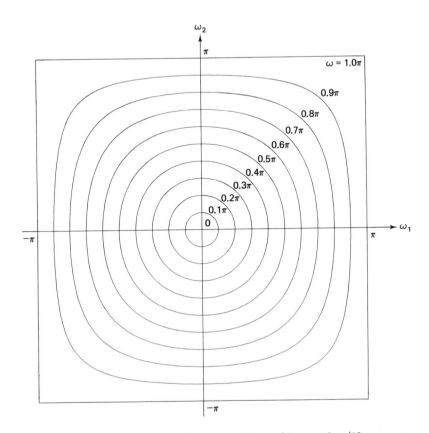

Figure 7.44   The contours obtained by cos $\omega = T(\omega_1, \omega_2)$ for $\omega = 0, \pi/10, \ldots, \pi$ for $T(\omega_1, \omega_2)$ given by Eq. (7.84).

7.7.4 Implementation of FIR Filters

In implementing a filter, the object is to realize a discrete space system with a specified unit sample response or transfer function. The simplest method of implementing an FIR filter is to use the convolution sum. Let $x(n_1, n_2)$ and $y(n_1, n_2)$ denote the input and output of the filter. Then $y(n_1, n_2)$ is related to $x(n_1, n_2)$ by

$$y(n_1, n_2) = \sum\sum_{(k_1, k_2) \in R_h} h(k_1, k_2) \cdot x(n_1 - k_1, n_2 - k_2) \qquad (7.85)$$

where R_h is the region of support of $h(n_1, n_2)$. From Eq. (7.85), the number of arithmetic operations required for each output point is about N multiplications and N additions, where N is the number of nonzero amplitudes in $h(n_1, n_2)$. As in the 1-D case, the realization can be improved by exploiting the symmetry of $h(n_1, n_2)$. Since $h(n_1, n_2) = h(-n_1, -n_2)$, by rewriting Eq. (7.85) and combining the two terms that

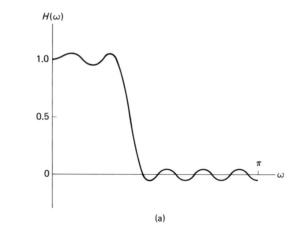

(a)

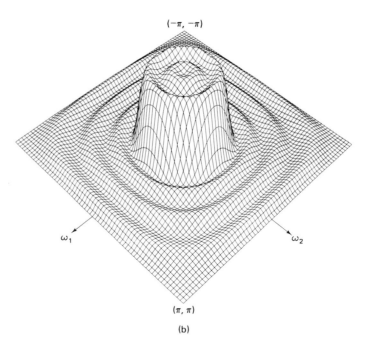

(b)

Figure 7.45   The frequency response of a 2-D filter designed by the transformation method. (a) Frequency response of the 1-D filter used in the design; (b) frequency response of the 2-D filter designed.

have the same value for $h(k_1, k_2)$, we can reduce the number of multiplications by about 50% without affecting the number of additions.

   If a filter is designed by using the transformation method discussed in Section 7.7.3, the number of arithmetic operations can be reduced significantly. If the 2-D filter is derived from a 1-D filter of length $2N + 1$ and a transformation sequence $t(n_1, n_2)$ in Eq. (7.83) of size $(2M_1 + 1) \times (2M_2 + 1)$, the resulting filter $h(n_1, n_2)$ is of size $(2M_1 \cdot N + 1) \times (2M_2 \cdot N + 1)$. If the filter is implemented by direct convolution, exploiting the property that $h(n_1, n_2) = h(-n_1, -n_2)$, then the number of arithmetic operations per output sample is around $[(2M_1 \cdot N + 1)(2M_2 \cdot N + 1)]/2$ multiplications and $(2M_1 \cdot N + 1) \times (2M_2 \cdot N + 1)$ additions, which are proportional to N^2. To achieve additional computational savings, we exploit the fact (Eq. 7.82) that $H(\omega_1, \omega_2)$ designed by the transformation method is of the following form:

$$H(\omega_1, \omega_2) = \sum_{n=0}^{N} b(n) \cdot [T(\omega_1, \omega_2)]^n \tag{7.86}$$

Equation (7.86) can be realized by the system shown in Fig. 7.46. Since $T(\omega_1, \omega_2)$ corresponds to a finite-extent sequence of size $(2M_1 + 1) \times (2M_2 + 1)$, the number of arithmetic operations per output point is about $(2M_1 + 1) \cdot (2M_2 + 1) \cdot N$ multiplications and $(2M_1 + 1) \cdot (2M_2 + 1) \cdot N$ additions, which are proportional to N. For large N, this represents considerable computational savings. When $N = 20$ and $M_1 = M_2 = 1$, direct convolution with only symmetry exploitation will require about 800 multiplications and 1600 additions per output point, while the realization by Fig. 7.46 will involve 180 multiplications and 180 additions.

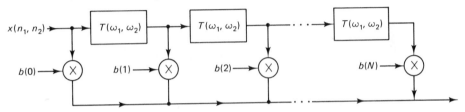

Figure 7.46   One implementation of a 2-D filter designed by the transformation method.

Any FIR filter can also be implemented by using an FFT algorithm. As we discussed in Section 7.6.1, the overlap-add method can be used to perform the filtering operation. In typical cases, this method reduces the number of arithmetic operations by a factor of 5–10 compared with realization by direct convolution.

7.8 INFINITE IMPULSE RESPONSE DIGITAL FILTERS

An infinite impulse response (IIR) filter has a unit sample response that is of infinite extent. As a result, an IIR filter differs from an FIR filter in some major ways.

An IIR filter with an arbitrary unit sample response $h(n_1, n_2)$ cannot be realized because computing each output sample may require an infinite number of arithmetic operations. As a result, in addition to requiring $h(n_1, n_2)$ to be real and stable, we require $h(n_1, n_2)$ to have a rational z-transform corresponding to a computational procedure with a wedge support output mask so that it is recursively computable. Specifically, we require $H(z_1, z_2)$, the z-transform of $h(n_1, n_2)$, to be a rational function of the following form:

$$H(z_1, z_2) = \frac{\sum\sum_{(k_1, k_2) \in R_B} b(k_1, k_2) \cdot z_1^{-k_1} \cdot z_2^{-k_2}}{\sum\sum_{(k_1, k_2) \in R_A} a(k_1, k_2) \cdot z_1^{-k_1} \cdot z_2^{-k_2}} \tag{7.87}$$

In addition, we require $h(n_1, n_2)$ to be a wedge support sequence. As we discussed in Section 7.4.5, a wedge support $h(n_1, n_2)$ with a rational z-transform can be realized by an LCCDE with proper boundary conditions.

One major difference between IIR and FIR filters is stability. An FIR filter is always stable as along as $h(n_1, n_2)$ is bounded (finite) for all (n_1, n_2), so stability is not an issue in design or implementation. With an IIR filter, however, ensuring stability is a major task. This imposes considerable restrictions on the design and implementation of IIR filters.

Zero phase is very easy to achieve for an FIR filter, and we discussed only zero-phase filters in Section 7.7. With an IIR filter, however, controlling the phase characteristics is very difficult. As a result, only the magnitude response is typically specified when an IIR filter is designed. The phase characteristic of the resulting filter is then regarded as acceptable phase. This lack of control over the phase characteristics also limits the usefulness of IIR digital filters.

7.8.1 Design of IIR Filters

The problem of designing an IIR filter is determining the coefficients of the system function. The magnitude specification that can be used is the tolerance scheme discussed in Section 7.7. In the 1-D case, there are two standard approaches to designing IIR filters. One is to design the filter from an analog filter system function, and the other is to design directly. In the 1-D IIR filter design, the first approach is typically much simpler and much more useful than the second approach. Using an elliptic analog filter system function and the bilinear transformation method, for example, optimal IIR lowpass, highpass, bandpass, and bandstop filters can be designed by following a finite fixed set of steps. Unfortunately, this approach is not useful in the 2-D case. In the 1-D case we exploit the availability of many simple methods to design 1-D analog filters that meet a given design specification. Simple methods do not exist in the design of 2-D analog filters.

In the second, direct method, an ideal unit sample response $h_d(n_1, n_2)$ or ideal magnitude response $|H_d(\omega_1, \omega_2)|$ is assumed known, and the coefficients of $H(z_1, z_2)$ are estimated so that $h(n_1, n_2)$ is closest to $h_d(n_1, n_2)$ or $|H(\omega_1, \omega_2)|$ is closest to $|H_d(\omega_1, \omega_2)|$ in some sense. The error criterion typically used is given by

$$\text{Error} = \sum_{n_1=-\infty}^{\infty} \sum_{n_2=-\infty}^{\infty} |h(n_1, n_2) - h_d(n_1, n_2)|^2 \tag{7.88}$$

or

$$\text{Error} = \int_{\omega_1=-\pi}^{\pi} \int_{\omega_2=-\pi}^{\pi} W(\omega_1, \omega_2)(|H(\omega_1, \omega_2)| - |H_d(\omega_1, \omega_2)|)^2 \cdot d\omega_1 \cdot d\omega_2 \tag{7.89}$$

where $W(\omega_1, \omega_2)$ is some weighting function. Minimization of Eq. (7.88) or Eq. (7.89) with respect to the coefficients of $H(z_1, z_2)$ is a highly nonlinear problem. To linearize the problem, many reasonable but ad hoc procedures have been considered. To illustrate the style in which these procedures were developed, we discuss one in particular.

Suppose $h_d(n_1, n_2)$ is given, and we wish to estimate the coefficients of $H(z_1, z_2)$ with assumed input and ouput mask shapes so that $h(n_1, n_2)$ is closest to $h_d(n_1, n_2)$. By considering the region of support of $h_d(n_1, n_2)$, the output and input mask shapes are chosen so that the resulting filter will have the same or a similar region of support. If $h_d(n_1, n_2)$ is a first-quadrant sequence, for instance, the output and input mask shapes shown in Fig. 7.47 will generate a filter with a first-quadrant support $h(n_1, n_2)$. The larger the size of the output and input masks, the better the resulting filter will be in general. The choice of the output and input mask shapes and sizes determines R_A and R_B in Eq. (7.87).

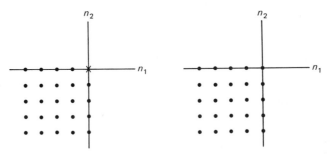

Figure 7.47   (a) Output mask of size 5×5 and (b) input mask of size 5×5 that will generate a filter with a first-quadrant $h(n_1, n_2)$.

The difference equation that corresponds to Eq. (7.87) is given by

$$\sum_{(k_1, k_2) \in R_A} a(k_1, k_2) \cdot y(n_1 - k_1, n_2 - k_2) = \sum_{(k_1, k_2) \in R_B} b(k_1, k_2) \cdot x(n_1 - k_1, n_2 - k_2)$$

$$(7.90)$$

In Eq. (7.90), when $x(n_1, n_2) = \delta(n_1, n_2)$, $y(n_1, n_2) = h(n_1, n_2)$. Therefore, from Eq. (7.90),

$$\sum_{(k_1, k_2) \in R_A} a(k_1, k_2) \cdot h(n_1 - k_1, n_2 - k_2) = \sum_{(k_1, k_2) \in R_B} b(k_1, k_2) \cdot \delta(n_1 - k_1, n_2 - k_2)$$

$$(7.91)$$

If we replace $h(n_1, n_2)$ with $h_d(n_1, n_2)$ in Eq. (7.91), we cannot expect the left-hand expression to equal the right-hand expression. Since we wish to have $h(n_1, n_2)$ as close as possible to $h_d(n_1, n_2)$, a reasonable error criterion is

$$\text{Error} = \sum_{n_1 = -\infty}^{\infty} \sum_{n_2 = -\infty}^{\infty} \left[\sum_{(k_1, k_2) \in R_A} a(k_1, k_2) \cdot h_d(n_1 - k_1, n_2 - k_2) \right.$$

$$(7.92)$$

$$\left. - \sum_{(k_1, k_2) \in R_B} b(k_1, k_2) \cdot \delta(n_1 - k_1, n_2 - k_2) \right]^2$$

Since the error in Eq. (7.92) is in the quadratic form of the unknown coefficients $a(n_1, n_2)$ and $b(n_1, n_2)$, minimization of error in Eq. (7.92) with respect to $a(n_1, n_2)$ and $b(n_1, n_2)$ requires solving a set of linear equations. An example of a filter designed by this method is shown in Fig. 7.48. The ideal unit sample response used is the circularly

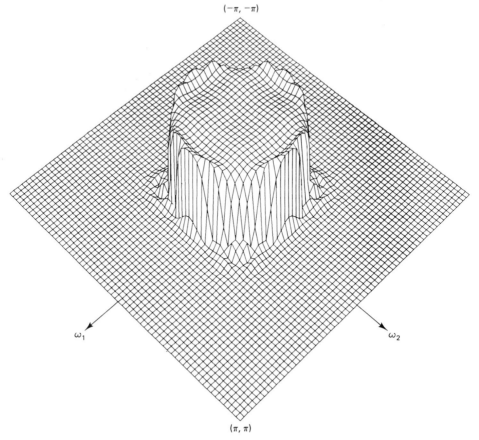

$(-\pi, -\pi)$

(π, π)

ω_1 ω_2

Figure 7.48   The frequency response of a 2-D zero-phase IIR filter designed.

symmetric ideal lowpass filter given by

$$h_d(n_1, n_2) = \frac{\omega_c}{2\pi\sqrt{n_1^2 + n_2^2}} \cdot J_1(\omega_c \cdot \sqrt{n_1^2 + n_2^2}) \tag{7.93}$$

To design a zero-phase filter $H(z_1, z_2)$, four one-quadrant filters $H_1(z_1, z_2)$, $H_2(z_1, z_2)$, $H_3(z_1, z_2)$, and $H_4(z_1, z_2)$ were designed. The first-quadrant filter $H_1(z_1, z_2)$ was obtained by using the output and input masks shown in Fig. 7.47 and the design method just discussed. The unit sample response used in the design of $H_1(z_1, z_2)$ is given by

$$h_d^1(n_1, n_2) = w'(n_1, n_2) \cdot h_d(n_1, n_2) = \begin{cases} h_d(n_1, n_2), & n_1, n_2 \geq 1 \\ \frac{1}{2}h_d(n_1, n_2), & n_1 = 0, n_2 \geq 1 \\ \frac{1}{2}h_d(n_1, n_2), & n_1 \geq 1, n_2 = 0 \\ \frac{1}{4}h_d(n_1, n_2), & n_1 = 0, n_2 = 0 \end{cases} \tag{7.94}$$

The window sequence $w'(n_1, n_2)$ used in Eq. (7.94) is shown in Fig. 7.49. The other three filters, $H_2(z_1, z_2)$, $H_3(z_1, z_2)$, and $H_4(z_1, z_2)$, were derived from $H_1(z_1, z_2)$. Specifically, the unit sample responses used in the design of $H_2(z_1, z_2)$, $H_3(z_1, z_2)$, and

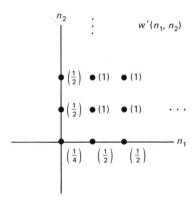

Figure 7.49 The window sequence $w'(n_1, n_2)$ in Eq. (7.94) used to design the filter in Fig. 7.48.

$H_4(z_1, z_2)$ are given by

$$h_d^2(n_1, n_2) = w'(-n_1, n_2) \cdot h_d(n_1, n_2) \qquad (7.95a)$$

$$h_d^3(n_1, n_2) = w'(-n_1, -n_2) \cdot h_d(n_1, n_2) \qquad (7.95b)$$

$$h_d^4(n_1, n_2) = w'(n_1, -n_2) \cdot h_d(n_1, n_2) \qquad (7.95c)$$

Since $h_d^1(n_1, n_2) = h_d^2(-n_1, n_2) = h_d^3(-n_1, -n_2) = h_d^4(n_1, -n_2)$, we can obtain $H_2(z_1, z_2)$, $H_3(z_1, z_2)$, and $H_4(z_1, z_2)$ from $H_1(z_1, z_2)$ by

$$H_2(z_1, z_2) = H_1(z_1^{-1}, z_2) \qquad (7.96a)$$

$$H_3(z_1, z_2) = H_1(z_1^{-1}, z_2^{-1}) \qquad (7.96b)$$

$$H_4(z_1, z_2) = H_1(z_1, z_2^{-1}) \qquad (7.96c)$$

Since $h_d(n_1, n_2) = h_d^1(n_1, n_2) + h_d^2(n_1, n_2) + h_d^3(n_1, n_2) + h_d^4(n_1, n_2)$, we can obtain the resulting filter $H(z_1, z_2)$ by

$$H(z_1, z_2) = H_1(z_1, z_2) + H_2(z_1, z_2) + H_3(z_1, z_2) + H_4(z_1, z_2) \qquad (7.97)$$

The filter can be implemented, therefore, by a parallel combination of four recursively computable computational procedures.

In addition to the IIR filter design method discussed above, there are many variations that require solving only sets of linear equations. All these methods, of course, are not optimal procedures, and precise control over frequency-domain parameters is not possible.

7.8.2 Implementation of IIR Filters

In the implementation of 1-D IIR filters, the standard methods are direct forms, cascade forms, and parallel forms. In the implementation of 2-D IIR filters, the only method that can be used for any recursively computable rational system function is the direct-form method. In the direct-form method, a difference equation is first obtained from the system function, and the difference equation is used to recursively compute the output.

In the realization of 1-D IIR filters, the cascade form is probably the most often used because of its relatively small sensitivity to coefficient quantization. In the 1-D cascade form, the system function $H(z)$ is expressed as

$$H(z) = A \cdot \prod_{k} H_k(z) \tag{7.98}$$

Since a 1-D polynomial can always be factored as a product of lower-order polynomials, $H(z)$ can always be expressed in the form of Eq. (7.98). In the case of 2-D IIR filters, the cascade form generally cannot be used. A 2-D polynomial cannot, in general, be factored as a product of lower-order polynomials, and $H(z_1, z_2)$ cannot generally be written in the form of

$$H(z_1, z_2) = A \cdot \prod_{k} H_k(z_1, z_2) \tag{7.99}$$

To use a cascade form in the realization of a 2-D IIR filter, therefore, the form in Eq. (7.99) should be used explicitly in the design step.

In the 1-D parallel form, the system function $H(z)$ is expressed as

$$H(z) = \sum_{k} H_k(z) \tag{7.100}$$

The parallel form also requires the factorization of the denominator polynomial and therefore cannot be used as a general procedure for realizing a 2-D IIR filter. Like the cascade form, the parallel form can be used when the form in Eq. (7.100) is used explicitly in the design step. In the IIR filter design example in the previous section the form of Eq. (7.100) was used explicitly in the filter design and therefore the parallel form could be used for its implementation. The parallel form is useful when a zero-phase IIR filter is desired, as in the design example in the previous section.

7.8.3 Comparison of FIR and IIR Filters

FIR filters have many advantages over IIR filters. Stability is not an issue in FIR filter design or implementation. For IIR filters, however, testing the filter stability and stabilizing an unstable filter without significantly affecting the magnitude response is a very big task. Zero phase is extremely easy to achieve for FIR filters. Designing zero-phase IIR filters is possible, but is more involved than designing zero-phase FIR filters. In addition, design methods are simpler for FIR filters than for IIR filters.

The main advantage of an IIR filter over an FIR filter is the reduction in the number of arithmetic operations when implemented in direct form. To meet the same magnitude specification, an FIR filter typically requires more arithmetic operations per output sample than an IIR filter. If an FIR filter is implemented exploiting the computational efficiency of FFT algorithms, this advantage of an IIR filter often disappears.

Because of the overwhelming advantage of FIR filters over IIR filters, FIR filters are much more common in practice.

7.9 APPLICATIONS

The theories discussed in the previous sections in this chapter can be applied to a number of practical problems, such as images, radar signals, and geophysical data. In this section, we show a few application examples derived from image processing problems. Our objective is not to give a comprehensive treatment of the image processing field, but just to show a few examples where digital signal processing techniques have been successfully applied. Since most of the signals that arise in practice are analog, we first briefly discuss issues related to digital processing of analog signals.

7.9.1 Digital Processing of Analog Signals

The issues that arise in digital processing of analog signals are essentially the same for the 1-D and the 2-D case, and therefore we simply summarize the 2-D results.

To differentiate an analog signal from a sequence, we denote the analog signal by $x_a(t_1, t_2)$. The continuous space Fourier transform of $x_a(t_1, t_2)$, $X_a(\Omega_1, \Omega_2)$, is related to $x_a(t_1, t_2)$ by

$$X_a(\Omega_1, \Omega_2) = \int_{t_1=-\infty}^{\infty} \int_{t_2=-\infty}^{\infty} x_a(t_1, t_2) \cdot e^{-j\Omega_1 t_1} \cdot e^{-j\Omega_2 t_2} \cdot dt_1 \cdot dt_2 \qquad (7.101)$$

$$x_a(t_1, t_2) = \frac{1}{(2\pi)^2} \int_{\Omega_1=-\infty}^{\infty} \int_{\Omega_2=-\infty}^{\infty} X_a(\Omega_1, \Omega_2) \cdot e^{j\Omega_1 t_1} \cdot e^{j\Omega_2 t_2} \cdot d\Omega_1 \cdot d\Omega_2 \qquad (7.102)$$

Suppose we obtain a discrete space signal $x(n_1, n_2)$ by sampling an analog signal $x_a(t_1, t_2)$ with sampling period (T_1, T_2) as follows:

$$x(n_1, n_2) = x_a(t_1, t_2)\big|_{t_1=n_1 T_1, t_2=n_2 T_2} \qquad (7.103)$$

Equation (7.103) represents the input-output relationship of an ideal analog-to-digital (A/D) converter. The relation between $X(\omega_1, \omega_2)$, the discrete space Fourier transform of $x(n_1, n_2)$, and $X_a(\Omega_1, \Omega_2)$, the continuous space Fourier transform of $x_a(t_1, t_2)$, is

$$X(\omega_1, \omega_2) = \frac{1}{T_1 \cdot T_2} \cdot \sum_{r_1=-\infty}^{\infty} \sum_{r_2=-\infty}^{\infty} X_a\left(\frac{\omega_1 - 2\pi r_1}{T_1}, \frac{\omega_2 - 2\pi r_2}{T_2}\right) \qquad (7.104)$$

Examples of $X_a(\Omega_1, \Omega_2)$ and $X(\omega_1, \omega_2)$ are shown in Fig. 7.50 for the case $1/T_1 > \Omega_c/\pi$ and $1/T_2 > \Omega_c'/\pi$, where Ω_c and Ω_c' are the cutoff frequencies of $X_a(\Omega_1, \Omega_2)$, as shown. From the figure, when $1/T_1 > \Omega_c/\pi$ and $1/T_2 > \Omega_c'/\pi$, $x_a(t_1, t_2)$ can be recovered from $x(n_1, n_2)$. This is the 2-D sampling theorem, which is a straightforward extension of the 1-D results.

An ideal digital-to-analog (D/A) converter recovers $x_a(t_1, t_2)$ from $x(n_1, n_2)$ when the sampling frequencies $1/T_1$ and $1/T_2$ are sufficiently high to satisfy the sampling

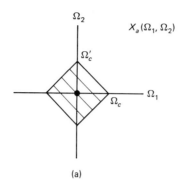

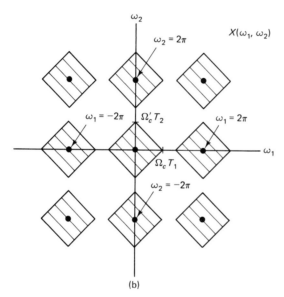

Figure 7.50 An example illustrating the relationship between $X_a(\Omega_1, \Omega_2)$ and $X(\omega_1, \omega_2)$ given by Eq. (7.104). (a) $X_a(\Omega_1, \Omega_2)$; (b) $X(\omega_1, \omega_2)$.

theorem. The output of the ideal D/A converter, $y_a(t_1, t_2)$, is given by

$$y_a(t_1, t_2) = \sum_{n_1=-\infty}^{\infty} \sum_{n_2=-\infty}^{\infty} x(n_1, n_2) \cdot \frac{\sin \frac{\pi}{T_1}(t_1 - n_1 \cdot T_1)}{\frac{\pi}{T_1}(t_1 - n_1 \cdot T_1)} \cdot \frac{\sin \frac{\pi}{T_2}(t_2 - n_2 \cdot T_2)}{\frac{\pi}{T_2}(t_2 - n_2 \cdot T_2)}$$

(7.105)

The function $y_a(t_1, t_2)$ is identical to $x_a(t_1, t_2)$ when the sampling frequencies used in the ideal A/D converter are sufficiently high. Otherwise, $y_a(t_1, t_2)$ is an aliased version of $x_a(t_1, t_2)$. Equation (7.105) is a straightforward extension of the 1-D results.

An analog signal can often be processed by digital signal processing techniques using the A/D and D/A converters just discussed. Digital processing of analog signals can, in general, be represented by the system in Fig. 7.51. The analog lowpass filter limits the bandwidth of the analog filter to reduce the effect of aliasing.

Figure 7.51   Digital processing of analog signals.

7.9.2 Examples in Image Processing Applications

An important application area of 2-D signal processing theories is image processing, which in recent years has received considerable attention. This is due in part to significant advances in hardware technology that allow sophisticated image processing algorithms to be implemented in real time and in part to a large number of applications of image processing in such diverse areas as medicine, communications, consumer electronics, defense, law enforcement, robotics, geophysics, and agriculture. Image processing can be classified broadly into four areas: image restoration, enhancement, coding, and understanding. In this section, we illustrate one example in each area.

Image restoration.   In image restoration, an image has been degraded in some manner and the objective is to reduce or eliminate the effect of degradation. Typical degradations that occur in practice include image blurring, additive random noise, quantization noise, multiplicative noise, and geometric distortion. Suppose an image $f(n_1, n_2)$ is degraded by additive random noise. Then the degraded image $g(n_1, n_2)$ can be expressed as

$$g(n_1, n_2) = f(n_1, n_2) + w(n_1, n_2) \tag{7.106}$$

where $w(n_1, n_2)$ is a random background noise. An example of an image degraded by white noise is shown in Fig. 7.52. Part (a) shows an undegraded original image of 256×256 pixels, and part (b) shows the degraded image.

An image processed by a signal processing algorithm to reduce additive noise in the image in Fig. 7.52(b) is shown in Fig. 7.52(c). The degraded image is filtered by a space-variant FIR filter. A new filter was designed at each pixel based on the amount of image detail in the neighborhood of the pixel to be processed. If there is a fair amount of detail, such as near edges, then little lowpass filtering is performed since details generally correspond to high-frequency components and therefore a high level of lowpass filtering can reduce the image details. In addition, the same amount of noise in high-detail image areas is less visible than in low-detail image areas. In low-detail image areas such as uniform background areas, a large amount of lowpass filtering is performed. Low details in images generally correspond to low-frequency components, and therefore the signal is not significantly degraded by a large amount of lowpass filtering. The background noise, on the other hand, is significantly reduced by a large amount of lowpass filtering.

Image enhancement.   Image enhancement is the processing of an image to improve its visual appearance to a human viewer or to enhance the performance of another image processing system. Methods and objectives vary with the application. When images are enhanced for human viewers, as in television, the objective may be to improve perceptual aspects: image quality, intelligibility, or visual appearance. In

(a)

(b)

(c)

Figure 7.52 (a) An original image of 256×256 pixels; (b) the image in part (a) degraded by additive white noise; (c) a restored image.

applications such as object identification by machine, an image may be preprocessed to aid machine performance. Because the objective of image enhancement is heavily dependent on the application context and the criteria for enhancement are often subjective or too complex to be easily converted to useful objective measures, image enhancement algorithms tend to be simple, qualitative, and ad hoc. Image enhancement is closely related to image restoration. When an image is degraded, restoration of the original image often results in enhancement. There are, however, some important differences. In image restoration, an ideal image has been degraded and the objective is to make the processed image resemble the original as much as possible. In image enhancement, the objective is to make the processed image better in some sense than the unprocessed image. To understand this difference, note that the original, undegraded image cannot be further restored but can be enhanced by increasing sharpness.

The visual appearance of an image can often be enhanced significantly by proper manipulation of the contrast and the overall dynamic range of the image. An example of this is illustrated in Fig. 7.53. Part (a) shows an image that was taken from an airplane. Because of the varying amounts of cloud cover, the details on the ground are not very visible. Part (b) shows an image obtained by processing the image in part (a). In the processing, different operations were performed for different regions of the image. By measuring the local average intensity in a particular region, we can estimate the approximate level of cloud cover. The high average intensity region generally corresponds to a higher level of cloud cover. In regions where cloud cover appears to be present, the image is highpass filtered to increase the contrast. In addition, the local average intensity is reduced so that the contrast increase will not be clipped because of dynamic range increase caused by the contrast increase. The amount of highpass filtering and reduction in the local average intensity is adapted to the estimated level of cloud cover. In this example, the highpass filter used is FIR.

Image coding.   The objective in image coding is to represent an image with as few bits as possible, preserving a certain level of image quality and intelligibility acceptable for a given application. Image coding can be used in reducing the bandwidth of a communication channel when an image is transmitted and in reducing the amount of required storage when an image needs to be retrieved at some future time. Image coding is related to image restoration and enhancement. If we can reduce the degradation such as quantization noise that results from an image coding algorithm or to enhance the visual appearance of the reconstructed image, for example, we can reduce the number of bits required to represent the image at a given level of image quality and intelligibility.

One approach to image coding is transform coding, in which an image is first transformed into a different domain and then the transformed image is coded. For transforms such as the Fourier transform, the energy of typical images is concentrated in a small region and therefore only the transform coefficients in the small region can be coded without significant distortion of an image. An example that illustrates the performance of a transform coding method is shown in Fig. 7.54. Part (a) shows an image of 256 × 256 pixels where each pixel value is represented by 8 bits of uniform quantization. Part (b) shows the coded image obtained by coding only a small number

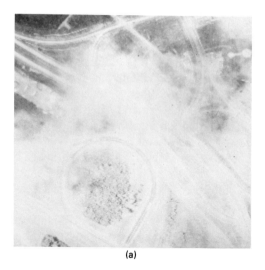

(a)

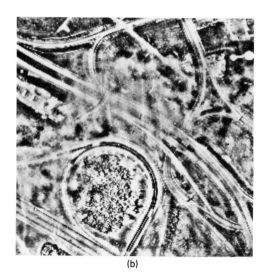

(b)

Figure 7.53 (a) An image of 256 × 256 pixels taken from an airplane; (b) the image in part (a) processed for enhancement.

of discrete cosine transform coefficients. The discrete cosine transform is closely related to the discrete Fourier transform. The number of bits used for this image is 0.7 bit/pixel. If a straightforward pulse code modulation (PCM) system that codes the intensity of the image were used, at least 2–4 bits/pixel would be necessary to obtain an image quality similar to the one in Fig. 7.54(b).

Image understanding. The objective in image understanding is to symbolically represent the contents of an image. Applications of image understanding include computer vision, robotics, and target identification. Image understanding differs from the other three areas in one major aspect. In image understanding, the input is an image, but the output is typically some symbolic representation of the contents of the image. Successful development of a system in this area generally requires both signal

(a)

(b)

Figure 7.54   (a) An original image of 256 × 256 pixels; (b) the image in part (a) coded by a discrete cosine transform method at 0.7 bit/pixel.

processing and artificial intelligence concepts. In a typical image understanding system, signal processing is used to perform lower-level processing such as reduction of degradation and extraction of image features such as edges, and artificial intelligence is used to perform higher-level processing such as symbol manipulation and knowledge-base management.

An example of edges detected by a simple signal processing algorithm is shown in Fig. 7.55. Part (a) shows an original image of 256 × 256 pixels. Part (b) shows the image where the edges or intensity discontinuities are shown. The edges were obtained by applying a bandpass filter to the original image in part (a) to emphasize intensity discontinuities and then applying a threshold test. Regions of the bandpass filtered image are declared to be edges when the pixel values are above a certain threshold.

(a)

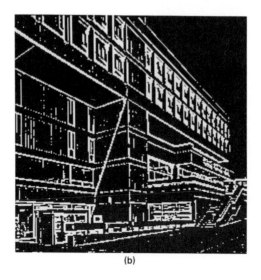

(b)

Figure 7.55 (a) An original image of 256 × 256 pixels; (b) the edge contour obtained from the image in part (a).

7.10 SUMMARY

In this chapter, we discussed the fundamentals of 2-D signal processing, including the Fourier transform, the z-transform, difference equations, discrete Fourier transform, fast Fourier transform, and design and implementation of digital filters. These are the same topics typically discussed in fundamentals of 1-D signal processing, but there are a number of differences between 1-D signal processing and 2-D signal processing, which we have attempted to show. Although our discussion in this chapter concentrated on 2-D signal processing, we note that the results can be extended in a straightforward manner to higher-dimensional signal processing. Those who wish to study 2-D signal processing in greater detail should refer to [4,5,6,7,8,9].

REFERENCES

1. J. O. Eklundh, "A Fast Computer Method for Matrix Transposing," *IEEE Trans. Computers,* Vol. 21, pp. 801–803, 1972.

2. T. W. Parks and J. H. McClellan, "Chebyshev Approximation for Nonrecursive Digital Filters with Linear Phase," *IEEE Trans. Circuit Theory,* Vol. 19, pp. 189–194, 1972.

3. R. M. Mersereau, W. F. G. Mecklenbrauken, and T. F. Quatieri, Jr., "McClellan Transformations for Two-Dimensional Digital Filtering: I, Design." *IEEE Trans. Circuits and Systems,* Vol. 23, pp. 405–414, July 1976.

4. J. S. Lim, *Two-Dimensional Signal Processing and Image Processing,* to be published.

5. D. E. Dudgeon and R. M. Mersereau, *Multidimensional Digital Signal Processing,* Prentice-Hall, Englewood Cliffs, NJ, 1983.

6. T. S. Huang, Ed., *Two-Dimensional Digital Signal Processing I,* in *Topics in Applied Physics,* Vol. 42, Springer-Verlag, Berlin, 1981.

7. T. S. Huang, Ed., *Two-Dimensional Digital Signal Processing II,* in *Topics in Applied Physics,* Vol. 43, Springer-Verlag, Berlin, 1981.

8. S. K. Mitra and M. P. Ekstron, Eds., *Two-Dimensional Digital Signal Processing,* Dowden, Hutchinson, and Ross, Stroudsburg, PA, 1978.

9. IEEE ASSP Society's MDSP Technical Committee, Ed., *Selected Papers in Multidimensional Digital Signal Processing,* IEEE Press, New York, 1986.

8

Some Advanced Topics in Filter Design

Hans Wilhelm Schüssler
Peter Steffen
University of Erlangen-Nuremberg

8.0 INTRODUCTION

The term *filter* most typically denotes a device having selective properties. In the ideal case, some parts of the spectrum of the incoming signal are passed without any change while other parts are suppressed completely. This process can be expressed by specifying a desired idealized transfer function $H_i(\omega)$. In the case of a lowpass filter, for example,

$$H_i(\omega) = \begin{cases} 1 & \text{in the passband} \\ 0 & \text{in the stopband} \end{cases}$$

In reality these properties cannot be achieved exactly. Fortunately, they are not necessary. A certain constant delay can always be tolerated, and some deviations from the desired behavior in the passband and stopband as well as in a transition band between both are permissible [1]. This leads to tolerance schemes for the magnitude and group delay of the system to be designed. An example of a lowpass filter is shown in Fig. 8.1. In (b) specifications for the group delay are given in the passband only.

The following remarks can be made about the filter shown in the figure:

1. The parameters ω_p, ω_s, δ_1, δ_2, and Δ as well as τ_0 depend on the particular application. In many cases of practical interest a phase distortion of the output signal is acceptable. Thus only the tolerance scheme for the magnitude in Fig. 8.1(a) has to be satisfied.

Hans Wilhelm Schüssler and Peter Steffen are with Lehrstuhl für Nachrichtentechnik, Universität Erlangen-Nürnberg, Cauerstrasse 7, 8520 Erlangen, West Germany.

416

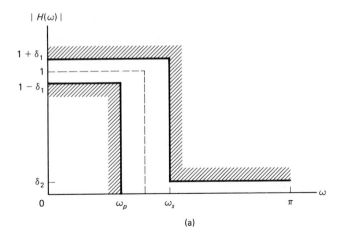

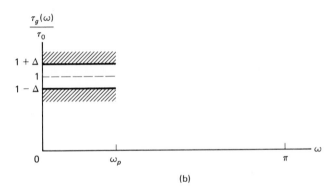

Figure 8.1 A tolerance scheme of a lowpass filter describing a desired magnitude and group delay.

2. We must expect that the degree of the required transfer function, and in that sense the cost (in some terms) of the system, will increase if the tolerances δ_1, δ_2, and Δ as well as the width $\Delta\omega = \omega_s - \omega_p$ of the transition band decrease.

3. We can also expect that a design is optimal, i.e., yields a transfer function of minimum degree, if the tolerance scheme is fully used. This leads to an equiripple approximation of the ideal behavior in the passband and the stopband. In this case the design must be such that, for the magnitude $|H(\omega)|$, the Chebyshev norm of the error function becomes a minimum:

$$\left\| |H_i(\omega)| - |H(\omega)| \right\|_\infty = \max \left| W(\omega)[|H_i(\omega)| - |H(\omega)|] \right|$$

$$= \max |E(\omega)|$$

Here $W(\omega)$ is a nonnegative weighting function, defined as

$$W(\omega) = \begin{cases} 1, & 0 \le |\omega| \le \omega_p \\ 0, & \omega_p < |\omega| < \omega_s \\ \dfrac{\delta_1}{\delta_2}, & \omega_s \le |\omega| \le \pi, \end{cases}$$

and consequently incorporates the parameters of the tolerance scheme.

The essential point is that the specifications for a selective system are expressed by a tolerance scheme and that the design leads to an optimal solution if the L_∞ norm of the error function is minimized. An example in the infinite impulse response (IIR) case is an elliptic (or Cauer) filter with minimum phase, and in the finite impulse response (FIR) case an equiripple filter with linear phase [1].

In this chapter the term *filter* will be used not only for selective systems but in a somewhat wider sense. We want to design systems having certain desired properties, at least approximately. The specifications can be expressed in the time domain (e.g., by a desired impulse response) or in the frequency domain (for example, in solving an equalization problem). The intended application might require specifications at certain points or over a large interval in the time or frequency domain. In the latter case we must formulate a norm for the error, the minimization of which leads to the desired filter. The choice of the error norm depends exclusively on the application. In the case of a selective system, the Chebyshev norm is generally the natural one, but, as we will see, other norms are preferred in other cases. Confusing different norms has led system designers down some erroneous paths. The following are two notable examples.

The easiest way to design an FIR filter is the Fourier series expansion of the desired ideal frequency response. Such a design minimizes the L_2 norm of the error function given by

$$\|H_i(\omega) - H(\omega)\|_2 = \left[\int_{-\pi}^{+\pi} |H_i(\omega) - H(\omega)|^2 \, d\omega\right]^{1/2}$$

If applied to the design of selective filters with discontinuous $H_i(\omega)$, this method yields a transfer function $H(\omega)$ showing the well-known Gibbs phenomenon [1]. The result is considered bad, and certain methods, especially windowing, have been tried to "improve" it. But the Fourier series expansion yields the optimal solution according to the L_2 norm. Its result has been considered poor only because the designer was in reality interested in satisfying a tolerance scheme and thus in minimizing the L_∞ norm but used instead the L_2 norm, since it is much simpler to apply. The well-known (and to some extent successful) improvement by the use of the Kaiser window [1] is in fact another example in which different norms are confused. The Kaiser window is an easily calculated approximation of the prolate spheroidal wave functions, which in turn are timelimited and have minimum energy outside some selected frequency interval. Obviously still another norm has been used to define these functions. More or less by accident, the Kaiser window turns out to be useful, especially since it provides a free parameter that can be fixed such that the tolerance scheme is satisfied [2]. The method leads to a quick and easy solution, but of course this solution is not optimal according to the L_∞ norm.

Another example of an erroneous path in system design is the attempt to design a selective filter by using an impulse-invariant transformation of a corresponding continuous system. The result is a discrete system such that $h_d(n) = h_c(nT)$, where $h_d(n)$ and $h_c(t)$ are the impulse responses of the discrete and continuous systems, respectively. Discrete systems of this type are extremely useful in certain applications, such as the simulation of continuous systems to be excited by a sequence of weighted unit impulses on a digital computer [3]. That means—in the limits of numerical accuracy—a precise numerical method for the solution of linear differential equations with constant coefficients. But it does not mean in general that the two corresponding frequency responses are similar, e.g., such that an optimum selective discrete system is achieved, if the continuous system is an elliptic filter. Only in the case of the L_2 norm does an optimization in one domain yield an optimal solution in the other domain, due to Parseval's relation [1].

The following general rules can be formulated for the design of systems, and, in fact, they hold correspondingly for any technical system.

1. The specifications for the system must be formulated according to the desired application. The essential general constraints such as causality and stability have to be taken into account to guarantee physical realizability. The specifications might be expressed in the time domain or the frequency domain (or both).

2. Usually an approximation problem must be formulated by prescribing a norm of the error function to be minimized. It might be necessary to regard additional side constraints. The error function and its norm depend on the particular application.

Remark. In practice it might be difficult to express some vague concept of the desired behavior in terms of an error function and in an appropriate norm. In fact, experience tells us that a careful and precise formulation of the problem based only on the application is already a major step toward the solution.

3. Efforts to find a solution in closed form are always worthwhile. If they are not successful, an appropriate numerical technique for the minimization of the prescribed error norm must be used. Questions about the existence of a solution and its uniqueness must be considered.

Remark. There is always a temptation to apply an easily handled design method to problems that really require more elaborate procedures. An easy method is tolerable only as long as solutions for the real problem are unknown. Violations of this rule have been and still are the reason for unsatisfying solutions.

In addition to presenting examples for these rules, in this chapter we will demonstrate the existence of interesting problems (and their optimal solutions) that are perhaps unfamiliar to the communications engineer but are nevertheless important for signal processing. Some are not formulated as a tolerance scheme in the frequency domain and thus do not require the minimization of the L_∞ norm. We will also show interesting, recent design procedures for old problems as well as other procedures for uncommon questions.

We start with problems formulated primarily in the time domain. First we consider the design of systems with prescribed impulse responses, using Prony's method, with different extensions and applications than those discussed in Chapter 1.

We also deal with the problem of reconstructing a function out of its samples and designing a realizable system for the task. Smoothing the numerical results of an experiment is a well-established problem discussed here in terms of both the classical approach and a signal processing approach. In all these cases the problem is stated primarily in the time domain, but we try to give some insight into the properties of the designed systems by showing their frequency responses for particular examples as well.

The classical problems of designing selective IIR or FIR filters is covered in most basic signal processing textbooks and therefore not treated here. However, for problems formulated in the frequency domain we consider a generalized L_2 design of FIR filters and the design of optimal minimum-phase FIR filters. Finally, we present a relatively new class of IIR filters consisting of allpass subsystems that yield interesting new approximation problems as well as favorable solutions and implementations.

8.1 DESIGN OF A SYSTEM WITH PRESCRIBED IMPULSE RESPONSE

The classical design problem in the time domain can be stated as follows:

Given a sequence $g(n)$, $n = 0, 1, \ldots, K$, design a digital system of prescribed degree such that its impulse response $h(n)$ approximates $g(n)$ as well as possible.

As is well known, this problem arises as a design task in the usual sense. It must also be solved if an unknown system is to be modeled by a system with a rational transfer function or identified as such a system. In this case the given values $g(n)$ might be samples of the output function of a continuous system. The modeling of the signal, treated in Section 1.3, is very important. Some other aspects as well as further applications are shown here.

The method we describe here is named for Prony, who developed it in 1795 for problems in hydro and gas mechanics (see Section 1.2) [4].

Let the transfer function $H(z)$ to be designed be

$$H(z) = \frac{\sum\limits_{k=0}^{p} b_k z^{-k}}{1 + \sum\limits_{k=1}^{p} a_k z^{-k}} = \sum\limits_{n=0}^{\infty} h(n)z^{-n}, \qquad p \in \mathbf{N}, \tag{8.1}$$

where $\mathbf{N}$ is the set of natural numbers. Here we assume the degree of the numerator and the denominator to be equal. In the first step we choose the number of given values $g(n)$ to be equal to the number of coefficients to be determined. At least one recursive system always exists, the impulse response of which satisfies exactly the condition

$$h(n) = g(n), \qquad n = 0, 1, \ldots, K \tag{8.2}$$

By multiplying Eq. (8.1) by the denominator and comparing the terms of equal order, we get, with Eq. (8.2),

$$
\begin{bmatrix} b_0 \\ b_1 \\ \vdots \\ b_{p-1} \\ b_p \\ \hline 0 \\ \vdots \\ 0 \end{bmatrix}
=
\begin{bmatrix}
g(0) & 0 & \cdots & & \cdots & & 0 \\
g(1) & g(0) & & & & & \\
\vdots & & & & & & \vdots \\
g(p-1) & \cdots & g(1) & & g(0) & & 0 \\
g(p) & \cdots & g(2) & & g(1) & & g(0) \\
\hline
g(p+1) & \cdots & \cdots & & & g(2) & g(1) \\
\vdots & & & & & & \vdots \\
g(2p) & \cdots & \cdots & & & g(p+1) & g(p)
\end{bmatrix}
\begin{bmatrix} 1 \\ a_1 \\ \vdots \\ a_{p-1} \\ a_p \end{bmatrix}
\tag{8.3}
$$

The indicated partition of Eq. (8.3) leads to the pair of matrix equations

$$\mathbf{b} = \mathbf{G}_1 \mathbf{a} \tag{8.4a}$$

and

$$\mathbf{0} = \mathbf{G}_2 \mathbf{a} \tag{8.4b}$$

where $\mathbf{G}_1$ is a $(p+1) \times (p+1)$ lower triangular Toeplitz matrix, and

$$\mathbf{G}_2 = [\mathbf{g}_1, \mathbf{g}_2, \ldots, \mathbf{g}_{p+1}] \tag{8.4c}$$

is a $p \times (p+1)$ rectangular matrix. Equation (8.4a) yields the vector $\mathbf{b}$ of the numerator coefficients for any denominator such that the impulse response has the desired values for $n = 0, 1, \ldots, p$.

To calculate the denominator, we write Eq. (8.4b) as

$$
\begin{aligned}
\mathbf{0} &= \mathbf{g}_1 + [\mathbf{g}_2, \ldots, \mathbf{g}_{p+1}]\mathbf{a}' \\
&= \mathbf{g}_1 + \mathbf{G}_3 \cdot \mathbf{a}'
\end{aligned}
$$

where $\mathbf{a}' = [a_1, a_2, \ldots, a_p]^T$ is the vector of the unknown coefficients.

If $\mathbf{G}_3$ has rank p, we obtain

$$\mathbf{a}' = -\mathbf{G}_3^{-1} \cdot \mathbf{g}_1 \tag{8.5}$$

Together with $\mathbf{b}$ from Eq. (8.4a) we then have all coefficients of $H(z)$.

For linearly dependent vectors $\mathbf{g}_n$, $n = 2, 3, \ldots, p + 1$, the matrix $\mathbf{G}_3$ has a rank $p' < p$. In this case the procedure will yield a system of degree p', the impulse response of which will exactly satisfy Eq. (8.2).

We add some remarks and an example.

1. The method just described solves the given problem precisely, but it does not satisfy any further conditions implicitly desired. For example, the system is not necessarily stable. And in general it is $h(n) \neq 0$ for $n > K$.
2. If the given values $g(n)$ are the first $K + 1$ samples of the impulse response of a discrete-time system, having degree $p \leq K/2$, we get precisely the transfer function of this system. A corresponding statement holds if the values $g(n)$ are

samples of the impulse response of an analog time-invariant system, described by a linear differential equation of order $p \le K/2$. Specifically, the resulting system will correspond to the impulse invariant transformation of that analog system.

3. It is interesting to note that $K + 1$ different solutions can be found if the number of unknown coefficients is equal to the number of independent values $g(n)$. This is true if the degrees q and p of the numerator and denominator polynomials are assumed to be not necessarily equal. The condition is $q + p + 1 = K + 1$. For $q = K$ we obtain a nonrecursive filter, and for $p = K$ an all-pole system.

Example 1

As a simple numerical example, let

$$\{g(n)\} = \{1; 18; 9; 2; 1; \tfrac{2}{9}; \tfrac{1}{9}\}$$

be the given sequence. As is easily confirmed, the matrix

$$\mathbf{G}_2 = \begin{bmatrix} 1 & 2 & 9 & 18 \\ \frac{2}{9} & 1 & 2 & 9 \\ \frac{1}{9} & \frac{2}{9} & 1 & 2 \end{bmatrix}$$

has rank 2. So a system with $q = p = 2$ will be sufficient. The described procedure, applied to the first five values of $g(n)$, yields the result

$$H(z) = \frac{1 + 18z^{-1} + \dfrac{80}{9}z^{-2}}{1 - \dfrac{1}{9}z^{-2}}$$

corresponding to

$$h(n) = \delta(n) + [22.5(\tfrac{1}{3})^{n-1} - 4.5(-\tfrac{1}{3})^{n-1}]u(n-1)$$

As we expect, $h(n) = g(n)$ for all seven values of $g(n)$; but in general $h(n) \ne 0$ for $n > 6$.

To demonstrate other possible solutions, we restrict ourselves for simplicity to five values of $g(n)$. Then corresponding solutions are as follows.

For $q = 3$, $p = 1$:

$$H_1(z) = \frac{1 + 17.5z^{-1} - 2.5z^{-3}}{1 - 0.5z^{-1}}$$

with $h_1(n) = \delta(n) + 18\delta(n-1) + 9\delta(n-2) + 2(0.5)^{n-3}u(n-3)$, where $h_1(n) \ne g(n)$, $n = 5$ and 6.

For $q = 1$, $p = 3$:

$$H_3(z) = \frac{1 + (18 + a_1)z^{-1}}{1 + a_1z^{-1} + a_2z^{-2} + a_3z^{-3}}$$

with $a_1 = -280/551$, $a_2 = 81/551$, and $a_3 = -40/551$. For $q = 0, p = 4$, we obtain an unstable system but still such that $h(n) = g(n)$, $n = 0, 1, \ldots, 4$. Finally, the nonrecursive case $q = 4$, $p = 0$ is obvious.

4. It is interesting to note that Prony's method can be applied to the analysis of a function

$$g_c(t) = \sum_{\ell=0}^{L} c_\ell \cos(\Omega_\ell t + \phi_\ell)$$

with unknown values c_ℓ, Ω_ℓ, and ϕ_ℓ. We assume that upper limits for Ω_ℓ and L are known so that the sampling interval T can be chosen small enough to avoid aliasing and that a sufficient number of samples $g(n) = g_c(t = nT)$, $n = 0, 1, \ldots, (4L + 4)$, can be used. The solution as described above yields a transfer function of degree $2L + 2$ with poles on the unit circle. The calculation of these poles and a partial fraction expansion is required for the determination of the unknown values c_ℓ, Ω_ℓ, and ϕ_ℓ. Generally, $4L + 5$ samples out of an arbitrary small interval of $g_c(t)$ provide enough information for this type of spectral analysis. But practical experience shows that the method becomes numerically sensitive and ill conditioned if the sampling interval T is decreased too much.

Now we consider the general case, characterized by $(K + 1) > (2p + 1)$, assuming $q = p$. No exact solution is possible in general. Equations such as (8.3) and (8.4) can be stated, but this time $\mathbf{G}_2$ is a $(K - p) \times (p + 1)$ matrix such that Eq. (8.4b) cannot be solved. Instead we consider

$$\mathbf{\Delta} = \mathbf{G}_2 \cdot \mathbf{a} \tag{8.6}$$

and determine the coefficient vector $\mathbf{a}$ such that $\mathbf{\Delta}^T\mathbf{\Delta}$, the mean quadratic error of the equations, becomes a minimum. With notations as before, we get

$$\mathbf{\Delta} = \mathbf{g}_1 + \mathbf{G}_3 \cdot \mathbf{a}'$$

where $\mathbf{G}_3$ is a $(K - p) \times p$ matrix now. Minimization of $\mathbf{\Delta}^T\mathbf{\Delta}$ yields the equation

$$\mathbf{G}_3^T\mathbf{G}_3\mathbf{a}' = -\mathbf{G}_3^T\mathbf{g}_1$$

If $\mathbf{G}_3$, and thus $\mathbf{G}_3^T\mathbf{G}_3$, has the rank p, we get

$$\mathbf{a}' = -(\mathbf{G}_3^T\mathbf{G}_3)^{-1}\mathbf{G}_3^T\mathbf{g}_1 \tag{8.7}$$

There are several different approaches to calculating the vector $\mathbf{b}$ of the numerator coefficients. If we use Eq. (8.4a) as before, we get a solution such that

$$h(n) = g(n), \ n = 0, 1, \ldots, p \tag{8.8}$$

Instead it might be of interest to calculate the numerator such that the error vanishes in another interval, starting at n_0. Such a solution can be found as follows. If we assume for simplicity that the transfer function has distinct poles, we obtain

$$H(z) = \frac{\sum\limits_{k=0}^{p} b_k z^{-k}}{\prod\limits_{k=1}^{p} (1 - z_k z^{-1})} = B_0 + \sum_{k=1}^{p} \frac{B_k}{1 - z_k z^{-1}} \tag{8.9a}$$

where the coefficients B_k, $k = 0, 1, \ldots, p$ are to be calculated. The impulse response becomes

$$h(n) = B_0 \delta(n) + \sum_{k=1}^{p} B_k z_k^n \cdot u(n) \qquad (8.9b)$$

If

$$h(n) = g(n), \qquad n = n_0, n_0 + 1, \ldots, (n_0 + p - 1) \qquad (8.10)$$

is prescribed, we get, for $n_0 > 0$,

$$
\begin{bmatrix}
z_1^{n_0} & z_2^{n_0} & \cdots & z_p^{n_0} \\
z_1^{n_0+1} & z_2^{n_0+1} & \cdots & z_p^{n_0+1} \\
\vdots & \vdots & & \vdots \\
z_1^{n_0+p-1} & z_2^{n_0+p-1} & \cdots & z_p^{n_0+p-1}
\end{bmatrix}
\begin{bmatrix}
B_1 \\ B_2 \\ \vdots \\ B_p
\end{bmatrix}
=
\begin{bmatrix}
g(n_0) \\ g(n_0 + 1) \\ \vdots \\ g(n_0 + p - 1)
\end{bmatrix}
\qquad (8.11)
$$

We can factor out a diagonal matrix, yielding the equation

$$\mathbf{V} \cdot \text{diag}[z_1^{n_0}, z_2^{n_0}, \ldots, z_p^{n_0}] \cdot \mathbf{B}' = \mathbf{g}_{n_0}$$

where $\mathbf{V}$ is the Vandermonde matrix of $z_1, \ldots, z_p$, the inverse of which always exists due to the assumption of simple poles. The solution becomes

$$\mathbf{B}' = \text{diag}[z_1^{-n_0}, z_2^{-n_0}, \ldots, z_p^{-n_0}]\mathbf{V}^{-1}\mathbf{g}_{n_0} \qquad (8.12a)$$

With

$$B_0 = g(0) - \sum_{k=1}^{p} B_k \qquad (8.12b)$$

we get in addition $h(0) = g(0)$.

A simple modification of Eq. (8.11) is required if $h(n) = g(n)$ is prescribed at points $n = n_0 + k \cdot \Delta n$, $k = 0, 1, \ldots, (p - 1)$, $\Delta n > 1$.

Finally, we show how $\mathbf{B} = [B_0, B_1, \ldots, B_p]^T$ can be calculated if the mean-square error for $n = 0, 1, \ldots, \infty$, given by

$$\boldsymbol{\epsilon}^T \boldsymbol{\epsilon} = \sum_{n=0}^{\infty} |h(n) - g(n)|^2, \qquad (8.13)$$

has to be minimized in addition to Eq. (8.7). This is of course possible only if the solution of Eq. (8.7) yields a stable polynomial.

Differentiating with respect to B_k yields

$$
\begin{bmatrix}
1 & 1 & 1 & \cdots & 1 \\
1 & w_{11} & w_{12} & \cdots & w_{1p} \\
1 & w_{21} & w_{22} & \cdots & w_{2p} \\
\vdots & \vdots & \vdots & & \vdots \\
1 & w_{p1} & w_{p2} & \cdots & w_{pp}
\end{bmatrix}
\begin{bmatrix}
B_0 \\ B_1 \\ B_2 \\ \vdots \\ B_p
\end{bmatrix}
=
\begin{bmatrix}
g(0) \\ G_1 \\ G_2 \\ \vdots \\ G_p
\end{bmatrix}
\qquad (8.14a)
$$

where

$$w_{ik} = \sum_{n=0}^{\infty} z_i^{*n} z_k^n = \frac{1}{1 - z_i^* z_k}, \qquad i, k = 1, 2, \ldots, p$$

$$G_k = \sum_{n=0}^{K} g(n) z_k^{*n}, \qquad k = 1, 2, \ldots, p$$

$$(8.14b)$$

As an example, we design a system whose impulse response approximates the desired sequence

$$g(n) = \frac{\sin[(n - 24)\pi/4]}{(n - 24)\pi}, \qquad n = 0, 1, \ldots, K$$

Clearly, $g(n)$ would be the impulse response of an idealized discrete lowpass filter with cutoff frequency $\omega_c = \pi/4$ and delay 24 if this equation $g(n)$ would hold for $n \in \mathbf{Z}$, where $\mathbf{Z}$ is the set of integers. But since we are designing a causal system and since Prony's method deals with sequences $g(n)$ of finite length, we have to restrict the interval as we did above. Figure 8.2 illustrates the problem and the results for $K = 100$ with a system of degree $p = 17$. As to be expected, the result is a nonminimum-phase system. Its transfer function has six zeros outside the unit circle.

The desired sequence is depicted in Fig. 8.2(a), while Fig. 8.2(b) shows the relative error for $0 \leq n \leq 100$ and Fig. 8.2(c) the impulse response for $n > 100$. A closer investigation shows that in this range, which was not included in the design procedure, $h(n)$ approximates the original sequence

$$g(n) = \frac{\sin[(n - 24)\pi/4]}{(n - 24)\pi}$$

such that the magnitude of the relative error remains smaller than 0.005. It is interesting to note that the dominant term in $h(n)$, which alone determines the impulse response for large values of n, consists of an exponentially decreasing sinusoidal sequence whose frequency is $\pi/4.032$ (instead of $\pi/4$). As we might expect, a larger value for K improves the approximation. For $K = 200$, for example, we get a dominant term having the frequency $\pi/4.018$.

For comparison, we examine the frequency response of the designed system. The result, depicted in Figs. 8.2(d) and (e), shows a rather large deviation in relation to the frequency response of an idealized lowpass filter. But first we could not expect such an ideal result because it was not our primary goal. Second and more important, it turns out that $H(\omega)$ is an excellent approximation to the spectrum of the causal sequence $g(n) = \sin[(n - 24)]\pi/4]/[(n - 24)\pi]$ for $n \geq 0$. That means the deviations from the idealized lowpass filter are due mainly to the required causality.

As another application of Prony's method, we show the design of a causal system such that its impulse response $h(n)$ has an autocorrelation sequence $v(\ell)$ that approximates a given sequence $g_v(\ell)$. This problem arises, for example, if we want to generate a random sequence with a prescribed autocorrelation sequence or power

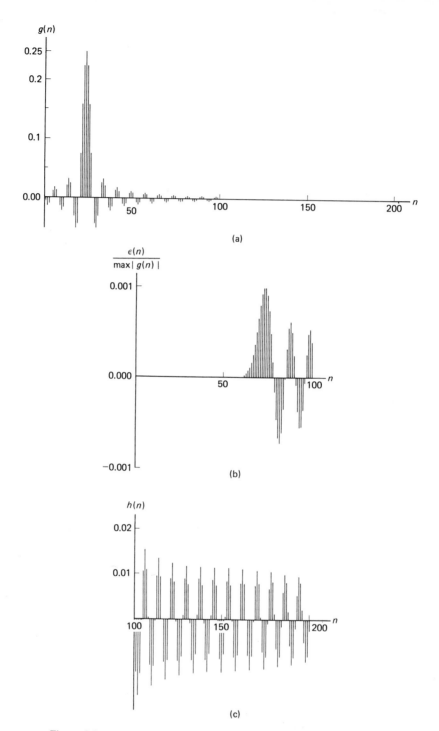

Figure 8.2 Approximation of an idealized lowpass filter, using Prony's method. (a) Desired impulse response $g(n) = \sin[(n - 24)\pi/4]/[(n - 24)\pi]$, $0 \leq n \leq 100$; (b) $\epsilon(n)/\max|g(n)| = [h(n) - g(n)]/\max|g(n)|$, $0 \leq n \leq 100$; (c) $h(n)$, $100 \leq n \leq 200$; (d) frequency response $|H(\omega)|$ of the designed system; (e) group delay $\tau_g(\omega)$ of the designed system.

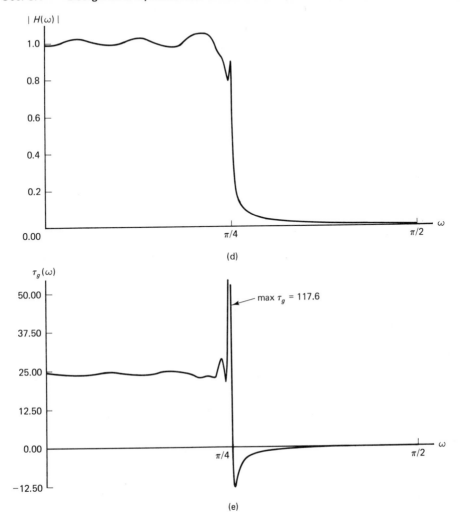

Figure 8.2 (*cont.*)

density spectrum. For convenience we cite the required well-known relations [1]:

$$v(\ell) = h(\ell) * h(-\ell) = \sum_{n=0}^{\infty} h(n)h(n + \ell) \tag{8.15a}$$

$$= \mathcal{Z}^{-1}\{H(z)H(z^{-1})\} \tag{8.15b}$$

$$= \frac{1}{2\pi} \int_{-\pi}^{+\pi} |H(\omega)|^2 \cos \ell\omega \, d\omega \tag{8.15c}$$

We are given a prescribed even sequence $g_v(\ell)$, $-K \leq \ell \leq K$, having the properties of an autocorrelation sequence. To satisfy this condition it is necessary and sufficient that

$$\mathcal{Z}\{g_v(\ell)\}\bigg|_{z=e^{j\omega}} = G_v(\omega) \geq 0 \qquad \text{for all } \omega \tag{8.16}$$

The system to be designed should be such that

$$v(\ell) \approx g_v(\ell) \tag{8.17a}$$

or, correspondingly,

$$|H(\omega)|^2 \approx G_v(\omega) \tag{8.17b}$$

For a solution in the time domain, we define the causal sequence

$$g(n) = 0.5 \, g_v(0) \cdot \delta(n) + g_v(n) \cdot u(n-1) \qquad n = 0, 1, \ldots, K \tag{8.18a}$$

Obviously we have

$$g_v(n) = g(n) + g(-n) \tag{8.18b}$$

and thus

$$\mathcal{Z}\{g_v(n)\} = G_v(z) = G(z) + G(z^{-1}) \tag{8.18c}$$

Using Prony's method appropriately with condition (8.13) for the calculation of the numerator coefficients, we design a stable system with the transfer function $F(z)$ such that its impulse response $f(n)$ satisfies $f(n) \approx g(n)$. Now, according to Eq. (8.18b), $g_v(n)$ is twice the even part of $g(n)$, and thus

$$2\mathcal{R}e\{G(\omega)\} = G_v(\omega)$$

So if the transfer function $F(z)$ is such that

$$2\mathcal{R}e\{F(\omega)\} = F(\omega) + F^*(\omega) \geq 0 \qquad \text{for all } \omega \tag{8.19}$$

the transfer function $H(z)$ of the desired system can be determined using the factorization

$$H(z)H(z^{-1}) = F(z) + F(z^{-1}) \tag{8.20}$$

obtained from a calculation of the poles and zeros of this function. For a stable solution we pick the poles inside the unit circle, which are the poles of $F(z)$, while we have the well-known ambiguity concerning the selection of zeros. We must check the validity of Eq. (8.19) numerically. If the test fails, we determine the absolute minimum

$$\min_{\forall \omega} \mathcal{R}e\{F(\omega)\} = \tfrac{1}{2}\min_{\forall \omega}[F(\omega) + F^*(\omega)] < 0$$

and introduce a modified transfer function

$$F_1(z) = F(z) - \min \mathcal{R}e\{F(\omega)\} \tag{8.21}$$

which means only a corresponding increase of $f(0)$. Proceeding with $F_1(z)$, as we explained above for $F(z)$, we see that only the zeros of $H(z)H(z^{-1})$ will be different.

8.2 SMOOTH INTERPOLATION OF SAMPLES

Quite often digital systems are embedded in an analog world. The necessary interfaces require analog-to-digital (A/D) and digital-to-analog (D/A) converters and in addition some filtering to avoid aliasing due to the sampling at the input or for smoothing at the output. In both cases analog filters can hardly be avoided, but their properties can be improved considerably by anticipating their presence in the design of the digital system.

This section deals with the design of an FIR filter to be used in front of a given cascade of a D/A converter and a smoothing filter to achieve a smooth output function. This corresponds to the classical problem of reconstructing a bandlimited function from its samples. But of course we are interested specifically in a realizable solution.

We first state the problem carefully. We introduce a family of idealized systems for the required interpolation. Let $x(n)$ be an infinite sequence of numbers at the output of a digital system. It can be interpreted as the result of sampling a bandlimited continuous function $x_c(t)$ such that $x(n) = x_c(nT)$, $n \in \mathbf{Z}$, where $\mathbf{Z}$ is the set of integers. If the sampling interval T is chosen such that

$$T = \frac{1}{f_s} \leq \frac{\pi}{\Omega_c} \tag{8.22a}$$

where Ω_c is the limit of the spectrum $X_c(j\Omega)$ of $x_c(t)$, this function can be expressed as

$$x_c(t) = \sum_{n=-\infty}^{+\infty} x(n) \frac{\sin \pi(t/T - n)}{\pi(t/T - n)} \tag{8.22b}$$

This well-known result shows how $x_c(t)$ can be reconstructed in principle. An idealized lowpass filter with a cutoff frequency $\Omega_s/2$ is required, the impulse response of which is

$$g_0(t) = \frac{\sin \pi t/T}{\pi t} \qquad \text{for all } t \tag{8.23a}$$

If excited by the generalized function

$$x_*(t) = T \sum_{n=-\infty}^{+\infty} x(n)\delta_c(t - nT) \tag{8.23b}$$

where $\delta_c(t)$ denotes the unit impulse, the output function of that lowpass filter is $x_c(t)$.

The impulse response $g_0(t)$ has the so-called interpolation property

$$g_0(t) = \begin{cases} 1/T, & t = 0 \\ 0, & t = nT, \ n \neq 0 \end{cases} \tag{8.23c}$$

which is convenient but not necessary. As is well known, the idealized lowpass filter is neither causal nor stable. Furthermore, the series of Eq. (8.22b) converges rather slowly if evaluated for any fixed value t since $g_0(t)$ decays only with $1/t$. Hence it seems to be worthwhile to look for other expansions, converging more rapidly as a starting point for the design of a realizable system.

As a first step, we weaken the basic condition in the sense that we assume the sampling frequency f_s to be

$$f_s = \frac{\beta \cdot \Omega_c}{\pi} = 2\beta f_c \tag{8.24a}$$

such that $x_c(t)$ is oversampled by a factor $\beta > 1$. Now $X_*(j\Omega)$, the spectrum of the function $x_*(t)$ given in Eq. (8.23b), shows gaps around the points $(2n + 1)\Omega_s/2$ of width

$$2\Delta\Omega = \Omega_s(1 - 1/\beta) := \alpha \cdot \Omega_s, \qquad 0 < \alpha < 1 \qquad (8.24b)$$

(See also Fig. 8.3.) Obviously any lowpass filter described by

$$G(j\Omega, \alpha) = \begin{cases} 1, & |\Omega| \le \Omega_c \\ \text{arbitrary}, & \Omega_c < |\Omega| < (1 + \alpha)\Omega_s/2 \\ 0, & |\Omega| \ge (1 + \alpha)\Omega_s/2 \end{cases} \qquad (8.24c)$$

will reconstruct $x_c(t)$. An example is given in Fig. 8.3. If we choose $G(j\Omega, \alpha)$ in the transition band such that it satisfies the second Nyquist condition,

$$G[j(\Omega_s/2 + \Omega'), \alpha] + G[j(\Omega_s/2 - \Omega'), \alpha] = 1 \qquad \text{for } |\Omega'| \le \alpha\Omega_s/2 \qquad (8.25a)$$

we can represent the corresponding impulse response by

$$g(t, \alpha) = \frac{\sin \pi t/T}{\pi t} \cdot f(t, \alpha) \qquad (8.25b)$$

Obviously the interpolation property of Eq. (8.23c) is satisfied if $f(0, \alpha) = 1$. Instead of Eq. (8.22b), we now have the expansion

$$x_c(t) = \sum_{n=-\infty}^{+\infty} x(n)g(t - nT, \alpha) \qquad (8.26)$$

This series will converge more rapidly than the series in Eq. (8.22b) if $f(t, \alpha)$ is chosen appropriately. The corresponding transfer function of the lowpass filter is now

$$G(j\Omega, \alpha) = \frac{1}{2\pi}G_0(j\Omega) * F(j\Omega, \alpha) \qquad (8.25c)$$

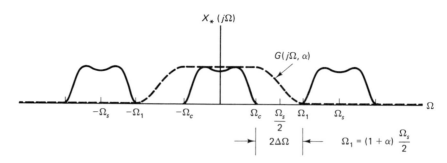

Figure 8.3 Reconstruction of oversampled signals according to Eq. (8.24c). [Adapted from Fig. 1 in H. W. Schüssler and P. Steffen, "A Hybrid System for the Reconstruction of a Smooth Function from Its Samples," *Circuits, Systems, and Signal Processing*, Vol. 3, No. 3, pp. 295–314, 1984.]

where

$$G_0(j\Omega) = \begin{cases} 1, & |\Omega| \le \Omega_s/2 \\ 0, & |\Omega| > \Omega_s/2 \end{cases}$$

is the frequency response of the idealized lowpass filter and $F(j\Omega, \alpha) = \mathcal{F}\{f(t, \alpha)\}$. An appropriate smoothness of $F(j\Omega, \alpha)$, i.e., a corresponding higher smoothness of $G(j\Omega, \alpha)$ in the transition range, yields the desired faster convergence of $g(t, \alpha)$.

Different functions $f(t, \alpha)$ have been proposed in the literature (e.g., [5]). Here we choose the class

$$f_m(t, \alpha) = \left[\frac{\sin \alpha\pi t/mT}{\alpha\pi t/mT} \right]^m, \qquad m = 1, 2, \ldots \qquad (8.27a)$$

The corresponding spectrum is

$$F_m(j\Omega, \alpha) = \frac{1}{(2\pi)^{m-1}} \cdot \underbrace{F_{1,m}(j\Omega, \alpha) * \cdots * F_{1,m}(j\Omega, \alpha)}_{m \text{ terms}} \qquad (8.27b)$$

where the expression on the right-hand side is an $(m - 1)$-fold convolution of the rectangular frequency function

$$F_{1,m}(j\Omega, \alpha) = \begin{cases} m\pi/\Delta\Omega, & |\Omega| \le \Delta\Omega/m \\ 0, & |\Omega| > \Delta\Omega/m \end{cases} \qquad (8.27c)$$

Figure 8.4(a) shows $G_m(j\Omega, \alpha) = (1/2\pi) \cdot G_0(j\Omega) * F_m(j\Omega, \alpha)$ as well as $F_m(j\Omega, \alpha)$ for $m = 1, 2$. The value $\alpha = 0.5$ has been chosen, indicating an oversampling factor $\beta = 2$. $G_0(j\Omega)$ is shown for comparison. In Fig. 8.4(b) the corresponding time functions $g_m(t, \alpha)$ are shown, with $g_0(t)$ plotted for comparison. As we can see, increasing m accelerates the convergence.

What we have achieved so far is an expansion shown in Eq. (8.26) that converges more rapidly than the original series in Eq. (8.22b). The corresponding lowpass filter is now stable, but it is not causal and thus not realizable. We will use it here as the basis for the design of a realizable system, the block diagram of which is shown in Fig. 8.5. The sequence $x(n)$ is used as the input to a digital filter, the design of which is the main goal. It works with a clock frequency $f_1 = 1/T_1 = r \cdot f_s$, where the integer $r \ge 1$ can be specified, thus yielding an interpolated sequence if $r > 1$. At the output of the D/A converter we get a staircase function, which in turn is smoothed by an analog filter. This entire system has to be designed such that the overall continuous impulse response

$$h_c(t) = \sum_{n=0}^{q} h(n) \cdot h_r(t - nT_1) \qquad (8.28a)$$

approximates a shifted and causal version of the desired time function

$$g_i(t) = g_m(t - \tau, \alpha) \cdot u_c(t) \qquad (8.28b)$$

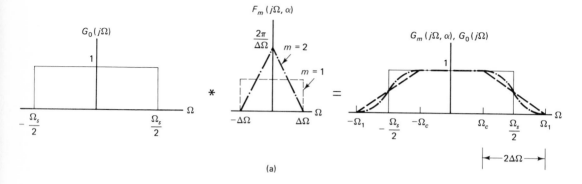

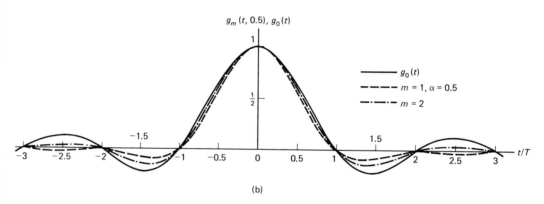

(b)

Figure 8.4 Transfer functions and impulse responses of interpolating systems with different convergence properties. (a) Illustration of Eq. (8.25c) together with Eqs. (8.27b, c); (b) interpolating time functions for $m = 1$, 2 and $\alpha = 0.5$ compared with $g_0(t)$. [Adapted from Fig. 2 in H. W. Schüssler and P. Steffen, "A Hybrid System for the Reconstruction of a Smooth Function from Its Samples," *Circuits, Systems, and Signal Processing*, Vol. 3, No. 3, pp. 295–314, 1984.]

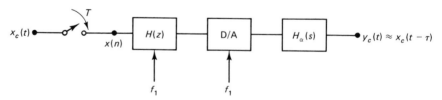

Figure 8.5 Block diagram of a realizable system for an approximate reconstruction of $x_c(t)$ out of its samples. [Adapted from Fig. 3 in H. W. Schüssler and P. Steffen, "A Hybrid System for the Reconstruction of a Smooth Function from Its Samples," *Circuits, Systems, and Signal Processing*, Vol. 3, No. 3, pp. 295–314, 1984.]

according to a criterion still to be chosen. In Eq. (8.28a) we assume a nonrecursive digital filter of degree q. The time function $h_r(t)$ describes the response of the continuous filter when excited by a rectangular impulse of length $T_1 = T/r$. Since $h_c(t)$ is a causal function, we have to accept a delay by a certain value τ in the desired function $g_m(t, \alpha)$. The value τ has to be chosen appropriately, e.g., such that $g_i(0) = 0$. The unit step function is denoted by $u_c(t)$.

The approximating nature of the solution has, beside others, the following consequences: While the bandlimited function $g_m(t, \alpha)$ is an entire function and, as such, has an infinite number of continuous derivatives, the smoothness of $h_c(t)$ is finite and is determined by the properties of the analog filter described by its rational transfer function $H_a(s)$. The impulse response $h_c(t)$ as well as its first k derivatives will be continuous if $p_a - q_a = k + 1$, where p_a and q_a are the degrees of the denominator and numerator polynomials of $H_a(s)$, respectively. In the following discussion we assume the analog smoothing filter to be given and thus the degree of achievable smoothness at the output to be fixed. So the problem reduces to the design of the discrete interpolation filter described by its impulse response $h(n)$. Since we are interested in the time function at the output, we present a design in the time domain. A frequency-domain solution has been considered in [6].

In a first step we determine $h(n)$ such that $h_c(t)$ and $g_i(t)$ coincide at equidistant points. With $T_1 = T/r$ and $\tau = L \cdot T$, we prescribe

$$h_c(\lambda T_1) = g_i(\lambda T_1), \qquad \lambda = 0, 1, \ldots, (q + 1) \tag{8.29}$$

Note that this condition is satisfied for $\lambda = 0$ in any case if $k \geq 0$, i.e., if the degree p_a of the denominator polynomial of $H_a(s)$ is larger than the degree q_a of the numerator polynomial.

The solution is straightforward. With

$$h_r(\lambda T_1) = a_{0\lambda}, \qquad \lambda = 1, 2, \ldots, (q + 1) \tag{8.30}$$

we immediately obtain a set of linear equations for the coefficients $h(n)$, $n = 0, 1, \ldots, q$, the matrix of which is a lower triangular Toeplitz matrix:

$$
\begin{bmatrix}
a_{01} & 0 & \cdots & & 0 \\
a_{02} & a_{01} & 0 & & \vdots \\
a_{03} & a_{02} & a_{01} & 0 & \vdots \\
a_{04} & a_{03} & a_{02} & a_{01} & \ddots \\
\vdots & & \ddots & \ddots & \ddots
\end{bmatrix}
\cdot
\begin{bmatrix}
h(0) \\
h(1) \\
h(2) \\
h(3) \\
\vdots
\end{bmatrix}
=
\begin{bmatrix}
g_i(T_1) \\
g_i(2T_1) \\
g_i(3T_1) \\
g_i(4T_1) \\
\vdots
\end{bmatrix}
\tag{8.31}
$$

Since $T_1 = T/r$, $h_c(t)$ has the interpolation property at the points $t_\ell = \ell \cdot T = r \cdot \ell T_1$.

To achieve maximum symmetry we will choose for the length of the nonrecursive filter $q + 1 = 2rL$. In addition to the interpolation condition, $h_c(t)$ agrees with $g_i(t)$ at $(r - 1)$ points in each of the intervals $(\ell - 1)T < t < \ell T$, $\ell = 1, 2, \ldots, 2L$; but $h_c(t)$ will not be smooth if $k = 0$.

To obtain a closer coincidence of the two functions, in a second approach we use an approximation in the sense of Hermite, i.e., such that the functions as well as their first k derivatives are equal at equidistant points. We explain the procedure for $k = 1$.

As we see from Fig. 8.6, the behavior at $t = 2T_1$ is determined by the two samples $h(0)$ and $h(1)$ only. If we proceed to the next point, just two new values of $h(n)$ gain influence. So we can write a sequence of paired conditions. We prescribe

$$h_c(2\lambda T_1) = g_i(2\lambda T_1) \tag{8.32a}$$

and

$$\left. \frac{dh_c(t)}{dt} \right|_{t=2\lambda T_1} = \left. \frac{dg_i(t)}{dt} \right|_{t=2\lambda T_1} := g'_i(2\lambda T_1), \qquad \lambda = 1, 2, \ldots, (q+1)/2 \tag{8.32b}$$

Using the notation

$$h_r(\lambda T_1) = a_{0\lambda}, \qquad \left. \frac{dh_r(t)}{dt} \right|_{t=\lambda T_1} = a_{1\lambda} \tag{8.33}$$

we obtain the set of equations

$$
\begin{bmatrix}
a_{02} & a_{01} & 0 & 0 & 0 & 0 & \cdots \\
a_{12} & a_{11} & 0 & 0 & 0 & 0 \\
a_{04} & a_{03} & a_{02} & a_{01} & 0 & 0 \\
a_{14} & a_{13} & a_{12} & a_{11} & 0 & 0 \\
\vdots & & & & & \vdots
\end{bmatrix}
\begin{bmatrix}
h(0) \\
h(1) \\
h(2) \\
h(3) \\
\vdots
\end{bmatrix}
=
\begin{bmatrix}
g_i(2T_1) \\
g'_i(2T_1) \\
g_i(4T_1) \\
g'_i(4T_1) \\
\vdots
\end{bmatrix}
\tag{8.34}
$$

which can be split up into a sequence of equations with only two unknowns and thus can be solved easily.

Figure 8.7 shows a result obtained for $m = 2$, $\alpha = 0.5$, $r = 4$, $L = 2$, and $q + 1 = 16$. A Bessel lowpass filter of second degree ($k = 1$) has been used for smoothing.

In [6] it has been shown that the method can be generalized such that $h_c(t)$ and $g_i(t)$ and their first k derivatives are equal at $M = (q+1)/(k+1)$ points.

It is interesting to note that a slight modification of the design yields an interpolating function $h_c(t)$ of precisely finite length. Suppose we have designed a digital filter as just described such that the desired function $g_i(t)$ is approximated over an interval of length $(q+1) \cdot T_1$. If the transfer function $H_a(s)$ has distinct poles, its partial fraction expansion is

$$H_a(s) = \sum_{\nu=1}^{p_a} \frac{B_\nu}{s - s_{\infty\nu}}, \qquad q_a < p_a \tag{8.35a}$$

The Laplace transform of the overall impulse response can be written as

$$H_c(s) = \sum_{\nu=1}^{p_a} B_\nu \frac{1 - e^{-sT_1}}{sT_1} \cdot H(e^{sT_1}) \cdot \frac{1}{s - s_{\infty\nu}} \tag{8.35b}$$

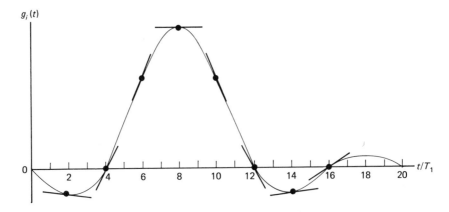

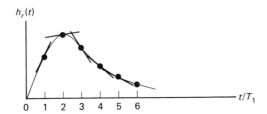

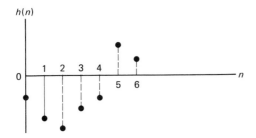

Figure 8.6 Illustration for setting up Eq. (8.34). [Adapted from Fig. 6 in H. W. Schüssler and P. Steffen, "A Hybrid System for the Reconstruction of a Smooth Function from Its Samples," *Circuits, Systems, and Signal Processing,* Vol. 3, No. 3, pp. 295–314, 1984.]

Thus the impulse response itself becomes

$$h_c(t) = \sum_{\nu=1}^{p_a} R_\nu \sum_{n=0}^{q} h(n) e^{-s_{\infty\nu} n T_1} e^{s_{\infty\nu} t}, \qquad t \ge (q+1)T \tag{8.36a}$$

where

$$R_\nu = B_\nu \cdot \frac{1 - e^{-s_{\infty\nu} T_1}}{s_{\infty\nu} T_1}, \qquad \nu = 1, 2, \dots, p_a \tag{8.36b}$$

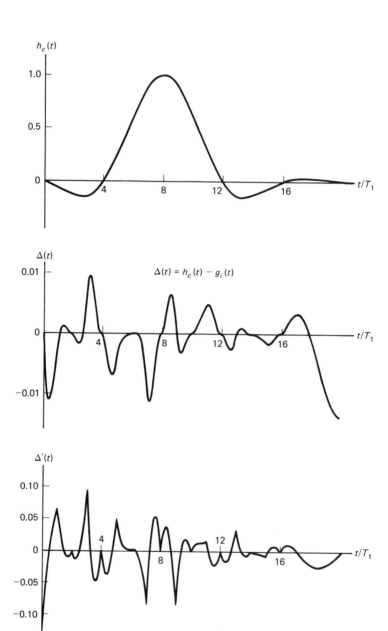

Figure 8.7 Result obtained for $m = 2$, $\alpha = 0.5$, $r = 4$, and $q = 15$. The figure shows $h_c(t)$, $\Delta(t) = h_c(t) - g_i(t)$, and $\Delta'(t) = h_c'(t) - g_i'(t)$. [Adapted from Fig. 7 in H. W. Schüssler and P. Steffen, "A Hybrid System for the Reconstruction of a Smooth Function from Its Samples," *Circuits, Systems, and Signal Processing*, Vol. 3, No. 3, pp. 295–314, 1984.]

To obtain an impulse response of finite length we increase the length of the digital filter by p_a samples, to be chosen such that the residues of $H_c(s)$ at all the poles $s_{\infty\nu}$ become zero. The resulting impulse response will have length $(q + 1 + p_a)T_1$. If

$$H(z) = \sum_{n=0}^{q} h(n)z^{-n}$$

is the transfer function of the digital system designed as explained above, we obtain with the notation $z_{\infty\nu} = e^{s_{\infty\nu}T}$, the conditions

$$H(z_{\infty\nu}) + \sum_{n=q+1}^{q+p_a} h(n)z_{\infty\nu}^{-nT_1} = 0, \qquad \nu = 1, 2, \ldots, p_a \qquad (8.37)$$

The solution of the linear equations

$$\begin{bmatrix} 1 & z_{\infty 1}^{-1} & \cdots & z_{\infty 1}^{-(p_a-1)} \\ 1 & z_{\infty 2}^{-1} & \cdots & z_{\infty 2}^{-(p_a-1)} \\ \vdots & \vdots & & \vdots \\ 1 & z_{\infty p_a}^{-1} & \cdots & z_{\infty p_a}^{-(p_a-1)} \end{bmatrix} \cdot \begin{bmatrix} h(q+1) \\ h(q+2) \\ \vdots \\ h(q+p_a) \end{bmatrix} = - \begin{bmatrix} z_{\infty 1}^{q+1} H(z_{\infty 1}) \\ z_{\infty 2}^{q+1} H(z_{\infty 2}) \\ \vdots \\ z_{\infty p_a}^{q+1} H(z_{\infty p_a}) \end{bmatrix} \qquad (8.38)$$

yields the additional coefficients of the digital filter. Since the coefficient matrix (8.38) is a Vandermonde matrix, a solution always exists. Thus we get a finite-length interpolating impulse response that is still the same close approximation of $g_i(t)$ in the interval $0 \le t \le (q + 1)T_1$.

8.3 SAVITZKY-GOLAY FILTERS

8.3.1 The Problem and an Elementary Solution

Again we consider the problem of smoothing. However, now our task is quite different from that in the preceding section. Here we assume that the sequence $x(n)$ is the result of a single nonrepeatable experiment, to be expressed as

$$x(n) = w(n) + r(n), \qquad n \in \mathbf{Z} \qquad (8.39)$$

where $w(n)$ is the unknown sequence of interest and $r(n)$ represents the error due to imperfections in the measurement. For simplicity we assume $r(n)$ to be white noise with variance σ_r^2. Our aim is the reduction of $r(n)$, combined with a minimum distortion of $w(n)$. Consequently, we want to design a system that, if excited by $x(n)$, yields

$$y(n) = w(n) + e(n) \qquad (8.40)$$

such that the error sequence $e(n)$ is minimized.

A very simple and well-known method for the solution of this problem is based on the use of smoothing polynomials, which we describe briefly, following closely the presentation in [7].

Each successive subset of $2K + 1$ input samples $x(n + k)$, $k = -K$, $(-K + 1), \ldots, K$, $n \in \mathbf{Z}$, is approximated by a polynomial $p_n(\xi)$ of degree L in the least-squares sense, thus leading to the functional

$$\epsilon_{K,L}(\mathbf{p}_n) = \sum_{k=-K}^{K} [p_n(n + k) - x(n + k)]^2, \qquad K \geq 0 \tag{8.41}$$

where

$$p_n(\xi) = \sum_{\ell=0}^{L} p_{\ell,n} \cdot (\xi - n)^\ell, \qquad \xi \in \mathbf{R}, \; L \geq 0, \tag{8.42a}$$

$$\mathbf{R} \text{ is the set of real numbers}$$

is the approximating polynomial in the nth interval and

$$\mathbf{p}_n = [p_{0,n}, p_{1,n}, \ldots, p_{L,n}]^T \tag{8.42b}$$

is the vector of coefficients of $p_n(\xi)$ (see Fig. 8.8).

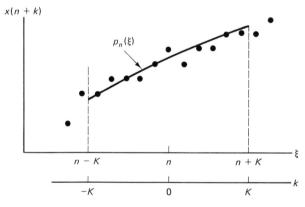

Figure 8.8 The disturbed sequence $x(n + k)$ and the smoothing polynomial $p_n(\xi)$.

The value of the smoothed sequence $y(n)$ is simply given by the value of the corresponding polynomial at $\xi = n$:

$$y(n) = p_n(n) = p_{0,n} \tag{8.43}$$

We will demonstrate the next step by an example, choosing $L = 2$ for the degree of the approximating polynomial. In this case the functional $\epsilon_{K,L}$ becomes

$$\epsilon_{K,2}(\mathbf{p}_n) = \sum_{k=-K}^{K} [p_{0,n} + p_{1,n} \cdot k + p_{2,n} \cdot k^2 - x(n + k)]^2$$

Partial differentiation of $\epsilon_{K,2}$ with respect to $p_{\ell,n}$ yields the linear system of equations

$$\begin{bmatrix} s_K(0) & s_K(1) & s_K(2) \\ s_K(1) & s_K(2) & s_K(3) \\ s_K(2) & s_K(3) & s_K(4) \end{bmatrix} \cdot \begin{bmatrix} p_{0,n} \\ p_{1,n} \\ p_{2,n} \end{bmatrix} = \begin{bmatrix} \tilde{x}_{0,n} \\ \tilde{x}_{1,n} \\ \tilde{x}_{2,n} \end{bmatrix}$$

where we have introduced the abbreviations

$$s_K(\ell) = \sum_{k=-K}^{K} k^\ell, \qquad \ell = 0, 1, 2, \ldots \tag{8.44a}$$

and

$$\tilde{x}_{\ell,n} = \sum_{k=-K}^{K} k^\ell \cdot x(n + k), \qquad \ell = 0, 1, 2, \ldots \qquad (8.44b)$$

Obviously the elements $s_K(\ell)$ of the matrix do not depend on the discrete-time variable n.

Since $s_K(1) = s_K(3) = 0$, we obtain immediately

$$p_{0,n} = \frac{\tilde{x}_{0,n} \cdot s_K(4) - \tilde{x}_{2,n} \cdot s_K(2)}{s_K(0) \cdot s_K(4) - s_K^2(2)} \qquad (8.45a)$$

Using the closed-form expressions

$$s_K(0) = 2K + 1, \qquad s_K(2) = \tfrac{1}{3} \cdot K \cdot (K + 1) \cdot (2K + 1) \qquad (8.45b)$$

$$s_K(4) = \tfrac{1}{15} \cdot K \cdot (K + 1) \cdot (2K + 1) \cdot (3K^2 + 3K - 1)$$

we are led to

$$y(n) = p_{0,n} = \frac{3 \cdot \sum\limits_{k=-K}^{K} x(n - k) \cdot [3K^2 + 3K - 1 - 5k^2]}{(2K - 1) \cdot (2K + 1) \cdot (2K + 3)} \qquad (8.45c)$$

$$= x(n) * h(n)$$

Here we have defined $h(n)$ as the impulse response of a nonrecursive filter of degree $q = 2K$:

$$h(n) = \frac{3[3K^2 + 3K - 1 - 5n^2]}{(2K - 1) \cdot (2K + 1) \cdot (2K + 3)}, \qquad n = -K, (-K + 1), \ldots, K \qquad (8.45d)$$

Note that we have chosen a noncausal representation with symmetry about $n = 0$. For example, for $K = 3$, we obtain

$$\{h(n)\} = \tfrac{1}{21} \cdot \{-2, 3, 6, 7, 6, 3, -2\}$$

As a second example, consider the case $L = 0$ or $L = 1$, for which we obtain the well-known moving-point average with the impulse response

$$h(n) = \begin{cases} 1/(2K + 1), & |n| \le K \\ 0, & |n| > K \end{cases}$$

The solution for arbitrary values of L is obtained by generalizing the procedure outlined so far for $L = 2$. Again the coefficients $p_{\ell,n}$ of $p_n(\xi)$ are determined by minimizing $\epsilon_{K,L}$, leading to the well-known "normal equations"

$$\sum_{\ell=0}^{L} p_{\ell,n} \cdot s_K(\ell + \lambda) = \sum_{k=-K}^{K} k^\lambda \cdot x(n + k), \qquad \lambda = 0, 1, \ldots, L \quad (8.44c)$$

Defining the Hankel matrix

$$\mathbf{S}_{K,L} = [s_K(\ell + \lambda)], \qquad \ell, \lambda = 0, 1, \ldots, L \qquad (8.44d)$$

we can rewrite Eq. (8.44c) in matrix notation:

$$\mathbf{S}_{K,L} \cdot \mathbf{p}_n = \tilde{\mathbf{x}}_n \tag{8.44e}$$

where

$$\tilde{\mathbf{x}}_n = [\tilde{x}_{0,n}, \ldots, \tilde{x}_{L,n}]^T, \qquad n \in \mathbf{Z} \tag{8.44f}$$

is the vector with the elements defined by Eq. (8.44b)

Since the matrix $\mathbf{S}_{K,L}$ does not depend on n, it would suffice to invert it once. However, it is interesting and surprising that there exist closed form solutions of the problem stated above. The first one was given by Sheppard in 1913 [8]. Another approach utilizing orthogonal polynomials was presented in [9] and is described in the next section.

On the other hand, if we would like to solve Eq. (8.44e) directly, we oberve that, from Eq. (8.43), the only coefficient that we need to calculate $y(n)$ is $p_{0,n}$. Therefore, we need only determine the first row of $\mathbf{S}_{K,L}^{-1}$, which we denote by $\mathbf{r}^T$. Since $\mathbf{S}_{K,L}$ is independent of n, $\mathbf{r}^T$ is as well, and

$$y(n) = p_{0,n} = \mathbf{r}^T \cdot \tilde{\mathbf{x}}_n \tag{8.46a}$$

where

$$\mathbf{r}^T = [r_0, \ldots, r_L] \tag{8.46b}$$

Now a simple manipulation of Eq. (8.46a) will lead us to the desired result.

By use of Eq. (8.44b) we obtain

$$y(n) = \sum_{\ell=0}^{L} r_\ell \cdot \tilde{x}_{\ell,n} = \sum_{\ell=0}^{L} r_\ell \cdot \sum_{k=-K}^{K} k^\ell \cdot x(n + k) \tag{8.47}$$

which can be written as

$$y(n) = \sum_{k=-K}^{K} h(k) \cdot x(n - k) \tag{8.48a}$$

where

$$h(n) = \begin{cases} \sum_{\ell=0}^{L} (-1)^\ell \cdot r_\ell \cdot n^\ell, & n = -K, (-K + 1), \ldots, K \\ 0, & |n| > K \end{cases} \tag{8.48b}$$

Obviously, Eq. (8.48a) describes the input-output relation of a nonrecursive filter of degree

$$q = 2 \cdot K \tag{8.49}$$

with impulse response $h(n)$, as in the example above, but now for arbitrary values of L.

Summarizing, we observe that the approach of fitting noisy data by an appropriate polynomial leads in a natural manner to the concept of digital filters. These filters are nonrecursive since we use only a total of $2K + 1$ values of $x(n)$ for the calculation of $y(n)$. Since, in addition, future values influence $y(n)$, the resulting systems are not causal. The corresponding transfer function is given by

$$H(z) = \sum_{n=-K}^{K} h(n)z^{-n} \tag{8.50}$$

Before we proceed with further investigations of this class of filters, we present a closed-form solution of the problem, utilizing the results of [7–9].

8.3.2 Explicit Solution by the Use of Orthogonal Polynomials

It is well known that the matrices $S_{K,L}$ tend to be ill conditioned if L increases [10]. Hence the direct calculation of $p_{0,n}$ from Eq. (8.44e) may lead to incorrect values or might even be impossible for numerical reasons. A simpler and more successful procedure is provided by the use of polynomials $q_\ell(\xi)$, which are orthogonal over $2K + 1$ equidistant points with respect to summation. They are defined by the conditions

$$\sum_{n=-K}^{K} q_\ell(n)q_\lambda(n) = \begin{cases} 0, & \ell \neq \lambda \\ \|q_\ell\|^2 > 0, & \ell = \lambda \end{cases} \ell, \lambda = 0, 1, \ldots, L \tag{8.51}$$

The degree of $q_\ell(\xi)$ is equal to ℓ. The value $\|q_\ell\|$ denotes the norm of $q_\ell(\xi)$ and is defined by

$$\|q_\ell\| = \left[\sum_{n=-K}^{K} q_\ell^2(n) \right]^{1/2}, \qquad \ell < 2K + 1 \tag{8.52a}$$

An explicit value can be calculated by

$$\|q_\ell\|^2 = 2^{-2\ell} \cdot \frac{1}{2\ell + 1} \cdot \prod_{\lambda=-\ell}^{\ell} (2K + 1 + \lambda), \qquad \ell \geq 0 \tag{8.52b}$$

or, equivalently,

$$\|q_{\ell+1}\|^2 = \frac{2\ell + 1}{2\ell + 3} \cdot \frac{2K + \ell + 2}{2} \cdot \frac{2K - \ell}{2} \cdot \|q_\ell\|^2, \qquad \ell \geq 0 \tag{8.52c}$$

where we obviously have for $\ell = 0$

$$\|q_0\|^2 = 2K + 1 \tag{8.52d}$$

It can be shown [11] that, if we start with

$$q_0(\xi) = 1, \qquad q_1(\xi) = \xi \tag{8.53a}$$

these polynomials can be obtained from the recurrence relation

$$q_{\ell+1}(\xi) = \frac{2\ell + 1}{\ell + 1} \cdot \xi \cdot q_\ell(\xi) - \frac{\ell}{\ell + 1} \cdot \frac{2K + 1 + \ell}{2} \cdot \frac{2K + 1 - \ell}{2} \cdot q_{\ell-1}(\xi), \ell \geq 1 \tag{8.53b}$$

For instance, we obtain with Eq. (8.53a)

$$q_2(\xi) = \tfrac{1}{2} \cdot [3\xi^2 - K \cdot (K + 1)] \tag{8.53c}$$

$$q_3(\xi) = \tfrac{1}{2} \cdot [5\xi^3 - (3K^2 + 3K - 1)\xi] \tag{8.53d}$$

The smoothing polynomial $p_n(\xi)$ in Eqs. (8.41)–(8.43) is now represented by the equivalent expression

$$p_n(\xi) = \sum_{\ell=0}^{L} a_{\ell,n} \cdot q_\ell(\xi - n) \tag{8.54a}$$

The minimum of the functional $\epsilon_{K,L}$ of Eq. (8.41) with respect to the parameters $a_{\ell,n}$ is achieved for

$$a_{\ell,n} = \frac{1}{\|q_\ell\|^2} \cdot \sum_{k=-K}^{K} x(n - k) \cdot q_\ell(-k), \qquad \ell = 0, 1, \ldots, L \tag{8.54b}$$

Here we have used the orthogonality of the polynomials $q_\ell(\xi)$ according to Eq. (8.51). The value of the smoothed output sequence is now calculated by

$$y(n) = p_n(n) = \sum_{\ell=0}^{L} a_{\ell,n} \cdot q_\ell(0)$$

$$= h(n) * x(n) \tag{8.54c}$$

where now

$$h(n) = \begin{cases} \sum_{\ell=0}^{L} \dfrac{1}{\|q_\ell\|^2} \cdot q_\ell(-n) \cdot q_\ell(0), & |n| \le K \\ 0, & |n| > K \end{cases} \tag{8.54d}$$

Of course, this impulse response has to be identical with that obtained by Eq. (8.48) since in either case $p_n(\xi)$ is the best smoothing polynomial with respect to Eq. (8.41).

The representation of $h(n)$ can be simplified rigorously if we make use of the Christoffel-Darboux summation [11]. Assuming L to be odd, i.e.,

$$L = 2M + 1, \quad M \in \mathbf{Z}, \quad M \ge 0 \tag{8.55a}$$

we obtain

$$h(n) = C_M \cdot \frac{1}{n} \cdot q_{2M+1}(n), \quad n \ne 0 \tag{8.55b}$$

where

$$C_M = (-1)^M \cdot \frac{(2M + 1)!}{[M!]^2} \cdot \frac{1}{\displaystyle\prod_{m=-M}^{M} (2K + 2m + 1)} \tag{8.55c}$$

For $n = 0$ we can use Eq. (8.54d) to obtain

$$h(0) = \sum_{\ell=0}^{2M+1} \frac{1}{\|q_\ell\|^2} \cdot q_\ell^2(0) = C_M \cdot q'_{2M+1}(0) \tag{8.55d}$$

If L is even, i.e.,

$$L = 2M, \qquad M \in \mathbf{Z}, \qquad M \geq 0 \tag{8.56a}$$

we use the special values

$$q_{2m+1}(0) = 0, \qquad m = 0, 1, \ldots, M \tag{8.57}$$

which can easily be derived from the recurrence relation (8.53a, b). Hence we can write for Eq. (8.54d)

$$h(n) = \sum_{\ell=0}^{2M} \frac{1}{\|q_\ell\|^2} q_\ell(-n) \cdot q_\ell(0) = \sum_{\ell=0}^{2M+1} \frac{1}{\|q_\ell\|^2} q_\ell(-n) \cdot q_\ell(0) \tag{8.56b}$$

and we obtain the same result as for $L = 2M + 1$ in Eqs. (8.55).

As a consequence we have:

For $L = 2M$ or $L = 2M + 1$, $h(n)$ is the truncated version of the even polynomial $C_M \cdot (1/n) \cdot q_{2M+1}(n)$, which is of degree $2M$.

Consider, for instance, the case $L = 3$, i.e., $M = 1$. First, with Eq. (8.55c), we obtain

$$C_1 = -\frac{6}{(2K - 1) \cdot (2K + 1) \cdot (2K + 3)} \tag{8.58a}$$

and, finally, with Eq. (8.53d),

$$h(n) = 3 \cdot \frac{3K^2 + 3K - 1 - 5n^2}{(2K - 1) \cdot (2K + 1) \cdot (2K + 3)}, \qquad |n| \leq K \tag{8.58b}$$

Of course, this result is identical with that of Eq. (8.45d), obtained by the direct approach.

8.3.3 An Equivalent Approach to Savitzky-Golay Filters

Up to now, we have performed the reconstruction process by the use of smoothing polynomials of degree L. A direct consequence of this approach is that a system being designed for $L = 2M$ will exactly reproduce the input $x(n)$ if it consists of samples of a polynomial of degree not larger than $2M + 1$. From a practical point of view this approach seems to be somewhat unsatisfying since real data generally cannot be described by polynomials. But it is surprising that a more direct reconstruction process will lead to identical results.

For this purpose let us consider again Eq. (8.40), i.e.,

$$y(n) = w(n) + e(n) \tag{8.40}$$

which can be written as

$$y(n) = h_1(n) * x(n) \tag{8.59a}$$

where $h_1(n)$ is the impulse response of the system to be designed according to completely different specifications. Using the representation of Eq. (8.39) we obtain

$$y(n) = h_1(n) * w(n) + h_1(n) * r(n) \tag{8.59b}$$

A direct comparison with Eq. (8.40) yields the total error sequence

$$e(n) = e_w(n) + e_r(n) \tag{8.60a}$$

which is divided into the signal error

$$e_w(n) = h_1(n) * w(n) - w(n)$$

$$= y_1(n) - w(n) \tag{8.60b}$$

and the remaining noise

$$e_r(n) = h_1(n) * r(n) \tag{8.60c}$$

The smoothing process should now reduce appropriate measures of $e_w(n)$ and $e_r(n)$ as much as possible. One possible measure would be the variance of the total error,

$$\text{VAR}[e] = \sigma_e^2 = E[e^2(n)] \tag{8.60d}$$

where E is the expectation operator. Here we assumed $e(n)$ to have zero mean. This definition of the error will lead us directly to the concept of Wiener filters (e.g., [12]).

For our purposes it is advantageous to be able to influence both parts of $e(n)$ separately, which means a special weighting of the errors subject to the special application. This can be achieved if we postulate the following:

1. The moments of $w(n)$ are conserved up to a desired order.
2. The variance of $e_r(n)$ should be minimum.

First, we consider the variance of $e_r(n)$. It is minimum if the energy of $h_1(n)$ is minimum. Hence the second condition is equivalent to the minimization of

$$W[h_1] = \sum_{n=-K}^{K} h_1^2(n) \tag{8.61}$$

where we presupposed the system to be nonrecursive.

The investigation of the first condition requires the definition of the moments of a sequence $v(n)$, which we assume to be of finite length. Let n_0 be an integer such that

$$v(n) = 0, \quad \text{for all } |n| > n_0 \tag{8.62a}$$

Then we call the value

$$m_\ell[v] = \sum_{n=-n_0}^{n_0} n^\ell \cdot v(n), \quad \ell = 0, 1, 2, \ldots \tag{8.62b}$$

the ℓth moment of $v(n)$. We can rewrite condition 1 more precisely as

$$m_\ell[y_1] = m_\ell[w], \quad \ell = 0, 1, \ldots, L \tag{8.63}$$

Since $y_1(n)$ is given by

$$y_1(n) = \sum_{k=-K}^{K} w(k) \cdot h_1(n - k) \tag{8.64a}$$

its moments can easily be expressed in terms of those of $w(n)$ and $h_1(n)$; i.e., we get

$$m_0[y_1] = m_0[w] \cdot m_0[h_1] \tag{8.64b}$$

and

$$m_\ell[y_1] = \sum_{\lambda=0}^{\ell} \binom{\ell}{\lambda} \cdot m_\lambda[w] \cdot m_{\ell-\lambda}[h_1], \qquad \ell \geq 1 \tag{8.64c}$$

Consequently, condition (8.63) is valid if $h_1(n)$ satisfies

$$m_0[h_1] = 1, \qquad m_\ell[h_1] = 0, \qquad \ell = 1, 2, \ldots, L \tag{8.65a}$$

Writing these equations in explicit form using Eq. (8.62b), we have

$$\sum_{n=-K}^{K} h_1(n) = 1, \qquad \sum_{n=-K}^{K} n^\ell \cdot h_1(n) = 0, \qquad \ell = 1, 2, \ldots, L \tag{8.65b}$$

To summarize: conditions 1 and 2 are equivalent to the minimization of $W[h_1]$ in Eq. (8.61) under the constraints of Eq. (8.65b).

The solution is obtained by introducing a new functional

$$\tilde{\epsilon}_{K,L}[\mathbf{h}_1, \mathbf{g}] = W[h_1] - 2 \cdot \{g_0 \cdot [m_0[h_1] - 1] + \sum_{\ell=1}^{L} g_\ell \cdot m_\ell[h_1]\} \tag{8.66a}$$

where

$$\mathbf{g} = [g_0, \ldots, g_L]^T \tag{8.66b}$$

is the vector of Lagrange multipliers. The factor -2 has been introduced to simplify later results.

The necessary and sufficient conditions for a minimum of $\tilde{\epsilon}_{K,L}$ are finally found to be

$$h_1(n) = \sum_{\ell=0}^{L} g_\ell \cdot n^\ell, \qquad n = -K, (-K + 1), \ldots, K \tag{8.67a}$$

$$\sum_{n=-K}^{K} h_1(n) = 1 \tag{8.67b}$$

$$\sum_{n=-K}^{K} n^\ell \cdot h_1(n) = 0, \qquad \ell = 1, 2, \ldots, L$$

Inserting Eq. (8.67a) into Eq. (8.67b) leads to the reduced system

$$\sum_{\ell=0}^{L} g_\ell s_K(\ell) = 1, \qquad \sum_{\ell=0}^{L} g_\ell s_K(\ell + \lambda) = 0, \qquad \lambda = 1, 2, \ldots, L \tag{8.67c}$$

where we have used the abbreviation of Eq. (8.44a). With the matrix $\mathbf{S}_{K,L}$ as defined by Eq. (8.44d), the result is

$$\mathbf{S}_{K,L} \cdot \mathbf{g} = \mathbf{e} \tag{8.67d}$$

with

$$\mathbf{e} = [1, 0, \ldots, 0]^T \tag{8.67e}$$

As in the classical approach (described in Section 8.3.1), we have the same reduction for the solution of the system. The special structure of $\mathbf{e}$ demands only the knowledge of the first column of $\mathbf{S}_{K,L}^{-1}$ being denoted by $\mathbf{s}$. From Eq. (8.67d) we deduce the solution

$$\mathbf{g} = \mathbf{s} \tag{8.67f}$$

Since $\mathbf{S}_{K,L}$ is symmetric, the same is true for $\mathbf{S}_{K,L}^{-1}$. Hence we have, with $\mathbf{r}$ from Eq. (8.46b),

$$\mathbf{s} = \mathbf{r} \tag{8.67g}$$

A closer look at the elements of the matrix $\mathbf{S}_{K,L}$ shows that

$$s_K(2\lambda - 1) = 0, \qquad \lambda \in \mathbf{N} \tag{8.67h}$$

where $\mathbf{N}$ is the set of natural numbers, thus yielding a nice checkerboard structure in $\mathbf{S}_{K,L}$ and $\mathbf{S}_{K,L}^{-1}$. As a direct consequence, we conclude

$$r_{2\mu+1} = 0 \qquad \mu \geq 0 \tag{8.67i}$$

and we finally obtain, comparing Eq. (8.67a) with Eq. (8.48b),

$$h_1(n) = h(n) \tag{8.68a}$$

due to the identity (8.67g). Hence the values g_ℓ are the coefficients of $h(n)$ if expanded in powers of n:

$$h(n) = \sum_{\ell=0}^{L} g_\ell \cdot n^\ell = \sum_{m=0}^{M} g_{2m} n^{2m}$$

$$= C_M \cdot \frac{1}{n} \cdot q_{2M+1}(n) \tag{8.68b}$$

where $L = 2M$ or $L = 2M + 1$.

Thus, we have the interesting result that smoothing by polynomials is equivalent to maximal noise reduction and to preserving the moments of the original input signal.

One property of these filters results directly from the identity (8.67i) or the representation in Eq. (8.68b):

$$h(n) = h(-n), \qquad n \in \mathbf{Z} \tag{8.68c}$$

Thus, since the impulse response is symmetric with respect to $n = 0$, they have (in this noncausal description) zero phase.

Now let us consider again the example for $L = 3$. Taking into account condition (8.67h), we are led to the following system of equations:

$$
\begin{bmatrix}
s_K(0) & 0 & s_K(2) & 0 \\
0 & s_K(2) & 0 & s_K(4) \\
s_K(2) & 0 & s_K(4) & 0 \\
0 & s_K(4) & 0 & s_K(6)
\end{bmatrix}
\cdot
\begin{bmatrix}
g_0 \\ g_1 \\ g_2 \\ g_3
\end{bmatrix}
=
\begin{bmatrix}
1 \\ 0 \\ 0 \\ 0
\end{bmatrix}
\tag{8.69a}
$$

with the solution

$$
g_1 = g_3 = 0
$$

$$
\begin{bmatrix} g_0 \\ g_2 \end{bmatrix} = \frac{1}{s_K(0) \cdot s_K(4) - s_K^2(2)} \begin{bmatrix} s_K(4) \\ -s_K(2) \end{bmatrix}
\tag{8.69b}
$$

Inserting this special result into Eq. (8.68b), we have

$$
h(n) = \frac{s_K(4) - s_K(2) \cdot n^2}{s_K(0) \cdot s_K(4) - s_K^2(2)}, \qquad |n| \le K
\tag{8.69c}
$$

a result that again is identical to those obtained in Eqs. (8.45d) and (8.58b).

8.3.4 Some Properties of the Filters

In the preceding section we separated the total error $e(n)$ (Eq. 8.40) into the signal error $e_w(n)$ and the remaining noise $e_r(n)$ according to Eq. (8.60). Obviously it is not possible to simultaneously make both parts arbitrarily small. For example, a vanishing signal error $e_w(n)$ would imply $h(n) = \delta(n)$, and hence no noise reduction would occur. On the other hand, a vanishing $e_r(n)$ would imply $h(n) = 0$ for all n and as a consequence $y(n) = 0$ for all n.

The distribution of the total error among $e_w(n)$ and $e_r(n)$ is controlled by the parameter L, the degree of the smoothing polynomial, or the order of moments to be preserved. More precise statements can be made with the knowledge of some properties of these filters. For this purpose we first consider condition 1, the conservation of moments. Condition 1 has been shown to be equivalent to Eqs. (8.65b). By use of the transfer function $H(z)$ as given in Eq. (8.50), we have for the ℓth derivative of $H(\omega)$ with respect to ω

$$
\frac{d^\ell H(\omega)}{d\omega^\ell} = (-j)^\ell \cdot \sum_{n=-K}^{K} n^\ell \cdot h(n) \cdot e^{-jn\omega}
\tag{8.70a}
$$

and especially at $\omega = 0$

$$
\left. \frac{d^\ell H(\omega)}{d\omega^\ell} \right|_{\omega=0} = (-j)^\ell \cdot \sum_{n=-K}^{K} n^\ell \cdot h(n)
\tag{8.70b}
$$

A comparison with Eq. (8.67b) immediately yields the equivalent conditions

$$H(\omega)\Big|_{\omega=0} = 1$$

$$\frac{d^\ell H(\omega)}{d\omega^\ell}\Big|_{\omega=0} = 0, \qquad \ell = 1, 2, \ldots, L$$

(8.70c)

The conservation of the moments turns out to be equivalent to a maximally flat approximation of an ideal lowpass filter at $\omega = 0$.

The lowpass behavior is achieved by the condition of minimum energy of the impulse response. The value of this minimum is one characteristic figure of the derived filters since it is identical with the noise reduction factor. We are able to calculate it explicitly, as we will show now. Inserting $h(n)$ from Eq. (8.68b) into Eq. (8.61), we obtain

$$W[h] = \sum_{n=-K}^{K}\left[\sum_{\ell=0}^{L} g_\ell \cdot n^\ell\right]^2$$

$$= \sum_{\ell=0}^{L} g_\ell \cdot \sum_{\lambda=0}^{L} g_\lambda \cdot s_K(\ell + \lambda)$$

and, according to Eq. (8.67c), we finally find the simple identity

$$W[h_{M,K}] = g_0 = h(0) = C_M \cdot q'_{2M+1}(0)$$

$$= \sum_{\ell=0}^{L} \frac{q_\ell^2(0)}{\|q_\ell\|^2}$$

(8.71a)

The impulse response that is obtained for any two values of M and K has now been denoted more precisely by $h_{M,K}(n)$ to indicate the dependence on these design parameters. A corresponding notation is used for the frequency responses:

$$H_{M,K}(\omega) = \sum_{n=-K}^{K} h_{M,K}(n) \cdot e^{-jn\omega}$$

(8.71b)

Figure 8.9 shows the frequency responses for $K = 11$ and different values of M. Obviously the passband increases with M growing, and therefore the noise suppression becomes worse. Note that these frequency responses would not satisfy any meaningful tolerance scheme. But nevertheless the systems are optimal for their purpose.

At a first glance the inspection of the systems in the frequency domain results in the same conclusion as before. The filter will reduce both signal and noise. However, just this fact suggests a solution, as we will describe now.

If there are no spectral components of $w(n)$ outside the region, where

$$H_{M,K}(\omega) \approx 1, \qquad |\omega| \leq \omega_c$$

(8.72a)

is valid, only a small amount of distortion of $w(n)$ will occur. On the other hand, in the region $\omega_c < |\omega| \leq \pi$, the spectral components of the noise can be reduced

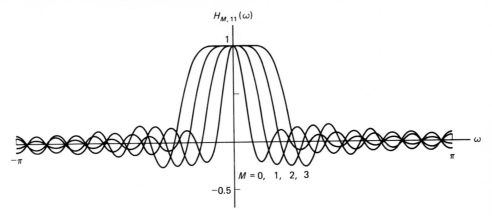

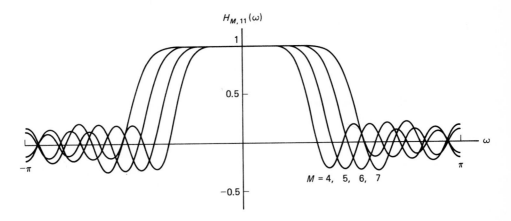

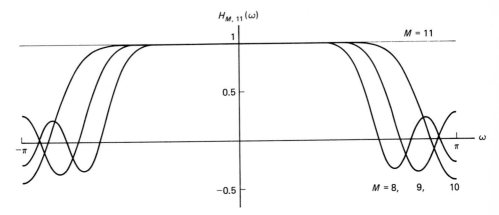

Figure 8.9 Frequency responses of the smoothing filters $K = 11, M = 0, 1, \ldots,$ 11. [Adapted from Fig. 3 in Peter Steffen, "On Digital Smoothing Filters: A Brief Review of Closed Form Solutions and Two New Filter Approaches," *Circuits, Systems, and Signal Processing*, Vol. 5, pp. 187–210, 1986.]

without affecting the signal. More precisely, we assume the power spectrum $P_w(\omega)$ of $w(n)$ to have (at least approximately) a limited bandwidth

$$P_w(\omega) = 0, \qquad 0 < \omega_c < |\omega| \le \pi \tag{8.72b}$$

We are led to such a condition if an originally bandlimited signal $w_0(t)$ is oversampled by a factor

$$\beta = \frac{\pi}{\omega_c} > 1 \tag{8.72c}$$

(see also Section 8.2). Figure 8.10 demonstrates this situation.

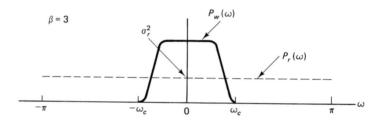

Figure 8.10 Fundamental situation when $w(n)$ results from an oversampling by the factor β.

Obviously, an ideal lowpass filter with cutoff frequency ω_c would reduce the total power of the noise by the factor β without any change of the power spectrum $P_w(\omega)$. The distortion within the range $|\omega| \le \omega_c$ cannot be removed without a change of the signal.

To demonstrate the performance of the filters, we investigate the cutoff frequency and the noise reduction factor as functions of M and K. The cutoff frequency was arbitrarily defined by

$$H_{M,K}(\omega) \le 0.99; \qquad \omega_c \le |\omega| \le \pi \tag{8.73}$$

Figure 8.11 shows $\omega_c(M, K)$ and $W[h_{M,K}]$ as a function of K and M, where M is a parameter. Since the degree of the filter is $q = 2K$, the value K is a direct measure of the cost of implementing the filter, i.e., the number of required arithmetic operations.

The combination of both results leads to a relation that can be used in the form

$$\frac{1}{W[h_{M,K}]} = f\left(\frac{\pi}{\omega_c}, M, K\right) \tag{8.74}$$

Note that the ideal lowpass filter will always yield the value $W^{-1}[h] = \pi/\omega_c$. Figure 8.12 shows the values that can be achieved with realizable filters.

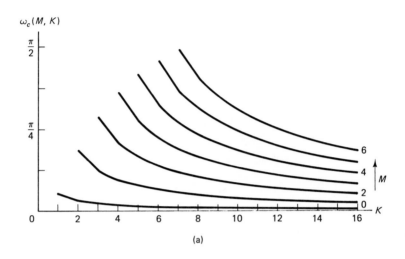

(a)

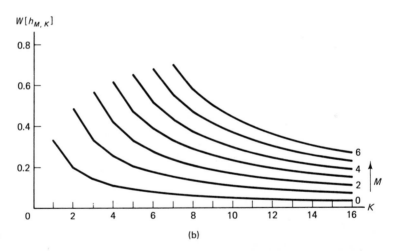

(b)

Figure 8.11   (a) Cutoff frequency ω_c (1% tolerated deviation in the passband); (b) noise-reduction factor. [Part (b) adapted from Fig. 6 in Peter Steffen, "On Digital Smoothing Filters: A Brief Review of Closed Form Solutions and Two New Filter Approaches," *Circuits, Systems, and Signal Processing*, Vol. 5, pp. 187–210, 1986.]

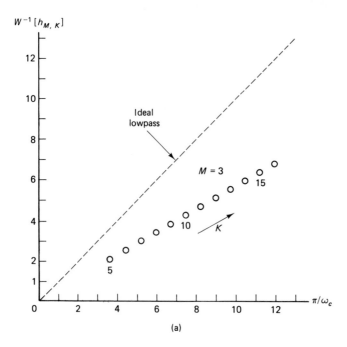

(a)

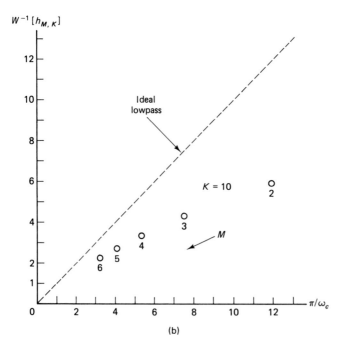

(b)

Figure 8.12 Inverse noise-reduction factor depending on the oversampling factor π/ω_c. (a) $M = 3$, $K = 5, 6, \ldots , 16$; (b) $K = 10$, $M = 2, 3, \ldots , 6$.

8.4 GENERALIZED L_2 DESIGN OF FIR FILTERS

8.4.1 Introduction

It is well known that some properties of FIR filters make them preferable to IIR filters in many cases. In addition, most design procedures for FIR filters are very simple and easy to implement on a computer. For example, the Fourier approximation, window methods, and the frequency sampling design will lead to *linear* problems or even explicit expressions for the determination of the desired impulse response.

The price for the simplicity of the design is often that the resulting systems do not have the wanted behavior at all points of interest. Hence, modifications of the classical methods seem to be appropriate for certain applications. But changes should try to preserve the simplicity of the design procedures as much as possible. One method that satisfies this condition is described in this section.

8.4.2 Statement of the Problem

To develop and illustrate the procedure, let us consider the design of a system whose frequency response approximates the ideal function

$$H_i(\omega) = |\omega|, \qquad |\omega| \le \pi \qquad (8.75a)$$

a function that frequently arises in geophysical applications [13]. It consists of a digital differentiator followed by a discrete Hilbert transformer. The Fourier approximation of length $q + 1 = 2K + 1$ is obtained directly by

$$H_K(z) = \sum_{n=-K}^{K} h(n)z^{-n} \qquad (8.75b)$$

with

$$\left. \begin{array}{l} h(0) = \dfrac{\pi}{2} \\[2ex] h(2\nu) = 0 \\[2ex] h(2\nu + 1) = -\dfrac{2}{\pi} \cdot \dfrac{1}{(2\nu + 1)^2} \end{array} \right\} \nu \in \mathbf{Z} \qquad (8.75c)$$

Since the function H_i is continuous, the convergence of the H_K for $K \to \infty$ will be uniform. However, for most applications the actual deviation

$$\Delta_K(\omega) = H_K(\omega) - H_i(\omega) \qquad (8.76)$$

is unacceptable in the neighborhood of $\omega = 0$ (and π). Figure 8.13 shows $H_5(\omega)$ and the corresponding error function $\Delta_5(\omega)$. Increasing the value of K to 15, for example, reduces the deviation at $\omega = 0$ to $\Delta_{15}(0) = -0.0397$ only, thus indicating the slow reduction of the deviation $\Delta_K(0)$ with growing values of K. Hence it seems desirable to fix the value $H_K(0)$ to be $H_K(0) = H_i(0) = 0$.

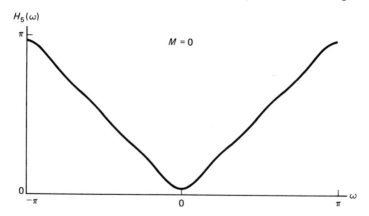

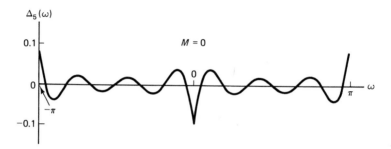

Figure 8.13 Fourier approximation of the function $|\omega|$ with $q = 2K = 10$. [Adapted from Fig. 1 in P. Steffen, "Fourier Approximation with Interpolative Constraints," *AEÜ*, Vol. 36, pp. 363–367, 1984.]

Let us proceed with more general prescriptions for the resulting frequency response. At a fixed set of points z_m,

$$z_m = e^{j\omega_m}, \qquad \omega_m \in \mathbf{R}, \quad m = 0, 1, \ldots, (M-1), \ M \in \mathbf{N} \qquad (8.77)$$

we prescribe values for the frequency response $H_K(\omega)$ and (if desired) its derivatives up to a certain order $\ell_m - 1$ by a set of linear constraints

$$C_{m,\lambda} = 0 \qquad (8.78a)$$

where

$$C_{m,\lambda} = \frac{d^\lambda H(\omega)}{d\omega^\lambda}\bigg|_{\omega = \omega_m} - A_{m,\lambda}
\begin{cases}
\lambda = 0, 1, \ldots, (\ell_m - 1) \\[4pt]
m = 0, 1, \ldots, (M-1) \\[4pt]
\ell_m \in \mathbf{N} \\[4pt]
A_{m,\lambda} \in \mathbf{C} \quad (\mathbf{C} \text{ the set of complex numbers})
\end{cases}
\qquad (8.78b)$$

Of course, the prescribed values $A_{m,\lambda}$ can originate—although not necessarily—from the ideal frequency response $H_i(\omega)$ and its derivatives at the points z_m:

$$A_{m,\lambda} = \frac{d^\lambda H_i(\omega)}{d\omega^\lambda}\bigg|_{\omega=\omega_m}, \quad \begin{cases} \lambda = 0, 1, \ldots, \ell_m - 1, \\ m = 0, 1, \ldots, (M-1) \end{cases} \qquad (8.78c)$$

Filters of this type can be found using the Hermite interpolation procedure. If just the values $H(\omega_m)$ and no derivatives are prescribed, the problem can be solved, e.g., by the Lagrange interpolation formula, a special case of which is the well-known frequency sampling method [14,15]. Another example is the maximally flat approximation of an idealized lowpass filter at the points $\omega_0 = 0$ and $\omega_1 = \pi$ as it has been presented in [16]. In all these cases a filter degree q has been assumed such that

$$q + 1 = P \qquad (8.79a)$$

where P is the total number of prescribed values, i.e.,

$$P = \sum_{m=0}^{M-1} \ell_m \qquad (8.79b)$$

In the following discussion we consider the far more interesting case

$$q + 1 > P \qquad (8.80)$$

The reason for increasing the degree is that the conditions (8.78) guarantee only a local approximation at or near the points z_m. In addition, we can now determine the remaining $q + 1 - P$ free parameters to obtain a global approximation subject to a norm that depends on the application.

For simplicity we choose the L_2 norm. (The Chebyshev norm, e.g., has been used in [17].)

Remark. Using the L_2 norm, the design of Savitzky-Golay filters is included here if we choose

$$H_i(\omega) = 0, \quad \text{for all } \omega$$

$$M = 1, \quad \omega_0 = 0, \quad \ell_0 = L + 1$$

$$A_{0,0} = 1, \quad A_{0,\lambda} = 0, \quad \lambda = 1, 2, \ldots, L$$

In the following discussion we do not assume linear phase of the realized transfer functions. Therefore, in general no simplification will result from a symmetric noncausal representation of the transfer function as in Eq. (8.75b). Consequently, if we use a causal notation,

$$H(z) = \sum_{n=0}^{q} h(n)z^{-n} \qquad (8.81)$$

we have to modify the error function as defined in Eq. (8.76). Particularly, we must tolerate a certain delay of r samples of the system actually realized, where r will be an additional parameter in the design procedure. The complex error function that has to be considered now is given by the expression

$$\Delta_{q;r}(\omega) = e^{jr\omega} \cdot H(\omega) - H_i(\omega), \qquad r \in \mathbf{Z} \tag{8.82a}$$

We will determine it as follows. For fixed values ρ we compute the minimum of the functional E with respect to the impulse response $h(n)$ of $H(z)$,

$$E_{q;\rho} = \min_{\mathbf{h}} \| \Delta_{q;\rho}(\omega) \|_2^2, \qquad \rho \in \mathbf{Z} \tag{8.82b}$$

subject to the desired constraints $C_{m,\lambda} = 0$ in Eq. (8.78). Then the minimum of E yields the necessary delay of r samples:

$$E_{q;r} = \min_{\rho \in \mathbf{Z}} E_{q;\rho} \tag{8.82c}$$

If we want to design a linear phase FIR filter of degree $q = 2K$, we can determine the value of r a priori. Obviously we always have $r = K$. Therefore, in the case of a desired linear phase there is no need to carry out the minimization of $E_{q;\rho}$.

8.4.3 The General Solution

The problem stated in the preceding section can always be solved by standard procedures using Lagrange multipliers (see also Section 8.3.3) leading to a linear system of equations of dimension $q + 1 + P$. Instead of proceeding this way, we present a special decomposition of $H(z)$ such that the constraints (8.78) are fulfilled a priori:

$$H(z) = H_0(z) \cdot [z^{-k} \cdot H_1(z) + H_2(z)], \qquad k \in \mathbf{Z} \tag{8.83a}$$

where

$$H_0(z) = z^{-P} \cdot \prod_{m=0}^{M-1} (z - z_m)^{\ell_m} \tag{8.83b}$$

As when we introduced the factor $e^{jr\omega}$ in Eq. (8.82a), it turns out to be useful to introduce an additional delay of k samples. Its value is again obtained by minimizing the functional E with respect to k. This delay by k samples has to be introduced even if a linear phase of $H(z)$ is desired. In this case, k can be determined a priori, as we will show later.

Besides $H_0(z)$, $H_1(z)$ can also be calculated explicitly. By use of Hermite interpolation, we choose $H_1(z)$ such that the product

$$\hat{H}_{01}(z) = z^{-k} \cdot H_{01}(z), \qquad \text{with } H_{01}(z) = H_0(z) \cdot H_1(z) \tag{8.83c}$$

satisfies the conditions (8.78). Therefore, we can state that the maximal degree of $H_{01}(z)$ will be $P - 1$.

Let the resulting degree of $H_{01}(z)$ be $\tilde{P} - 1$; then we always have

$$\tilde{P} \leq P \tag{8.83d}$$

and hence the value of k will be bounded by

$$0 \leq k \leq q + 1 - \tilde{P} \tag{8.83e}$$

If a linear phase has to be realized, we can calculate the value of k directly. Assuming q to be even, $q = 2K$, and P to be odd, $P = 2P_0 + 1$, we have the optimal delay given by

$$k = q + 1 - P_0 \qquad (8.83f)$$

The special choice of $H_0(z)$ guarantees that $H_2(z)$ does not influence the behavior of $H(z)$ at the points z_m, provided $H_2(z)$ has no poles at these points. Since we are dealing with FIR filters, this will always be the case. Therefore, only

$$q_2 + 1 = q + 1 - P \qquad (8.84a)$$

remaining parameters can be used to optimize the global behavior of the final solution. The partial transfer function will be used in the direct form,

$$H_2(z) = \sum_{n=0}^{q_2} h_2(n) z^{-n} \qquad (8.84b)$$

Its parameters are determined subject to Eqs. (8.82). Note that the constraints (8.78) are satisfied for any choice of $h_2(n)$. This fact leads to an interesting property of the proposed decomposition. Defining the vector

$$\mathbf{h}_2 = [h_2(0), \ldots , h_2(q_2)]^T \qquad (8.85a)$$

we can write for the error functional, from Eq. (8.82),

$$E_{q,P;r,k}[\mathbf{h}_2] = \| \hat{H}_{01}(\omega) + H_0(\omega) \cdot H_2(\omega) - e^{-jr\omega} \cdot H_i(\omega) \|_2^2$$
$$= \| H_0(\omega) \cdot [H_2(\omega) - \tilde{H}_i(\omega)] \|_2^2 \qquad (8.85b)$$

where we have introduced a modified ideal function

$$\tilde{H}_i(\omega) = \frac{e^{-jr\omega} \cdot H_i(\omega) - \hat{H}_{01}(\omega)}{H_0(\omega)} \qquad (8.85c)$$

that has to be approximated by $H_2(\omega)$, while $H_0(\omega)$ acts as a weighting function. In addition, the dependence on P and k is indicated in the functional E.

The representation of Eq. (8.85) is of great interest since it proves directly that the remaining parameters can be easily determined without any constraints.

Let us consider two extremes in the problem under consideration. First we prescribe $H(\omega) = A_{m,0}$ at M points z_m, where $M > 1$, but we do not prescribe values for the derivatives:

$$M > 1, \qquad \ell_m = 1, \qquad m = 0, 1, \ldots , (M - 1), \qquad P = M$$

This special case is identical with the Lagrange interpolation method [14], which includes the well-known frequency sampling method, if $z_m = e^{j2\pi m/M}$ (see Section 8.4.4). Second, all prescriptions can be stated at one single point, leading to the specifications

$$M = 1, \qquad \ell_0 > 1, \qquad P = \ell_0 - 1$$

As already mentioned, the design of the Savitzky-Golay filters is a special application of this second case. We proceed with the calculation of $H_0(z)$ and $\hat{H}_{01}(z)$ for this class of filters, assuming that their transfer function $H(z)$ is known. We see that $H_0(z)$ and $\hat{H}_{01}(z)$ are obtained as

$$H_0(z) = [1 - z^{-1}]^{\ell_0}, \qquad \ell_0 = L + 1$$
$$\hat{H}_{01}(z) = z^{-K}$$

assuming $q = 2K$ even.

If desired, we can calculate the remaining function $H_2(z)$ from

$$H_2(z) = \sum_{n=0}^{2K-\ell_0} h_2(n) \cdot z^{-n} = \frac{H(z) - \hat{H}_{01}(z)}{H_0(z)}$$

The conditions (8.70c) of maximal flatness guarantee that $H_2(z)$ is nonrecursive as needed. A more direct closed-form solution for $H_2(z)$ can be found as well [8].

Figure 8.14 shows a block diagram for the realization of $H(z)$. Note that only $q_2 + 1$ multiplications are necessary to obtain one output value showing the advantage of the decomposition according to Eqs. (8.83).

Finally, let us reconsider the example of Section 8.3.3, where we used the values $q = 2K = 6$ and $L = 3$. Using the notation in Eqs. (8.81) and (8.84b), we obtain

$$H_0(z) = (1 - z^{-1})^4, \qquad \hat{H}_{01}(z) = z^{-3}, \qquad H_2(z) = -\tfrac{1}{21} \cdot (2 + 5z^{-1} + 2z^{-2})$$

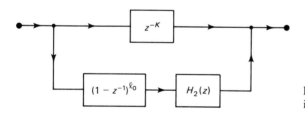

Figure 8.14   Block diagram for the realization of Savitzky-Golay filters.

8.4.4 Systems with Transfer Functions Having Prescribed Values at Equally Spaced Points

We now consider the special case of M equidistant points,

$$z_m = e^{j2\pi m/M}, \qquad m = 0, 1, \ldots, (M - 1) \tag{8.86a}$$

In addition we assume

$$\ell_m = 1, \qquad m = 0, 1, \ldots, (M - 1) \tag{8.86b}$$

which means that no derivatives are prescribed. The restrictions $C_{m,\lambda}$ in Eq. (8.78) now read

$$C_{m,0} = H\left(\frac{2\pi m}{M}\right) - A_{m,0} = 0, \qquad m = 0, 1, \ldots, (M - 1) \tag{8.86c}$$

Note that we do not assume linear phase. This special case permits a closed-form solution [18]. As will be shown later, the general case leads to an explicit solution as well.

The total transfer function is

$$H(z) = H_0(z) \cdot [z^{-k} \cdot H_1(z) + H_2(z)] \tag{8.83a}$$

where now

$$H_0(z) = 1 - z^{-M} \tag{8.87a}$$

and

$$H_1(z) = \sum_{m=0}^{M-1} B_m \cdot \frac{z}{z - z_m} \tag{8.87b}$$

with the values

$$B_m = \frac{1}{M} \cdot z_m^k \cdot A_{m,0}, \qquad m = 0, 1, \ldots, (M-1) \tag{8.87c}$$

We obtain the transfer function $\hat{H}_{01}(z)$ by evaluating Eq. (8.83c) with the special results of Eq. (8.87):

$$\hat{H}_{01}(z) = z^{-k} \cdot \sum_{n=0}^{M-1} h_{01}(n) \cdot z^{-n} \tag{8.88a}$$

Alternatively, corresponding to the representation in Eq. (8.83c) we can write

$$H_{01}(z) = \sum_{n=0}^{M-1} h_{01}(n) \cdot z^{-n} \tag{8.88b}$$

where the values of the corresponding impulse response are obtained by inverse DFT due to the special choice of the points z_m:

$$h_{01}(n) = \frac{1}{M} \sum_{m=0}^{M-1} z_m^k \cdot A_{m,0} \cdot z_m^n \tag{8.88c}$$
$$= \mathrm{DFT}_M^{-1}\{z_m^k \cdot A_{m,0}\}$$

Using Eq. (8.87a), we obtain the overall transfer function

$$H(z) = z^{-k} \cdot H_{01}(z) + (1 - z^{-M}) \cdot H_2(z) \tag{8.89}$$

and the functional E of Eq. (8.85b) explicitly reads

$E_{q,P;r,k}[\mathbf{h}_2]$

$$= \frac{1}{2\pi} \int_{-\pi}^{\pi} |e^{-jk\omega} \cdot H_{01}(\omega) + (1 - e^{-jM\omega}) \cdot H_2(\omega) - e^{-jr\omega} \cdot H_i(\omega)|^2 \, d\omega \tag{8.90}$$

The minimization with respect to the remaining parameters $h_2(n)$ leads to the set of linear equations

$$\sum_{n=0}^{q_2} h_2(n) \cdot \frac{1}{2\pi} \int_{-\pi}^{\pi} [-e^{j(\lambda-M-n)\omega} + 2e^{j(\lambda-n)\omega} - e^{j(\lambda+M-n)\omega}] \, d\omega$$

$$= \frac{1}{2\pi} \int_{-\pi}^{\pi} H_i(\omega) \cdot [e^{j(\lambda-r)\omega} - e^{j(M+\lambda-r)\omega}] \, d\omega \qquad (8.91a)$$

$$-\frac{1}{2\pi} \int_{-\pi}^{\pi} H_{01}(\omega) \cdot [e^{j(\lambda-k)\omega} - e^{j(M+\lambda-k)\omega}] \, d\omega, \qquad \lambda = 0, 1, \ldots, q_2$$

Using the Fourier series expansion of the desired frequency response $H_i(\omega)$, we have

$$H_i(\omega) = \sum_{n=-\infty}^{+\infty} h_i(n) \cdot e^{-jn\omega}$$

$$h_i(n) = \frac{1}{2\pi} \int_{-\pi}^{\pi} H_i(\omega) \cdot e^{jn\omega} d\omega, \qquad n \in \mathbf{Z}$$

and using the representation of $H_{01}(z)$ in Eq. (8.88b), we obtain a simpler version of Eqs. (8.91a):

$$-h_2(\lambda - M) + 2h_2(\lambda) - h_2(\lambda + M)$$

$$= h_i(\lambda - r) - h_i(M + \lambda - r) - h_{01}(\lambda - k) + h_{01}(M + \lambda - k),$$

$$\lambda = 0, 1, \ldots, q_2 \qquad (8.91b)$$

For the solution of this system, we distinguish two cases:

$$Case\ 1: \qquad q_2 < M \qquad (8.92a)$$

$$Case\ 2: \qquad M \le q_2 \qquad (8.93a)$$

The first case permits a closed-form solution, which follows immediately from the fact that

$$h_2(n) = 0, \qquad n < 0, \qquad n > q_2 \qquad (8.92b)$$

Consequently, we have, with Eq. (8.92a),

$$h_2(\lambda \pm M) = 0, \qquad \lambda = 0, 1, \ldots, q_2 \qquad (8.92c)$$

The left-hand side of Eq. (8.91b) reduces to the single term $2h_2(\lambda)$. Hence we obtain the solution

$$h_2(\lambda) = \frac{1}{2} \cdot [h_i(\lambda - r) - h_i(M + \lambda - r) - h_{01}(\lambda - k) + h_{01}(M + \lambda - k)],$$

$$\lambda = 0, 1, \ldots, q_2 \qquad (8.92d)$$

The second case permits a closed-form solution as well. To demonstrate this we take into account the special structure of the system of equations (8.91). Using matrix notation, we can write

$$\mathbf{Ah}_2 = \mathbf{g}(r, k) \qquad (8.93b)$$

where we have defined the matrix of dimension $(q_2 + 1) \times (q_2 + 1)$

$$
\mathbf{A} = \begin{bmatrix}
2 & 0 & \cdots & 0 & -1 & 0 & \cdots & 0 \\
0 & & & & & & & \vdots \\
\vdots & & & & & & & 0 \\
0 & & & & & & & -1 \\
-1 & & & & & & & 0 \\
0 & & & & & & & \vdots \\
\vdots & & & & & & & 0 \\
0 & \cdots & 0 & -1 & 0 & \cdots & 0 & 2 \\
0 & \cdots & \cdots & M & \cdots & \cdots & \cdots & q_2
\end{bmatrix}
\tag{8.93c}
$$

and the vector

$$
\mathbf{g}(r, k) = [g_{r,k}(0), \ldots, g_{r,k}(q_2)]
\tag{8.93d}
$$

with the elements

$$
g_{r,k}(\lambda) = h_i(\lambda - r) - h_i(M + \lambda - r) - h_{01}(\lambda - k) + h_{01}(M + \lambda - k),
$$
$$
\lambda = 0, 1, \ldots, q_2
\tag{8.93e}
$$

The simple structure of $\mathbf{A}$ (which is independent of r and k) induces a special partitioning of the system of equations. To illustrate this, let us consider an example. Let $q_2 = 6$ and $M = 3$. We can reorder the equations and obtain

$$
\begin{bmatrix} 2 & -1 & 0 \\ -1 & 2 & -1 \\ 0 & -1 & 2 \end{bmatrix} \begin{bmatrix} h_2(0) \\ h_2(3) \\ h_2(6) \end{bmatrix} = \begin{bmatrix} g_{r,k}(0) \\ g_{r,k}(3) \\ g_{r,k}(6) \end{bmatrix}
$$

$$
\begin{bmatrix} 2 & -1 \\ -1 & 2 \end{bmatrix} \begin{bmatrix} h_2(1) \\ h_2(4) \end{bmatrix} = \begin{bmatrix} g_{r,k}(1) \\ g_{r,k}(4) \end{bmatrix}
$$

$$
\begin{bmatrix} 2 & -1 \\ -1 & 2 \end{bmatrix} \begin{bmatrix} h_2(2) \\ h_2(5) \end{bmatrix} = \begin{bmatrix} g_{r,k}(2) \\ g_{r,k}(5) \end{bmatrix}
$$

The general case will be solved by using the decomposition of $q_2 + 1$ with respect to M:

$$
q_2 + 1 = Q \cdot M + m, \qquad Q \geq 0, \qquad 0 \leq m < M
\tag{8.94a}
$$

The system of equations described by Eq. (8.93) separates into m systems of dimension $Q + 1$ and $M - m$ systems of dimension Q. In particular, we obtain

$$
\mathbf{T}_{Q+1} \mathbf{h}_{2,\mu} = \mathbf{g}_\mu(r, k), \qquad \mu = 1, 2, \ldots, m
\tag{8.94b}
$$

$$
\mathbf{T}_Q \mathbf{h}_{2,\mu} = \mathbf{g}_\mu(r, k), \qquad \mu = (m + 1), (m + 2), \ldots, M
\tag{8.94c}
$$

Here we have defined the tridiagonal matrices $\mathbf{T}_p$ of dimension $p \times p$

$$\mathbf{T}_p = \begin{bmatrix} 2 & -1 & 0 & \cdots & & 0 \\ -1 & & & & & \\ 0 & & & & & 0 \\ & & & & & -1 \\ 0 & \cdots & 0 & -1 & 2 \end{bmatrix}, \quad p \in \mathbf{N} \qquad (8.94d)$$

and the corresponding vectors that are obtained by subsampling the original ones

$$\mathbf{h}_{2,\mu} = [h_2(\mu), h_2(\mu + M), \ldots]^T$$

$$\mathbf{g}_\mu(r, k) = [g_{r,k}(\mu), g_{r,k}(\mu + M), \ldots]^T, \quad \mu = 0, 1, \ldots, (M-1) \qquad (8.94e)$$

Note that the first m vectors have length $Q + 1$, while the remaining $M - m$ vectors are of dimension Q.

The advantage in using the matrices $\mathbf{T}_p$ is that they can be inverted in closed form. It is well known that they are diagonalized by the orthonormal matrices

$$\mathbf{P}_p = \mathbf{P}_p^{-1} = \sqrt{\frac{2}{p+1}} \cdot \left(\sin \frac{\mu\nu\pi}{p+1}\right), \quad \mu, \nu = 1, 2, \ldots, p \qquad (8.95a)$$

Hence we can write

$$\mathbf{T}_p = \mathbf{P}_p \mathbf{D}_p \mathbf{P}_p \qquad (8.95b)$$

with the diagonal matrix

$$\mathbf{D}_p = \text{diag}[d_{p,1}, \ldots, d_{p,p}] \qquad (8.95c)$$

and the eigenvalues

$$d_{p,\mu} = 2 \cdot \left(1 - \cos \frac{\mu\pi}{p+1}\right), \quad \mu = 1, 2, \ldots, p \qquad (8.95d)$$

Consequently, the solution of the systems in Eqs. (8.94) is simply given by

$$\mathbf{h}_{2,\mu} = \mathbf{P}_p \mathbf{D}_p^{-1} \mathbf{P}_p \mathbf{g}_\mu(r, k) \qquad (8.96)$$

where $p = Q + 1$ for $1 \le \mu \le m$ and $p = Q$ for $m < \mu \le M$. By rearranging the $\mathbf{h}_{2,\mu}$, we find the wanted solution $\mathbf{h}_2$. We will demonstrate the design procedure by continuing the example of Eq. (8.75), where

$$H_i(\omega) = |\omega|, \quad |\omega| \le \pi \qquad (8.75a)$$

Figure 8.15 shows the resulting frequency responses that are obtained for the parameters $q = 2K = 10$ and $M = 1, 2$. A comparison with Fig. 8.13 shows that the maximum values of the error functions are all about the same size. As required, the error functions have zeros at $\omega = 0$ or $\omega = 0$ and $\omega = \pi$. According to the linear phase of $H_i(\omega) = |\omega|$, the resulting frequency responses have linear phase too. In both cases we obtain $r = K = 5$. The first case leads to $H_{01}(z) = 0$ and therefore there is no need to specify k. The second example results in

$$H_{01}(z) = \frac{\pi}{2} \cdot (1 - z^{-1})$$

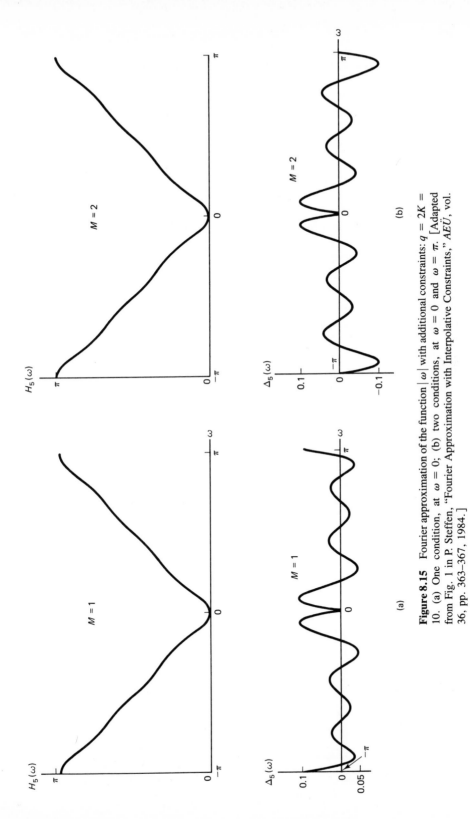

Figure 8.15 Fourier approximation of the function $|\omega|$ with additional constraints: $q = 2K = 10$. (a) One condition, at $\omega = 0$; (b) two conditions, at $\omega = 0$ and $\omega = \pi$. [Adapted from Fig. 1 in P. Steffen, "Fourier Approximation with Interpolative Constraints," *AEÜ*, vol. 36, pp. 363–367, 1984.]

8.5 DESIGN OF MINIMUM-PHASE FREQUENCY-SELECTIVE FIR FILTERS

8.5.1 Introduction

The design of linear phase frequency-selective digital filters has a well-established solution based on the work of Parks and McClellan in 1972 [19]. However, in several applications the delay of these filters is intolerably high. In addition, in many cases linear phase is not required at all. A smaller delay could be realized with equivalent minimum-phase FIR filters. In this context, the equivalence of different filters is interpreted to mean that they are satisfying the same tolerances with respect to the magnitude. It can be expected in general that the degree of equivalent minimum-phase filters will be less than that of the corresponding linear phase filters because linear phase would represent an additional requirement. Hence the use of minimum-phase filters is potentially advantageous in several applications, and their design has been treated extensively in the literature [e.g., 20]. They all utilize the fact that attenuation and phase constitute a pair of Hilbert transforms [e.g., 1].

As in the design of recursive filters, we could use a direct approximation. However, there exists a very simple but nevertheless elegant method to derive equiripple minimum-phase FIR filters from certain equiripple linear phase filters that can easily be determined with the aid of the Parks-McClellan program.

8.5.2 Statement of the Problem

We explain the design procedure with an example of a lowpass filter, given the tolerance scheme for the magnitude response of the minimum-phase filter in Fig. 8.16. The frequencies ω_p and ω_s define the passband and the stopband, respectively, while δ_p and δ_s describe the corresponding allowable deviations. Our goal is the calculation of a FIR filter transfer function $H(z)$ with minimum phase:

$$H(z) = \sum_{n=0}^{q} h(n) \cdot z^{-n}, \qquad q \in \mathbf{N} \tag{8.97a}$$

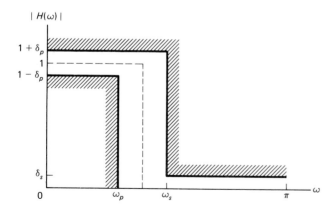

Figure 8.16 Tolerance scheme for the desired minimum-phase filter.

such that the constraints

$$|1 - |H(\omega)|| \le \delta_p, \qquad |\omega| \le \omega_p$$
$$|H(\omega)| \le \delta_s, \qquad \omega_s \le |\omega| \le \pi \tag{8.98}$$

are satisfied and q is as small as possible. The condition of minimum phase allows us to write

$$H(z) = h(0) \cdot \prod_{n=1}^{q} (1 - z_{0n} \cdot z^{-1}) \tag{8.97b}$$

where all the zeros z_{0n} are located within the closed unit circle:

$$|z_{0n}| \le 1, \qquad n = 1, 2, \ldots, q \tag{8.97c}$$

This representation is the key to deriving FIR filters with minimum phase from those with linear phase.

8.5.3 Transformation of Linear Phase FIR Filters into Minimum-Phase FIR Filters

The procedure we use involves three essential steps, repeated here for convenience [20,1]. The first step consists of designing an appropriate lowpass filter with linear phase of degree $2q$. Its transfer function is

$$H_{\mathrm{LP}}(z) = \sum_{n=0}^{2q} h_{\mathrm{LP}}(n) \cdot z^{-n} \tag{8.99a}$$

and on the unit circle we have

$$H_{\mathrm{LP}}(\omega) = e^{-jq\omega} \cdot \tilde{H}_{\mathrm{LP}}(\omega) \tag{8.99b}$$

with the real-valued function

$$\tilde{H}_{\mathrm{LP}}(\omega) = h_{\mathrm{LP}}(q) + 2 \sum_{n=1}^{q} h_{\mathrm{LP}}(q - n) \cdot \cos n\omega \tag{8.99c}$$

due to the symmetry property of the impulse response of $H_{\mathrm{LP}}(z)$:

$$h_{\mathrm{LP}}(n) = h_{\mathrm{LP}}(2q - n), \qquad n \in \mathbf{Z} \tag{8.99d}$$

$\tilde{H}_{\mathrm{LP}}(\omega)$ is the Chebyshev solution of the real approximation problem defined by

$$|1 - \tilde{H}_{\mathrm{LP}}(\omega)| \le \delta_1, \qquad |\omega| \le \omega_p$$
$$|\tilde{H}_{\mathrm{LP}}(\omega)| \le \delta_2, \qquad \omega_s \le |\omega| \le \pi \tag{8.100}$$

The tolerances $\delta_{1,2}$ can be determined uniquely by δ_p and δ_s of Eq. (8.98), as we will derive later (see Eq. 8.104c). Figure 8.17(a) shows an example for the result of this first step. We can observe two properties that are both a consequence of the Chebyshev approximation. First, the zeros of $\tilde{H}_{\mathrm{LP}}$ are simple. Second, the solution possesses equiripple behavior that is of special interest in the stopband. It will be used in the second step of the indicated transformation.

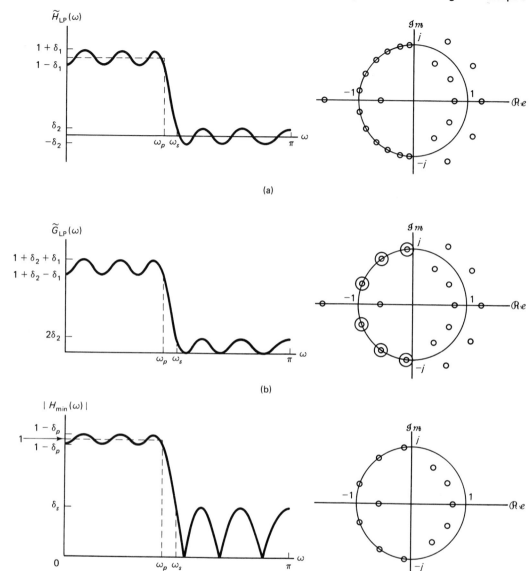

Figure 8.17   Simple example for the transformation of a linear phase filter of degree 24 into a minimum-phase FIR filter of degree 12.

This second step simply consists of adding δ_2 to $\tilde{H}_{LP}(\omega)$, giving the new trigonometric polynomial $\tilde{G}_{LP}(\omega)$, an example of which is shown in Fig. 8.17(b):

$$\tilde{G}_{LP}(\omega) = \delta_2 + \tilde{H}_{LP}(\omega) \tag{8.101a}$$

Equivalently, we can use the transfer function

$$G_{LP}(z) = z^{-q} \cdot \delta_2 + H_{LP}(z) \tag{8.101b}$$

The special construction of these auxiliary functions guarantees three important facts:

1. G_{LP} still has the linear phase $q \cdot \omega$.
2. $\tilde{G}_{LP}(\omega)$ is nonnegative:

$$\tilde{G}_{LP}(\omega) \geq 0, \text{ for all } \omega \qquad (8.101c)$$

More precisely, we have

$$1 - \delta_1 + \delta_2 \leq \tilde{G}_{LP}(\omega) \leq 1 + \delta_1 + \delta_2, \qquad |\omega| \leq \omega_p$$
$$0 \leq \tilde{G}_{LP}(\omega) \leq 2\delta_2, \qquad \omega_s \leq |\omega| \leq \pi \qquad (8.101d)$$

3. $\tilde{G}_{LP}$ still possesses the equiripple behavior, and consequently all zeros on the unit circle have multiplicity 2.

From these properties we can deduce the representation

$$G_{LP}(z) = \lambda^2 \cdot \prod_{n=1}^{q} (z^{-1} - \tilde{z}_{0n}) \cdot (1 - \tilde{z}_{0n} \cdot z^{-1}) \qquad (8.102a)$$

where $\tilde{z}_{0n}$ are the zeros of G_{LP} with a magnitude not exceeding 1. The constant parameter λ is determined by

$$\lambda^2 = \frac{H_{LP}(\omega = 0) + \delta_2}{\prod\limits_{n=1}^{q} (1 - \tilde{z}_{0n})^2} = \frac{h_{LP}(0)}{\prod\limits_{n=1}^{q} (-\tilde{z}_{0n})} \qquad (8.102b)$$

The second equality follows from the fact that only $h_{LP}(q)$ in H_{LP} is changed to obtain G_{LP}.

The third step is initiated by the observation that we can write

$$\Phi(z) = b^2 \cdot \tilde{G}_{LP}(z) = H_{min}(z) \cdot H_{min}(z^{-1}) \qquad (8.103a)$$

where b is a scaling factor to be determined later and $H_{min}(z)$ is the minimum-phase part of $\Phi(z)$. We obtain it by taking simple zeros on the unit circle and all zeros with magnitude less than 1; i.e., we have

$$|\tilde{z}_{0n}| \leq 1, \qquad n = 1, 2, \ldots, q \qquad (8.103b)$$

The basic situation for this condition is demonstrated in Fig. 8.17(c).

From Eq. (8.103a) it follows that we have

$$b \cdot \sqrt{1 - \delta_1 + \delta_2} \leq |H_{min}(\omega)| \leq b \cdot \sqrt{1 + \delta_1 + \delta_2}, \qquad |\omega| \leq \omega_p$$
$$0 \leq |H_{min}(\omega)| \leq b \cdot \sqrt{2\delta_2}, \qquad \omega_s \leq |\omega| \leq \pi \qquad (8.104a)$$

By identifying these equations with those of Eq. (8.98), we obtain the desired solution. The system of equations for the unknowns b, δ_1, and δ_2

$$1 + \delta_p = b \cdot \sqrt{1 + \delta_1 + \delta_2}$$
$$1 - \delta_p = b \cdot \sqrt{1 - \delta_1 + \delta_2}$$
$$\delta_s = b \cdot \sqrt{2\delta_2} \qquad (8.104b)$$

is solved by

$$b = \sqrt{\frac{2 + 2\delta_p^2 - \delta_s^2}{2}}$$

$$\delta_1 = \frac{4\delta_p}{2 + 2\delta_p^2 - \delta_s^2} \qquad (\approx 2\delta_p, \text{ if } \delta_p, \delta_s \ll 1) \qquad (8.104c)$$

$$\delta_2 = \frac{\delta_s^2}{2 + 2\delta_p^2 - \delta_s^2} \qquad (\approx \tfrac{1}{2}\delta_s^2, \text{ if } \delta_p, \delta_s \ll 1)$$

The tolerances $\delta_{1,2}$ for the design of $H_{\mathrm{LP}}(z)$ are now determined such that after transformation, $H_{\min}(z)$ will satisfy the original tolerance scheme. Therefore, we can identify $H(z)$ and $H_{\min}(z)$. It follows from Eqs. (8.97) and (8.102)–(8.104) that

$$H(z) = H_{\min}(z) = h(0) \cdot \prod_{n=1}^{q} (1 - z_{0n} \cdot z^{-1}) \qquad (8.105a)$$

where

$$h(0) = b \cdot \lambda \qquad (8.105b)$$

and

$$z_{0n} = \tilde{z}_{0n}, \qquad n = 1, 2, \ldots, q \qquad (8.105c)$$

Thus it has been shown that the transfer function $H(z)$ is the theoretical solution of the original design problem stated by Eqs. (8.97) and (8.98). In practice, however, there are several difficulties.

The first difficulty arises in the design of the linear phase filter. It is obvious from Eq. (8.104c) that δ_2 may become very small, even for moderate values of δ_s. Consequently, this could possibly lead to numerical problems since, in general, these cases will lead to a high degree of $H_{\mathrm{LP}}(z)$. A second problem is the decomposition of $\Phi(z)$ according to Eq. (8.103), which we treat in the next section.

8.5.4 Separation of the Minimum-Phase Component

The problem we are concerned with is the decomposition of the function $\Phi(z)$ of Eq. (8.103a) into its minimum-phase and maximum-phase components. The most direct method to solve this problem consists of determining all zeros of $\Phi(z)$ in Eq. (8.103a). If we associate all zeros inside and simple zeros on the unit circle with the minimum-phase component, this component is then known except for a constant, which can easily be obtained. This solution was used in [20,21,25]. But experience shows that this method is limited in efficiency because of the issues of determination of the zeros if $H_{\mathrm{LP}}(z)$ is of high degree.

Another approach to the solution of this problem starts with Eq. (8.103a), evaluated for $z = e^{j\omega}$. For simplicity we identify here and in the following $H_{\min}(z)$ with $H(z)$ as in Eq. (8.105a).

$$\Phi(\omega) = H(\omega) \cdot H(-\omega) = |H(\omega)|^2$$

Equivalently, we can write

$$|H(\omega)| = [\Phi(\omega)]^{1/2}$$

stating that the magnitude of $H(\omega)$ is known. If we introduce the function

$$G(\omega) = \ln H(\omega) = -a(\omega) - j\theta(\omega) \tag{8.106a}$$

we observe that its real part can be calculated directly from Eq. (8.103a):

$$-a(\omega) = \ln |H(\omega)| = \frac{1}{2}\ln \Phi(\omega) \tag{8.106b}$$

The minimum-phase property now permits us to calculate the corresponding phase function $\theta(\omega)$ since in this case a and θ constitute a pair of Hilbert transforms.

It is possible to determine the phase function by

$$\theta(\omega) = \frac{1}{2\pi} \int_{-\pi}^{\pi} a(\eta) \cot \frac{\eta - \omega}{2} \, d\eta \tag{8.106c}$$

However, at least for filters of higher degree, this integral relation is not appropriate for a numerical evaluation.

Another approach to solving the decomposition problem was proposed in [23]. It utilizes the cepstrum $\hat{h}(n)$ of a sequence $h(n)$ that is defined [1] by

$$\hat{h}(n) = Z^{-1}\{\hat{H}(z)\} \tag{8.107a}$$

where

$$\hat{H}(z) = \ln[Z\{h(n)\}] = \ln H(z) \tag{8.107b}$$

A comparison with the function G in Eq. (8.106a) leads to the identity

$$\hat{H}(\omega) = G(\omega) \tag{8.107c}$$

where the existence of $\hat{H}$ for $z = e^{j\omega}$ is guaranteed in the case of minimum phase. Only the zeros of H have to be excluded where $\hat{H}$ has logarithmic singularities. Moreover, minimum phase is the key to the solution of the original problem. It forces the cepstral sequence $\hat{h}(n)$ to be causal:

$$\hat{h}(n) = 0, \quad n < 0 \tag{8.108a}$$

Hence we have the representation

$$\hat{H}(z) = \sum_{n=0}^{\infty} \hat{h}(n)z^{-n} \tag{8.108b}$$

and, especially for $z = e^{j\omega}$, we obtain the trigonometric expansions for the attenuation and the phase:

$$a(\omega) = -\sum_{n=0}^{\infty} \hat{h}(n) \cos n\omega \tag{8.108c}$$

$$\theta(\omega) = \sum_{n=1}^{\infty} \hat{h}(n) \sin n\omega \tag{8.108d}$$

Consequently, if the cosine expansion of $a(\omega)$ in Eq. (8.108c) is known, the corresponding sine expansion of the phase can easily be obtained, thus completing the theoretical solution of the original problem.

The practical solution of the problem, however, suffers from the fact that the expansions in Eqs. (8.108) will always be infinite, even if $H(z)$ is nonrecursive. This can be seen by calculating $\hat{H}(z)$ using the representation of $H(z)$ in Eq. (8.105a):

$$\hat{H}(z) = \ln h(0) - \sum_{m=1}^{q} \sum_{n=1}^{\infty} \frac{1}{n} z_{0m}^{n} \cdot z^{-n} \tag{8.109}$$

Therefore, errors will occur in practice. The approximate computation of the coefficients $\hat{h}(n)$ will be performed by means of an FFT with appropriate length. Since this length will always be finite, aliasing errors will be introduced.

Another approach utilizes the fact that all zeros with unity magnitude are known. This is a consequence of the Chebyshev approximation, which yields the set of all extremal frequencies. For our purpose we need only those M extremal frequencies that are located in the stopband and, in addition, where $\tilde{H}_{\text{LP}}$ takes on its minimal value:

$$\tilde{H}_{\text{LP}}(\omega_{en}) = \tilde{H}_{\text{LP}}(-\omega_{en}) = -\delta_2 \tag{8.110a}$$

Defining

$$z_{en} = e^{j\omega_{en}}, \qquad n = 1(1)M \tag{8.110b}$$

we can set up the "stopband function"

$$\Phi_s(z) = H_s(z) \cdot H_s(z^{-1}) \tag{8.111a}$$

which contains all the twofold zeros with unity magnitude:

$$\Phi_s(z) = \Phi_s(z^{-1}) = z^{-M} \cdot \prod_{n=1}^{M} (z - z_{en})^2$$

$$= \prod_{n=1}^{M} (z - 2\cos\omega_{en} + z^{-1}) \tag{8.111b}$$

The function $H_s(z)$ in Eq. (8.111a) is directly obtained as

$$H_s(z) = \prod_{n=1}^{M} (1 - z_{en}z^{-1}) \tag{8.111c}$$

It constitutes that part of $H_{\min}(z)$ that can be determined *a priori* since the zeros z_{en} are known as extremal points of the linear phase filter. All other (unknown) zeros of $\Phi(z)$ are collected to yield the "passband function"

$$\Phi_p(z) = H_p(z) \cdot H_p(z^{-1}) \tag{8.112}$$

where $H_p(z)$ is the minimum-phase component of $\Phi_p(z)$. Finally, the implicit partition of $\Phi(z)$,

$$\Phi(z) = \Phi_p(z) \cdot \Phi_s(z), \tag{8.113a}$$

permits the calculation of $\Phi_p(z)$:

$$\Phi_p(z) = \frac{\Phi(z)}{\Phi_s(z)} \qquad (8.113b)$$

Note that Φ_p contains no zeros on the unit circle. Therefore, the corresponding cepstrum $\hat{\Phi}_p$ is well defined. A closer look at the series in Eq. (8.109) shows that the coefficients of $\hat{\Phi}_p$ decay exponentially, while those of $\hat{\Phi}_s$ decay only as $1/n$ as $n \to \infty$. Fortunately, the phase $\theta_s(\omega)$ of this slowly decaying part can be calculated analytically since $H_s(z)$ is known explicitly. Using the common representation

$$H_s(\omega) = |H_s(\omega)| \cdot e^{-j\theta_s(\omega)} \qquad (8.114a)$$

it is easy to show that the phase is given by

$$\begin{aligned}
\theta_s(\omega) &= \sum_{n=1}^{M} \tan^{-1}\left[\frac{\sin(\omega_{en} - \omega)}{1 - \cos(\omega_{en} - \omega)}\right] \\
&= \frac{M}{2} \cdot \omega - \frac{\pi}{2} \cdot \sum_{n=1}^{M} \text{sign}\left(\sin\frac{\omega - \omega_{en}}{2}\right)
\end{aligned} \qquad (8.114b)$$

where, in addition to the linear phase component, the discontinuities (which are caused by the zeros z_{0n} located on the unit circle) are represented by the second term.

The remaining task is the calculation of the phase $\theta_p(\omega)$, which belongs to $H_p(z)$. Inverse z-transformation of the function

$$\begin{aligned}
\hat{\Phi}_p(z) &= \ln \Phi_p(z) \\
&= \sum_{n=-\infty}^{+\infty} \hat{\phi}_p(n) z^{-n}
\end{aligned} \qquad (8.115a)$$

leads us to the desired result. From Eq. (8.112) it follows that

$$\begin{aligned}
\hat{\Phi}_p(z) &= \ln H_p(z) + \ln H_p(z^{-1}) \\
&= \hat{H}_p(z) + \hat{H}_p(z^{-1})
\end{aligned} \qquad (8.115b)$$

If we now take into account that the minimum-phase property of $H_p(z)$ results in a causal cepstral sequence, as was stated in Eq. (8.108) for the general case, we conclude that

$$\hat{h}_p(n) = 0, \qquad n < 0 \qquad (8.115c)$$

and hence

$$\hat{H}_p(z) = \sum_{n=0}^{\infty} \hat{h}_p(n) z^{-n} \qquad (8.115d)$$

A comparison with the sequence $\hat{\phi}_p(n)$ yields

$$\hat{h}_p(0) = \frac{1}{2}\hat{\phi}_p(0)$$

$$\hat{h}_p(n) = \hat{\phi}_p(n), \qquad n \in \mathbf{N} \qquad (8.115e)$$

These results permit us to calculate the attenuation and especially the phase belonging to H_p:

$$a_p(\omega) = -\sum_{n=0}^{\infty} \hat{h}_p(n) \cos n\omega \tag{8.116a}$$

$$\theta_p(\omega) = \sum_{n=1}^{\infty} \hat{h}_p(n) \sin n\omega \tag{8.116b}$$

These equations will be of central interest for the numerical computation of $H_{min}(z)$, which we consider later.

The next step is to obtain the "passband function" $H_p(z)$ by exponentiation, i.e.,

$$H_p(z) = e^{\hat{H}_p(z)} \tag{8.117}$$

and the total minimum-phase transfer function is then known based on the decomposition

$$H_{min}(z) = H_p(z) \cdot H_s(z) \tag{8.118}$$

thus completing the theoretical derivation.

As mentioned already, the practical realization of this procedure will be performed by the FFT, the length K of which must be sufficiently large. Following the preceding development, we may assume that the samples of the frequency responses $|H_{min}|$, $|H_s|$, and $|H_p|$ at $z = e^{j2k\pi/K}$ are known, for simplicity written as

$$|H_{min,k}|, \qquad |H_{s,k}|, \qquad |H_{p,k}|$$

respectively.

The calculation of the corresponding attenuation values

$$a_{min,k}, \qquad a_{s,k}, \qquad a_{p,k}$$

will be the next step. Up to this point, all operations were performed exactly, except for possible numerical errors. The following step, however, will always introduce an aliasing error, the size of which depends on the length K chosen for the FFT. We have to calculate the sequence

$$\hat{\phi}_{p,a}(n) = \text{DFT}^{-1}\{a_{p,k}\} \tag{8.119a}$$

by the inverse DFT as an approximation for $\hat{\phi}_p(n)$, as has been indicated by the index a. The transformation length K has to be chosen subject to the condition

$$\hat{\phi}_p(n) \approx 0, \qquad |n| \geq \frac{K}{2} \tag{8.119b}$$

The correspondence of Eqs. (8.115c, e) has to be realized by

$$\hat{h}_{p,a}(0) = \frac{1}{2} \hat{\phi}_{p,a}(0)$$

$$\hat{h}_{p,a}(n) = \hat{\phi}_{p,a}(n), \qquad 0 < n < \frac{K}{2} \tag{8.120a}$$

$$\hat{h}_{p,a}(n) = 0, \qquad \frac{K}{2} \leq n < K$$

At this point we would be able to calculate

$$\hat{H}_{p,a}(z) = \mathcal{Z}\{\hat{h}_{p,a}(n)\} \qquad (8.120b)$$

Instead, we calculate the phase function

$$\theta_{p,a,k} = \text{DFT}\{\hat{h}_{p,a,0}(n)\} \qquad (8.121a)$$

where $\hat{h}_{p,a,0}(n)$ is the odd part of $\hat{h}_{p,a}(n)$:

$$\hat{h}_{p,a,0}(n) = \begin{cases} 0, & n = 0, K/2 \\ \hat{h}_{p,a}(n), & 1 \le n < K/2 \end{cases} \qquad (8.121b)$$

$$\hat{h}_{p,a,0}(n) = -h_{p,a,0}(K-n), \qquad 1 \le n < K$$

We obtain the values

$$H_{p,a,k} = |H_{p,k}| \cdot e^{-j\theta_{p,a,k}} \qquad (8.122a)$$

as an approximation for $H_p[(2\pi k)/K]$. The reason that we prefer the representation of Eq. (8.122a) to that of Eq. (8.120b) combined with Eq. (8.117) is that the magnitude is still exact and the aliasing errors affect only the phase. The minimum-phase function we finally use in Eq. (8.118) is given by

$$H_{p,a}(z) = \sum_{n=0}^{q-M} h_{p,a}(n) z^{-n} \qquad (8.122b)$$

where the impulse response is obtained by an inverse DFT of $H_{p,a,k}$ in Eq. (8.122a):

$$h_{p,a}(n) = \text{DFT}^{-1}\{H_{p,a,k}\} \qquad (8.122c)$$

We make two additional remarks concerning accuracy:

1. Some problems occasionally arise when $|H_p|$ is calculated according to Eq. (8.113) because the zeros of $\Phi_s(z)$ are located on the unit circle. As described in [23], this difficulty can be easily circumvented.
2. The influence of the aliasing errors can be measured directly by

$$E_{\text{aliasing}} = \sum_{n=q+1-M}^{K-1} |h_{p,a}(n)|^2 \qquad (8.123a)$$

since the correct impulse response $h_p(n)$ has the exact length $q + 1 - M$, i.e., we have

$$h_p(n) = 0, \qquad n \ge q + 1 - M \qquad (8.123b)$$

Consequently, if K was chosen large enough, E_{aliasing} will be of negligible size.

The last step in the design procedure is the calculation of the total minimum-phase transfer function as an approximation of that in Eq. (8.118):

$$H_{\min,a}(z) = H_{p,a}(z) \cdot H_s(z) \qquad (8.124)$$

thus completing the practical determination of the desired system.

For a design example we use the specifications

$$\delta_p = 0.05, \qquad \delta_s = 0.01, \qquad \omega_p = 0.44\pi, \qquad \omega_s = 0.5\pi$$

For a filter with linear phase, a degree of 50 is required to meet these specifications, while the procedure explained here yields a minimum-phase filter of degree 42. The value of the group delay $\omega = 0$ was calculated to be 3.604 compared with 25.0 in the linear phase case. The magnitude of the corresponding frequency response is shown in Fig. 8.18.

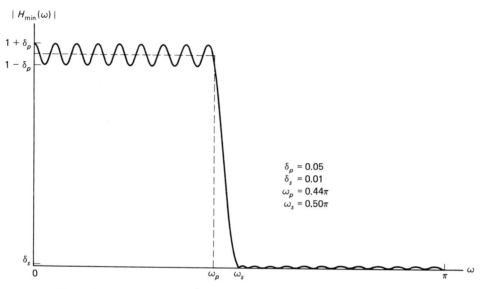

Figure 8.18   Magnitude of the frequency response of a minimum-phase FIR filter of degree 42.

8.6 FILTERS CONSISTING OF ALLPASS SUBSYSTEMS

8.6.1 Introduction

The design of a filter is not completed with the solution of the approximation problem treated so far. Finding an optimal system for the implementation is an essential part of the design as well. Here we call a structure optimal if it yields a canonical realization of the desired transfer function with a minimum number of arithmetic operations such that the sensitivity due to the quantization of coefficients as well as to the noise caused by rounding the variables is low. Though problems of implementation and the effects of limited wordlength are not topics of this chapter, we will discuss a rather new idea for a structure that leads to new approximation problems as well.

In analog circuit design it is well known that selective filters can be implemented using a bridge or lattice structure. While this possibility is totally impractical for accuracy reasons in the continuous case, the method can be used with advantage in a digital implementation. Consequently, the idea has been developed starting from an analog lattice structure leading to wave digital lattice filters [26]. Recently several authors presented independently an attractive solution without referring to analog circuits [27,28]. Its essential points will be described here briefly.

8.6.2 Analysis of the General Structure

Suppose we are given two stable allpass systems having the transfer functions

$$A_i(z) = \frac{z^{-p_i} + \sum_{k=1}^{p_i} a_{ki} z^{k-p_i}}{1 + \sum_{k=1}^{p_i} a_{ki} z^{-k}}, \qquad i = 1, 2 \tag{8.125a}$$

$$= \frac{z^{-p_i} D_i(z^{-1})}{D_i(z)} \tag{8.125b}$$

We assume that the poles of the two allpass systems are different. A connection as shown in Fig. 8.19(a) yields two transfer functions $H_1(z)$ and $H_2(z)$, the properties of which we investigate first.

$$H_1(z) = \frac{Y_1(z)}{X(z)} = \frac{1}{2}[A_1(z) + A_2(z)] = \frac{N_1(z)}{D(z)} \tag{8.126a}$$

$$H_2(z) = \frac{Y_2(z)}{X(z)} = \frac{1}{2}[A_1(z) - A_2(z)] = \frac{N_2(z)}{D(z)} \tag{8.126b}$$

Obviously, the denominator polynomial is $D(z) = D_1(z) \cdot D_2(z)$. Under the assumptions made in Eq. (8.125), it has the degree $p = p_1 + p_2$. The numerator polynomials turn out to be

$$N_{1,2}(z) = \frac{1}{2}[z^{-p_1} D_1(z^{-1}) D_2(z) \pm z^{-p_2} D_2(z^{-1}) D_1(z)] \tag{8.127}$$

Here and in the following equations, the first index (1) refers to the upper sign (+), the second to the lower one.

As we can readily confirm, the polynomials have the properties

$$z^{-(p_1+p_2)} \cdot N_1(z^{-1}) = N_1(z) \tag{8.128a}$$

$$z^{-(p_1+p_2)} \cdot N_2(z^{-1}) = -N_2(z) \tag{8.128b}$$

That means $N_1(z)$ is a mirror-image polynomial, while $N_2(z)$ is an anti-mirror-image polynomial, both with the center of symmetry at $p/2$. In general, their degree is p.

We also investigate the interdependence of $H_1(z)$ and $H_2(z)$. Using Eq. (8.126), we get the important result

$$H_1(z)H_1(z^{-1}) + H_2(z)H_2(z^{-1}) = 1 \tag{8.129a}$$

and, especially, on the unit circle

$$|H_1(\omega)|^2 + |H_2(\omega)|^2 = 1 \tag{8.129b}$$

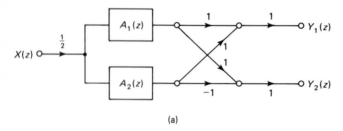

(a)

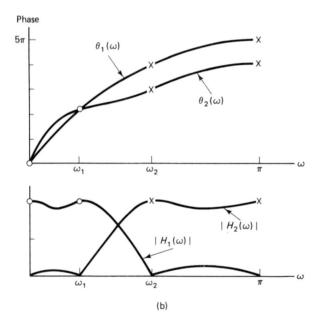

(b)

Figure 8.19 (a) The basic structure for a complementary pair of transfer functions; (b) phase responses and the resulting magnitude functions.

Because of this property the functions $H_1(z)$ and $H_2(z)$ are called *lossless bounded-real*. The energy of the input signal $x(n)$ is distributed to the outputs 1 and 2 without loss. Before we interpret this property further we derive a related result. With $A_i(\omega) = e^{-j\theta_i(\omega)}$, Eqs. (8.126) lead to

$$H_{1,2}(\omega) = \frac{1}{2}e^{-j\theta_1(\omega)}[1 \pm e^{-j\Delta\theta(\omega)}] \tag{8.130a}$$

where $\Delta\theta(\omega) = \theta_2(\omega) - \theta_1(\omega)$. Using

$$|H_1(\omega)| = \frac{1}{2}|1 + e^{-j\Delta\theta(\omega)}| = \left|\cos\frac{\Delta\theta(\omega)}{2}\right| \tag{8.130b}$$

and

$$|H_2(\omega)| = \frac{1}{2}|1 - e^{-j\Delta\theta(\omega)}| = \left|\sin\frac{\Delta\theta(\omega)}{2}\right| \tag{8.130c}$$

we can confirm Eq. (8.129b) readily. These results yield the following consequences: If for a particular frequency ω_λ we have $\Delta\theta(\omega_\lambda) = 0$, i.e., $\theta_2(\omega_\lambda) = \theta_1(\omega_\lambda)$, we get

$$|H_1(\omega_\lambda)| = 1 \quad \text{and} \quad H_2(\omega_\lambda) = 0$$

while for $\Delta\theta(\omega_\lambda) = \pi$ the result is

$$H_1(\omega_\lambda) = 0 \quad \text{and} \quad |H_2(\omega_\lambda)| = 1$$

Figure 8.19(b) illustrates this interdependence. Another very important consequence of Eq. (8.129b) is the "built-in" insensitivity in the passband. Suppose the coefficients of the two allpass systems in the basic structure are somewhat incorrect, for example due to the required limitation of their wordlength. Thus in Eq. (8.125a) we have $a_{ki} + \Delta a_{ki}$ instead of a_{ki}. Furthermore, we assume that the structure of the allpass systems is such that their main property $|A_i(\omega)| = 1$ for all ω as well as the stability of the system is not affected by the erroneous coefficients. This condition can be satisfied easily, as we show in Section 8.6.3. Under these assumptions Eq. (8.129b) still holds.

Now let ω_λ be a point where $|H_i(\omega_\lambda)| = 1$ with ideal coefficients. From Eq. (8.129b), any change of any coefficients a_{ki} will result in a decrease of $|H_i(\omega_\lambda)|$ regardless of the sign of Δa_{ki}. Thus the slope of $|H_i(\omega)|$ as a function of any a_{ki} is precisely zero at $\omega = \omega_\lambda$. So the first-order sensitivity, defined as

$$S_{ki} = \frac{a_{ki}}{|H_i(\omega)|} \cdot \frac{\partial |H_i(\omega)|}{\partial a_{ki}}$$

is zero at all points ω_λ, where $|H_i(\omega)|$ is unity. Thus, under rather mild conditions we can expect a small sensitivity of the system.

Furthermore, we consider the quotient

$$F(z) = \frac{H_2(z)}{H_1(z)} = \frac{N_2(z)}{N_1(z)} \tag{8.131a}$$

on the unit circle $z = e^{j\omega}$. With Eq. (8.130a) we readily get

$$F(\omega) = \frac{1 - e^{-j\Delta\theta(\omega)}}{1 + e^{-j\Delta\theta(\omega)}} = j\tan\frac{\Delta\theta(\omega)}{2} \tag{8.131b}$$

We conclude that for any input signal, the phase shift between the two output signals is precisely $-\pi/2$.

Finally we calculate

$$H_1(\omega) + jH_2(\omega) = H_1(\omega)\left[1 - \tan\frac{\Delta\theta(\omega)}{2}\right] \tag{8.132}$$

If we assume $\Delta\theta(\omega)$ to be $-\pi/2 \cdot \text{sign}[\omega]$ for all ω, the two signals would be a Hilbert transform pair, i.e., the real and imaginary part of an analytic signal, the spectrum of which is zero for $\omega < 0$. We return to this property in Section 8.6.5.

8.6.3 Allpass Structures

We now consider the issue of effective implementations for the structure depicted in Fig. 8.19. Because an allpass system of p_ith degree described by Eq. (8.125a) can be built with p_i multipliers only, we conclude that a transfer function $H_i(z)$ of pth degree satisfying the constraints given by Eqs. (8.128) and (8.129) can be implemented with only p multiplications per output sample. This is the case, for example, with a large class of selective systems, the transfer functions of which have zeros only on the unit circle (see Section 8.6.4). If for comparison we consider the usual implementation of such a system of pth degree as a cascade of blocks of second order, possibly with a further block of first order, we end up with roughly twice as many multiplications per output sample. Furthermore, the complementary system (e.g., the corresponding highpass filter if $H_i(z)$ describes a lowpass filter) can be achieved with just one additional summation.

The favorable sensitivity of the system considered in the previous section required subsystems having the structural allpass property, i.e., they remain allpass systems after a change of their coefficients in a certain range. In addition we desire canonic structures with a minimum number of multiplications. Quite a few allpass systems are well known that satisfy these conditions (e.g., [29]). Here we consider a cascade of allpass systems of first and second order and restrict ourselves to one example each. Corresponding signal flow graphs are shown in Fig. 8.20. Their transfer functions are

$$A_{1k}(z) = \frac{z^{-1} + \alpha_k}{1 + \alpha_k z^{-1}}, \qquad |\alpha_k| < 1 \tag{8.133a}$$

$$A_{2k}(z) = \frac{z^{-2} + \beta_k(1 + \alpha_k)z^{-1} + \alpha_k}{1 + \beta_k(1 + \alpha_k)z^{-1} + \alpha_k z^{-2}}, \qquad |\alpha_k|, \quad |\beta_k| < 1 \tag{8.133b}$$

Obviously the structures are canonic, and they do have the structural allpass property. Furthermore controlling the coefficients to guarantee stability is easily done.

If the numerator polynomials N_1 and N_2 are functions of z^2 only, the two allpass systems can be built as cascades of blocks of second and fourth order yielded by replacing z^{-1} by z^{-2} in Fig. 8.20. In this case, of interest for quadrature mirror filters and for certain Hilbert transformers (considered in Section 8.6.5), the total number of coefficients for both systems is further reduced by a factor of 2.

8.6.4 Design of Selective Filters

Suppose we are given the problem of designing a lowpass filter according to the tolerance scheme for the magnitude as given in Fig. 8.21(a) and having the structure considered here. First we assume the approximation problem to be solved already, yielding a transfer function $H_1(z) = N_1(z)/D(z)$ with the properties found by the analysis of the circuit in Fig. 8.19. For convenience the properties are repeated here [28]:

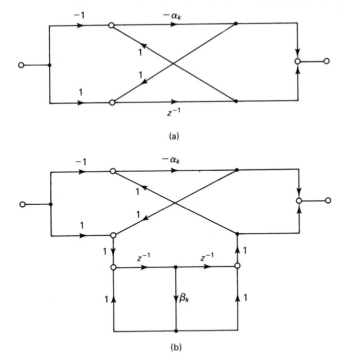

(a)

(b)

Figure 8.20   Canonic structures of allpass systems of first and second order with minimum number of multipliers.

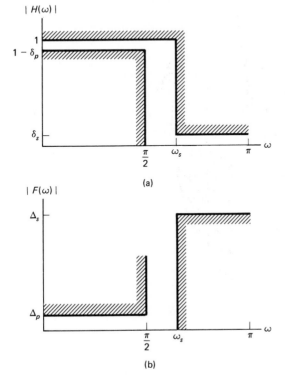

(a)

(b)

Figure 8.21   Tolerance schemes for (a) $|H(\omega)|$ and (b) $|F(\omega)|$ of a normalized lowpass filter.

1. $H_1(z)$ is bounded real, specifically, $|H_1(\omega)| \leq 1$.
2. $N_1(z)$ is a mirror-image polynomial.
3. There exists a transfer function $H_2(z) = N_2(z)/D(z)$ that is complementary to $H_1(z)$ such that $|H_1(\omega)|^2 + |H_2(\omega)|^2 = 1$.
4. In addition, $N_2(z)$ has to be an anti-mirror-image polynomial.

Obviously the last two conditions lead to some additional restrictions for $H_1(z)$: Since an anti-mirror-image polynomial has to have a zero of odd multiplicity at $z = 1$ and Eq. (8.129) has to be satisfied, we conclude that

$$|H_1(\omega = 0)| = 1$$

$$\left. \frac{\partial^\lambda |H_1(\omega)|}{\partial \omega^\lambda} \right|_{\omega=0} = 0, \qquad \lambda = 1, 2, \ldots, (2\ell - 1) \tag{8.134a}$$

are additional constraints, permitting $H_2(z)$ to have a zero at $z = 1$ of multiplicity ℓ, where ℓ is odd.

Furthermore, $N_2(z)$ can be of even order only if it has a zero of odd order at $z = -1$ as well. Now if we assume $H_1(z)$ to be a transfer function of a lowpass filter, $H_2(z)$ has to describe the corresponding highpass filter (see Fig. 8.19b) and thus cannot have a zero at $z = -1$. We conclude that, at least in the case of highpass and lowpass filters, the complementary pair of transfer functions $H_1(z)$ and $H_2(z)$ has to be of odd order.

All these conditions are fulfilled, e.g., with one of the classical closed-form solutions of the approximation problem, where maximally flat or Chebyshev behavior is obtained in one or both subbands as long as we restrict ourselves to functions of odd order in the lowpass or highpass case. Now if a complementary pair $H_1(z)$ and $H_2(z)$ with the required properties is known, the desired allpass transfer functions are

$$A_1(z) = H_1(z) + H_2(z), \qquad A_2(z) = H_1(z) - H_2(z) \tag{8.135}$$

As an example we consider the design of a normalized Butterworth lowpass filter of fifth degree, having its 3-dB cutoff frequency at $\omega_p = \pi/2$. The well-known design (e.g., [1]) yields

$$H_1(z) = \frac{0.052786(1 + z^{-1})^5}{(1 + 0.105573z^{-2})(1 + 0.527864z^{-2})}$$

which has a flat frequency response at $\omega = 0$. For the complementary function $H_2(z)$ we get

$$H_2(z) = \frac{0.052786(1 - z^{-1})^5}{D(z)}$$

Using Eq. (8.135) we obtain

$$A_1(z) = \frac{z^{-2} + 0.105573}{1 + 0.105573z^{-2}}$$

$$A_2(z) = z^{-1} \cdot \frac{z^{-2} + 0.527864}{1 + 0.527864z^{-2}}$$

Since the second allpass system is a cascade of a delay and an allpass system of second order, this system can be implemented with only two multiplications per output sample. Together with the corresponding system described by $H_2(z)$, it is an implementation of a quadrature mirror filter. Note that any other Butterworth lowpass (or highpass) filter of the same degree can be obtained by applying an appropriate lowpass-lowpass transformation yielding two general allpass systems of second and third degree, respectively, which can be implemented with a total of five multiplications per output sample.

We investigate the inherent approximation problem somewhat further from a different point of view [27]. Multiplying the desired transfer function $H_1(z)$ in the numerator and denominator by $A_1(z) = H_1(z) + H_2(z)$, we get

$$H_1(z) = A_1(z) \cdot \frac{H_1(z)}{H_1(z) + H_2(z)}$$

With $F(z) = H_2(z)/H_1(z) = N_2(z)/N_1(z)$, as introduced in Eq. (8.131a), we obtain

$$H_1(z) = A_1(z) \cdot \frac{1}{1 + F(z)} \qquad (8.136a)$$

and thus, since $F(\omega)$ is always imaginary (see Eq. 8.131b),

$$|H_1(\omega)|^2 = \frac{1}{1 + |F(\omega)|^2} \qquad (8.136b)$$

Obviously, the poles of $F(\omega)$ correspond with the zeros of $|H_1(\omega)|$, while the zeros of $F(\omega)$ determine the points, where $|H_1(\omega)| = 1$. Equations (8.136) have a counterpart in the classical method of designing selective analog filters. Essentially $F(z)$ plays the role of the characteristic function, which controls the behavior of the transfer functions to establish its bounded real property. The standard solutions of the approximation problem are found by selecting specific expressions for $F(z)$.

Before we proceed with an example for picking $F(z)$ appropriately, we finish the description of the design procedure. Suppose we selected a function $F(z) = N_2(z)/N_1(z)$ such that the tolerance scheme given for $|H_1(\omega)|$ is satisfied. To find the required $D(z)$ we start with

$$N_1(z)N_1(z^{-1}) + N_2(z)N_2(z^{-1}) = D(z)D(z^{-1}), \qquad (8.137a)$$

an equation equivalent to Eq. (8.129a). According to Eq. (8.128), we have

$$N_{1,2}(z^{-1}) = \pm z^p N_{1,2}(z) \qquad (8.137b)$$

Thus the left-hand side of Eq. (8.137a) can be written as

$$z^p[N_1^2(z) - N_2^2(z)] = z^p[N_1(z) + N_2(z)][N_1(z) - N_2(z)] \qquad (8.138)$$

Using Eq. (8.137b) again, we find

$$N_1(z) - N_2(z) = z^{-p}[N_1(z^{-1}) + N_2(z^{-1})] \qquad (8.139)$$

and conclude that the zeros of $N_1(z) - N_2(z)$ are reciprocal to those of $N_1(z) + N_2(z)$.

Due to stability they cannot lie on the unit circle. If z_k denote the zeros of $N_1(z) + N_2(z)$ such that $|z_k| < 1$, $k = 1, 2, \ldots, r$, and $|z_k| > 1$, $k = (r + 1)$, $(r + 2), \ldots, p$, we get

$$N_1(z) + N_2(z) = \prod_{k=1}^{r} (1 - z^{-1}z_k) \cdot \prod_{k=r+1}^{p} (z^{-1} - z_k^{-1}) \tag{8.140a}$$

and

$$N_1(z) - N_2(z) = \prod_{k=1}^{r} (z^{-1} - z_k) \cdot \prod_{k=r+1}^{p} (1 - z^{-1}z_k^{-1}) \tag{8.140b}$$

Dividing by

$$D(z) = \prod_{k=1}^{r} (1 - z^{-1}z_k) \cdot \prod_{k=r+1}^{p} (1 - z^{-1}z_k^{-1}) \tag{8.140c}$$

yields finally, according to Eq. (8.135),

$$A_1(z) = \frac{\displaystyle\prod_{k=r+1}^{p} (z^{-1} - z_k^{-1})}{\displaystyle\prod_{k=r+1}^{p} (1 - z^{-1}z_k^{-1})} \tag{8.141a}$$

and

$$A_2(z) = \frac{\displaystyle\prod_{k=1}^{r} (z^{-1} - z_k^{-1})}{\displaystyle\prod_{k=1}^{r} (1 - z^{-1}z_k^{-1})} \tag{8.141b}$$

Referring to the conditions formulated with $H_1(z)$ and $H_2(z)$ at the beginning of this section we conclude that there is a simpler way to express the conditions in terms of $N_1(z)$ and $N_2(z)$:

A pair of mirror-image and anti-mirror-image polynomials $N_1(z)$ and $N_2(z)$ of the same degree p with uncommon zeros always yields a pair of complementary bounded real transfer functions $H_1(z)$ and $H_2(z)$ to be implemented with two allpass subsystems.

Example 2

Suppose we are given a tolerance scheme for $|H(\omega)|$ as shown in Fig. 8.21(a). The passband cutoff frequency has been normalized such that $\omega_p = \pi/2$. Since we want to satisfy it by choosing an appropriate function $F(z)$, we transform it into a tolerance scheme for $|F(\omega)|$. Using Eq. (8.136b), we obtain in the passband $0 \le |\omega| \le \pi/2$

$$1 \ge |H(\omega)| \ge 1 - \delta_p \longrightarrow 0 \le |F(\omega)| \le \Delta_p = \frac{\sqrt{2\delta_p - \delta_p^2}}{1 - \delta_p}$$

and in the stopband $\omega_s \le |\omega| \le \pi$

$$|H(\omega)| \le \delta_s \longrightarrow |F(\omega)| \ge \Delta_s = \frac{\sqrt{1 - \delta_s^2}}{\delta_s}$$

Figure 8.21(b) shows the tolerance scheme for $|F(\omega)|$, which in turn leads to a tolerance scheme for $\Delta\theta(\omega)$, according to Eq. (8.131b). As a simple example, we design a lowpass filter with Chebyshev behavior in the passband, satisfying the specifications

$$\delta_p = 0.1 \longrightarrow \Delta_p = 0.484322 \qquad \text{(corresponding to } |\Delta\theta(\omega)| \leq 0.287\pi)$$

$$\delta_s = 0.01 \longrightarrow \Delta_s = 99.995, \qquad |\Delta\theta(\omega) - \pi| \leq 0.00637\pi$$

$$\omega_s = 0.7\pi$$

If we transfer the well-known equations for the design from the analog into the digital domain, we obtain for the required degree

$$p \geq \frac{\cosh^{-1}[\Delta_s/\Delta_p]}{\cosh^{-1}[\tan \omega_s/2]} = 4.6508$$

Thus we pick $p = 5$ as the next odd integer, yielding a reduced stopband cutoff frequency of $\omega_s = 0.68\pi$. In general the function $F(\omega)$ for a Chebyshev behavior is obtained by bilinear transformation of the appropriate Chebyshev polynomial with $\Delta_p = \max |F(\omega)|$ as a factor. We get

$$|F(\omega)| = \Delta_p \cdot \left| T_p\left(\tan \frac{\omega}{2}\right) \right|$$

where T_p is the Chebyshev polynomial of pth degree. The corresponding function $F(z)$ can be written as

$$F(z) = \Delta_p \cdot K_p \cdot \frac{\prod_{\lambda=1}^{p} (1 - z^{-1}z_\lambda)}{(1 + z^{-1})^p} = \frac{N_2(z)}{N_1(z)}$$

where $z_\lambda = e^{-j\omega_\lambda}$, with $\omega_\lambda = 2\tan^{-1}[\cos(2\lambda - 1)(\pi/2p)]$, are the zeros of the Chebyshev polynomial after the transformation and thus the zeros of $N_2(z)$. With

$$K_p = 0.5(1 + \sqrt{2})^p[1 - (1 - \sqrt{2})^{2p}]$$

for odd values p we get $|F(j)| = \Delta_p$, as required.

Proceeding with our example and $p = 5$, we have to find the zeros of the polynomials

$$(1 + z^{-1})^5 \pm \Delta_p K_p \prod_{\lambda=1}^{5} (1 - z^{-1}z_\lambda)$$

A separation, as explained by Eq. (8.140), finally yields

$$A_1(z) = \frac{z^{-2} - 0.58900z^{-1} + 0.49569}{1 - 0.58900z^{-1} + 0.49569z^{-2}}$$

and

$$A_2(z) = \frac{z^{-1} - 0.5395}{1 - 0.5395z^{-1}} \cdot \frac{z^{-2} - 0.005z^{-1} + 0.829927}{1 - 0.005z^{-1} + 0.829927z^{-2}}$$

As a somewhat unusual illustration associated with a selective filter, Fig. 8.22 shows the phase difference $\Delta\theta(\omega) = \text{phase}[A_2(\omega)/A_1(\omega)]$. The Chebyshev behavior in the range $0 \leq |\omega| \leq \pi/2$, as well as the monotonic increase for $|\omega| > \pi/2$, can be seen, corresponding to the desired properties of $|F(\omega)|$ and $|H(\omega)|$. The zeros ω_λ of $\Delta\theta(\omega)$ are indicated. As indicated previously, at these points, $|H(\omega)| = 1$.

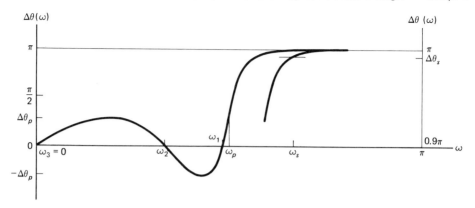

Figure 8.22   Phase difference $\Delta\theta(\omega) = \theta_2(\omega) - \theta_1(\omega)$ in the case of a Chebyshev lowpass filter of degree 5.

8.6.5 Recursive Phase Splitter and Hilbert Transformers

As we pointed out in Section 8.6.2, the structure under study can be used to generate approximately an analytic signal out of any given signal $x(n)$ We require two allpass systems whose phase responses differ nearly by $-\pi/2$. From analog circuit theory the design of a phase splitter is well known. It yields an optimal solution in the sense that the phase difference of the two allpass systems approximates $\pi/2$ inside a desired frequency band in the Chebyshev sense. Elliptic functions are required for a closed-form solution. As explained in [30], the bilinear transformation of an analog phase splitter yields a digital one since the transformation changes neither the allpass property nor the important feature of interest, the difference of the phase responses.

   In the usual application of a phase splitter, the output signals of the allpass subsystems are used directly. Thus they have the same magnitude, while their phase difference just approximates $\pi/2$. If we use the structure of Fig. 8.19 instead, i.e., besides multiplication by $1/2$ add and subtract the output signals of the two allpass systems first, we obtain precisely a phase shift of $\pi/2$, but an imperfect magnitude. Let $\Delta\theta(\omega) = \pi/2 - \epsilon(\omega)$, where $\epsilon(\omega)$ describes the deviation from the desired value $\pi/2$. Using Eq. (8.131b) we get

$$\frac{|H_2(\omega)|}{|H_1(\omega)|} = \frac{1 - \tan \epsilon/2}{1 + \tan \epsilon/2} \approx 1 - \epsilon \qquad \text{for } |\epsilon| \ll 1$$

Since the phase difference is precisely $\pi/2$, this structure might be compared with the well-known FIR Hilbert transformer [1]. However, there is an important difference: it does not perform the Hilbert transform of the input signal $x(n)$ but of a phase-distorted version of it. This is to be seen with Eq. (8.132), which shows that the phase response of $H_1(\omega)$ has to be taken into account.

   There are two possible ways to avoid this undesired effect and thus to design approximately a true recursive Hilbert transformer. In [31] a "noncausal" solution has been described. Its essential point is that the response of the first allpass system of a phase splitter is processed in reverse order by the second one. Its output after a second reversal is an approximation of the response of the noncausal system. Details about handling sequences of potentially unlimited length and reducing the unavoidable truncation error are given in [32].

Here we present another possibility [33]. We assume the first allpass system to be a delay element of order $p_1 = p - 1$, i.e., $A_1(z) = z^{-(p-1)}$. The problem is to design an allpass system $A_2(z)$ whose phase approximates the desired function

$$\theta_d(\omega) = (p - 1)\omega + \pi/2 \ \text{sign}[\omega] \tag{8.142}$$

for $\omega_\ell \leq |\omega| \leq \omega_u$ in a Chebyshev sense (see Fig. 8.23). For a simpler presentation we restrict the following considerations to even values of p and assume the second

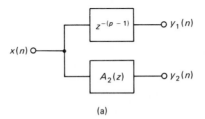

(a)

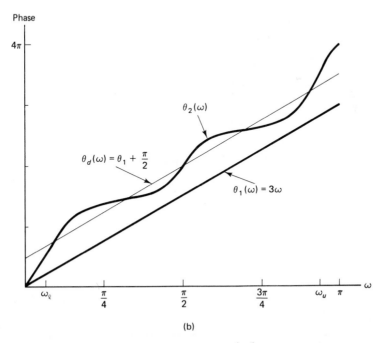

(b)

Figure 8.23   (a) Special structure with $A_1(z) = z^{-(p-1)}$; (b) optimal phase response for $p = 4$.

allpass system to be of pth order. Its transfer function on the unit circle is, according to Eq. (8.125a),

$$A_2(\omega) = e^{-jp\omega} \cdot \frac{1 + \sum\limits_{k=1}^{p} a_k e^{jk\omega}}{1 + \sum\limits_{k=1}^{p} a_k e^{-jk\omega}}$$

and its phase

$$\theta_2(\omega) = p\omega - 2 \tan^{-1} \frac{\sum\limits_{k=1}^{p} a_k \sin k\omega}{1 + \sum\limits_{k=1}^{p} a_k \cos k\omega} \tag{8.143}$$

As $\theta_2(0) = 0$, $\theta_2(\pi) = p\pi$, and, since the slope of $\theta_2(\omega)$—being the group delay of an allpass system—is always positive, the desired phase response will look like the one sketched in Fig. 8.23(b) for $p = 4$ and the case $\omega_u = \pi - \omega_\ell$. The error function $\epsilon(\omega) = \theta_d(\omega) - \theta_2(\omega)$ shows $p + 2$ alternations, as required for an optimal solution. It can be found by the well-known Remez exchange algorithm (e.g., [1]), if an interpolation method is available to find the coefficients of an allpass system such that its phase has prescribed (and permissible) values at given points ω_i, $i = 1, 2, \ldots, p$. We show this procedure in general terms.

Let $\omega_{i+1} > \omega_i$ and thus $\theta(\omega_{i+1}) > \theta(\omega_i)$ for all i. We prescribe

$$\theta(\omega_i) = p\omega_i - 2 \tan^{-1} \frac{\sum\limits_{k=1}^{p} a_k \sin k\omega_i}{1 + \sum\limits_{k=1}^{p} a_k \cos k\omega_i}$$

With

$$\beta_i = \frac{1}{2}[\theta(\omega_i) - p\omega_i] \tag{8.144a}$$

we get

$$\left[1 + \sum\limits_{k=1}^{p} a_k \cos k\omega_i\right] \tan \beta_i = -\sum\limits_{k=1}^{p} a_k \sin k\omega_i$$

and thus

$$\sum\limits_{k=1}^{p} a_k[\tan \beta_i \cos k\omega_i + \sin k\omega_i] = -\tan \beta_i$$

This yields, after elementary manipulations,

$$\sum\limits_{k=1}^{p} a_k \sin[k\omega_i + \beta_i] = -\sin \beta_i, \qquad i = 1, 2, \ldots, p \tag{8.144b}$$

Letting $\mathbf{a} = [a_1, a_2, \ldots, a_p]^T$ be the vector of unknown coefficients, $\mathbf{A}$ the matrix with the elements $a_{ik} = \sin[k\omega_i + \beta_i]$, and $\mathbf{b} = [b_1, b_2, \ldots, b_p]^T$ with $b_i = -\sin \beta_i$

the vector of known values on the right-hand side, then

$$\mathbf{a} = \mathbf{A}^{-1} \cdot \mathbf{b} \tag{8.144c}$$

is the solution of the problem of designing an allpass system with prescribed phase values at a given set of points. Using this result, we can apply the Remez exchange algorithm such that an equiripple solution is found in the desired interval $[\omega_\ell, \omega_u]$. The maximum deviation $\max|\epsilon(\omega)| = \max|\theta_d(\omega) - \theta_2(\omega)|$ cannot be prescribed in advance. It turns out that an approximation interval with $\omega_u = \pi - \omega_\ell$ being symmetrical to $\pi/2$ yields a very efficient solution. For details, see [33].

Example 3

With $\omega_\ell = 0.07\pi$ (and $\omega_u = 0.93\pi$) and degree $p = 6$ we get $\max|\epsilon(\omega)| = 5.99 \cdot 10^{-2} \triangleq 3.43°$. Figure 8.24(a) shows $\epsilon(\omega)$. The transfer function of the resulting allpass system can be written as

$$A_2(z) = \frac{z^{-2} + \alpha_1}{1 + \alpha_1 z^{-2}} \cdot \frac{z^{-4} + \beta_2(1 + \alpha_2)z^{-2} + \alpha_2}{1 + \beta_2(1 + \alpha_2)z^{-2} + \alpha_2 z^{-4}}$$

where $\alpha_1 = -0.751622$, $\alpha_2 = 0.091969$, and $\beta_2 = 0.240721$.

    A cascade of the two structures shown in Fig. 8.20 can be used if all delay elements are doubled (i.e., z^{-1} replaced by z^{-2}). Thus only three multiplications per output sample are required. This favorable property is a consequence of the symmetrical approximation interval.

    For comparison, an FIR Hilbert transformer has been designed with Chebyshev behavior of its amplitude response in the same interval (see Fig. 8.24b). An impulse response of length 19 is required, yielding an amplitude deviation $|A(\omega)| \leq 6.49 \cdot 10^{-2}$, a figure indicating performance similar to that of the recursive Hilbert transformer. Here five multiplications per output sample are required if we make use of the fact that the impulse response is an odd sequence. The total delay is nine steps, compared with five in the recursive case.

    Finally we show the properties of a phase splitter designed for the same approximation interval. With two allpass systems of degree $p_1 = 3$ and $p_2 = 2$, we get a deviation $|\epsilon(\omega)| \leq 5.655 \cdot 10^{-2} \triangleq 3.24°$ (see Fig. 8.24c). The systems are described by

$$A_1(z) = z^{-1} \cdot \frac{z^{-2} - 0.76275}{1 - 0.76275z^{-2}}$$

$$A_2(z) = \frac{z^{-2} - 0.28453}{1 - 0.28453z^{-2}}$$

and thus the phase splitter can be implemented by two multiplications per output sample only. Again this is due to the symmetry of the approximation interval. Figure 8.24(d) illustrates the drawback of this solution. The resulting group delay in either channel is not constant. Thus the two output signals are phase-distorted versions of $x(n)$, but approximately Hilbert transforms of each other.

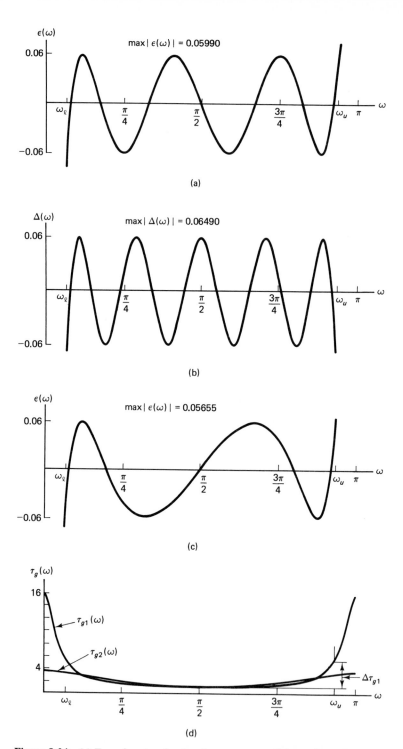

Figure 8.24 (a) Error function for the phase response of the optimal IIR Hilbert transformer; $p = 6$. (b) Error function of the magnitude for an FIR Hilbert transformer of length 19. (c) Error function of the phase for a phase splitter; $p_1 = 3$, $p_2 = 2$. (d) Group delay functions of the phase splitter solution of part (c).

8.7 SUMMARY

In this chapter we discussed some problems arising in the design of digital filters. We did not attempt to give a complete picture of all the design problems solved in recent years. Instead we dealt with some general aspects of design, emphasizing that in all cases the actual problem has to be stated very carefully according to the desired application before an appropriate design procedure is used. Consequently the systems obtained are tailored for their particular purposes and thus may not be suitable for other applications. Furthermore, we explained different methods for the solution of the approximation problem with emphasis on closed-form solutions wherever they are known.

We considered in some detail the design of filters that satisfy specifications in the time domain. Beside Prony's method and some of its extensions, the old problem of reconstructing a smooth function out of its samples was considered in Section 8.2, yielding an efficient and generally applicable solution. Savitzky-Golay filters, designed for optimal smoothing of measured data, are of particular interest in physics and chemistry. But since they are not particularly selective in the usual sense, they are not as familiar to communication engineers. Nevertheless, their specification and the design procedures are very interesting.

In some applications a certain behavior of the frequency response at distinct points is of interest corresponding to global constraints described by an appropriate norm of an error. This problem is considered for FIR filters in a general formulation in Section 8.4. Methods for the design of optimal minimum-phase FIR filters are also presented, including a more recent approach that can be applied in case of high-order filters.

Formulating and solving the approximation problem was the main objective. An optimal design requires a careful choice of the structure for the implementation as well. The last section deals with a filter that consists of allpass subsystems, a recently introduced, interesting candidate for solutions of different problems. Beside an analysis of the structure, its use for the implementation of selective filters and recursive Hilbert transformers is shown.

The design of digital systems is still an open field. New applications will call for new solutions. In addition it is necessary to include implementation considerations into the design procedure such that a really optimal system is achieved.

REFERENCES

1. A. V. Oppenheim and R. W. Schafer, *Digital Signal Processing,* Prentice-Hall, Englewood Cliffs, NJ, 1975.

2. J. F. Kaiser, "Nonrecursive Digital Filter Design Using the I_0-sinh Window Function," *Proc. IEEE International Symposium on Circuits and Systems,* pp. 20–23, 1974; reprinted in *Selected Papers in Digital Signal Processing II,* IEEE Press, New York, 1976, pp. 123–126.

3. H. W. Schüssler, "A Signal Processing Approach to Simulation," *Frequenz,* Vol. 35, pp. 174–184, 1981.

4. C. S. Burrus and T. W. Parks, "Time Domain Design of Recursive Digital Filters," *IEEE Trans. Audio Electroacoustics,* Vol. 18, pp. 137–141, 1970; reprinted in *Digital Signal Processing,* IEEE Press (Selected Reprint Series), New York, pp. 138–142, 1972.

5. H. D. Helms and J. B. Thomas, "Truncation Error of Sampling Theorem Expansions." *Proc. IRE,* Vol. 50, pp. 179–184, 1962.

6. H. W. Schüssler and P. Steffen, "A Hybrid System for the Reconstruction of a Smooth Function from Its Samples," *Circuits, Systems, and Signal Processing,* Vol. 3, pp. 295–314, 1984.

7. P. Steffen, "On Digital Smoothing Filters: A Brief Review of Closed Form Solutions and Two New Filter Approaches," *Circuits, Systems, and Signal Processing,* Vol. 5, pp. 187–210, 1986.

8. W. F. Sheppard, "Reduction of Errors by Means of Negligible Differences," *Proc. 5th International Congress of Mathematicians,* Vol. 2, Cambridge, pp. 348–384, 1913.

9. U. Bromba and H. Ziegler, "Explicit Formula for Filter Function of Maximally Flat Nonrecursive Digital Filters," *Electronics Letters,* No. 16, pp. 905–906, 1980.

10. J. F. Kaiser and K. Steiglitz, "Design of FIR Filters with Flatness Constraints," *Proc. ICASSP-83,* pp. 197–200, 1983.

11. F. B. Hildebrand, *Introduction to Numerical Analysis.* McGraw-Hill, New York, 1956.

12. A. Papoulis, *Probability, Random Variables and Stochastic Processes,* 2nd ed., McGraw-Hill, New York, 1984.

13. P. Steffen, "Two-Dimensional Digital Filters for Geophysical Applications: A Simple Design Method Leading to an Easy Implementation," *Signal Processing,* Vol. 7, pp. 293–320, 1984.

14. H. W. Schüssler, "On Structures for Nonrecursive Digital Filters," *AEÜ,* Vol. 26, pp. 255–258, 1972; reprinted in *Selected Papers in Digital Signal Processing II,* IEEE Press, New York, 1976, pp. 506–509.

15. L. R. Rabiner, B. Gold, and C. A. McGonegal, "An Approach to the Approximation Problem for Nonrecursive Digital Filters," *IEEE Trans. AU,* Vol. 18, pp. 83–106. Reprinted in *Digital Signal Processing,* IEEE Press (Selected Reprint Series), New York, pp. 158–181, 1972.

16. O. Herrmann, "On the Approximation Problem in Nonrecursive Digital Filter Design," *IEEE Trans. Circuit Theory,* Vol. 18, pp. 411–413, 1971. Reprinted in *Digital Signal Processing,* IEEE Press (Selected Reprint Series), New York, pp. 202–203, 1972.

17. H. W. Schüssler and P. Steffen, "An Approach for Designing Systems with Prescribed Behavior at Distinct Frequencies Regarding Additional Constraints," *Proc. ICASSP-85,* pp. 2.6.1–2.6.4, 1985.

18. P. Steffen, "Fourier Approximation with Interpolative Constraints," *AEÜ,* Vol. 36, pp. 363–367, 1984.

19. T. Parks and J. McClellan, "Chebyshev Approximation for Nonrecursive Digital Filters with Linear Phase," *IEEE Trans. Circuit Theory,* Vol. 19, pp. 189–194, 1972.

20. O. Herrmann and H. W. Schüssler, "Design of Nonrecursive Digital Filters with Minimum Phase," *Electronics Letters,* Vol. 6, pp. 329–330, 1970. Reprinted in *Digital Signal Processing,* IEEE Press (Selected Reprint Series), New York, pp. 185—186, 1972.

21. P. Leistner and T. Parks, "On the Design of FIR Digital Filters with Optimum Magnitude and Minimum Phase," *AEÜ,* Vol. 29, pp. 270–274, 1975.

22. H. Göckler, "A General Approach to the Design of Sampled-Data FIR Filters with Optimum Magnitude and Minimum Phase," *Proc. EUSIPCO,* pp. 679–685, 1980.

23. R. Boite and H. Leich, "A New Procedure for the Design of High Order Minimum Phase FIR Digital or CCD Filters," *Signal Processing,* Vol. 3, pp. 101–108, 1981.

24. G. Mian and A. Nainer, "A Fast Procedure to Design Equiripple Minimum Phase FIR Filters," *IEEE Trans. Circuits and Systems,* Vol. 29, pp. 327–331, 1982.

25. X. Chen and T. W. Parks, "Design of Optimal Minimum Phase FIR Filters by Direct Factorization," *Signal Processing,* Vol. 10, pp. 369–383, 1986.

26. A. Fettweis, H. Levin, and A. Sedlmeyer, "Wave Digital Lattice Filters," *Int. J. of Circuit Theory and Applications,* pp. 203–211, June 1974.

27. R. Ansari and B. Liu, "A Class of Low-Noise Computationally Efficient Recursive Digital Filters with Applications to Sampling Rate Alterations, *IEEE Trans. Accoustics, Speech, and Signal Processing,* Vol. 33, pp. 90–97, 1985.

28. P. P. Vaidyanathan, S. K. Mitra, and Y. Neuvo, "A New Approach to the Realization of Low-Sensitivity IIR Digital Filters," *IEEE Trans. Acoustics, Speech and Signal Processing,* Vol. 34, pp. 350–361, 1986.

29. S. K. Mitra and K. Hirano, "Digital All-Pass Networks," *IEEE Trans. Circuits and Systems,* Vol. 21, pp. 688–700, 1974.

30. B. Gold and C. B. Rader, *Digital Processing of Signals,* McGraw-Hill, New York, 1969.

31. R. Czarnach, H. W. Schüssler, and G. Röhrlein, "Linear Phase Recursive Digital Filters for Special Applications," *Proc. ICASSP-82,* pp. 1825–1828, 1982.

32. R. Czarnach, "Recursive Processing by Noncausal Digital Filters," *IEEE Trans. Acoustics, Speech, and Signal Processing,* Vol. 30, pp. 363–370, 1982.

33. H. W. Schüssler and J. Weith, "On the Design of Recursive Hilbert Transformers," *Proc. ICASSP-87,* pp. 876–879, 1987.

Appendix:
Fundamentals of Digital
Signal Processing

A.0 INTRODUCTION

Since all the chapters in this book discuss advanced topics in signal processing, we assume that the reader has a background in signal processing fundamentals. However, the notation used and the amount of material covered often varies greatly. This Appendix summarizes the signal processing fundamentals with which we assume the reader is familiar and also establishes the notation used throughout this book.

A.1 SIGNALS AND SYSTEMS

A.1.1 Signals

Signals are scalar-valued functions of one or more independent variables. Often for convenience, when the signals are one-dimensional, the independent variable is referred to as "time." The independent variable may be continuous or discrete. Signals that are continuous in both amplitude and time (often referred to as continuous-time or analog signals) are the most commonly encountered in signal processing contexts. Discrete-time signals are typically associated with sampling of continuous-time signals. In a digital implementation of a signal processing system, quantization of signal amplitude is also required. Although not precisely correct in every context, discrete-time signal processing is often referred to as digital signal processing.

Discrete-time signals, also referred to as sequences, are denoted by functions whose arguments are integers. For example, $x(n)$ represents a sequence that is defined for all integer values of n and undefined for noninteger values of n. The notation $x(n)$ refers to the discrete-time function x or to the value of the function x at a specific value of n. The distinction between these two will be obvious from the context.

Some sequences and classes of sequences play a particularly important role in discrete-time signal processing. These are summarized below.

The unit sample sequence, denoted by $\delta(n)$, is defined as

$$\delta(n) = \begin{cases} 1, & n = 0 \\ 0, & \text{otherwise} \end{cases} \tag{A.1}$$

The sequence $\delta(n)$ plays a role similar to an impulse function in analog system analysis.

492

The unit step sequence, denoted by $u(n)$, is defined as

$$u(n) = \begin{cases} 1, & n \geq 0 \\ 0, & \text{otherwise} \end{cases} \tag{A.2}$$

Exponential sequences of the form

$$x(n) = A\alpha^n \tag{A.3}$$

play a role in discrete-time signal processing similar to the role played by exponential functions in continuous-time signal processing. Specifically, they are eigenfunctions of discrete-time linear systems and for that reason form the basis for transform analysis techniques. When $|\alpha| = 1$, $x(n)$ takes the form of a complex exponential sequence typically expressed in the form

$$x(n) = Ae^{j\omega n} \tag{A.4}$$

Because the variable n is an integer, complex exponential sequences separated by integer multiples of 2π in ω (frequency) are identical sequences, i.e.,

$$e^{j(\omega + k2\pi)n} = e^{j\omega n} \tag{A.5}$$

This fact forms the core of many of the important differences between the representation of discrete-time signals and systems and that of continuous-time signals and systems.

A general sinusoidal sequence can be expressed as

$$x(n) = A\cos(\omega_0 n + \phi) \tag{A.6}$$

where A is the amplitude, ω the frequency, and ϕ the phase. In contrast with continuous-time sinusoids, a discrete-time sinusoidal signal is not necessarily periodic and if it is, the period is $2\pi/\omega_0$ only when $2\pi/\omega_0$ is an integer. In both continuous time and discrete time, the importance of sinusoidal signals lies in the facts that a broad class of signals can be represented as a linear combination of sinusoidal signals and that the response of linear time-invariant systems to a sinusoidal signal is sinusoidal with the same frequency and with a change in only the amplitude and phase.

A.1.2 Systems

In general, a system maps an input signal $x(n)$ to an output signal $y(n)$ through a system transformation $T\{\cdot\}$. This definition of a system is very broad. Without some restrictions, the characterization of a system requires a complete input-output relationship—knowing the output of a system to a certain set of inputs does not allow us to determine the output of the system to other sets of inputs. Two types of restrictions that greatly simplify the characterization and analysis of a system are linearity and time invariance, alternatively referred to as shift invariance. Fortunately, many systems can often be approximated in practice by a linear and time-invariant system.

The linearity of a system is defined through the principle of superposition:

$$\text{Linearity} \longleftrightarrow T\{ax_1(n) + bx_2(n)\} = ay_1(n) + by_2(n) \tag{A.7}$$

where $T\{x_1(n)\} = y_1(n)$, $T\{x_2(n)\} = y_2(n)$, and a and b are any scalar constants.

Time invariance of a system is defined as

$$\text{Time invariance} \longleftrightarrow T\{x(n - n_0)\} = y(n - n_0) \qquad (A.8)$$

where $y(n) = T\{x(n)\}$ and n_0 is any integer. Linearity and time invariance are independent properties, i.e., a system may have one but not the other property, both or neither.

For a linear and time-invariant (LTI) system, the system response $y(n)$ is given by

$$y(n) = \sum_{K=-\infty}^{+\infty} x(k)h(n - k) = x(n) * h(n) \qquad (A.9)$$

where $x(n)$ is the input and $h(n)$ is the response of the system when the input is $\delta(n)$. Equation (A.9) is the convolution sum.

As with continuous-time convolution, the convolution operator in Eq. (A.9) is commutative and associative and distributes over addition:

Commutative:

$$x(n) * y(n) = y(n) * x(n) \qquad (A.10)$$

Associative:

$$[x(n) * y(n)] * w(n) = x(n) * [y(n) * w(n)] \qquad (A.11)$$

Distributive:

$$x(n) * [y(n) + w(n)] = [x(n) * y(n) + x(n) * w(n)] \qquad (A.12)$$

In continuous-time systems, convolution is primarily an analytical tool. For discrete-time systems, the convolution sum, in addition to being important in the analysis of LTI systems, is important as an explicit mechanism for implementing a specific class of LTI systems, namely those for which the impulse response is of finite length (FIR systems).

Two additional system properties that are referred to frequently are the properties of stability and causality. A system is considered stable in the bounded input–bounded output (BIBO) sense if and only if a bounded input always leads to a bounded output. A necessary and sufficient condition for an LTI system to be stable is that its unit sample response $h(n)$ be absolutely summable. For an LTI system,

$$\text{Stability} \longleftrightarrow \sum_{n=-\infty}^{\infty} |h(n)| < \infty \qquad (A.13)$$

Because of Eq. (A.13), an absolutely summable sequence is often referred to as a stable sequence.

A system is referred to as causal if and only if, for each value of n, say n_0, $y(n)$ does not depend on values of the input for $n > n_0$. A necessary and sufficient condition for an LTI system to be causal is that its unit sample response $h(n)$ be zero for $n < 0$. For an LTI system,

$$\text{Causality} \longleftrightarrow h(n) = 0 \text{ for } n < 0 \qquad (A.14)$$

Because of Eq. (A.14), a sequence that is zero for $n < 0$ is often referred to as a causal sequence.

A.2 FREQUENCY-DOMAIN REPRESENTATION OF SIGNALS

In this section, we summarize the representation of sequences as linear combinations of complex exponentials, first for periodic sequences using the discrete-time Fourier series, next for stable sequences using the discrete-time Fourier transform, then through a generalization of the discrete-time Fourier transform, namely, the z-transform, and finally for finite-extent sequences using the discrete Fourier transform. In Section A.3, we review the use of these representations in characterizing LTI systems.

A.2.1 Discrete-Time Fourier Series

Any periodic sequence $\tilde{x}(n)$ with period N can be represented through the discrete-time Fourier series (DFS) pair in Eqs. (A.15) and (A.16).

$$\text{\textit{Synthesis equation:}} \qquad \tilde{x}(n) = \frac{1}{N} \sum_{k=0}^{N-1} \tilde{X}(k) e^{j(2\pi/N)nk} \qquad (A.15)$$

$$\text{\textit{Analysis equation:}} \qquad \tilde{X}(k) = \frac{1}{N} \sum_{n=0}^{N-1} \tilde{x}(n) e^{-j(2\pi/N)nk} \qquad (A.16)$$

The synthesis equation expresses the periodic sequence as a linear combination of harmonically related complex exponentials. The choice of interpreting the DFS coefficients $\tilde{X}(k)$ either as zero outside the range $0 \le k \le (N-1)$ or as periodically repeated does not in any way affect Eq. (A.15). It is a commonly accepted convention, however, to interpret $\tilde{X}(k)$ as periodic to maintain a duality between the analysis and synthesis equations.

A.2.2 Discrete-Time Fourier Transform

Any stable sequence $x(n)$ (i.e., one that is absolutely summable) can be represented as a linear combination of complex exponentials. For aperiodic stable sequences, the synthesis equation takes the form of Eq. (A.17), and the analysis equation takes the form of Eq. (A.18).

$$\text{\textit{Synthesis equation:}} \qquad x(n) = \frac{1}{2\pi} \int_{-\pi}^{\pi} X(\omega) e^{j\omega n} d\omega \qquad (A.17)$$

$$\text{\textit{Analysis equation:}} \qquad X(\omega) = \sum_{r=-\infty}^{+\infty} x(n) e^{-j\omega n} \qquad (A.18)$$

To relate the discrete-time Fourier transform and the discrete-time Fourier series, consider a stable sequence $x(n)$ and the periodic signal $\tilde{x}_1(n)$ formed by time-aliasing $x(n)$, i.e.,

$$\tilde{x}_1(n) = \sum_{r=-\infty}^{+\infty} x(n + rN) \qquad (A.19)$$

Then the DFS coefficients of $\tilde{x}_1(n)$ are proportional to samples spaced by $2\pi/N$ of the

Fourier transform $x(n)$. Specifically,

$$\tilde{X}_1(k) = \frac{1}{N} X(\omega) \Big|_{\omega=(2\pi k/N)} \tag{A.20}$$

Among other things, this implies that the DFS coefficients of a periodic signal are proportional to the discrete-time Fourier transform of one period.

A.2.3 z-Transform

A generalization of the Fourier transform, the z-transform, permits the representation of a broader class of signals as a linear combination of complex exponentials, for which the magnitudes may or may not be unity.

The z-transform analysis and synthesis equations are as follows:

$$\textit{Synthesis equation:} \quad x(n) = \frac{1}{2\pi j} \oint_C X(z) z^{n-1} dz \tag{A.21}$$

$$\textit{Analysis equation:} \quad X(z) = \sum_{n=-\infty}^{+\infty} x(n) z^{-n} \tag{A.22}$$

From Eqs. (A.18) and (A.22), $X(\omega)$ is related to $X(z)$ by $X(\omega) = X(z)|_{z=e^{j\omega}}$, i.e., for a stable sequence, the Fourier transform $X(\omega)$ is the z-transform evaluated on the contour $|z| = 1$, referred to as the unit circle.

Equation (A.22) converges only for some values of z and not others. The range of values of z for which $X(z)$ converges, i.e., the region of convergence (ROC), corresponds to the values of z for which $x(n)z^{-n}$ is absolutely summable. We summarize the properties of the ROC in more detail later. Complete specification of the z-transform requires specification not only of the algebraic expression for the z-transform but also of the ROC. For example, the two sequences $a^n u(n)$ and $-a^n u(-n - 1)$ have z-transforms that are identical algebraically and that differ only in the ROC.

The synthesis equation as expressed in Eq. (A.21) is a contour integral with the contour encircling the origin and contained within the region of convergence. While this equation provides a formal means for obtaining $x(n)$ from $X(z)$, its evaluation requires contour integration. Such an integral can be evaluated using complex residues, but the procedure is often tedious and usually unnecessary. When $X(z)$ is a rational function of z, a more typical approach is to expand $X(z)$ using a partial fraction equation. The inverse z-transform of the individual simpler terms can usually then be recognized by inspection.

When $X(z)$ is a ratio of polynomials in z, it is often described to within a scale factor by a pole-zero plot marking the location in the complex z-plane of the roots of the numerator polynomial (the zeros) and the denominator polynomial (the poles). The complex z-plane is indicated in Fig. A.1. On the unit circle the z-transform reduces to the Fourier transform.

There are a number of important properties of the ROC that, together with properties of the time-domain sequence, permit implicit specification of the ROC. These properties are summarized as follows:

Property 1. The ROC is a connected region.

Property 2. For a rational z-transform, the ROC does not contain any poles and is bounded by poles.

Property 3. If $x(n)$ is a right-sided sequence and if the circle $|z| = r_0$ is in the ROC, then all finite values of z for which $|z| > r_0$ will also be in the ROC.

Property 4. If $x(n)$ is a left-sided sequence and if the circle $|z| = r_0$ is in the ROC, then all values of z for which $0 < |z| < r_0$ will also be in the ROC.

Property 5. If $x(n)$ is a stable and causal sequence with a rational z-transform, then all the poles of $X(z)$ are inside the unit circle.

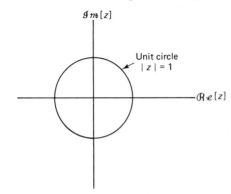

Figure A.1   Unit circle in the z-plane.

A.2.4 Discrete Fourier Transform

In Section A.2.1, we discussed the representation of periodic sequences in terms of the discrete Fourier series. With the correct interpretation, the same representation can be applied to finite-duration sequences. The resulting Fourier representation for finite-duration sequences is referred to as the discrete Fourier transform (DFT).

The DFT analysis and synthesis equations are

Analysis equation:
$$X(k) = \sum_{n=0}^{N-1} x(n)e^{-j(2\pi/N)kn}, \qquad 0 \le k \le N-1$$
(A.23)

Synthesis equation:
$$x(n) = \frac{1}{N}\sum_{k=0}^{N-1} X(k)e^{j(2\pi/N)kn}, \qquad 0 \le n \le N-1$$
(A.24)

The fact that $X(k) = 0$ for k outside the interval $0 \le k \le N-1$ and that $x(n) = 0$ for k outside the interval $0 \le k \le N-1$ is implied but not always stated explicitly.

The DFT is used in a variety of signal processing applications, so it is of considerable interest to efficiently compute the DFT and inverse DFT. A straightforward computation of the N-point DFT or inverse DFT requires on the order of N^2 arithmetic operations (multiplications and additions). The number of arithmetic operations required is significantly reduced through the set of fast Fourier transform (FFT) algorithms. Most FFT algorithms are based on the simple principle that an N-point

DFT can be computed by computing two $(N/2)$-point DFTs, or three $(N/3)$-point DFTs, etc. Computation of the N-point DFT or inverse DFT using FFT algorithms requires on the order of $N \log N$ arithmetic operations. Chapter 4 discusses in detail the problem of efficiently computing the DFT.

A.2.5 Transform Properties

In Section A.2, we have summarized four frequency domain representations for signals. Particular properties of these representations are heavily exploited in signal and system analysis. In Tables A.1, A.2, A.3, and A.4 these properties are summarized for the discrete-time Fourier series, the discrete-time Fourier transform, the z-transform, and the discrete Fourier transform.

TABLE A.1 PROPERTIES OF THE DISCRETE-TIME FOURIER SERIES

Property 1: Linearity	$a\tilde{x}(n) + b\tilde{y}(n) \longleftrightarrow a\tilde{X}(k) + b\tilde{Y}(k)$
Property 2: Periodic convolution	$\tilde{x}(n) * \tilde{y}(n) \longleftrightarrow \tilde{X}(k) \cdot \tilde{Y}(k)$
Property 3: Modulation	$\tilde{x}(n) \cdot \tilde{y}(n) \longleftrightarrow \dfrac{1}{N}\tilde{X}(k) * \tilde{Y}(k) = \sum\limits_{\ell=0}^{N-1} \tilde{X}(\ell)\tilde{Y}(k - \ell)$
Property 4: Time and frequency shift	**(a)** $\tilde{x}(n - n_0) \longleftrightarrow \tilde{X}(k) \cdot e^{-j(2\pi/N)k \cdot n_0}$
	(b) $\tilde{x}(n)e^{j(2\pi k_0 n/N)} \longleftrightarrow \tilde{X}(k - k_0)$
Property 5: Symmetry properties	**(a)** $\tilde{x}^*(n) \longleftrightarrow \tilde{X}^*(-k)$
	(b) $\tilde{x}(-n) \longleftrightarrow \tilde{X}^*(-k)$

TABLE A.2 PROPERTIES OF THE DISCRETE-TIME FOURIER TRANSFORM

Property 1: Linearity	$ax_1(n) + bx_2(n) \longleftrightarrow aX_1(\omega) + bX_2(\omega)$
Property 2: Convolution	$x(n) * h(n) \longleftrightarrow X(\omega)H(\omega)$
Property 3: Modulation	$x(n)y(n) \longleftrightarrow \dfrac{1}{2\pi}\displaystyle\int_{\theta-\pi}^{\pi} X(\theta)Y(\omega - \theta)\,d\theta$
Property 4: Time and frequency shift	**(a)** $x(n - n_0) \longleftrightarrow X(\omega)e^{-j\omega n_0}$
	(b) $e^{-j\omega_0 n} \cdot x(n) \longleftrightarrow X(\omega - \omega_0)$
Property 5: Differentiation	$-jnx(n) \longleftrightarrow \dfrac{dX(\omega)}{d\omega}$
Property 6: Symmetry property	**(a)** $x(n)$: real $\longleftrightarrow X(\omega) = X^*(-\omega)$
	$X_R(\omega), \lvert X(\omega)\rvert$: even
	$X_I(\omega), \theta_x(\omega)$: odd
	(b) $x(n)$: real and even $\longleftrightarrow X(\omega)$: real and even
	(c) $x(n)$: real and odd $\longleftrightarrow X(\omega)$: pure imaginary and odd

TABLE A.3 PROPERTIES OF THE z-TRANSFORM

Property 1: Linearity	$ax(n) + by(n) \longleftrightarrow aX(z) + bY(z)$ $\text{ROC}: R_- < \lvert z \rvert < R_+$ $R_- = \max[R_{x-}, R_{y-}]$ $R_+ = \min[R_{x+}, R_{y+}]$
Property 2: Convolution	$x(n) * h(n) \longleftrightarrow X(z)H(z)$ $\text{ROC}: R_- < \lvert z \rvert < R_+$
Property 3: Shift of a sequence	$x(n - n_0) \longleftrightarrow X(z)z^{-n_0}$
Property 4: Differentiation	$nx(n) \longleftrightarrow -z\dfrac{dX(z)}{dz}n$
Property 5: Symmetry property	**(a)** $x^*(n) \longleftrightarrow X^*(z^*)$ **(b)** $x(-n) \longleftrightarrow X(z^{-1}), \quad \text{ROC}: \dfrac{1}{R_{X+}} < \lvert z \rvert < \dfrac{1}{R_{x-}}$

TABLE A.4 PROPERTIES OF THE DISCRETE FOURIER TRANSFORM

For all the properties listed, the expressions given specify $x(n)$ and $y(n)$ for $0 \le n \le N - 1$ and $X(k)$ and $Y(k)$ for $0 \le k \le N - 1$. The four sequences $x(n)$, $y(n)$, $X(k)$, and $Y(k)$ are all zero outside those ranges. The expression $x((n))_N$ refers to $x(n)$ modulo N.

Property 1: Linearity	$ax(n) + by(n) \longleftrightarrow aX(k) + bY(k)$
Property 2: Circular convolution	$\left. \begin{aligned} x(n) * y(n) &= \sum_{m=0}^{N-1} x((m))_N \cdot y((n - m))_N \\ &= \sum_{m=0}^{N-1} x(m) \cdot y((n - m))_N \end{aligned} \right\} \longleftrightarrow X(k) \cdot Y(k)$
Property 3: Modulation	$x(n) \cdot y(n) \longleftrightarrow \dfrac{1}{N}\left[\sum_{\ell=0}^{N-1} X((\ell))_N \cdot Y((k - \ell))_N\right]$ $= \dfrac{1}{N}\left[\sum_{\ell=0}^{N-1} X(\ell) \cdot Y((k - \ell))_N\right]$
Property 4: Circular shift of a sequence	**(a)** $x((n - n_0))_N \longleftrightarrow e^{-j(2\pi/N)kn_0} \cdot X(k)$ **(b)** $x(n)e^{-j(2\pi/N)kn_0} \longleftrightarrow X((k - k_0))_N$
Property 5: Symmetry property	**(a)** $x^*(n) \longleftrightarrow X^*((-k))_N$ **(b)** $x((-n))_N \longleftrightarrow X((-k))_N$

A.3 FREQUENCY-DOMAIN REPRESENTATION OF LTI SYSTEMS

In Section A.2 we reviewed the representation of signals as a linear combination of complex exponentials of the form $e^{j\omega n}$ or, more generally, z^n. For linear systems, the response is then the same linear combination of the responses to the individual complex exponentials. If in addition the system is time-invariant, the complex exponentials are eigenfunctions. Consequently, the system can be characterized by the spectrum of eigenvalues, corresponding to the frequency response if the signal decomposition is in terms of complex exponentials with unity magnitude or, more generally, to the system function in the context of the more general complex exponentials z^n.

The eigenfunction property follows directly from the convolution sum and states that with $x(n) = z^n$, the output $y(n)$ has the form

$$y(n) = H(z)z^n \tag{A.25}$$

where

$$H(z) = \sum_{k=-\infty}^{+\infty} h(k)z^{-n} \tag{A.26}$$

The system function $H(z)$ is the eigenvalue associated with the eigenfunction z^n. Also, from Eq. (A.26), $H(z)$ is the z-transform of the system unit sample response. When $z = e^{j\omega}$, it corresponds to the Fourier transform of the unit sample response.

Since Eq. (A.17) or (A.21) corresponds to a decomposition of $x(n)$ as a linear combination of complex exponentials, we can obtain the response $y(n)$, using linearity and the eigenfunction property, by multiplying the amplitudes of the eigenfunctions z^n in Eq. (A.22) by the eigenvalues $H(z)$, i.e.,

$$y(n) = \frac{1}{2\pi j} \int_C H(z)X(z)z^{n-1}dz \tag{A.27}$$

Equation (A.27) then becomes the synthesis equation for the output, i.e.,

$$Y(z) = H(z)X(z) \tag{A.28}$$

Equation (A.28) corresponds to the z-transform convolution property included in Table A.3.

A.4 SYSTEMS CHARACTERIZED BY LINEAR CONSTANT-COEFFICIENT DIFFERENCE EQUATIONS

A particularly important class of discrete-time systems are those characterized by linear constant-coefficient difference equations (LCCDE) of the form

$$\sum_{k=0}^{N} a_k y(n-k) = \sum_{k=0}^{M} b_k x(n-k) \tag{A.29}$$

where the a_k's and the b_k's are constants. Equation (A.29) is typically referred to as an Nth-order difference equation.

A system characterized by an Nth-order difference equation of the form in Eq. (A.29) represents a linear time-invariant system only under an appropriate choice of the homogeneous solution, i.e., linearity and time invariance are additional constraints to the equation itself. Even under these additional constraints, the system is not restricted to be causal.

A.4.1 Solution of Linear Constant-Coefficient Difference Equations

Assuming the system is linear, time-invariant, and causal, the response of a system characterized by Eq. (A.29) can be obtained recursively. Specifically, we can rewrite Eq. (A.29) as

$$a_0 y(n) = \sum_{k=0}^{M} b_k x(n-k) - \sum_{k=1}^{N} a_k y(n-k) \qquad (A.30)$$

Since we are assuming that the system is linear and causal, if $x(n) = 0$ for $n < n_0$ then $y(n) = 0$ for $n < n_0$. With this assumed zero state, $y(n)$ can be generated recursively from Eq. (A.30). This recursion is illustrated in a linear signal flow graph form in Fig. A.2, where z^{-1} represents a unit delay. While the recursion in Eq. (A.30) will generate the correct output sequence and, in fact, represents a specific algorithm for computing the output, the result will not be in an analytically convenient form. A convenient procedure to obtain the solution analytically is through the use of the z-transform. Specifically, applying the z-transform to both sides of Eq. (A.29) and using the linearity and time-shifting properties, and after appropriate algebraic manipulation, we obtain

$$H(z) = \frac{Y(z)}{X(z)} = \frac{\sum_{k=0}^{M} b_k z^{-k}}{\sum_{k=0}^{N} a_k z^{-k}} \qquad (A.31)$$

Equation (A.31) specifies the algebraic expression for the system function, which we note is a rational function of z. It does not, however, explicitly specify the ROC. If we assume that the system is causal, then the ROC associated with Eq. (A.31) will be the region outside a circle passing through the outermost pole of $H(z)$. If we do not

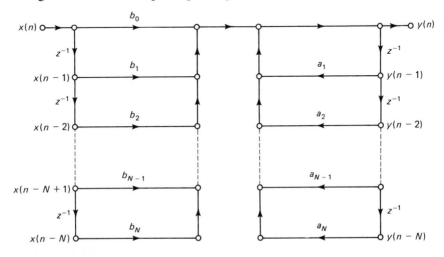

Figure A.2 Direct form I realization of an Nth-order difference equation. For convenience N and M are assumed equal.

impose causality, then in general there are many choices for the ROC and correspondingly for the system impulse response.

A.4.2 Finite Impulse Response and Infinite Impulse Response Systems

When N in Eq. (A.29) or (A.31) is greater than zero, the system impulse response is a linear combination of exponentials and consequently is of infinite length. In this case, the system is referred to as an infinite impulse response (IIR) system. When $N = 0$ (assuming a_0 is normalized to unity), the difference equation becomes

$$y(n) = \sum_{k=0}^{M} b_k x(n - k) \tag{A.32}$$

The unit sample response of the system is then

$$h(n) = b_n, \qquad 0 \le n \le M \tag{A.33}$$

and is zero for $n < 0$ and $n > M$. Such systems are referred to as finite impulse response (FIR) systems.

Generally in designing a discrete-time filter to meet certain prescribed specifications, a design utilizing both poles and zeroes, i.e., an IIR design, results in far fewer overall delays and coefficients than an FIR design. On the other hand, there are certain situations in which FIR filters are preferable to IIR filters. One major advantage of FIR filters is that they can be designed such that the impulse response is symmetric, i.e., so that $h(n) = h(M - 1 - n)$ where M is the filter length, in which case the frequency response is real to within a linear phase factor $e^{-j[2\pi(M-1)/2]\omega}$. Causal IIR filters cannot be designed to have a symmetric impulse response. Also, since one realization of FIR filters is directly through Eq. (A.32) as the sum of weighted taps on a delay line, implementation of FIR filters is well matched to certain technologies, in particular the use of CCDs and other charge transfer devices and surface acoustic wave devices.

A.4.3 Linear Signal Flow Graph Representation of Linear Constant-Coefficient Difference Equations

The flow graph in Fig. A.2 is generally referred to as direct form I realization of the Nth-order difference equation. This realization can be viewed as a cascade of two systems, one implementing the zeroes of $H(z)$ and the other the poles. Since each of these is an LTI system, the order in which they are cascaded can be reversed. After doing so, and combining the two chains of delays into a single chain of delays, the linear signal flow graph in Fig. A.3 results. This representation is typically referred to as the direct form II realization of the difference equation.

The rational function $H(z)$ in Eq. (A.31) can also be rearranged in a number of other ways leading to other realizations of the Nth-order difference equation. For example, $H(z)$ can be expressed as a product of second-order factors in the form

$$H(z) = \frac{b_0}{a_0} \prod_{k=1}^{N/2} \frac{1 + \beta_{1k} z^{-1} + \beta_{2k} z^{-2}}{1 + \alpha_{1k} z^{-1} + \alpha_{2k} z^{-2}} \tag{A.34}$$

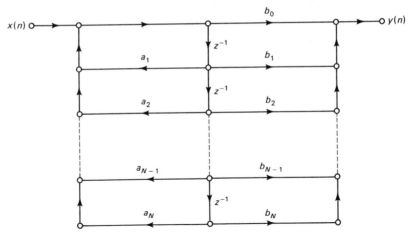

Figure A.3 Direct form II realization of an Nth-order difference equation with $N = M$.

where we have assumed that $N = M$ and N is even. If this is not the case, we can simply include terms with zero coefficients. The signal flow graph structure suggested by Eq. (A.34) is the cascade structure shown in Fig. A.4, illustrated for $N = 6$. In this figure we have represented each of the second-order sections in direct II form. The parallel form structure is obtained by expanding $H(z)$ in a partial fraction expansion. For example, if we expand $H(z)$ in second-order terms, again assuming N is even, we can write $H(z)$ as

$$H(z) = \frac{b_N}{a_N} \sum_{k=1}^{N/2} \frac{1 + \gamma_{0k} + \gamma_{1k}z^{-1}}{1 + \alpha_{1k}z^{-1} + \alpha_{2k}z^{-2}} \tag{A.35}$$

Equation (A.35) then corresponds to the parallel form realization shown in Fig. A.5, where each section is implemented in direct II form.

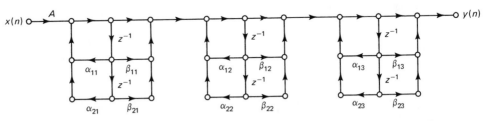

Figure A.4 Cascade structure with a direct form II realization of each second-order subsystem.

There are, of course, many variations on the cascade and parallel forms, such as the use of both first- and second-order sections, different ordering of sections, the specific form used for each of the sections, etc. There are also many forms in addition to the basic ones that we have summarized here.

The direct form II, cascade, and parallel structures, including those based on applying the signal flow graph transposition theorem to Figs. A.3, A.4, and A.5, are perhaps the most commonly encountered, although in specialized situations certain

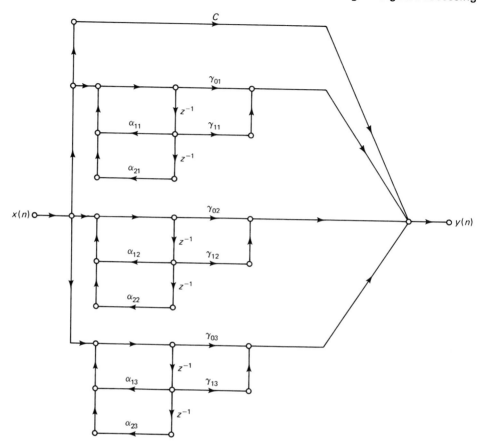

Figure A.5 Parallel-form realization with the real and complex poles grouped in pairs.

other structures are preferable. The choice of structure for realization of a given transfer function in a specific application is closely associated with issues of implementation: modularity, considerations of parallel computation, effects of coefficient inaccuracies, roundoff noise, dynamic range, etc. For example, it is well known that, in general, the direct-form structures tend to be more sensitive to coefficient inaccuracies than either the cascade or parallel structures since in the direct-form structures, pole and zero locations are controlled through the coefficients of high-order polynomials, whereas in the cascade and parallel forms they are controlled through the coefficients of first- and second-order polynomials.

The basic structures just described were discussed in the context of a general difference equation with both poles and zeros, i.e., IIR filters. For FIR filters, both the direct form I and II structures reduce to the tapped delay line structure in Fig. A.6. As indicated earlier, one of the potential advantages of FIR filters lies in the fact that they can be designed and implemented to have exactly linear phase. Since linear phase FIR filters have a symmetric impulse response, i.e., $h(n) = h(N - 1 - n)$, the direct-form structure in Fig. A.6 can be rearranged in this case to reduce the number

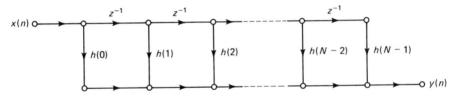

Figure A.6   Direct-form realization of an FIR system.

of multiplies by first adding terms with identical coefficients and then carrying out the multiplication.

FIR filters can also be implemented as a cascade of first- and/or second-order zeros. Generally, however, the most typical implementation of FIR filters is in direct form since technologies such as charge transfer devices are best suited to implementing a direct tapped delay line structure.

A.5 DIGITAL FILTER DESIGN

The design of digital filters can be thought of as involving two stages: (1) the specification of the desired properties of the filter and (2) the approximation of the specifications using a discrete-time system. Although these two steps are not independent, it is usually convenient to treat them separately.

A typical form for the specification is depicted in Fig. A.7 for a lowpass filter in which δ_1 and δ_2 represent the allowable passband and stopband tolerance and ω_p and ω_s, respectively, denote the passband and stopband edge frequencies.

Many filters used in practice are specified by such a tolerance scheme, with no constraints on the phase response other than those imposed by stability and causality requirements; i.e., the poles of the system function must lie inside the unit circle. In designing FIR digital filters, we more often impose the constraint that the phase be linear, thereby again removing phase from consideration in the design process.

Given a set of specifications, the filter design problem becomes a problem in functional approximation to obtain a discrete-time linear system whose frequency response falls within the prescribed tolerances. IIR systems imply approximation by

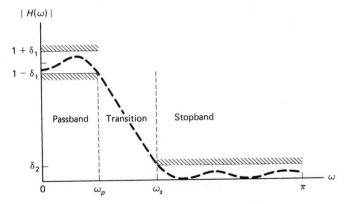

Figure A.7   A typical form for digital lowpass filter specification.

a rational function of z, while FIR systems imply polynomial approximation. There are a variety of design techniques for both types of filters, ranging from closed-form procedures, which involve only substitution of design specifications into design formulas, to algorithmic techniques, where, for example, a solution may be obtained by an iterative procedure.

A.5.1 Design of IIR Digital Filters from Analog Filters

The classical approach to the design of IIR digital filters involves the transformation of an analog filter into a digital filter meeting prescribed specifications. This approach is motivated principally by the fact that the art of analog IIR filter design is highly advanced and, since useful results can be achieved, it is advantageous to utilize the design procedures already developed for analog filters.

In transforming an analog system to a digital system we generally require that the essential properties of the analog frequency response be preserved in the frequency response of the resulting digital filter. Loosely speaking, this implies that we want the imaginary axis of the s-plane to map onto the unit circle in the z-plane and that a stable analog filter be transformed to a stable digital filter. These constraints are basic to the techniques discussed in this section.

In impulse invariant design, an analog filter impulse response $h_a(t)$ is mapped to a digital filter impulse response $h(n)$ through the relation

$$h(n) = Th_a(nT) \tag{A.36}$$

where T denotes a sampling period. The frequency response of the digital filter $H(\omega)$ is related to the frequency response of the analog filter $H_a(\Omega)$ as

$$H(\omega) = \sum_{k=-\infty}^{+\infty} H_a\left(\frac{\omega}{T} + \frac{2\pi k}{T}\right) \tag{A.37}$$

Any practical analog filter will not be bandlimited, and consequently there is interference between successive terms in Eq. (A.37) (i.e., aliasing), as indicated in Fig. A.8. However, if the analog filter approaches zero at high frequencies sufficiently rapidly, the aliasing may be negligibly small and a useful digital filter can result from the sampling of the impulse response of an analog filter.

In structuring the design problem, we begin with discrete-time specifications that are then mapped to corresponding analog specifications, and the resulting analog filter is then mapped back to discrete time. From this point of view, the parameter T in the impulse invariant design procedure has no effect and often for convenience is chosen as unity.

The basis for impulse invariance as just described is to choose an impulse response for the digital filter that is similar in some sense to the impulse response of the analog filter. The use of this often is motivated not so much by a desire to maintain the impulse response shape but by the knowledge that if the analog filter is bandlimited, then the digital filter frequency response will closely approximate the analog frequency response. However, in some filter design problems, a primary objective may be to control some aspect of the time response such as the impulse response or

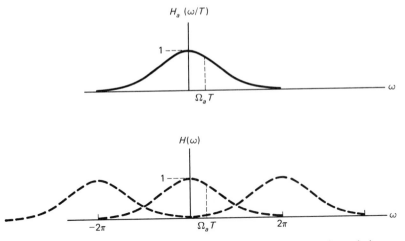

Figure A.8 Graphical representation of aliasing in the impulse-invariance design technique.

the step response. In such cases a natural approach would be to design the digital filter by impulse invariance or a step invariance procedure.

Impulse invariance as a procedure for mapping analog filter designs to digital filter designs has the limitation that the analog filter must (at least approximately) have a bandlimited frequency response. As an alternative, the bilinear transformation for mapping analog to digital designs maps the entire $j\Omega$-axis in the s-plane to once around the unit circle in the z-plane, and consequently there is no aliasing. By necessity, however, the mapping from analog to digital frequency is nonlinear. Consequently, in contrast to impulse invariance, the bilinear transformation is useful only for mapping analog filter designs that have piecewise-constant frequency characteristics.

The bilinear transformation is given by

$$s = \frac{2}{T}\left(\frac{1 - z^{-1}}{1 + z^{-1}}\right) \tag{A.38}$$

The parameter T has again been included in the mapping since this commonly appears in some texts. However, just as with impulse invariance, and with the point of view that we begin the process with specifications in terms of discrete-time frequency, the parameter T will cancel.

With the use of the bilinear transformation, discrete-time frequency ω is related to continuous-time frequency Ω by

$$\omega = 2\,\tan^{-1}(\Omega T/2) \tag{A.39}$$

This relationship is depicted in Fig. A.9.

Although the bilinear transformation can be used effectively in mapping a piecewise-constant magnitude characteristic from the s-plane to the z-plane, the distortion in the frequency axis will manifest itself in terms of distortion in the phase characteristic associated with the filter. If, for example, we were interested in a digital

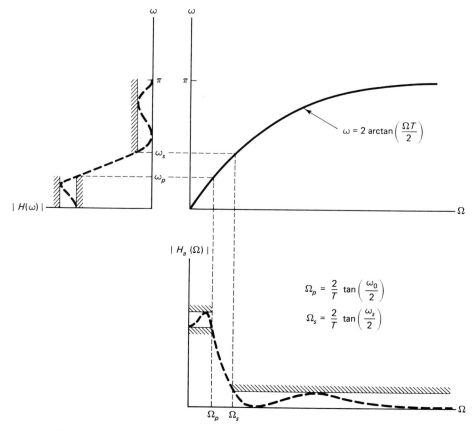

Figure A.9 Frequency warping encountered in transforming an analog lowpass filter to a digital lowpass filter using the bilinear transformation. To achieve the desired digital cutoff frequency, the analog cutoff frequencies must be prewarped as indicated.

lowpass filter with a linear phase characteristic, we could not obtain such a filter by applying the bilinear transformation to an analog lowpass filter with a linear phase characteristic.

A.5.2 Computer-Aided Design of IIR Digital Filters

In the previous section we have seen that digital filters can be designed by transforming an appropriate analog filter design into a digital filter design. This approach is reasonable when we can take advantage of analog designs that are given in terms of formulas or extensive design tables, e.g., frequency-selective filters such as Butterworth, Chebyshev, or elliptic filters. In general, however, analytic formulas do not exist for the design of either analog or digital filters to match arbitrary frequency response specifications or other types of specifications. In these more general cases, design procedures have been developed that are algorithmic, generally relying on the use of a computer to solve sets of linear or nonlinear equations. In most cases the computer-aided design techniques apply equally well to the design of either analog or digital filters with only minor modification. Therefore, nothing is gained by first obtaining an analog design and then transforming this design to a digital filter.

Most algorithmic design procedures for IIR filters take the following form:

1. $H(z)$ is assumed to be a rational function. It can be represented as a ratio of polynomials in z (or z^{-1}), as a product of numerator and denominator factors (zeros and poles), or as a product of second-order factors.
2. The orders of the numerator and denominator of $H(z)$ are fixed.
3. An ideal desired frequency response and a corresponding approximation error criterion is chosen.
4. By a suitable optimization algorithm, the free parameters (numerator and denominator coefficients, zeros and poles, etc.) are varied in a systematic way to minimize the approximation error according to the assumed error criterion.
5. The set of parameters that minimize the approximation error is the desired result.

A.5.3 Design of FIR Filters

In contrast to IIR filters, there is not available a set of continuous-time FIR design techniques that can be exploited in the design of discrete-time FIR filters. Consequently, FIR design is most typically carried out directly in the discrete-time domain.

The most straightforward approach to FIR filter design is to obtain a finite-length impulse response by truncating an infinite-duration impulse response sequence. If we suppose that $H_d(\omega)$ is an ideal desired frequency response, then

$$H_d(\omega) = \sum_{n=-\infty}^{\infty} h_d(n)e^{-j\omega n} \tag{A.40}$$

where $h_d(n)$ is the corresponding impulse response sequence, i.e.,

$$h_d(n) = \frac{1}{2\pi}\int_{-\pi}^{\pi} H_d(\omega)e^{j\omega n} d\omega \tag{A.41}$$

In general, $H_d(\omega)$ for a frequency-selective filter may be piecewise constant with discontinuities at the boundaries between bands. In such cases the sequence $h_d(n)$ is of infinite duration and it must be truncated to obtain a finite-duration impulse response $h(n)$. The truncation corresponds to representing $h(n)$ as the product of the desired impulse response and a finite-duration "window" $w(n)$, i.e.,

$$h(n) = h_d(n)w(n) \tag{A.42}$$

Correspondingly, in the frequency domain,

$$H(\omega) = \frac{1}{2\pi}\int_{-\pi}^{\pi} H_d(\theta)W(\theta - \omega) d\theta \tag{A.43}$$

That is, $H(\omega)$ is the periodic convolution of the desired frequency response with the Fourier transform of the window. Thus the frequency response $H(\omega)$ will be a "smeared" version of the desired response $H_d(\omega)$. The choice of the window is governed by the desire to have $w(n)$ as short as possible in duration so as to minimize computation in the implementation of the filter, while having $W(\omega)$ highly concentrated in the frequency domain so as to faithfully reproduce the desired frequency response. These are, of course, conflicting requirements. Examples of some commonly used windows are Hanning, Hamming, Blackman and Kaiser windows.

While FIR filters obtained by windowing are relatively easy to design, they are not optimum, i.e., do not have the minimum passband or stopband ripple for a given filter order and transition bandwidth. The Parks-McClellan algorithm provides a procedure for designing linear phase FIR filters that are optimal in the sense of minimizing the maximum weighted error over a specified set of frequency bands. The algorithm is based on expressing the frequency response of a linear phase FIR filter as a linear phase factor times a weighted trigonometric polynomial. For example, for a causal linear phase FIR filter of length $N = 2M + 1$, the impulse response has the symmetry property

$$h(n + M) = h(-n + M) \tag{A.44}$$

and $H(\omega)$ is of the form

$$H(\omega) = e^{-jwM}\left[h(0) + \sum_{k=1}^{M} 2h(n)\cos(\omega n)\right]$$
$$= e^{-j\omega M}\left[\sum_{k=0}^{M} a_k(\cos \omega)^k\right] \tag{A.45}$$

Given the desired specifications for $|H(\omega)|$, determination of the coefficients a_k and equivalently the impulse response values $h(n)$ becomes a problem in polynomial approximation. An efficient algorithm to solve this polynomial approximation problem exists and is widely used to design optimal FIR filters.

Index